Analytic Number Theory

Developments in Mathematics

VOLUME 6

Series Editor:

Krishnaswami Alladi, *University of Florida, U.S.A.* VOLUME 3

Series Editor:

Krishnaswami Alladi, *University of Florida, U.S.A.*

Aims and Scope

Developments in Mathematics is a book series publishing

(i) Proceedings of Conferences dealing with the latest research advances,

(ii) Research Monographs, and

(iii) Contributed Volumes focussing on certain areas of special interest.

Editors of conference proceedings are urged to include a few survey papers for wider appeal. Research monographs which could be used as texts or references for graduate level courses would also be suitable for the series. Contributed volumes are those where various authors either write papers or chapters in an organized volume devoted to a topic of special/current interest or importance. A contributed volume could deal with a classical topic which is once again in the limelight owing to new developments.

Analytic Number Theory

Edited by

Chaohua Jia
Academia Sinica,
China

and

Kohji Matsumoto
Nagoya University,
Japan

KLUWER ACADEMIC PUBLISHERS
DORDRECHT / BOSTON / LONDON

A C.I.P. Catalogue record for this book is available from the Library of Congress.

ISBN 978-1-4419-5214-1

Published by Kluwer Academic Publishers,
P.O. Box 17, 3300 AA Dordrecht, The Netherlands.

Sold and distributed in North, Central and South America
by Kluwer Academic Publishers,
101 Philip Drive, Norwell, MA 02061, U.S.A.

In all other countries, sold and distributed
by Kluwer Academic Publishers,
P.O. Box 322, 3300 AH Dordrecht, The Netherlands.

Printed on acid-free paper

Printed in the Netherlands.

Contents

Preface

From September 13 to 17 in 1999, the First China-Japan Seminar on Number Theory was held in Beijing, China, which was organized by the Institute of Mathematics, Academia Sinica jointly with Department of Mathematics, Peking University. Ten Japanese Professors and eighteen Chinese Professors attended this seminar. Professor Yuan Wang was the chairman, and Professor Chengbiao Pan was the vice-chairman. This seminar was planned and prepared by Professor Shigeru Kanemitsu and the first-named editor. Talks covered various research fields including analytic number theory, algebraic number theory, modular forms and transcendental number theory. The Great Wall and acrobatics impressed Japanese visitors.

From November 29 to December 3 in 1999, an annual conference on analytic number theory was held in Kyoto, Japan, as one of the conferences supported by Research Institute of Mathematical Sciences (RIMS), Kyoto University. The organizer was the second-named editor. About one hundred Japanese scholars and some foreign visitors coming from China, France, Germany and India attended this conference. Talks covered many branches in number theory. The scenery in Kyoto, Arashiyama Mountain and Katsura River impressed foreign visitors. An informal report of this conference was published as the volume 1160 of Sūrikaiseki Kenkyūsho Kōkyūroku (June 2000), published by RIMS, Kyoto University.

The present book is the Proceedings of these two conferences, which records mainly some recent progress in number theory in China and Japan and reflects the academic exchanging between China and Japan.

In China, the founder of modern number theory is Professor Lookeng Hua. His books "Introduction to Number Theory", "Additive Prime Number Theory" and so on have influenced not only younger generations in China but also number theorists in other countries. Professor Hua created the strong tradition of analytic number theory in China. Professor Jingrun Chen did excellent works on Goldbach's conjecture. The report literature of Mr. Chi Xu "Goldbach Conjecture" made many people out of the circle of mathematicians to know something on number theory.

In Japan, the first internationally important number theorist is Professor Teiji Takagi, one of the main contributors to class field theory. His books "Lectures on Elementary Number Theory" and "Algebraic Number Theory" (written in Japanese) are still very useful among Japanese number theorists. Under the influence of Professor Takagi, a large part of research of the first generation of Japanese analytic number theorists such as Professor Zyoiti Suetuna, Professor Tikao Tatuzawa and Professor Takayoshi Mitsui were devoted to analytic problems on algebraic number fields.

Now mathematicians of younger generations have been growing in both countries. It is natural and necessary to exchange in a suitable scale between China and Japan which are near in location and similar in cultural background. In his visiting to Academia Sinica twice, Professor Kanemitsu put forward many good suggestions concerning this matter and pushed relevant activities. This is the initial driving force of the project of the First China-Japan Seminar. Here we would like to thank sincerely Japanese Science Promotion Society and National Science Foundation of China for their great support, Professor Yuan Wang for encouragement and calligraphy, Professor Yasutaka Ihara for his support which made the Kyoto Conference realizable, Professor Shigeru Kanemitsu and Professor Chengbiao Pan for their great effort of promotion.

Since many attendants of the China-Japan Seminar also attended the Kyoto Conference, we decided to make a plan of publishing the joint Proceedings of these two conferences. It was again Professor Kanemitsu who suggested the way of publishing the Proceedings as one volume of the series "Developments in Mathematics", Kluwer Academic Publishers, and made the first contact to Professor Krishnaswami Alladi, the series editor of this series. We greatly appreciate the support of Professor Alladi. We are also indebted to Kluwer for publishing this volume and to Mr. John Martindale and his assistant Ms. Angela Quilici for their constant help.

These Proceedings include 23 papers, most of which were written by participants of at least one of the above conferences. Professor Akio Fujii, one of the invited speakers of the Kyoto Conference, could not attend the conference but contributed a paper. All papers were refereed. We sincerely thank all the authors and the referees for their contributions. Thanks are also due to Dr. Masami Yoshimoto, Dr. Hiroshi Kumagai, Dr. Jun Furuya, Dr. Yumiko Ichihara, Mr. Hidehiko Mishou, Mr. Masatoshi Suzuki, and especially Dr. Yuichi Kamiya for their effort of making files of Kluwer LaTeX style. The contents include several survey or half-survey articles (on prime numbers, divisor problems and Diophantine equations) as well as research papers on various aspects of analytic number theory such as additive problems, Diophantine approximations and the theory of zeta and L-functions. We believe that the contents of the Proceedings reflect well the main body of mathematical activities of the two conferences.

The Second China-Japan Seminar was held from March 12 to 16, 2001, in Iizuka, Fukuoka Prefecture, Japan. The description of this conference will be found in the coming Proceedings. We hope that the prospects of the exchanging on number theory between China and Japan will be as beautiful as Sakura and plum blossom.

April 2001

CHAOHUA JIA AND KOHJI MATSUMOTO (EDITORS)

LIST OF PARTICIPANTS (THE FIRST CHINA-JAPAN SEMINAR, BEIJING)

S. Akiyama	S. Kanemitsu	C. B. Pan
T. Cai	H. Li	Z. Sun
Y. Cai	C. Liu	Y. Tanigawa
X. Cao	J. Liu	I. Wakabayashi
Y. Chen	M. Lu	W. Wang
S. Egami	K. Matsumoto	Y. Wang
X. Gao	Z. Meng	G. Xu
M. Hata	K. Miyake	W. Zhai
C. Jia	L. Murata	W. Zhang
	Y. Nakai	

LIST OF PARTICIPANTS (Kyoto)
(This is only the list of participants who signed the sheet on the desk
at the entrance of the lecture room.)

T. Adachi	K. Kawada	Y. Ohkubo
S. Akiyama	H. Kawai	Y. Ohno
M. Amou	Y. Kitaoka	T. Okano
K. Azuhata	I. Kiuchi	R. Okazaki
J. Brüdern	Takao Komatsu	Y. Okuyama
K. Chinen	Toru Komatsu	Y. Ônishi
S. Egami	Y. Koshiba	T. P. Peneva
J. Furuya	Y. Koya	K. Saito
Y. Gon	S. Koyama	A. Sankaranarayanan
K. Goto	H. Kumagai	Y. Sano
Y. Hamahata	M. Kurihara	H. Sasaki
T. Harase	T. Kuzumaki	R. Sasaki
M. Hata	S. Louboutin	I. Shiokawa
K. Hatada	K. Matsumoto	M. Sudo
T. Hibino	H. Mikawa	T. Sugano
M. Hirabayashi	H. Mishou	M. Suzuki
Y. Ichihara	K. Miyake	S. Suzuki
Y. Ihara	T. Mizuno	I. Takada
M. Ishibashi	R. Morikawa	R. Takeuchi
N. Ishii	N. Murabayashi	A. Tamagawa
Hideaki Ishikawa	L. Murata	J. Tamura
Hirofumi Ishikawa	K. Nagasaka	T. Tanaka
S. Ito	M. Nagata	Y. Tanigawa
C. Jia	H. Nagoshi	N. Terai
T. Kagawa	D. Nakai	T. Toshimitsu
Y. Kamiya	Y. Nakai	Y. Uchida
M. Kan	M. Nakajima	A. Umegaki
S. Kanemitsu	K. Nakamula	I. Wakabayashi
H. Kangetu	I. Nakashima	A. Yagi
T. Kano	M. Nakasuji	M. Yamabe
T. Kanoko	K. Nishioka	S. Yasutomi
N. Kataoka	T. Noda	M. Yoshimoto
M. Katsurada	J. Noguchi	W. Zhang

中日友好

萬古永存

祝中日首屆數論會議在北京召開

庚辰 王元

ON ANALYTIC CONTINUATION
OF MULTIPLE *L*-FUNCTIONS
AND RELATED ZETA-FUNCTIONS

Shigeki AKIYAMA

Department of Mathematics, Faculty of Science, Niigata University, Ikarashi 2-8050, Niigata 950-2181, Japan

akiyama@math.sc.niigata-u.ac.jp

Hideaki ISHIKAWA

Graduate school of Natural Science, Niigata University, Ikarashi 2-8050, Niigata 950-2181, Japan

isikawah@qed.sc.niigata-u.ac.jp

Keywords: Multiple *L*-function, Multiple Hurwitz zeta function, Euler-Maclaurin summation formula

Abstract   A multiple *L*-function and a multiple Hurwitz zeta function of Euler-Zagier type are introduced. Analytic continuation of them as complex functions of several variables is established by an application of the Euler-Maclaurin summation formula. Moreover location of singularities of such zeta functions is studied in detail.

1991 Mathematics Subject Classification: Primary 11M41; Secondary 32Dxx, 11Mxx, 11M35.

1.   INTRODUCTION

Analytic continuation of Euler-Zagier's multiple zeta function of two variables was first established by F. V. Atkinson [3] with an application to the mean value problem of the Riemann zeta function. We can find recent developments in [8], [7] and [5]. From an analytic point of view, these results suggest broad applications of multiple zeta functions. In [9] and [10], D. Zagier pointed out an interesting interplay between positive integer values and other areas of mathematics, which include knot theory and mathematical physics. Many works had been done according to his motivation but here we restrict our attention to the analytic contin-

1

C. Jia and K. Matsumoto (eds.), Analytic Number Theory, 1–16.
© *2002 Kluwer Academic Publishers. Printed in the Netherlands.*

uation. T. Arakawa and M. Kaneko [2] showed an analytic continuation with respect to the last variable. To speak about the analytic continuation with respect to all variables, we have to refer to J. Zhao [11] and S. Akiyama, S. Egami and Y. Tanigawa [1]. In [11], an analytic continuation and the residue calculation were done by using the theory of generalized functions in the sense of I. M. Gel'fand and G. E. Shilov. In [1], they gave an analytic continuation by means of a simple application of the Euler-Maclaurin formula. The advantage of this method is that it gives the complete location of singularities. This work also includes some study on the values at non positive integers.

In this paper we consider a more general situation, which seems important for number theory, in light of the method of [1]. We shall give an analytic continuation of multiple Hurwitz zeta functions (Theorem 1) and also multiple *L*-functions (Theorem 2) defined below. In special cases, we can completely describe the whole set of singularities, by using a property of zeros of Bernoulli polynomials (Lemma 4) and a non vanishing result on a certain character sum (Lemma 2).

We explain notations used in this paper. The set of rational integers is denoted by $\mathbb{Z}$, the rational numbers by $\mathbb{Q}$, the complex numbers by $\mathbb{C}$ and the positive integers by $\mathbb{N}$. We write $\mathbb{Z}_{\leq \ell}$ for the integers not greater than ℓ. Let χ_i $(i = 1, 2, \ldots, k)$ be Dirichlet characters of the same conductor $q \geq 2$ and β_i $(i = 1, 2, \cdots, k)$ be real numbers in the half open interval $[0, 1)$. The principal character is denoted by χ_0. Then multiple Hurwitz zeta function and multiple *L*-function are defined respectively by:

$$\zeta_k(s_1, \ldots, s_k \mid \beta_1, \ldots, \beta_k)$$

$$= \sum_{0 < n_1 < \cdots < n_k} \frac{1}{(n_1 + \beta_1)^{s_1}(n_2 + \beta_2)^{s_2} \ldots (n_k + \beta_k)^{s_k}} \quad (1)$$

and

$$L_k(s_1, \ldots, s_k \mid \chi_1, \ldots, \chi_k) = \sum_{0 < n_1 < \cdots < n_k} \frac{\chi_1(n_1)}{n_1^{s_1}} \frac{\chi_2(n_2)}{n_2^{s_2}} \cdots \frac{\chi_k(n_k)}{n_k^{s_k}}, \quad (2)$$

where $n_i \in \mathbb{N}$ $(i = 1, \ldots, k)$. If $\Re(s_i) \geq 1$ $(i = 1, 2, \ldots, k - 1)$ and $\Re(s_k) > 1$, then these series are absolutely convergent and define holomorphic functions of k complex variables in this region. In the sequel we write them by $\zeta_k(s \mid \beta)$ and $L_k(s \mid \chi)$, for abbreviation. The Hurwitz zeta function $\zeta(s, \alpha)$ in the usual sense for $\alpha \in (0, 1)$ is written as

$$\zeta(s, \alpha) = \sum_{n=0}^{\infty} \frac{1}{(n + \alpha)^s} = \frac{1}{\alpha^s} + \zeta_1(s \mid \alpha),$$

by the above notation.

We shall state the first result. Note that $\beta_j - \beta_{j+1} = 1/2$ for some j implies $\beta_{j-1} - \beta_j \neq 1/2$, since $\beta_j \in [0, 1)$.

Theorem 1. *The multiple Hurwitz zeta function* $\zeta_k(s \mid \beta)$ *is meromorphically continued to* $\mathbb{C}^k$ *and has possible singularities on:*

$$s_k = 1, \quad \sum_{i=1}^{j} s_{k-i+1} \in \mathbb{Z}_{\leq j} \ (j = 2, 3, \ldots, k).$$

Let us assume furthermore that all β_i ($i = 1, \ldots, k$) *are rational. If* $\beta_{k-1} - \beta_k$ *is not 0 nor 1/2, then the above set coincides with the set of whole singularities. If* $\beta_{k-1} - \beta_k = 1/2$ *then*

$$s_k = 1$$
$$s_{k-1} + s_k = 2, 0, -2, -4, -6, \ldots$$
$$\sum_{i=1}^{j} s_{k-i+1} \in \mathbb{Z}_{\leq j} \quad \text{for } j = 3, 4, \ldots, k$$

forms the set of whole singularities. If $\beta_{k-1} - \beta_k = 0$ *then*

$$s_k = 1$$
$$s_{k-1} + s_k = 2, 1, 0, -2, -4, -6, \ldots$$
$$\sum_{i=1}^{j} s_{k-i+1} \in \mathbb{Z}_{\leq j} \quad \text{for } j = 3, 4, \ldots, k$$

forms the set of whole singularities.

For the simplicity, we only concerned with special cases and determined the whole set of singularities in Theorem 1. The reader can easily handle the case when all $\beta_i - \beta_{i+1}$ ($i = 1, \ldots, k - 1$) are not necessary rational and fixed. So we have enough information on the location of singularities of multiple Hurwitz zeta functions. For the case of multiple L-functions, our knowledge is rather restricted.

Theorem 2. *The multiple L-function* $L_k(s \mid \chi)$ *is meromorphically continued to* $\mathbb{C}^k$ *and has possible singularities on:*

$$s_k = 1, \quad \sum_{i=1}^{j} s_{k-i+1} \in \mathbb{Z}_{\leq j} \ (j = 2, 3, \ldots, k).$$

Especially for the case $k = 2$, we can state the location of singularities in detail as follows:

Corollary 1. *We have a meromorphic continuation of $L_2(s \,|\, \chi)$ to $\mathbb{C}^2$.*
$L_2(s \,|\, \chi)$ is holomorphic in

$$
\begin{cases}
\{(s_1, s_2) \in \mathbb{C}^2 \mid s_1 + s_2 \notin \mathbb{Z}_{\leq 2}, \quad s_2 \neq 1\} \\
\qquad\qquad\qquad\qquad\qquad if \quad \chi_1 = \chi_0, \quad \chi_2 = \chi_0 \\
\{(s_1, s_2) \in \mathbb{C}^2 \mid s_1 + s_2 \notin \mathbb{Z}_{\leq 1}, \quad s_2 \neq 1\} \\
\qquad\qquad\qquad\qquad\qquad if \quad \chi_1 \neq \chi_0, \quad \chi_2 = \chi_0 \\
\{(s_1, s_2) \in \mathbb{C}^2 \mid s_1 + s_2 \notin \mathbb{Z}_{\leq 1}\} \qquad if \quad \chi_2 \neq \chi_0,
\end{cases}
\tag{3}
$$

*where the excluded sets are possible singularities. Suppose that χ_1 and χ_2
are primitive characters with $\chi_1\chi_2 \neq \chi_0$. Then $L_2(s \,|\, \chi)$ is a holomorphic
function in*

$$
\begin{cases}
\{(s_1, s_2) \in \mathbb{C}^2 \mid s_1 + s_2 \neq 0, -2, -4, -6, -8, \dots\} \\
\qquad\qquad\qquad\qquad\qquad if \quad \chi_1\chi_2(-1) = 1, \\
\{(s_1, s_2) \in \mathbb{C}^2 \mid s_1 + s_2 \neq 1, -1, -3, -5, -7, \dots\} \\
\qquad\qquad\qquad\qquad\qquad if \quad \chi_1\chi_2(-1) = -1,
\end{cases}
\tag{4}
$$

where the excluded set forms the whole set of singularities.

Unfortunately the authors could not get the complete description of
singularities of multiple L-function for $k \geq 3$.

2.   PRELIMINARIES

Let $N_1, N_2 \in \mathbb{N}$ and η be a real number. Suppose that a function
$f(x)$ is $l+1$ times continuously differentiable. By using Stieltjes integral
expression, we see

$$
\sum_{N_1+\eta < n \leq N_2} f(n) = \int_{N_1+\eta}^{N_2} f(x)d[x]
$$

$$
= \int_{N_1+\eta}^{N_2} f(x)dx - [f(x)\tilde{B}_1(x)]_{N_1+\eta}^{N_2}
$$

$$
+ \int_{N_1+\eta}^{N_2} f'(x)\tilde{B}_1(x)dx
$$

where $\tilde{B}_j(x) = B_j(x - [x])$ is the j-th *periodic Bernoulli polynomial*.
Here j-th Bernoulli polynomial $B_j(x)$ is defined by

$$
\frac{te^{xt}}{e^t - 1} = \sum_{j=0}^{\infty} \frac{B_j(x)}{j!} t^j
$$

and $[x]$ is the largest integer not exceeding x. Define the Bernoulli number B_r by the value $B_r = B_r(0)$. Repeating integration by parts,

$$\sum_{N_1+\eta<n\leq N_2} f(n) = \int_{N_1+\eta}^{N_2} f(x)dx + \frac{1}{2}f(N_2) + f(N_1+\eta)\tilde{B}_1(\eta)$$

$$+\sum_{r=1}^{l} \frac{(-1)^{r+1}}{(r+1)!} \left(B_{r+1}f^{(r)}(N_2) - f^{(r)}(N_1+\eta)\tilde{B}_{r+1}(\eta) \right)$$

$$-\frac{(-1)^{l+1}}{(l+1)!} \int_{N_1+\eta}^{N_2} f^{(l+1)}(x)\tilde{B}_{l+1}(x)dx. \tag{5}$$

When $\eta = 0$, the formula (5) is nothing but the standard Euler-Maclaurin summation formula. This slightly modified summation formula by a parameter η works quite fine in studying our series (1) and (2).

Lemma 1. *Let*

$$\Phi_l(s \mid N_1+\eta, \alpha) = \frac{(s)_{l+1}}{(l+1)!} \int_{N_1+\eta}^{\infty} \frac{\tilde{B}_{l+1}(x)}{(x+\alpha)^{s+l+1}}dx$$

and

$$(s)_r = \begin{cases} s(s+1)(s+2)\dots(s+r-1) & \text{if} \quad r \geq 1 \\ 1 & \text{if} \quad r = 0 \\ (s-1)^{-1} & \text{if} \quad r = -1 \end{cases}.$$

Then it follows that

$$\sum_{N_1+\eta<n} \frac{1}{(n+\alpha)^s} = \sum_{r=-1}^{l} \frac{\tilde{B}_{r+1}(\eta)}{(r+1)!} \frac{(s)_r}{(N_1+\alpha+\eta)^{s+r}} - \Phi_l(s \mid N_1+\eta, \alpha)$$

with

$$\Phi_l(\dot{s} \mid N_1+\eta, \alpha) \ll N_1^{-(\Re s+l+1)}.$$

Proof. Put $f(x) = (x+\alpha)^{-s}$. Then we have $f^{(r)}(x) = (-1)^r(s)_r(x+\alpha)^{-s-r}$. So from (5),

$$\sum_{N_1+\eta<n\leq N_2}\frac{1}{(n+\alpha)^s}=\left[\frac{1}{1-s}\frac{1}{(x+\alpha)^{s-1}}\right]_{N_1+\eta}^{N_2}$$

$$+\frac{1}{2}\frac{1}{(N_2+\alpha)^s}+\frac{\tilde{B}_1(\eta)}{(N_1+\eta+\alpha)^s}$$

$$-\sum_{r=1}^{l}\frac{(s)_r}{(r+1)!}\left(\frac{B_{r+1}}{(N_2+\alpha)^{s+r}}-\frac{\tilde{B}_{r+1}(\eta)}{(N_1+\eta+\alpha)^{s+r}}\right)$$

$$-\frac{1}{(l+1)!}\int_{N_1+\eta}^{N_2}\tilde{B}_{l+1}(x)\frac{(s)_{l+1}}{(x+\alpha)^{s+l+1}}dx.$$

When $\Re s > 1$, we have

$$\sum_{N_1+\eta<n}\frac{1}{(n+\alpha)^s}=\frac{1}{s-1}\frac{1}{(N_1+\alpha+\eta)^{s-1}}+\frac{\tilde{B}_1(\eta)}{(N_1+\eta+\alpha)^s}$$

$$+\sum_{r=1}^{l}\frac{\tilde{B}_{r+1}(\eta)}{(r+1)!}\frac{(s)_r}{(N_1+\alpha+\eta)^{s+r}}-\Phi_l(s\mid N_1+\eta,\alpha)$$

as $N_2\to\infty$. When $\Re s\leq 1$, if we take a sufficiently large l, the integral in the last term $\Phi_l(s\mid N_1+\eta,\alpha)$ is absolutely convergent. Thus this formula gives an analytic continuation of the series of the left hand side. Performing integration by parts once more and comparing two expressions, it can be easily seen that $\Phi_l(s\mid N_1+\eta,\alpha)\ll N_1^{-(\Re s+l+1)}$. $\square$

Let $A_{\chi_1,\chi_2}(j)$ be the sum

$$\sum_{a_1=1}^{q-1}\sum_{a_2=1}^{q-1}\chi_1(a_1)\chi_2(a_2)\tilde{B}_j(\frac{a_1-a_2}{q}).$$

Lemma 2. *Suppose χ_1 and χ_2 are primitive characters modulo q with $\chi_1\chi_2\neq\chi_0$. Then we have: for $1\leq j$*

$$A_{\chi_1,\chi_2}(j)=\begin{cases}-2\dfrac{j!}{(2\pi i)^j}\tau(\chi_1)\overline{\tau(\overline{\chi_2})}L(j,\overline{\chi_1\chi_2}) & \text{if } (-1)^j\chi_1\chi_2(-1)=1,\\[2mm]0 & \text{if } (-1)^j\chi_1\chi_2(-1)=-1,\end{cases}$$

where $\tau(\chi)$ is the Gauss sum defined by $\tau(\chi)=\sum_{u=0}^{q-1}\chi(u)e^{2\pi iu/q}$.

Proof. Recall the Fourier expansion of Bernoulli polynomial:

$$\tilde{B}_j(y)=-j!\lim_{M\to\infty}\sum_{\substack{n=-M\\n\neq 0}}^{M}\frac{e^{2\pi iny}}{(2\pi in)^j}\qquad(6)$$

for $1 \leq j$, $0 \leq y < 1$ except $(j, y) = (1, 0)$. First suppose $j \geq 2$, then the right hand side of (6) is absolutely convergent. Thus it follows from (6) that

$$A_{\chi_1,\chi_2}(j) = \sum_{a_1=1}^{q-1} \sum_{a_2=1}^{q-1} \chi_1(a_1)\chi_2(a_2)\left(-j! \sum_{\substack{n=-\infty \\ n\neq 0}}^{\infty} \frac{\exp[2\pi i n \frac{a_1-a_2}{q}]}{(2\pi i n)^j}\right).$$

Since

$$\sum_{u=0}^{q-1} \chi(u)e^{2\pi i n u/q} = \bar{\chi}(n)\tau(\chi)$$

for a primitive character χ, we have

$$A_{\chi_1,\chi_2}(j) = -j! \sum_{\substack{n\neq 0 \\ n=-\infty}}^{\infty} \frac{\bar{\chi}_1(n)\bar{\chi}_2(-n)\tau(\chi_1)\tau(\chi_2)}{(2\pi i n)^j},$$

from which the assertion follows immediately by the relation $\overline{\tau(\bar{\chi})} = \chi(-1)\tau(\chi)$. Next assume that $j = 1$. Divide $A_{\chi_1,\chi_2}(1)$ into

$$A_{\chi_1,\chi_2}(1) = \sideset{}{'}\sum \chi_1(a_1)\chi_2(a_2)\tilde{B}_1\left(\frac{a_1-a_2}{q}\right) + \sum_{a_1=1}^{q-1} \chi_1\chi_2(a_1)B_1 \quad (7)$$

where $\sum'$ taken over all the terms $1 \leq a_1, a_2 \leq q - 1$ with $a_1 \neq a_2$. The second sum in (7) is equal to 0 by the assumption. By using (6), the first sum is

$$\sideset{}{'}\sum \chi_1(a_1)\chi_2(a_2)\tilde{B}_1\left(\frac{a_1-a_2}{q}\right)$$

$$= \sideset{}{'}\sum \chi_1(a_1)\chi_2(a_2) \lim_{M\to\infty}\left(-\sum_{\substack{n=-M \\ n\neq 0}}^{M} \frac{\exp[2\pi i n\frac{a_1-a_2}{q}]}{2\pi i n}\right)$$

$$= -\lim_{M\to\infty} \sum_{\substack{n=-M \\ n\neq 0}}^{M} \frac{\displaystyle\sum_{a_1=1}^{q-1}\sum_{a_2=1}^{q-1} \chi_1(a_1)\chi_2(a_2)\exp[2\pi i n\frac{a_1-a_2}{q}] - \sum_{a_1=1}^{q-1} \chi_1\chi_2(a_1)}{2\pi i n}.$$

Using $\sum_{a_1=1}^{q-1} \chi_1\chi_2(a_1) = 0$ again, we have

$$\sideset{}{'}\sum \chi_1(a_1)\chi_2(a_2)\tilde{B}_1\left(\frac{a_1 - a_2}{q}\right)$$

$$= -\frac{1}{2\pi i} \lim_{M\to\infty} \sum_{\substack{n=-M \\ n\neq 0}}^{M} \frac{\bar{\chi}_1(n)\bar{\chi}_2(-n)\tau(\chi_1)\tau(\chi_2)}{n}.$$

Hence we get the result. □

We recall the classical theorem of von Staudt & Clausen.

Lemma 3.

$$B_{2n} + \sum_{p-1\,|\,2n} \frac{1}{p} \qquad \text{is an integer.}$$

Here the summation is taken over all primes p such that $p-1$ divides $2n$.

Extending the former results of D. H. Lehmer and K. Inkeri, the distribution of zeros of Bernoulli polynomials is extensively studied in [4], where one can find a lot of references. On rational zeros, we quote here the result of [6].

Lemma 4. *Rational zeros of Bernoulli polynomial $B_n(x)$ must be $0, 1/2$ or 1. These zeros occur when and only when in the following cases:*

$$\begin{cases} B_n(0) = B_n(1) = 0 & n \text{ is odd} \quad n \geq 3 \\ B_n(1/2) = 0 & n \text{ is odd} \quad n \geq 1. \end{cases} \tag{8}$$

We shall give its proof, for the convenience of the reader.

Proof. First we shall show that if $B_n(\gamma) = 0$ with $\gamma \in \mathbb{Q}$ then $2\gamma \in \mathbb{Z}$. The Bernoulli polynomial is explicitly written as

$$B_n(x) = \sum_{k=0}^{n} \binom{n}{k} B_k x^{n-k}.$$

Let $\gamma = P/Q$ with $P, Q \in \mathbb{Z}$ and P, Q are coprime. Then we have

$$-\frac{P^n}{Q} = nB_1 P^{n-1} + \sum_{k=2}^{n} \binom{n}{k} B_k P^{n-k} Q^{k-1}.$$

Assume that there exists a prime factor $q \geq 3$ of Q. Then the right hand side is q-integral. Indeed, we see that $B_1 = -1/2$ and qB_k is q-integral since the denominator of B_k is always square free, which is an easy consequence of Lemma 3. But the left hand side is not q-integral, we get a contradiction. This shows that Q must be a power of 2. Let $Q = 2^m$ with a non negative integer m. Then we have

$$-\frac{P^n}{2^{m-1}} = -nP^{n-1} + \sum_{k=2}^{n} \binom{n}{k} B_k P^{n-k} 2^{m(k-1)+1}.$$

If $m \geq 2$ then we get a similar contradiction. Thus Q must divide 2, we see $2\gamma \in \mathbb{Z}$. Now our task is to study that values of Bernoulli polynomials at half integers. Since $B_0(x) = 1$ and $B_1(x) = x - 1/2$, the assertion is obvious if $n < 2$. Assume that $n \geq 2$ and even. Then by Lemma 3, the denominator of B_n is divisible by 3. Recalling the relation

$$B_n(x+1) - B_n(x) = nx^{n-1}, \tag{9}$$

we have for any integer m

$$B_n(m) \quad B_n \quad (\text{mod } \mathbb{Z}).$$

This shows $B_n(m) \not/ 0$. The relation

$$B_n(1/2) = (2^{1-n} - 1)B_n \tag{10}$$

implies that

$$B_n(1/2) \neq 0 \tag{11}$$

for $n \geq 2$ and even. We see that $B_n(1/2)$ is not 3 integral from Lemma 3 and the relation (10). Combining (9), (11), we have for any integer m and any even integer $n \geq 2$

$$B_n(1/2 + m) \neq 0.$$

It is easy to show the assertion for the remaining case when $n \geq 2$ is odd, by using (9) and (10). $\qquad\qquad\square$

3. ANALYTIC CONTINUATION OF MULTIPLE HURWITZ ZETA FUNCTIONS

This section is devoted to the proof of Theorem 1. First we treat the double Hurwitz zeta function. By Lemma 1, we see

$$\sum_{n_1 + (\beta_1 - \beta_2) < n_2} \frac{1}{(n_2 + \beta_2)^{s_2}}$$

$$= \sum_{r=-1}^{l} \frac{\tilde{B}_{r+1}(\beta_1 - \beta_2)}{(r+1)!} \frac{(s_2)_r}{(n_1 + \beta_1)^{s_2+r}} - \Phi_l(s_2 \mid n_1 + \beta_1 - \beta_2, \beta_2). \tag{12}$$

Suppose first that $\beta_1 \geq \beta_2$. Then the sum $\sum_{n_1+\beta_1-\beta_2<n_2}$ means $\sum_{n_1<n_2}$, so it follows from (12) that

$$
\begin{aligned}
\zeta_2(s \mid \beta) &= \sum_{n_1=1}^{\infty} \frac{1}{(n_1 + \beta_1)^{s_1}} \sum_{n_1<n_2} \frac{1}{(n_2 + \beta_2)^{s_2}} \\
&= \sum_{n_1=1}^{\infty} \frac{1}{(n_1 + \beta_1)^{s_1}} \sum_{n_1+(\beta_1-\beta_2)<n_2} \frac{1}{(n_2 + \beta_2)^{s_2}} \\
&= \sum_{n_1=1}^{\infty} \frac{1}{(n_1 + \beta_1)^{s_1}} \left\{ \sum_{r=-1}^{l} \frac{\tilde{B}_{r+1}(\beta_1 - \beta_2)}{(r+1)!} \frac{(s_2)_r}{(n_1 + \beta_1)^{s_2+r}} \right. \\
&\qquad \left. - \Phi_l(s_2 \mid n_1 + \beta_1 - \beta_2, \beta_2) \right\}.
\end{aligned}
$$

Suppose that $\beta_1 < \beta_2$. We consider

$$
\zeta_2(s \mid \beta) = \sum_{n_1=1}^{\infty} \frac{1}{(n_1 + \beta_1)^{s_1}} \left\{ \frac{-1}{(n_1 + \beta_2)^{s_2}} + \sum_{n_1 \leq n_2} \right\}. \tag{13}
$$

Noting that the sum $\sum_{n_1+\beta_1-\beta_2<n_2}$ means $\sum_{n_1 \leq n_2}$, we apply (12) to the second term in the braces. For the first term in the braces, we use the binomial expansion:

$$
\frac{1}{(n_1 + \beta_2)^{s_2}} = \frac{1}{(n_1 + \beta_1)^{s_2}} \left(\sum_{m=0}^{v} \frac{(-1)^m (s_2)_m}{m!} \left(\frac{\beta_2 - \beta_1}{n_1 + \beta_1} \right)^m + R_{v+1} \right) \tag{14}
$$

with $R_{v+1} \ll n_1^{v+1}$. By applying (12) and (14) to (13), we have

$$
\begin{aligned}
\zeta_2(s \mid \beta) &= \sum_{n_1=1}^{\infty} \frac{1}{(n_1 + \beta_1)^{s_1}} \left\{ \frac{1}{s_2 - 1} \frac{1}{(n_1 + \beta_1)^{s_2-1}} \right. \\
&\qquad + \sum_{r=0}^{l} \left(\frac{\tilde{B}_{r+1}(\beta_1 - \beta_2)}{(r+1)!} - \frac{(\beta_1 - \beta_2)^r}{r!} \right) \frac{(s_2)_r}{(n_1 + \beta_1)^{s_2+r}} \\
&\qquad \left. - \Phi_l(s_2 \mid n_1 + \beta_1 - \beta_2, \beta_2) - \frac{R_{l+1}}{(n_1 + \beta_1)^{s_2}} \right\}.
\end{aligned}
$$

Recalling the relation (9) and combining the cases $\beta_1 \leq \beta_2$ and $\beta_1 > \beta_2$, we have

$$
\zeta_2(s \mid \beta) = \frac{1}{s_2 - 1} \zeta_1(s_1 + s_2 - 1 \mid \beta_1)
$$
$$
+ \sum_{r=0}^{l} \frac{B_{r+1}(\beta_1 - \beta_2)}{(r+1)!} (s_2)_r \zeta_1(s_1 + s_2 + r \mid \beta_1)
$$
$$
- \sum_{n_1=1}^{\infty} \frac{\Phi_l^*(s_2 \mid n_1 + \beta_1 - \beta_2, \beta_2)}{(n_1 + \beta_1)^{s_1}} \tag{15}
$$

where

$$
\Phi_l^*(s_2 \mid n_1 + \beta_1 - \beta_2, \beta_2) = \begin{cases} \Phi_l(s_2 \mid n_1 + \beta_1 - \beta_2, \beta_2) & \text{if } \beta_1 \geq \beta_2 \\[2mm] \Phi_l(s_2 \mid n_1 + \beta_1 - \beta_2, \beta_2) + \dfrac{R_{l+1}}{(n_1 + \beta_1)^{s_2}} \\ \qquad\qquad\qquad\qquad \text{if } \beta_1 < \beta_2. \end{cases}
$$

The right hand side in (15) has meromorphic continuation except the last term. The last summation is absolutely convergent, and hence holomorphic, in $\Re(s_1 + s_2 + l) > 0$. Thus we now have a meromorphic continuation to $\Re(s_1 + s_2 + l) > 0$. Since we can choose arbitrary large l, we get a meromorphic continuation of $\zeta_2(s \mid \beta)$ to $\mathbb{C}^2$, holomorphic in

$$
\{(s_1, s_2) \in \mathbb{C}^2 \mid s_2 \neq 1, \quad s_1 + s_2 \neq 2, 1, 0, -1, -2, -3, \dots \}.
$$

The exceptions in this set are the possible singularities occurring in $(s_2 - 1)^{-1}$ and

$$
\frac{B_{r+1}(\beta_1 - \beta_2)}{(r+1)!} (s_2)_r \zeta_1(s_1 + s_2 + r \mid \beta_1). \tag{16}
$$

Whether they are 'real singularity' or not depends on the choice of parameters β_i $(i = 1, 2)$. For the case of multiple Hurwitz zeta functions with k variables,

$$\zeta_k(s_1, \ldots, s_k \mid \beta_1, \ldots, \beta_k)$$

$$= \sum_{n_1=1}^{\infty} \frac{1}{(n_1 + \beta_1)^{s_1}} \sum_{n_1 < n_2} \frac{1}{(n_2 + \beta_2)^{s_2}} \cdots$$

$$\cdots \sum_{n_{k-2} < n_{k-1}} \frac{1}{(n_{k-1} + \beta_{k-1})^{s_{k-1}}} \left\{ \frac{1}{s_k - 1} \frac{1}{(n_{k-1} + \beta_{k-1})^{s_k - 1}} \right.$$

$$+ \sum_{r=0}^{l} \frac{B_{r+1}(\beta_{k-1} - \beta_k)}{(r+1)!} \frac{(s_k)_r}{(n_{k-1} + \beta_{k-1})^{s_k + r}}$$

$$\left. - \Phi_l^*(s_k \mid n_{k-1} + \beta_{k-1} - \beta_k, \beta_k) \right\}$$

$$= \frac{1}{s_k - 1} \zeta_{k-1}(s_1, \ldots, s_{k-2}, s_{k-1} + s_k - 1 \mid \beta_1, \ldots, \beta_{k-1})$$

$$+ \sum_{r=0}^{l} \frac{B_{r+1}(\beta_{k-1} - \beta_k)}{(r+1)!} (s_k)_r$$

$$\times \zeta_{k-1}(s_1, \ldots, s_{k-2}, s_{k-1} + s_k + r \mid \beta_1, \ldots, \beta_{k-1})$$

$$- \sum_{0 < n_1 < n_2 < \cdots < n_{k-1}} \frac{\Phi_l^*(s_k \mid n_{k-1} + \beta_{k-1} - \beta_k, \beta_k)}{(n_1 + \beta_1)^{s_1} \cdots (n_{k-1} + \beta_{k-1})^{s_{k-1}}}.$$

Since

$$\sum_{0 < n_1 < \cdots < n_{k-1}} \frac{\Phi_l^*(s_k \mid n_{k-1} + \beta_{k-1} - \beta_k, \beta_k)}{(n_1 + \beta_1)^{s_1}(n_2 + \beta_2)^{s_2} \ldots (n_{k-1} + \beta_{k-1})^{s_{k-1}}}$$

$$\ll \sum_{n_{k-1}} \frac{n_{k-1}^{-l - \Re(s_k) + k - 3}}{n_{k-1}^L}$$

with $L = \Re(s_{k-1}) + \sum_{1 \le j \le k-2, \Re(s_i) \le 0} \Re(s_i)$, the last summation is convergent absolutely in

$$l - k + 2 + \Re(s_{k-1}) + \Re(s_k) + \sum_{\substack{1 \le j \le k-2 \\ \Re(s_i) \le 0}} \Re(s_i) > 0.$$

Since l can be taken arbitrarily large, we get an analytic continuation of $\zeta_k(s \mid \beta)$ to $\mathbb{C}^k$. Now we study the set of singularities more precisely. The 'singular part' of $\zeta_2(s \mid \beta)$ is

$$\frac{\zeta_1(s_1 + s_2 - 1 \mid \beta_1)}{s_2 - 1} + \sum_{r=0}^{\infty} \frac{(s_2)_r}{s_1 + s_2 + r - 1} \frac{B_{r+1}(\beta_1 - \beta_2)}{(r+1)!}.$$

Note that this sum is by no means convergent and just indicates local singularities. From this expression we see

$$s_2 = 1, \quad s_1 + s_2 \in \{2, 1, 0, -1, -2, -3, -4, \dots\}$$

are possible singularities and the second assertion of Theorem 1 for $k = 2$ is now clear with the help of Lemma 4. We wish to determine the whole singularities when all β_i $(i = 1, \dots, k)$ are rational numbers by an induction on k. Let us consider the case of k variables,

$$\zeta_k(s \mid \beta) = \sum_{r=-1}^{l} \frac{B_{r+1}(\beta_{k-1} - \beta_k)}{(r+1)!} (s_k)_r$$

$$\times \zeta_{k-1}(s_1, \dots, s_{k-2}, s_{k-1} + s_k + r \mid \beta_1, \dots \beta_{k-1})$$

$$- \sum_{0 < n_1 < \cdots < n_{k-1}} \frac{\Phi_l^*(s_k \mid n_{k-1} + \beta_{k-1} - \beta_k, \beta_k)}{(n_1 + \beta_1)^{s_1} \dots (n_{k-1} + \beta_{k-1})^{s_{k-1}}}.$$

We shall only prove the case when $\beta_{k-1} - \beta_k = 0$. Other cases are left to the reader. By the induction hypothesis and Lemma 4 the singularities lie on, at least for $r = -1, 0, 1, 3, 5, 7, \dots$,

$$s_k = 1, \quad s_{k-1} + s_k + r = 1,$$

$$s_{k-2} + s_{k-1} + s_k + r = 2, 0, -2, -4, -6, \dots$$

and

$$s_{k-j} + s_{k-j+1} + \cdots + s_k + r \in \mathbb{Z}_{\leq j}, \quad \text{for } j \geq 3$$

in any three cases; $\beta_{k-2} - \beta_{k-1} = 0, 1/2$, and otherwise.
Thus

$$s_k = 1, \quad s_{k-1} + s_k = 2, 1, 0, -2, -4, -6, \dots$$

and

$$s_{k-j+1} + s_{k-j+2} + \cdots + s_k \in \mathbb{Z}_{\leq j}, \quad \text{for } j \geq 3$$

are the possible singularities, as desired. Note that the singularities of the form

$$s_{k-2} + s_{k-1} + s_k + r = 1, -1, -3, -5, \dots$$

may appear. However, these singularities don't affect our description. Next we will show that they are the 'real' singularities. For example, the singularities of the form $s_{k-2} + s_{k-1} + s_k = \eta$ occurs in several ways for a fixed η. So our task is to show that no singularities defined by one of the above equations will identically vanish in the summation process. This can be shown by a small trick of replacing variables:

$$s_1 = u_1, \dots, s_{k-2} = u_{k-2}, \quad s_{k-1} + s_k = u_{k-1}, \quad s_k = u_k$$

In fact, we see that the singularities of

$$\zeta_k(u_1, \ldots, u_{k-2}, u_{k-1} - u_k, u_k \mid \beta_1, \ldots, \beta_k)$$

appear in

$$\sum_{r=-1}^{l} \frac{B_r(\beta_{k-1} - \beta_k)}{(r+1)!} (u_k)_r \zeta_{k-1}(u_1, \ldots, u_{k-2}, u_{k-1} + r \mid \beta_1, \ldots \beta_{k-1}).$$

By this expression we see that the singularities of $\zeta_{k-1}(u_1, \ldots, u_{k-1} + r \mid \beta_1, \ldots \beta_{k-1})$ are summed with functions of u_k of *different* degree. Thus these singularities, as weighted sum by another variable u_k, will not vanish identically. This argument seems to be an advantage of [1], which clarify the exact location of singularities. The Theorem is proved by the induction.   □

4.   ANALYTIC CONTINUATION OF MULTIPLE *L*-FUNCTIONS

Proof of Theorem 2. When $\Re s_i > 1$ for $i = 1, 2, \ldots, k$, the series is absolutely convergent. Rearranging the terms,

$$\sum_{n_1 < \cdots < n_k} \frac{\chi_1(n_1)}{n_1^{s_1}} \cdots \frac{\chi_k(n_k)}{n_k^{s_k}}$$

$$= \frac{1}{q^{s_1 + \cdots + s_k}} \sum_{a_1=1}^{q-1} \sum_{a_2=1}^{q-1} \cdots \sum_{a_k=1}^{q-1} \chi_1(a_1)\chi_2(a_2) \ldots \chi_k(a_k)$$

$$\times \left[\sum_{m_1=0}^{\infty} \frac{1}{(m_1 + \frac{a_1}{q})^{s_1}} \sum_{m_1 + \frac{a_1 - a_2}{q} < m_2} \frac{1}{(m_2 + \frac{a_2}{q})^{s_2}} \cdots \right.$$

$$\left. \cdots \sum_{m_{k-2} + \frac{a_{k-2} - a_{k-1}}{q} < m_{k-1}} \frac{1}{(m_{k-1} + \frac{a_{k-1}}{q})^{s_{k-1}}} \sum_{m_{k-1} + \frac{a_{k-1} - a_k}{q} < m_k} \frac{1}{(m_k + \frac{a_k}{q})^{s_k}} \right].$$

By this expression, it suffices to show that the series in the last brace has the desirable property. When $a_i - a_{i+1} \geq 0$ holds for $i = 1, \ldots, k-1$, this is clear form Theorem 1, since this series is just a multiple Hurwitz zeta function. Proceeding along the same line with the proof of Theorem 1, other cases are also easily deduced by recursive applications of Lemma 1. Since there are no need to use binomial expansions, this case is easier than before.   □

Proof of Corollary 1. Considering the case $k = 2$ in Theorem 2, we see

$$L_2(s \mid \chi) = \frac{1}{q} \frac{\sum\limits_{a_2=1}^{q-1} \chi_2(a_2)}{s_2 - 1} L(s_1 + s_2 - 1, \chi_1)$$

$$+ \frac{1}{q^{s_1+s_2}} \sum_{a_1=1}^{q-1} \sum_{a_2=1}^{q-1} \chi_1(a_1)\chi_2(a_2)$$

$$\times \left\{ \sum_{r=0}^{l} \frac{\tilde{B}_{r+1}(\frac{a_1-a_2}{q})}{(r+1)!} (s_2)_r \zeta(s_1 + s_2 + r, \frac{a_1}{q}) \right.$$

$$\left. - \sum_{m_1=0}^{\infty} \frac{\Phi_l(s_2 \mid m_1 + \frac{a_1-a_2}{q}, \frac{a_2}{q})}{(m_1 + \frac{a_1}{q})^{s_1}} \right\}.$$

We have a meromorphic continuation of $L_2(s \mid \chi)$ to $\mathbb{C}^2$, which is holo-morphic in the domain (3). Note that the singularities occur in

$$\frac{\sum\limits_{a_2=1}^{q-1} \chi_2(a_2)}{s_2 - 1} L(s_1 + s_2 - 1, \chi_1)$$

and

$$\sum_{a_1=1}^{q} \sum_{a_2=1}^{q} \chi_1(a_1)\chi_2(a_2) \frac{\tilde{B}_{r+1}(\frac{a_1-a_2}{q})}{(r+1)!} (s_2)_r \zeta(s_1 + s_2 + r, \frac{a_1}{q}).$$

If χ_2 is not principal then the first term vanishes and we see the 'singular part' is

$$\sum_{r=0}^{\infty} \frac{A_{\chi_1,\chi_2}(r+1)}{(r+1)!} \frac{(s_2)_r}{s_1 + s_2 + r - 1}.$$

Thus we get the result by using Lemma 2 and the fact:

$$L(n, \chi) \neq 0$$

for $n \geq 1$ and a non principal character χ.   □

   As we stated in the introduction, we do not have a satisfactory answer to the problem of describing whole sigularities of multiple L-functions in the case $k \geq 3$, at present. For example when $k = 3$, what we have to show is the non vanishing of the sum:

$$\sum_{a_1=1}^{q-1} \sum_{a_2=1}^{q-1} \sum_{a_3=1}^{q-1} \chi_1(a_1)\chi_2(a_2)\chi_3(a_3)\tilde{B}_{r_1+1}\left(\frac{a_1 - a_2}{q}\right)\tilde{B}_{r_2+1}\left(\frac{a_2 - a_3}{q}\right),$$

apart from trivial cases.

References

[1] S. Akiyama, S. Egami, and Y. Tanigawa , An analytic continuation of multiple zeta functions and their values at non-positive integers, *Acta Arith.* **98** (2001), 107–116.

[2] T. Arakawa and M. Kaneko, Multiple zeta values, poly-Bernoulli numbers, and related zeta functions, *Nagoya Math. J.* **153** (1999), 189–209.

[3] F. V. Atkinson, The mean value of the Riemann zeta-function, *Acta Math.*, **81** (1949), 353–376.

[4] K. Dilcher, Zero of Bernoulli, generalized Bernoulli and Euler polynomials, *Mem. Amer. Math. Soc.*, Number 386, 1988.

[5] S. Egami, *Introduction to multiple zeta function*, Lecture Note at Niigata University (in Japanese).

[6] K. Inkeri, The real roots of Bernoulli polynomials, *Ann. Univ. Turku. Ser. A I* **37** (1959), 20pp.

[7] M. Katsurada and K. Matsumoto, Asymptotic expansions of the mean values of Dirichlet *L*-functions. *Math. Z.*, **208** (1991), 23–39.

[8] Y. Motohashi, A note on the mean value of the zeta and *L*-functions. I, *Proc. Japan Acad., Ser. A Math. Sci.* **61** (1985), 222–224.

[9] D. Zagier, Values of zeta functions and their applications, *First European Congress of Mathematics*, Vol. II, Birkhäuser, 1994, pp.497–512.

[10] D. Zagier, Periods of modular forms, traces of Hecke operators, and multiple zeta values, *Research into automorphic forms and L functions (in Japanese) (Kyoto, 1992)*, *Sūrikaisekikenkyūsho Kōkyūroku*, **843** (1993), 162–170.

[11] J. Zhao, Analytic continuation of multiple zeta functions, *Proc. Amer. Math. Soc.*, **128** (2000), 1275–1283.

ON THE VALUES OF CERTAIN
q-HYPERGEOMETRIC SERIES II

Masaaki AMOU

Department of Mathematics, Gunma University, Tenjin-cho 1-5-1, Kiryu 376-8515, Japan

amou@math.sci.gunma-u.ac.jp

Masanori KATSURADA

Mathematics, Hiyoshi Campus, Keio University, Hiyoshi 4-1-1, Kohoku-ku, Yokohama 223-8521, Japan

masanori@math.hc.keio.ac.jp

Keijo VÄÄNÄNEN

Department of Mathematics, University of Oulu, P. O. Box 3000, SF-90014 Oulu, Finland

kvaanane@sun3.oulu.fi

Keywords: Irrationality, Irrationality measure, q-hypergeometric series, q-Bessel function, S-unit equation

Abstract   As a continuation of the previous work by the authors having the same title, we study the arithmetical nature of the values of certain q-hypergeometric series $\phi(z; q)$ with a rational or an imaginary quadratic integer q with $|q| > 1$, which is related to a q-analogue of the Bessel function $J_0(z)$. The main result determines the pairs (q, α) with $\alpha \in K$ for which $\phi(\alpha; q)$ belongs to K, where K is an imaginary quadratic number field including q.

2000 Mathematics Subject Classification. Primary: 11J72; Secondary: 11J82.

The first named author was supported in part by Grant-in-Aid for Scientific Research (No. 11640009), Ministry of Education, Science, Sports and Culture of Japan.

The second named author was supported in part by Grant-in-Aid for Scientific Research (No. 11640038), Ministry of Education, Science, Sports and Culture of Japan.

C. Jia and K. Matsumoto (eds.), Analytic Number Theory, 17–25.

1.   INTRODUCTION

Throughout this paper except in the appendix, we denote by q a rational or an imaginary quadratic integer with $|q| > 1$, and K an imaginary quadratic number field including q. Note that K must be of the form $K = \mathbf{Q}(q)$ if q is an imaginary quadratic integer. For a positive rational integer s and a polynomial $P(z) \in K[z]$ of degree s such that $P(0) \neq 0, P(q^{-n}) \neq 0$ for all integers $n \geq 0$, we define an entire function

$$\phi(z;q) = \sum_{n=0}^{\infty} \frac{q^{-s\binom{n}{2}}}{P(1)P(q^{-1})\cdots P(q^{-(n-1)})} z^n. \qquad (1.1)$$

Concerning the values of $\phi(z;q)$, as a special case of a result of Bézivin [3], we know that if $\phi(\alpha;q) \in K$ for nonzero $\alpha \in K$, then $\alpha = a_s q^n$ with some integer n, where a_s is the leading coefficient of $P(z)$. He used in the proof a rationality criterion for power series. Recently, the present authors [1] showed that n in Bézivin's result must be positive. Hence we know that, for nonzero $\alpha \in K$,

$$\phi(\alpha;q) \in K \Longrightarrow \alpha \in \{a_s q^n \mid n \in \mathbf{N}\}. \qquad (1.2)$$

In case of $P(z) = a_s z^s + a_0$, it was also proved in [1] that $\phi(a;q) \in K$ for nonzero $\alpha \in K$ if and only if $\alpha = a_s q^{sn}$ with some $n \in \mathbf{N}$.

In this paper we are interested in the particular case $s = 2, P(z) = (z - q)^2$ of (1.1), that is,

$$\phi(z;q) = \sum_{n=0}^{\infty} \frac{z^n}{(1-q)^2(1-q^2)^2\cdots(1-q^n)^2}. \qquad (1.3)$$

We note that the function $J(z;q) := \phi(-z^2/4;q)$ satisfies

$$\lim_{q \to 1} J((1-q)z;q) = \sum_{n=0}^{\infty} \frac{(-1)^n}{(n!)^2} \left(\frac{z}{2}\right)^{2n},$$

where the right-hand side is the Bessel function $J_0(z)$. In this sense $J(z;q)$ is a q-analogue of $J_0(z)$. The main purpose of this paper is to determine the pairs (q, α) with $\alpha \in K$ for which $\phi(\alpha;q)$ belong to K. In this direction we have the following result (see Theorems 2 and 3 of [2]): $\phi(\alpha;q)$ does not belong to K for all nonzero $\alpha \in K$ except possibly when q is equal to

$$-3, \pm 1 \pm 2\sqrt{-1}, \pm 2 \pm \sqrt{-7}, \pm 4 \pm \sqrt{-7},$$
$$(\pm 1 \pm \sqrt{-7})/2, (\pm 3 \pm \sqrt{-7})/2, b\sqrt{-D},$$

where b is a nonzero rational integer and D is a square free positive integer satisfying

$$D \equiv 3 \pmod 4, \quad 1 + b^2 D = 2^j \ (\exists j \in \mathbf{N}). \tag{1.4}$$

We now state our main result which completes the above result.

Theorem. *Let q be a rational or an imaginary quadratic integer with $|q| > 1$, and K an imaginary quadratic number field including q. Let $\phi(z; q)$ be the function (1.3). Then, for nonzero $\alpha \in K$, $\phi(\alpha; q)$ does not belong to K except when*

$$(q, \alpha) = (-3, -27), \ ((-1 \pm \sqrt{-7})/2, (1 \pm 3\sqrt{-7})/2),$$

where the order of each $\pm$ sign is taken into account.
    Moreover, α is a zero of $\phi(z; q)$ in each of these exceptional cases.

For the proof, we recall in the next section a method developed in [1] and [2]. In particular, we introduce a linear recurrence $c_n = c_n(q)$ ($n \in \mathbf{N}$) having the property that $\phi(q^n; q) \in K$ if and only if $c_n(q) = 0$. Then the proof of the theorem will be carried out in the third section by determining the cases for which $c_n(q) = 0$. In the appendix we remark that one of our previous results (see Theorem 1 of [2]) can be made effective.

The authors would like to thank the referee for valuable comments on refinements of an earlier version of the present paper.

2.   A LINEAR RECURRENCE c_n

Let $\phi(z; q)$ be the function (1.1). Then, for nonzero $\alpha \in K$, we define a function

$$f(z) = f(z; \alpha) = \sum_{n=0}^{\infty} \frac{q^{-s\binom{n+1}{2}} z^{sn}}{P(q^{-1}z) \cdots P(q^{-n}z)} \alpha^n,$$

which is holomorphic at the origin and meromorphic on the whole complex plane. Since $\phi(\alpha; q) = f(q)$, we may study $f(q)$ arithmetically instead of $\phi(\alpha; q)$. An advantage in treating $f(z)$ is the fact that it satisfies the functional equation

$$P(z)f(qz) = \alpha z^s f(z) + P(z), \tag{2.1}$$

which is simpler than the functional equation of $\phi(z) = \phi(z; q)$ such as

$$\{P(q\Delta) - z\Delta^s\}\phi(z) = P(q),$$

where Δ is a q^{-1}-difference operator acting as $(\Delta\phi)(z) = \phi(q^{-1}z)$. In fact, as a consequence of the result of Duverney [4], we know that $f(q)$ does not belong to K when $f(z)$ is not a polynomial (see also [1]). Since the functional equation (2.1) has the unique solution in $K[[z]]$, a polynomial solution of (2.1) must be in $K[z]$. Let $\mathcal{E}_q(P)$ be the set consisting of all elements $\alpha \in K$ for which the functional equation (2.1) has a polynomial solution. Then we see that, for $\alpha \in K$,

$$\phi(\alpha; q) \in K \Longleftrightarrow \alpha \in \mathcal{E}_q(P). \tag{2.2}$$

Note that $f(z) \equiv 1$ is the unique solution of (2.1) with $\alpha = 0$, and that no constant functions satisfy (2.1) with nonzero α.

In view of (1.2), $\alpha \in \mathcal{E}_q(P)\setminus\{0\}$ implies that $\alpha = a_s q^n$ with some positive integer n. Indeed, we can see that if (2.1) has a polynomial solution of degree $n \in \mathbf{N}$, then α must be of the form $\alpha = a_s q^n$. Hence, by (2.2), our main task is to determine the pairs (q, n) for which the functional equation (2.1) with $s = 2, P(z) = (z - q)^2$, and $\alpha = q^n$ has a polynomial solution of degree $n \in \mathbf{N}$. To this end we quote a result from Section 2 of [2] with a brief explanation.

Let $f(z)$ be the unique solution in $K[[z]]$ of (2.1) with $s = 2, P(z) = (z - q)^2$, and $\alpha = q^n$ ($n \in \mathbf{N}$). It is easily seen that $f(z)$ is a polynomial of degree n if and only if $f(z)/P(z)$ is a polynomial of degree $n - 2$. By setting

$$f(z)/P(z) = \sum_{i=0}^{n-2} b_i z^i$$

with unknown coefficients b_i, we have a system of $n - 1$ linear equations of the form

$$A_n \mathbf{b} = \mathbf{c},$$

where

$$A_n = \begin{pmatrix} q^2 & & & & & \\ -2q & q^2 & & & & \\ 1 - q^{n-2} & -2q & q^2 & & & \\ & \cdots & \cdots & \cdots & & \\ & & & 1 - q^2 & -2q & q^2 \\ & & & & 1 - q & -2q \end{pmatrix},$$

$$\mathbf{b} = \begin{pmatrix} b_0 \\ b_1 \\ \vdots \\ \vdots \\ b_{n-2} \end{pmatrix}, \quad \mathbf{c} = \begin{pmatrix} 1 \\ 0 \\ \vdots \\ \vdots \\ 0 \end{pmatrix}.$$

Let $\mathbf{B}_n$ be an $n \times n$ matrix which is $\mathbf{A}_n$ with $\mathbf{c}$ as the last column. Since $\mathbf{A}_n$ has the rank $n - 1$, this system of linear equations has a solution if and only if $\mathbf{B}_n$ has the same rank $n - 1$, so that $\det \mathbf{B}_n = 0$. We can show for $\det \mathbf{B}_n$ ($n \in \mathbf{N}$) the recursion formula

$$\det \mathbf{B}_{n+2} = 2q \det \mathbf{B}_{n+1} - q^2(1 - q^n) \det \mathbf{B}_n,$$

with the convention $\det \mathbf{B}_0 = 0, \det \mathbf{B}_1 = 1$. For simplicity let us introduce a sequence $c_n = c_n(q)$ to be $c_n = q^{-(n-1)} \det \mathbf{B}_n$, for which $c_1 = 1, c_2 = 2$, and

$$c_{n+2} = 2c_{n+1} - (1 - q^n)c_n \quad (n \in \mathbf{N}). \tag{2.3}$$

Then we can summarize the argument above as follows: *The functional equation (2.1) with $s = 2, P(z) = (z - q)^2$, and $\alpha = q^n$ ($n \in \mathbf{N}$) has a polynomial solution $f(z)$ if and only if $c_n(q) = 0$.* We wish to show in the next section that $c_n(q) = 0$ if and only if

$$(q, n) = (-3, 3), ((-1 \pm \sqrt{-7})/2, 4), \tag{2.4}$$

which correspond to the cases given in the theorem.

3. PROOF OF THE THEOREM

Let $c_n = c_n(q)$ ($n \in \mathbf{N}$) be the sequence defined in the previous section. The following is the key lemma for our purpose.

Lemma 1. *Let δ be a positive number. If the inequalities*

$$|c_{n+1}| > \delta|c_n|, \quad (|q|^n - 1)|c_n| > (2 + \delta)|c_{n+1}|, \tag{3.1}$$

and

$$(|q| - (2 + \delta)\delta^{-1})|q|^n \geq 2(3 + \delta + \delta^{-1}) \tag{3.2}$$

hold for some $n = m$, then (3.1) is valid for all $n \geq m$.

Proof. We show the assertion by induction on n. Suppose that the desired inequalities hold for n with $n \geq m$. By the recursion formula (2.3) and the second inequality of (3.1), we obtain

$$\begin{aligned}
|c_{n+2}| &\geq (|q|^n - 1)|c_n| - 2|c_{n+1}| \\
&> (2 + \delta)|c_{n+1}| - 2|c_{n+1}| \\
&= \delta|c_{n+1}|,
\end{aligned}$$

which is the first inequality of (3.1) with $n + 1$ instead of n.

We next show the second inequality of (3.1) with $n+1$ instead of n. By the recursion formula (2.3) and the first inequality of (3.1), we obtain

$$\begin{aligned}
|c_{n+2}| &\leq 2|c_{n+1}| + (|q|^n + 1)|c_n| \\
&< 2|c_{n+1}| + (|q|^n + 1)\delta^{-1}|c_{n+1}| \\
&= (2 + \delta^{-1}(|q|^n + 1))|c_{n+1}|.
\end{aligned}$$

Noting that $(2 + \delta)(2 + \delta^{-1}(|q|^n + 1))$ is equal to

$$|q|^{n+1} - 1 - \{(|q| - (2+\delta)\delta^{-1})|q|^n - 2(3 + \delta + \delta^{-1})\},$$

and that the inequality (3.2) for $n = m$ implies the same inequality for all $n \geq m$, we get the desired inequality. This completes the proof. $\square$

In view of the fact mentioned in the introduction, we may consider the sequences $c_n = c_n(q)$ $(n \in \mathbf{N})$ only for q given just before the statement of the Theorem. In the next lemma we consider the sequence $c_n(q)$ for these q excluding $b\sqrt{-D}$.

Lemma 2. *Let q be one of the numbers*

$$-3, \pm 1 \pm 2\sqrt{-1}, \pm 2 \pm \sqrt{-7}, \pm 4 \pm \sqrt{-7}, (\pm 1 \pm \sqrt{-7})/2, (\pm 3 \pm \sqrt{-7})/2.$$

Then, for the sequence $c_n = c_n(q)$ $(n \in \mathbf{N})$, $c_n = 0$ if and only if (2.4) holds.

Moreover, for the exceptional cases (2.4), $\phi(q^3; q) = 0$ if $q = -3$, and $\phi(q^4; q) = 0$ if $q = (-1 \pm \sqrt{-7})/2$, where $\phi(z; q)$ is the function (1.3).

Proof. Since

$$c_3 = 3 + q, \quad c_4 = 2(q^2 + q + 2), \tag{3.3}$$

we see that $c_n = 0$ in the cases (2.4). By using computer, we have the following table which ensures the validity of (3.1) and (3.2) with these values:

q	-3	$\pm 1 \pm 2\sqrt{-1}$	$\pm 2 \pm \sqrt{-7}$	$\pm 4 \pm \sqrt{-7}$	$(\pm 1 \pm \sqrt{-7})/2$	$(\pm 3 \pm \sqrt{-7})/2$
δ	3	2	2	2	7	3
m	5	5	3	3	15	7

It follows from Lemma 1 that, in each of the sequences, $c_n(q) \neq 0$ for all $n \geq m$. By using computer again, we can see the non-vanishing of the remaining terms except for the cases (2.4).

As we noted in the previous section, if the functional equation (2.1) has a polynomial solution $f(z; \alpha)$, it is divisible by $P(z)$. Hence we have $\phi(q^3; q) = f(q; q^3) = 0$ if $q = -3$, and $\phi(q^4; q) = f(q; q^4) = 0$ if $q = (-1 \pm \sqrt{-7})/2$. The lemma is proved. $\square$

We next consider the case where $q = b\sqrt{-D}$ without (1.4).

Lemma 3. *Let b be a nonzero integer, and D a positive integer such that $b^2 D \geq 5$. Then, for $q = b\sqrt{-D}$, the sequence $c_n = c_n(q)$ $(n \in \mathbf{N})$ does not vanish for all n.*

Proof. By (3.3), c_3 and c_4 are nonzero for the present q. Let us set $A := b^2 D$. To prove $c_n \neq 0$ for all $n \geq 5$, we show (3.1) and (3.2) with $\delta = 3, n = 4$. Indeed, by straightforward calculations, we obtain

$$|c_5|^2 - 9|c_4|^2 = A^4 + A^3 - 28A^2 + 77A - 119$$

and

$$(|q|^4 - 1)^2 |c_4|^2 - 25|c_5|^2 = 4A^6 - 12A^5 - 17A^4 - A^3 - 228A^2 + 763A - 609.$$

Since these values are positive whenever $A \geq 5$, (3.1) with $\delta = 3, n = 4$ holds. Moreover,

$$(|q| - 5/3)|q|^4 = (\sqrt{A} - 5/3)A^2 > 14$$

holds whenever $A \geq 5$. Hence (3.2) with $\delta = 3, n = 4$ also holds. Hence the desired assertion follows from Lemma 1. This completes the proof. $\qquad \square$

By this lemma there remains the consideration of the case where $q = b\sqrt{-D}$ with (1.4) and $b^2 D < 5$, that is the case $q = \pm\sqrt{-3}$. In this case, by using computer, we can show that (3.1) and (3.2) with $\delta = 4, n = 8$ are valid. Hence, by Lemma 1, $c_n = c_n(\pm\sqrt{-3}) \neq 0$ for all $n \geq 8$. We see also that $c_n \neq 0$ for all $n < 8$ by using computer again. Thus we have shown the desired assertion, and this completes the proof of the theorem.

Appendix

Here we consider an arbitrary algebraic number field K, and we denote by $\mathcal{O}_K$ the ring of integers in K. Let $d, h,$ and R be the degree over $\mathbf{Q}$, the class number, and the regulator of K, respectively. Let s be a positive integer, q a nonzero element of K, and $P(z)$ a polynomial in $K[z]$ of the form

$$P(z) = \sum_{i=0}^{s} a_i z^i, \qquad a_s a_0 \neq 0.$$

Then, as in Section 2, we define a set $\mathcal{E}_q(P)$ to be the set consisting of all $\alpha \in K$ for which the functional equation (2.1) has a polynomial solution. In this appendix we remark that the following result concerning the set $\mathcal{E}_q(P)$ holds. Hereafter, for any $\alpha \in K$, we denote by $H(\alpha)$ the ordinary height of α, that is, the maximum of the absolute values of the coefficients for the minimal polynomial of α over $\mathbf{Z}$.

Theorem A. *Let s be a positive integer with $s \geq 2$, and q a nonzero element of K with $q \in \mathcal{O}_K$ or $q^{-1} \in \mathcal{O}_K$. Let $a_i(x) \in \mathcal{O}_K[x]$, $i = 0, 1, ..., s$, be such that*

$$\gamma := a_{s-1}(1)a_{s-1}(-1)(a_{s-1}^2(-1) - 2a_s(-1)a_{s-2}(-1)) \neq 0.$$

Let $S = \{w_1, ..., w_t\}$ be the set of finite places of K for which $|\gamma|_{w_i} < 1$, and B an upper bound of the prime numbers $p_1, ..., p_t$ with $|p_i|_{w_i} < 1$. Let $P(z) = P(z; q)$ be a polynomial as above, where $a_i = a_i(q)$ $(i = 0, 1, ..., s)$. Then there exists a positive constant C, which is effectively computable from quantities depending only on d, h, R, t, and B, such that if $\mathcal{E}_q(P) \neq \{0\}$, then $H(q) \leq C$.

Note that we already proved the assertion of this theorem with a non-effective constant C (see Theorem 1 of [2]). In that proof we applied a generalized version of Roth's theorem (see Chapter 7, Corollary 1.2 of Lang [5]), which is not effective. However, as we see below, it is natural in our situation to apply a result on S-unit equations, which is effective.

Proof of Theorem A. We first consider the case where $q \in \mathcal{O}_K$. It follows from the Proposition of [2] that if $\mathcal{E}_q(P) \neq \{0\}$, then $1 \pm q$ are S-units. Since

$$(1 - q) + (1 + q) = 2,$$

$(1-q, 1+q)$ is a solution of the S-unit equation $x_1 + x_2 = 2$. Hence, by a result on S-unit equations in two variables (see Corollary 1.3 of Shorey and Tijdeman [6]), $H(1 \pm q)$ is bounded from above by an effectively computable constant depending only on the quantities given in the theorem. Since the minimal polynomial of q over $\mathbf{Z}$ is $Q(x + 1)$ if that of $1 + q$ is $Q(x)$, $H(q)$ is also bounded from above by a similar constant.

For the case where $q^{-1} \in \mathcal{O}_K$, by the Proposition of [2] again, we can apply the same argument replacing q by q^{-1}. Hence, by noting that $H(q^{-1}) = H(q)$, the desired assertion holds in this case. This completes the proof. □

References

[1] M. Amou, M. Katsurada, and K. Väänänen, Arithmetical properties of the values of functions satisfying certain functional equations of Poincaré, Acta Arith., to appear.

[2] M. Amou, M. Katsurada, and K. Väänänen, On the values of certain q-hypergeometric series, in "Proceedings of the Turku Symposium on Number Theory in Memory of Kustaa Inkeri" (M. Jutila and T. Metsänkylä, eds.), Walter de Gruyter, Berlin 2001, 5–17.

[3] J.-P. Bézivin, Indépendance linéaire des valeurs des solutions transcendantes de certaines équations fonctionnelles, Manuscripta Math. **61** (1988), 103–129.

[4] D. Duverney, Propriétés arithmétiques des solutions de certaines équations fonctionnelles de Poincaré, J. Théorie des Nombres Bordeaux **8** (1996), 443–447.

[5] S. Lang, Fundamentals of Diophantine Geometry, Springer-Verlag, 1983.

[6] T. N. Shorey and R. Tijdeman, Exponential Diophantine Equations, Cambridge Univ. Press, Cambridge, 1986.

THE CLASS NUMBER ONE PROBLEM FOR SOME NON-NORMAL SEXTIC CM-FIELDS

Gérard BOUTTEAUX
and Stéphane LOUBOUTIN
Institut de Mathématiques de Luminy, UPR 9016, 163 avenue de Luminy, Case 907,
13288 Marseille Cedex 9, France
loubouti@iml.univ-mrs.fr

Keywords: CM-field, relative class number, cyclic cubic field

Abstract   We determine all the non-normal sextic CM-fields (whose maximal totally real subfields are cyclic cubic fields) which have class number one.
There are 19 non-isomorphic such fields.

1991 Mathematics Subject Classification: Primary 11R29, 11R42 and 11R21.

1.   INTRODUCTION

Lately, great progress have been made towards the determination of
all the normal CM-fields with class number one. Due to the work of
various authors, all the normal CM-fields of degrees less than 32 with
class number one are known. In contrast, up to now the determination
of all the non-normal CM-fields with class number one and of a given
degree has only been solved for quartic fields (see [LO]). The present
piece of work is an abridged version of half the work to be completed
in [Bou] (the PhD thesis of the first author under the supervision of
the second auhtor): the determination of all the non-normal sextic CM-
fields with class number one, regardless whether their maximal totally
real subfield is a real cyclic cubic field (the situation dealt with in the
present paper) or a non-normal totally real cubic field.

Let **K** range over the non-normal sextic CM-fields whose maximal
totally real subfields are cyclic cubic fields. In the present paper we
will prove that the relative class number of **K** goes to infinity with the
absolute value of its discriminant (see Theorem 4), we will characterize

C. Jia and K. Matsumoto (eds.), Analytic Number Theory, 27–37.
© 2002 *Kluwer Academic Publishers. Printed in the Netherlands.*

these $\mathbf{K}$'s of odd class numbers (see Theorem 8) and we will finally determine all these $\mathbf{K}$'s of class number one (see Theorem 10).

Throughout this paper $\mathbf{K} = \mathbf{F}(\sqrt{-\delta})$ denotes a non-normal sextic CM-field whose maximal totally real subfield $\mathbf{F}$ is a cyclic cubic field, where δ is a totally positive algebraic element of $\mathbf{F}$. Let $\delta_1 = \delta$, δ_2 and δ_3 denote the conjugates of δ in $\mathbf{F}$ and let $\mathbf{N} = \mathbf{F}\left(\sqrt{-\delta_1}, \sqrt{-\delta_2}, \sqrt{-\delta_3}\right)$ denote the normal closure of $\mathbf{K}$. Then, $\mathbf{N}$ is a CM-field with maximal totally real subfield $\mathbf{N}^+ = \mathbf{F}\left(\sqrt{\delta_1\delta_2}, \sqrt{\delta_1\delta_3}\right)$ and $\mathbf{k} = \mathbf{Q}(\sqrt{-d})$ is an imaginary quadratic subfield of $\mathbf{N}$, where $d = \delta_1\delta_2\delta_3 = N_{\mathbf{F}/\mathbf{Q}}(\delta)$.

2.   EXPLICIT LOWER BOUNDS ON RELATIVE CLASS NUMBERS OF SOME NON-NORMAL SEXTIC CM-FIELDS

Let $h_{\mathbf{N}}^- = h_{\mathbf{K}}/h_{\mathbf{F}}$ and $Q_{\mathbf{N}} \in \{1,2\}$ denote the relative class number and Hasse unit index of $\mathbf{K}$, respectively. We have

$$h_{\mathbf{K}}^- = \frac{Q_{\mathbf{K}}}{4\pi^3}\sqrt{\frac{d_{\mathbf{K}}}{d_{\mathbf{F}}}}\frac{\mathrm{Res}_{s=1}(\zeta_{\mathbf{K}})}{\mathrm{Res}_{s=1}(\zeta_{\mathbf{F}})}, \tag{1}$$

where $d_{\mathbf{E}}$ and $\mathrm{Res}_{s=1}(\zeta_{\mathbf{E}})$ denote the absolute value of the discriminant and the residue at $s = 1$ of the Dedekind zeta function $\zeta_{\mathbf{E}}$ of the number field $\mathbf{E}$. The aim of this section is to obtain an explicit lower bound on $h_{\mathbf{K}}^-$ (see Theorem 4).

2.1.   FACTORIZATIONS OF DEDEKIND ZETA FUNCTIONS

Proposition 1. *Let $\mathbf{F}$ be a real cyclic cubic field, $\mathbf{K}$ be a non-normal CM-sextic field with maximal totally real subfield $\mathbf{F}$ and $\mathbf{N}$ be the normal closure of $\mathbf{K}$. Then, $\mathbf{N}$ is a CM-field of degree 24 with Galois group $\mathrm{Gal}(\mathbf{N}/\mathbf{Q})$ isomorphic to the direct product $\mathcal{A}_4 \times C_2$, $\mathbf{N}^+$ is a normal subfield of $\mathbf{N}$ of degree 12 and Galois group $\mathrm{Gal}(\mathbf{N}^+/\mathbf{Q})$ isomorphic to $\mathcal{A}_4$ and the imaginary cyclic sextic field $\mathbf{A} = \mathbf{F}\mathbf{k}$ is the maximal abelian subfield of $\mathbf{N}$. Finally, we have the following factorization of Dedekind zeta functions :*

$$\zeta_{\mathbf{N}}/\zeta_{\mathbf{N}^+} = (\zeta_{\mathbf{A}}/\zeta_{\mathbf{F}})(\zeta_{\mathbf{K}}/\zeta_{\mathbf{F}})^3. \tag{2}$$

Proof. Let us only prove (2). Set $\mathbf{K}_0 = \mathbf{A} = \mathbf{F}\mathbf{k}$ and $\mathbf{K}_i = \mathbf{F}(\sqrt{-\delta_i})$, $1 \le i \le 3$. Since the Galois group of the abelian extension $\mathbf{N}/\mathbf{F}$ is the elementary 2-group $C_2 \times C_2 \times C_2$, using abelian L-functions we easily obtain $\zeta_{\mathbf{N}}/\zeta_{\mathbf{N}^+} = \prod_{i=0}^{3}(\zeta_{\mathbf{K}_i}/\zeta_{\mathbf{F}})$. Finally, as the three $\mathbf{K}_i$'s with $1 \le i \le$

3 are isomorphic to $\mathbf{K}$, we have $\zeta_{\mathbf{K}_i} = \zeta_{\mathbf{K}}$ for $1 \leq i \leq 3$, and we obtain the desired result. •

2.2.   ON THE NUMBER OF REAL ZEROS OF DEDEKIND ZETA FUNCTIONS

For the reader's convenience we repeat the statement and proof of [LLO, Lemma 15]:

Lemma 2. *If the absolute value $d_{\mathbf{M}}$ of the discriminant of a number $\mathbf{M}$ satisfies $d_{\mathbf{M}} > \exp(2(\sqrt{m+1} - 1))$, then its Dedekind zeta function $\zeta_{\mathbf{M}}$ has at most m real zeros in the range $s_m = 1 - (2(\sqrt{m+1} - 1)^2/\log d_{\mathbf{M}}) \leq s < 1$. In particular, $\zeta_{\mathbf{M}}$ has at most two real zeros in the range $1 - (1/\log d_{\mathbf{M}}) \leq s < 1$.*

Proof. Assume $\zeta_{\mathbf{M}}$ has at least $m + 1$ real zeros in the range $[s_m, 1[$. According to the proof of [Sta, Lemma 3] for any $s > 1$ we have

$$f(s) = \frac{m+1}{s - s_m} - \frac{1}{s - 1} \leq \left(\sum_{\rho \in]0,1], \ \zeta_{\mathbf{M}}(\rho)=0} \frac{1}{s - \rho} \right) - \frac{1}{s - 1}$$

$$< \frac{1}{2} \log d_{\mathbf{M}} + h(s)$$

where

$$h(s) = \left(\frac{1}{s} - \frac{n}{2} \log \pi \right) + \frac{r_1}{2} \frac{\Gamma'}{\Gamma}(s/2) + r_2 \left(\frac{\Gamma'}{\Gamma}(s) - \log 2 \right)$$

where $n > 1$ is the degree of $\mathbf{M}$ and where ρ ranges over all the real zeros in $]0, 1[$ of $\zeta_{\mathbf{M}}$. Setting

$$t_m = \frac{\sqrt{m+1} - s_m}{\sqrt{m+1} - 1} = 1 + \frac{1 - s_m}{\sqrt{m+1} - 1} = 1 + \frac{2(\sqrt{m+1} - 1)}{\log d_{\mathbf{M}}} < 2,$$

we obtain

$$f(t_m) = \frac{\sqrt{m+1} - 1}{t_m - 1} = \frac{(\sqrt{m+1} - 1)^2}{1 - s_m} = \frac{1}{2} \log d_{\mathbf{M}} < \frac{1}{2} \log d_{\mathbf{M}} + h(t_m),$$

and since $h(t_m) < h(2) < 0$ we have a contradiction.
Indeed, let $\gamma = 0.577 \cdots$ denote Euler's constant. Since $h'(s) > 0$ for $s > 0$ (use $(\Gamma'/\Gamma)'(s) = \sum_{k \geq 0}(k + s)^{-2}$), we do have $h(t_m) < h(2) = (1 - n(\gamma + \log \pi))/2 + r_1(1 - \log 2) \leq (1 - n(\gamma + \log \pi - 1 + \log 2)/2 < 0$.
•

2.3.   BOUNDS ON RESIDUES

Lemma 3.

1. *Let* **K** *be a sextic CM-field. Then,* $\frac{1}{2} \leq 1 - (1/a \log d_{\mathbf{K}}) \leq \beta < 1$ *and* $\zeta_{\mathbf{K}}(\beta) \leq 0$ *imply*

$$\operatorname{Res}_{s=1}(\zeta_{\mathbf{K}}) \geq \epsilon_{\mathbf{K}} \frac{1 - \beta}{e^{1/2a}} \quad where \quad \epsilon_{\mathbf{K}} := 1 - (6\pi e^{1/6a}/d_{\mathbf{K}}^{1/6}). \quad (3)$$

2. *If* **F** *is real cyclic cubic field of conductor* $f_{\mathbf{F}}$ *then*

$$\operatorname{Res}_{s=1}(\zeta_{\mathbf{F}}) \leq \frac{1}{4}(\log f_{\mathbf{F}} + 0.05)^2, \quad (4)$$

and $\frac{1}{2} \leq \beta < 1$ *and* $\zeta_{\mathbf{F}}(\beta) = 0$ *imply*

$$\operatorname{Res}_{s=1}(\zeta_{\mathbf{F}}) \leq \frac{1 - \beta}{48} \log^3 d_{\mathbf{F}}. \quad (5)$$

Proof. To prove (3), use [Lou2, Proposition A]. To prove (4), use [Lou1]. For the proof of (5) (which stems from the use of [Lou4, bound (31)] and the ideas of [Lou5]) see [Lou6]. •

Notice that the residue at its simple pole $s = 1$ of any Dedekind zeta function $\zeta_{\mathbf{K}}$ is positive (use the analytic class number formula, or notice that from its definition we get $\zeta_{\mathbf{K}}(s) \geq 1$ for $s > 1$). Therefore, we have $\lim_{\substack{s \to 1 \\ s < 1}} \zeta_{\mathbf{K}}(s) = -\infty$ and $\zeta_{\mathbf{K}}(1 - (1/a \log d_{\mathbf{K}})) \leq 0$ if $\zeta_{\mathbf{K}}$ does not have any real zero in the range $1 - (1/a \log d_{\mathbf{K}}) \leq s < 1$.

2.4.   EXPLICIT LOWER BOUNDS ON RELATIVE CLASS NUMBERS

Theorem 4. *Let* **K** *be a non-normal sextic CM-field with maximal totally real subfield a real cyclic cubic field* **F** *of conductor* $f_{\mathbf{F}}$. *Let* **N** *denote the normal closure of* **K**. *Set* $\epsilon_{\mathbf{K}} := 1 - (6\pi e^{1/24}/d_{\mathbf{K}}^{1/6})$. *We have*

$$h_{\mathbf{K}}^- \geq \epsilon_{\mathbf{K}} \frac{\sqrt{d_{\mathbf{K}}/d_{\mathbf{F}}}}{e^{1/8}\pi^3(\log f_{\mathbf{F}} + 0.05)^2 \log d_{\mathbf{N}}} \quad (6)$$

and $d_{\mathbf{N}} \leq d_{\mathbf{K}}^{12}$. *Therefore,* $h_{\mathbf{K}}^-$ *goes to infinity with* $d_{\mathbf{K}}$ *and there are only finitely many non-normal CM sextic fields* **K** *(whose maximal totally real subfields are cyclic cubic fields) of a given relative class number.*

Proof. There are two cases to consider.

First, assume that $\zeta_{\mathbf{F}}$ has a real zero β in $[1 - (1/\log d_{\mathbf{N}}), 1[$. Then, $\beta \geq 1 - (1/4 \log d_{\mathbf{K}})$ (since $[\mathbf{N} : \mathbf{K}] = 4$, we have $d_{\mathbf{N}} \geq d_{\mathbf{K}}^4$). Since $\zeta_{\mathbf{K}}(\beta) = 0 \leq 0$, we obtain

$$\text{Res}_{s=1}(\zeta_{\mathbf{K}}) \geq \epsilon_{\mathbf{K}} \frac{1 - \beta}{e^{1/8}} \tag{7}$$

(use Lemma 3 with $a = 4$). Using (1), (5), (7) and $Q_{\mathbf{K}} \geq 1$, we get

$$h_{\mathbf{K}}^- \geq \epsilon_{\mathbf{K}} \frac{12\sqrt{d_{\mathbf{K}}/d_{\mathbf{F}}}}{e^{1/8}\pi^3 \log^3 d_{\mathbf{F}}} \geq \epsilon_{\mathbf{K}} \frac{24\sqrt{d_{\mathbf{K}}/d_{\mathbf{F}}}}{e^{1/8}\pi^3 (\log^2 f_{\mathbf{F}})(\log d_{\mathbf{N}})} \tag{8}$$

(since $[\mathbf{N} : \mathbf{F}] = 8$, we have $d_{\mathbf{N}} \geq d_{\mathbf{F}}^8 = f_{\mathbf{F}}^{16}$).
Second assume that $\zeta_{\mathbf{F}}$ does not have any real zero in $[1 - (1/\log d_{\mathbf{N}}), 1[$. Since according to (2) any real zero β in $[1 - (1/\log d_{\mathbf{N}}), 1[$ of $\zeta_{\mathbf{K}}$ would be a triple zero of $\zeta_{\mathbf{N}}$, which would contradict Lemma 2, we conclude that $\zeta_{\mathbf{K}}$ does not have any real zero in $[1 - (1/\log d_{\mathbf{N}}), 1[$, and $\beta :=$ $1 - (1/\log d_{\mathbf{N}})) \geq 1 - (1/4 \log d_{\mathbf{K}}))$ satisfies $\zeta_{\mathbf{F}}(\beta) \leq 0$. Therefore,

$$\text{Res}_{s=1}(\zeta_{\mathbf{K}}) \geq \epsilon_{\mathbf{K}} \frac{1}{e^{1/8} \log d_{\mathbf{N}}}. \tag{9}$$

(use Lemma 3 with $a = 4$). Using (1), (4), (9) and $Q_{\mathbf{K}} \geq 1$, we get (6). Since the lower bound (8) is always better than the lower bound (6), the worst lower bound (6) always holds.

Now, let $\delta \in \mathbf{F}$ be any totally positive element such that $\mathbf{K} = \mathbf{F}(\sqrt{-\delta})$. Let $\delta_1 = \delta$, δ_2 and δ_3 denote the three conjugates of δ in $\mathbf{F}$ and set $\mathbf{K}_i = \mathbf{F}(\sqrt{-\delta_i})$. Let $\mathbf{N}$ denote the normal closure of $\mathbf{K}$. Since $\mathbf{N} = \mathbf{K}_1\mathbf{K}_2\mathbf{K}_3$ and since the three $\mathbf{K}_i$'s are pairwise isomorphic then $d_{\mathbf{N}}$ divides $d_{\mathbf{K}}^{12}$ (see [Sta, Lemma 7]), and $d_{\mathbf{N}} \leq d_{\mathbf{K}}^{12}$. Finally, since $f_{\mathbf{F}} = d_{\mathbf{F}}^{1/2} \leq d_{\mathbf{K}}^{1/4}$ and since $\sqrt{d_{\mathbf{K}}/d_{\mathbf{F}}} \geq d_{\mathbf{K}}^{1/4}$, we do have $h_{\mathbf{K}}^- \longrightarrow \infty$ as $d_{\mathbf{K}} \longrightarrow \infty$ •

3. CHARACTERIZATION OF THE ODDNESS OF THE CLASS NUMBER

Lemma 5. *Let* $\mathbf{K}$ *be a non-normal sextic CM-field whose maximal totally real subfiel* $\mathbf{F}$ *is a cyclic cubic field. The class number* $h_{\mathbf{K}}$ *of* $\mathbf{K}$ *is odd if and only if the narrow class number* $h_{\mathbf{F}}^+$ *of* $\mathbf{F}$ *is odd and exactly one prime ideal* $\mathcal{Q}$ *of* $\mathbf{F}$ *is ramified in the quadratic extension* $\mathbf{K}/\mathbf{F}$.

Proof. Assume that $h_{\mathbf{K}}$ is odd. If the quadratic extension $\mathbf{K}/\mathbf{F}$ were unramified at all the finite places of $\mathbf{F}$ then 2 would divide $h_{\mathbf{F}}^+$. Since the narrow class number of a real cyclic cubic field is either equal to its wide class number or equal to four times its wide class number, we would have $h_{\mathbf{F}}^+ \equiv 4 \pmod 8$ and the narrow Hilbert 2-class field of $\mathbf{F}$ would be a normal number field of degree 12 containing $\mathbf{K}$, hence containing the normal closure $\mathbf{N}$ of $\mathbf{K}$ which is of degree 24 (see Proposition 1). A contradiction. Hence, $\mathbf{N}/\mathbf{F}$ is ramified at at least one finite place, which implies $\mathbf{H}_{\mathbf{F}}^+ \cap \mathbf{K} = \mathbf{F}$ where $\mathbf{H}_{\mathbf{F}}^+$ denotes the narrow Hilbert class field of $\mathbf{F}$. Consequently, the extension $\mathbf{KH}_{\mathbf{F}}^+/\mathbf{K}$ is an unramified extension of degree $h_{\mathbf{F}}^+$ of $\mathbf{K}$. Hence $h_{\mathbf{F}}^+$ divides $h_{\mathbf{K}}$, and the oddness of the class number of $\mathbf{K}$ implies that $h_{\mathbf{F}}^+$ is odd, which implies $h_{\mathbf{F}}^+ = h_{\mathbf{F}}$.

Conversely, assume that $h_{\mathbf{F}}^+$ is odd. Since $\mathbf{K}$ is a totally imaginary number field which is a quadratic extension of the totally real number field $\mathbf{F}$ of odd narrow class number, then, the 2-rank of the ideal class group of $\mathbf{K}$ is equal to $t-1$, where t denotes the number of prime ideals of $\mathbf{F}$ which are ramified in the quadratic extension $\mathbf{K}/\mathbf{F}$ (see [CH, Lemma 13.7]). Hence $h_{\mathbf{K}}$ is odd if and only if exactly one prime ideal $\mathcal{Q}$ of $\mathbf{F}$ is ramified in the quadratic extension $\mathbf{K}/\mathbf{F}$. •

Lemma 6. *Let $\mathbf{K}$ be a non-normal sextic CM-field whose maximal totally real subfield $\mathbf{F}$ is a cyclic cubic field. Assume that the class number $h_{\mathbf{K}}$ of $\mathbf{K}$ is odd, stick to the notation introduced in Lemma 5 and let $q \geq 2$ denote the rational prime such that $\mathcal{Q} \cap \mathbf{Z} = q\mathbf{Z}$. Then, (i) the Hasse unit $Q_{\mathbf{K}}$ of $\mathbf{K}$ is equal to one, (ii) $\mathbf{K} = \mathbf{F}(\sqrt{-\alpha_{\mathcal{Q}}})$ for any totally positive algebraic element $\alpha_{\mathcal{Q}} \in \mathbf{F}$ such that $\mathcal{Q}^{h_{\mathbf{F}}^+} = (\alpha_{\mathcal{Q}})$, (iii) there exists $e \geq 1$ odd such that the finite part of the conductor of the quadratic extension $\mathbf{K}/\mathbf{F}$ is given by $\mathcal{F}_{\mathbf{K}/\mathbf{F}} = \mathcal{Q}^e$, and (iv) q splits completely in $\mathbf{F}$.*

Proof. Let $U_{\mathbf{F}}$ and $U_{\mathbf{F}}^+$ denote the groups of units and totally positive units of the ring of algebraic integers of $\mathbf{F}$, respectively. Since $h_{\mathbf{F}}^+$ is odd we have $U_{\mathbf{F}}^+ = U_{\mathbf{F}}^2$. We also set $h = h_{\mathbf{F}}^+$, which is odd (Lemma 5).

1. If we had $Q_{\mathbf{K}} = 2$ then there would exist some $\epsilon \in U_{\mathbf{F}}^+$ such that $\mathbf{K} = \mathbf{F}(\sqrt{-\epsilon})$. Since $U_{\mathbf{F}}^+ = U_{\mathbf{F}}^2$, we would have $\mathbf{K} = \mathbf{F}(\sqrt{-1})$ and $\mathbf{K}$ would be a cyclic sextic field. A contradiction. Hence, $Q_{\mathbf{K}} = 1$.

2. Let α be any totally positive algebraic integer of $\mathbf{F}$ such that $\mathbf{K} = \mathbf{F}(\sqrt{-\alpha})$. Since $\mathcal{Q}$ is the only prime ideal of $\mathbf{F}$ which is ramified in the quadratic extension $\mathbf{K}/\mathbf{F}$, there exists some integral ideal $\mathcal{I}$ of $\mathbf{F}$ such that $(\alpha) = \mathcal{I}^2\mathcal{Q}^l$, with $l \in \{0,1\}$, which implies $\alpha^h = \epsilon \alpha_{\mathcal{I}}^2 \alpha_{\mathcal{Q}}^l$ for a totally positive generator $\alpha_{\mathcal{I}}$ of $\mathcal{I}^h$ and some $\epsilon \in U_{\mathbf{F}}^+$. Since $U_{\mathbf{F}}^+ = U_{\mathbf{F}}^2$, we have $\epsilon = \eta^2$ for some $\eta \in U_{\mathbf{F}}$ and $\mathbf{K} = \mathbf{F}(\sqrt{-\alpha}) = \mathbf{F}(\sqrt{-\alpha^h}) = \mathbf{F}(\sqrt{-\alpha_{\mathcal{Q}}^l})$. If we had $l = 0$ then $\mathbf{K} = \mathbf{F}(\sqrt{-1})$ would

be a cyclic sextic field. A contradiction. Therefore, $l = 1$ and $\mathbf{K} = \mathbf{F}(\sqrt{-\alpha_{\mathcal{Q}}})$.

3. Since $\mathcal{Q}$ is the only prime ideal of $\mathbf{F}$ ramified in the quadratic extension $\mathbf{K}/\mathbf{F}$, there exists $e \geq 1$ such that $\mathcal{F}_{\mathbf{K}/\mathbf{F}} = \mathcal{Q}^e$. Since $\mathbf{K}/\mathbf{F}$ is quadratic, for any totally positive algebraic integer $\alpha \in \mathbf{F}$ such that $\mathbf{K} = \mathbf{F}(\sqrt{-\alpha})$ there exists some integral ideal $\mathcal{I}$ of $\mathbf{F}$ such that $(4\alpha) = \mathcal{I}^2 \mathcal{F}_{\mathbf{K}/\mathbf{F}}$ (see [LYK]). In particular, there exists some integral ideal $\mathcal{I}$ of $\mathbf{F}$ such that $(4\alpha_{\mathcal{Q}}) = (2)^2 \mathcal{Q}^{h_{\mathbf{F}}^+} = \mathcal{I}^2 \mathcal{F}_{\mathbf{K}/\mathbf{F}} = \mathcal{I}^2 \mathcal{Q}^e$ and e is odd.

4. If $(q) = \mathcal{Q}$ were inert in $\mathbf{F}$, we would have $\mathbf{K} = \mathbf{F}(\sqrt{-q^h}) = \mathbf{F}(\sqrt{-q})$. If $(q) = \mathcal{Q}^3$ were ramified in $\mathbf{F}$, we would have $\mathbf{K} = \mathbf{F}(\sqrt{-\alpha_{\mathcal{Q}}}) = \mathbf{F}(\sqrt{-\alpha_{\mathcal{Q}}^3}) = \mathbf{F}(\sqrt{-q^{3h}}) = \mathbf{F}(\sqrt{-q})$. In both cases, $\mathbf{K}$ would be abelian. A contradiction. Hence, q splits in $\mathbf{F}$. $\bullet$

3.1.   THE SIMPLEST NON-NORMAL SEXTIC CM-FIELDS

Throughout this section, we assume that $h_{\mathbf{F}}^+$ is odd. We let $\mathbf{A_F}$ denote the ring of algebraic integers of $\mathbf{F}$ and for any non-zero $\alpha \in \mathbf{F}$ we let $\nu(\alpha) = \pm 1$ denote the sign of $N_{\mathbf{F}/\mathbf{Q}}(\alpha)$. Let us first set some notation. Let $q \geq 2$ be any rational prime which splits completely in a real cyclic cubic field $\mathbf{F}$ of odd narrow class number $h_{\mathbf{F}}^+$, let $\mathcal{Q}$ be any one of the three prime ideals of $\mathbf{F}$ above q and let $\alpha_{\mathcal{Q}}$ be any totally positive generator of the principal (in the narrow sense) ideal $\mathcal{Q}^{h_{\mathbf{F}}^+}$. We set $\mathbf{K_{F,Q}} := \mathbf{F}(\sqrt{-\alpha_{\mathcal{Q}}})$ and notice that $\mathbf{K_{F,Q}}$ is a non-normal CM-sextic field with maximal totally real subfield the cyclic cubic field $\mathbf{F}$. Clearly, $\mathcal{Q}$ is ramified in the quadratic extension $\mathbf{K}/\mathbf{F}$. However, this quadratic extension could also be ramified at primes ideals of $\mathbf{F}$ above the rational prime 2. We thus define **a simplest non-normal sextic CM-field** as being a $\mathbf{K_{F,Q}}$ such that $\mathcal{Q}$ is the only prime ideal of $\mathbf{F}$ ramified in the quadratic extension $\mathbf{K_{F,Q}}/\mathbf{F}$. Now, we would like to know when is $\mathbf{K_{F,Q}}$ a simplest non-normal sextic field. According to class field theory, there is a bijective correspondence between the simplest non-normal sextic CM-fields of conductor $\mathcal{Q}^e$ (e odd) and the primitive quadratic characters χ_0 on the multiplicative groups $(\mathbf{A_F}/\mathcal{Q}^e)^*$ which satisfy $\chi_0(\epsilon) = \nu(\epsilon)$ for all $\epsilon \in \mathbf{U_F}$ (which amounts to asking that $\mathcal{I} \mapsto \chi(\mathcal{I}) = \nu(\alpha_{\mathcal{I}})\chi_0(\alpha_{\mathcal{I}})$ be a primitive quadratic character on the unit ray class group of $\mathbf{L}$ for the modulus $\mathcal{Q}^e$, where $\alpha_{\mathcal{I}}$ is any totally positive generator of $\mathcal{I}^{h_{\mathbf{F}}^+}$). Notice that χ must be odd for it must satisfy $\chi(-1) = \nu(-1) = (-1)^3 = -1$. Since we have a canonical isomorphism from $\mathbf{Z}/q^e\mathbf{Z}$ onto $\mathbf{A_F}/\mathcal{Q}^e$, there exists an odd primitive quadratic character on the multiplicative group $(\mathbf{A_F}/\mathcal{Q}^e)^*$, e odd, if and only if there exists an odd primitive quadratic character on the multiplicative group $(\mathbf{Z}/q^e\mathbf{Z})^*$, e odd, hence if and only if $[q = 2$ and $e = 3]$ or $[q \equiv 3 \pmod{4}$ and $e = 1]$, in which cases there exists only one such odd primitive quadratic character modulo $\mathcal{Q}^e$ which we denote by $\chi_\mathcal{Q}$. The values of $\chi_\mathcal{Q}$ are very easy to compute: for $\alpha \in \mathbf{A_L}$ there exists $a_\alpha \in \mathbf{Z}$ such that $\alpha \equiv a_\alpha \pmod{\mathcal{Q}^e}$ and we have $\chi_\mathcal{Q}(\alpha) = \chi_q(a_\alpha)$ where χ_q denote the odd quadratic character associated with the imaginary quadratic field $\mathbf{Q}(\sqrt{-q})$. In particular, we obtain:

Proposition 7. $\mathbf{K} = \mathbf{F}(\sqrt{-\alpha_\mathcal{Q}})$ *is a simplest non-normal sextic CM-field if and only if $q \not\equiv 1 \pmod{4}$ and the odd primitive quadratic character $\chi_\mathcal{Q}$ satisfies $\chi_\mathcal{Q}(\epsilon) = N_{\mathbf{F}/\mathbf{Q}}(\epsilon)$ for the three units ϵ of any system of fundamental units of the unit group $\mathbf{U_F}$ of $\mathbf{F}$. In that case the finite part of the conductor of the quadratic extension $\mathbf{K_{F,\mathcal{Q}}}/\mathbf{F}$ is given by*

$$\mathcal{F}_{\mathbf{K_{F,\mathcal{Q}}}/\mathbf{F}} = \tilde{\mathcal{Q}} := \begin{cases} \mathcal{Q} & \text{if } q > 2, \\ \mathcal{Q}^3 & \text{if } q = 2 \end{cases}$$

3.2.   THE CHARACTERIZATION

Now, we are in a position to give the main result of this third section:

Theorem 8. *A number field $\mathbf{K}$ is a non-normal sextic CM-field of odd class number and of maximal totally real subfield a cyclic cubic field $\mathbf{F}$ if and only if the narrow class number $h_\mathbf{F}^+$ of $\mathbf{F}$ is odd and $\mathbf{K} = \mathbf{K_{F,\mathcal{Q}}} = \mathbf{F}(\sqrt{-\alpha_\mathcal{Q}})$ is a simplest non-normal sextic CM-field associated with $\mathbf{F}$ and a prime ideal $\mathcal{Q}$ of $\mathbf{F}$ above a positive prime $q \not\equiv 1 \pmod{4}$ which splits completely in $\mathbf{F}$ and such that the odd primitive quadratic character $\chi_\mathcal{Q}$ satisfies $\chi_\mathcal{Q}(\epsilon) = N_{\mathbf{F}/\mathbf{Q}}(\epsilon)$ for the three units ϵ of any system of fundamental units of the unit group $\mathbf{U_F}$ of $\mathbf{F}$.*

4.   THE DETERMINATION

Since the $\mathbf{K_{F,\mathcal{Q}}}$'s are isomorphic when $\mathcal{Q}$ ranges over the three prime ideals of $\mathbf{F}$ above a split prime q and since we do not want to distinguish isomorphic number fields, we let $\mathbf{K_{F,q}}$ denote any one of these $\mathbf{K_{F,\mathcal{Q}}}$. We will set $\tilde{q} := N_{\mathbf{F}/\mathbf{Q}}(\mathcal{F}_{\mathbf{K}/\mathbf{F}})$. Hence, $\tilde{q} = q$ if $q > 2$ and $\tilde{q} = 2^3$ if $q = 2$.

Theorem 9. *Let $\mathbf{K} = \mathbf{K_{F,q}}$ be any simplest non-normal sextic CM-field. Then, $Q_\mathbf{K} = 1$, $d_\mathbf{K} = \tilde{q}d_\mathbf{F}^2 = \tilde{q}f_\mathbf{F}^4$, $d_\mathbf{N} = \tilde{q}^{12}d_\mathbf{F}^8 = \tilde{q}^8 d_\mathbf{K}^4 = \tilde{q}^{12}f_\mathbf{F}^{16}$, and (6) yields :*

$$h_{\mathbf{K}}^{-} \geq \epsilon_{\mathbf{K}} \frac{\sqrt{\tilde{q}} f_{\mathbf{F}}^2}{e^{1/8}\pi^3 (\log f_{\mathbf{F}} + 0.05)^2 \log(\tilde{q}^{12} f_{\mathbf{F}}^{16})} \qquad (10)$$

(*where* $\epsilon_{\mathbf{K}} := 1 - (6\pi e^{1/24}/d_{\mathbf{K}}^{1/6}) \geq \epsilon_{\mathbf{F}} := 1 - (6\pi e^{1/24}/3^{1/6} f_{\mathbf{F}}^{2/3}))$. *In particular, if* $h_{\mathbf{K}} = 1$ *then* $f_{\mathbf{F}} \leq 9 \cdot 10^5$ *and for a given* $f_{\mathbf{F}} \leq 9 \cdot 10^5$ *we can use* (10) *to compute a bound on the* $\tilde{q}$'s *for which* $h_{\mathbf{K}}^{-} = 1$. *For example* $h_{\mathbf{K}} = 1$ *and* $f_{\mathbf{F}} = 7$ *imply* $\tilde{q} \leq 5 \cdot 10^7$, $h_{\mathbf{K}} = 1$ *and* $f_{\mathbf{F}} > 1700$ *imply* $\tilde{q} \leq 10^5$, $h_{\mathbf{K}} = 1$ *and* $f_{\mathbf{F}} > 7200$ *imply* $\tilde{q} \leq 10^4$ *and* $h_{\mathbf{K}} = 1$ *and* $f_{\mathbf{F}} > 30000$ *imply* $\tilde{q} \leq 10^3$.

Proof. Noticing that the right hand side of (10) increases with $\tilde{q} \geq 3$, we do obtain that $f_{\mathbf{F}} > 9 \cdot 10^5$ implies $h_{\mathbf{K}}^{-} > 1$. •

Assume that $h_{\mathbf{K}_{\mathbf{F},q}} = 1$. Then $h_{\mathbf{F}}^{+} = 1$, $f_{\mathbf{F}} \equiv 1 \pmod{6}$ is prime or $f_{\mathbf{F}} = 9$, $f_{\mathbf{F}} \leq 9 \cdot 10^5$, and we can compute $B_{\mathbf{F}}$ such that (10) yields $h_{\mathbf{K}_{\mathbf{F},q}}^{-} > 1$ for $q > B_{\mathbf{F}}$ (and we get rid of all the $q \leq B_{\mathbf{F}}$ for which either $q \equiv 1 \pmod{4}$ or q does not split in $\mathbf{F}$ (see Theorem 8)). Now, the key point is to use powerful necessary conditions for the class number of $\mathbf{K}_{\mathbf{F},q}$ to be equal to one, the ones given in [LO, Theorem 6] and in [Oka, Theorem 2]. Using these powerful necessary conditions, we get rid of most of the previous pairs $(q, f_{\mathbf{F}})$ and end up with a very short list of less than two hundred pairs $(q, f_{\mathbf{F}})$ such that any simplest non-normal sextic number fields with class number one must be associated with one of these less than two hundred pairs. Moreover, by getting rid of the pairs $(q, f_{\mathbf{F}})$ for which the modular characters $\chi_{\mathcal{Q}}$ do not satisfy $\chi_{\mathcal{Q}}(\epsilon) = N_{\mathbf{F}/\mathbf{Q}}(\epsilon)$ for the three units ϵ of any system of fundamental units of the unit group $\mathbf{U}_{\mathbf{F}}$, we end up with less than forty number fields $\mathbf{K}_{\mathbf{F},q}$ for which we have to compute their (relative) class numbers. Now, for a given $\mathbf{F}$ of narrow class number one and a given $\mathbf{K}_{\mathbf{F},q}$, we use the method developed in [Lou3] for computing $h_{\mathbf{K}_{\mathbf{F},q}}^{-}$. To this end, we pick up one ideal $\mathcal{Q}$ above q and notice that we may assume that the primitive quadratic character χ on the ray class group of conductor $\tilde{\mathcal{Q}}$ associated with the quadratic extension $\mathbf{K}_{\mathbf{F},q}/\mathbf{F}$ is given by $(\alpha) \mapsto \chi(\alpha) = \nu(\alpha)\chi_{\mathcal{Q}}(\alpha)$ where $\nu(\alpha)$ denotes the sign of the norm of α and where $\chi_{\mathcal{Q}}$ has been defined in subsection 3.1. According to our computation, we obtained:

Theorem 10. *There are 19 non-isomorphic non-normal sextic CM-fields* $\mathbf{K}$ *(whose maximal totally real subfields are cyclic cubic fields* $\mathbf{F}$) *which have class number one: the 19 simplest non-normal sextic CM-fields* $\mathbf{K}_{\mathbf{F},q}$ *given in the following Table:*

Table

$f_\mathbf{F}$	q	$P_\mathbf{F}(X)$	$d_\mathbf{K}$	$\rho_\mathbf{K} = d_\mathbf{K}^{1/6}$
7	167	$X^3 - 18X^2 + 101X - 167$	$7^4 \cdot 167$	$8.587\cdots$
7	239	$X^3 - 19X^2 + 118X - 239$	$7^4 \cdot 239$	$9.115\cdots$
7	251	$X^3 - 22X^2 + 145X - 251$	$7^4 \cdot 251$	$9.190\cdots$
7	379	$X^3 - 26X^2 + 181X - 379$	$7^4 \cdot 379$	$9.844\cdots$
7	491	$X^3 - 26X^2 + 209X - 491$	$7^4 \cdot 491$	$10.278\cdots$
7	547	$X^3 - 27X^2 + 222X - 547$	$7^4 \cdot 547$	$10.464\cdots$
7	1051	$X^3 - 34X^2 + 341X - 1051$	$7^4 \cdot 1051$	$10.668\cdots$
9	71	$X^3 - 30X^2 + 117X - 71$	$9^4 \cdot 71$	$8.804\cdots$
9	199	$X^3 - 39X^2 + 318X - 199$	$9^4 \cdot 199$	$10.454\cdots$
9	379	$X^3 - 30X^2 + 237X - 379$	$9^4 \cdot 379$	$11.639\cdots$
9	523	$X^3 - 57X^2 + 507X - 523$	$9^4 \cdot 523$	$12.281\cdots$
9	739	$X^3 - 33X^2 + 315X - 739$	$9^4 \cdot 739$	$13.009\cdots$
13	47	$X^3 - 15X^2 + 62X - 47$	$13^4 \cdot 47$	$10.502\cdots$
13	79	$X^3 - 14X^2 + 61X - 79$	$13^4 \cdot 79$	$11.452\cdots$
19	31	$X^3 - 11X^2 + 34X - 31$	$19^4 \cdot 31$	$12.620\cdots$
19	83	$X^3 - 18X^2 + 89X - 83$	$19^4 \cdot 83$	$14.871\cdots$
31	2	$X^3 - 12X^2 + 17X - 2$	$31^4 \cdot 2^3$	$13.955\cdots$
37	11	$X^3 - 10X^2 + 21X - 11$	$37^4 \cdot 11$	$16.558\cdots$
61	3	$X^3 - 15X^2 + 14X - 3$	$61^4 \cdot 3$	$18.609\cdots$

In this Table, $f_\mathbf{F}$ is the conductor of $\mathbf{F}$ and $\mathbf{F}$ is also defined as being the splitting field of an unitary cubic polynomial $P_\mathbf{F}(X) = X^3 - aX^2 + bX - c$ with integral coefficients and constant term $c = q$ which is the minimal polynomial of an algebraic element $\alpha_q \in \mathbf{F}$ of norm q such that $\mathbf{K}_{\mathbf{F},q} = \mathbf{F}(\sqrt{-\alpha_q})$. Therefore, $\mathbf{K}_{\mathbf{F},q}$ is generated by one of the complex roots of the sextic polynomial $P_{\mathbf{K}_{\mathbf{F},q}}(X) = -P_\mathbf{F}(-X^2) = X^6 + aX^4 + bX^2 + c$.

References

[Bou]   G. Boutteaux. Détermination des corps à multiplication complexe, sextiques, non galoisiens et principaux. *PhD Thesis,* in preparation.

[CH]   P.E. Conner and J. Hurrelbrink. *Class number parity.* Series in Pure Mathematics. Vol. **8**. Singapore etc.: World Scientific. xi, 234 p. (1988).

[LLO]   F. Lemmermeyer, S. Louboutin and R. Okazaki. The class number one problem for some non-abelian normal CM-fields of degree 24. *J. Théor. Nombres Bordeaux* **11** (1999), 387–406.

[LO]   S. Louboutin and R. Okazaki. Determination of all non-normal quartic CM-fields and of all non-abelian normal octic CM-fields with class number one. *Acta Arith.* **67** (1994), 47–62.

[Lou1]   S. Louboutin. Majorations explicites de $|L(1,\chi)|$. *C. R. Acad. Sci. Paris* **316** (1993), 11–14.

[Lou2]   S. Louboutin. Lower bounds for relative class numbers of CM-fields. *Proc. Amer. Math. Soc.* **120** (1994), 425–434.

[Lou3]   S. Louboutin. Computation of relative class numbers of CM-fields. *Math. Comp.* **66** (1997), 173–184.

[Lou4]   S. Louboutin. Upper bounds on $|L(1,\chi)|$ and applications. *Canad. J. Math.* **50** (1999), 794–815.

[Lou5]   S. Louboutin. Explicit bounds for residues of Dedekind zeta functions, values of L-functions at $s = 1$ and relative class numbers. *J. Number Theory,* **85** (2000), 263–282.

[Lou6]   S. Louboutin. Explicit upper bounds for residues of Dedekind zeta functions and values of L-functions at $s = 1$, and explicit lower bounds for relative class numbers of CM-fields. *Preprint Univ. Caen,* January 2000.

[LYK]   S. Louboutin, Y.-S. Yang and S.-H. Kwon. The non-normal quartic CM-fields and the dihedral octic CM-fields with ideal class groups of exponent ≤ 2. *Preprint* (2000).

[Oka]   R. Okazaki. Non-normal class number one problem and the least prime power-residue. In *Number Theory and Applications (series: Develoments in Mathematics Volume 2), edited by S. Kanemitsu and K. Györy from Kluwer Academic Publishers* (1999) pp. 273–289.

[Sta]   H.M. Stark. Some effective cases of the Brauer-Siegel theorem. *Invent. Math.* **23** (1974), 135–152.

[Wa]   L.C. Washington. *Introduction to Cyclotomic Fields.* Grad. Texts Math. **83**, second edition, Springer-Verlag (1997).

TERNARY PROBLEMS IN ADDITIVE PRIME NUMBER THEORY

Jörg BRÜDERN

Mathematisches Institut A, Universität Stuttgart, D-70511 Stuttgart, Germany
bruedern@mathematik.uni-stuttgart.de

Koichi KAWADA

Department of Mathematics, Faculty of Education, Iwate University, Morioka, 020-8550
Japan
kawada@iwate-u.ac.jp

Keywords: primes, almost primes, sums of powers, sieves

Abstract   We discuss the solubility of the ternary equations $x^2 + y^3 + z^k = n$ for an integer k with $3 \leq k \leq 5$ and large integers n, where two of the variables are primes, and the remaining one is an almost prime. We are also concerned with related quaternary problems. As usual, an integer with at most r prime factors is called a P_r-number. We shall show, amongst other things, that for almost all odd n, the equation $x^2 + p_1^3 + p_2^5 = n$ has a solution with primes p_1, p_2 and a P_{15}-number x, and that for every sufficiently large even n, the equation $x + p_1^2 + p_2^3 + p_3^4 = n$ has a solution with primes p_i and a P_2-number x.

1991 Mathematics Subject Classification: 11P32, 11P55, 11N36, 11P05.

1.   INTRODUCTION

The discovery of the circle method by Hardy and Littlewood in the 1920ies has greatly advanced our understanding of additive problems in number theory. Not only has the method developed into an indispensable tool in diophantine analysis and continues to be the only widely applicable machinery to show that a diophantine equation has many solutions, but also it has its value for heuristical arguments in this area.

---

[1]Written while both authors attended a conference at RIMS Kyoto in December 1999. We express our gratitude to the organizer for this opportunity to collaborate.

C. Jia and K. Matsumoto (eds.), Analytic Number Theory, 39–91.
© 2002 *Kluwer Academic Publishers. Printed in the Netherlands.*

This was already realized by its inventors in a paper of 1925 (Hardy and Littlewood [14]) which contains many conjectures still in a prominent chapter of the problem book. For example, one is lead to expect that the additive equation

$$\sum_{i=1}^{s} x_i^{k_i} = n \tag{1.1}$$

with fixed integers $k_i \geq 2$, is soluble in natural numbers x_i for all sufficiently large n, provided only that

$$\sum_{i=1}^{s} \frac{1}{k_i} > 1, \tag{1.2}$$

and that the allied congruences

$$\sum_{i=1}^{s} x_i^{k_i} \equiv n \pmod{q}$$

have solutions for all moduli q. In this generalization of Waring's problem, particular attention has been paid to the case where only three summands are present in (1.1). Leaving aside the classical territory of sums of three squares there remain the equations

$$x^2 + y^2 + z^k = n \qquad (k \geq 3), \tag{1.3}$$
$$x^2 + y^3 + z^k = n \qquad (3 \leq k \leq 5). \tag{1.4}$$

For none of these equations, it has been possible to confirm the result suggested by a formal application of the Hardy-Littlewood method. It is known, however, that for almost all[2] natural numbers n satisfying the congruence conditions, the equations (1.3) and (1.4) have solutions. Rather than recalling the extensive literature on this problem, we content ourselves with mentioning that Vaughan [28] and Hooley [16] independently added the missing case $k = 5$ of (1.4) to the otherwise complete list provided by Davenport and Heilbronn [6, 7] and Roth [24]. It came to a surprise when Jagy and Kaplansky [21] exhibited infinitely many n not of the form $x^2 + y^2 + z^9$, for which nonetheless the congruence conditions are satisfied.

In this paper, we are mainly concerned with companion problems in additive prime number theory. The ultimate goal would be to solve

---

[2] We use almost all in the sense usually adopted in analytic number theory: a statement is true for almost all n if the number of $n \leq N$ for which the statement is false, is $o(N)$ as $N \to \infty$.

(1.3) and (1.4) with all variables restricted to prime numbers. With existing technology, we can, at best, hope to establish this for almost all n satisfying necessary congruence conditions. A result of this type is indeed available for the equations (1.3). Although the authors are not aware of any explicit reference except for the case $k = 2$ (see Schwarz [26]), a standard application of the circle method yields that for any $k \geq 2$ and any fixed $A > 0$, all but $O(N/(\log N)^A)$ natural numbers $n \leq N$ satisfying the relevant congruence conditions[3] are of the form $n = p_1^2 + p_2^2 + p_3^k$, where p_i denotes a prime variable.

If only one square appears in the representation, the picture is less complete. Halberstam [10, 11] showed that almost all n can be written as

$$p_1^2 + p_2^3 + z^3 = n, \tag{1.5}$$

and also as

$$x^2 + y^3 + p^4 = n. \tag{1.6}$$

Hooley [16] gave a new proof of the latter result, and also found a similar result where the biquadrate in (1.6) is replaced by a fifth power of a prime. In his thesis, the first author [1] was able to handle the equations

$$x^2 + p_1^3 + p_2^k = n \qquad (3 \leq k \leq 5) \tag{1.7}$$

for almost all n. The replacement of the remaining variable in (1.5) or (1.7) by a prime has resisted all attacks so far. It is possible, however, to replace such a variable by an almost prime. Our results are as follows, where an integer with at most r prime factors, counted according to multiplicity, is called a P_r-number, as usual.

Theorem 1. *For almost all odd n the equation $x^2 + p_1^3 + p_2^5 = n$ has a solution with a P_{15}-number x and primes p_1, p_2.*

Theorem 2. *Let $\mathbb{N}_1$ be the set of all odd natural numbers that are not congruent to 2 modulo 3.*
(i) *For almost all $n \in \mathbb{N}_1$, the equation $x^2 + p_1^3 + p_2^4 = n$ has solutions with a P_6-number x and primes p_1, p_2.*
(ii) *For almost all $n \in \mathbb{N}_1$, the equation $p_1^2 + y^3 + p_2^4 = n$ has solutions with a P_4-number y and primes p_1, p_2.*

---

[3]The condition on n here is that the congruences $x^2 + y^2 + z^k \equiv n \pmod{q}$ have solutions with $(xyz, q) = 1$ for all moduli q. On denoting by q_k the product of all primes $p > 3$ such that $(p-1)|k$ and $p \equiv 3 \pmod 4$, this condition is equivalent to (i) $n \equiv 1$ or $3 \pmod 6$ when k is odd, (ii) $n \equiv 3 \pmod{24}$, $n \not\equiv 0 \pmod 5$ and $(n-1, q_k) = 1$ when k is even but $4 \nmid k$, (iii) $n \equiv 3 \pmod{24}$, $n \not\equiv 0, 2 \pmod 5$ and $(n-1, q_k) = 1$ when $4|k$. It is easy to see that almost all n violating this congruence condition cannot be written in the proposed manner.

Theorem 3. *Let $\mathbb{N}_2$ be the set of all odd natural numbers that are not congruent to 5 modulo 7.*
(i) For almost all $n \in \mathbb{N}_2$, the equation $x^2 + p_1^3 + p_2^3 = n$ has solutions with a P_3-number x and primes p_1, p_2.
(ii) For almost all $n \in \mathbb{N}_2$, the equation $p_1^2 + y^3 + p_2^3 = n$ has solutions with a P_3-number y and primes p_1, p_2.

It will be clear from the proofs below that in Theorems 1–3 we actually obtain a somewhat stronger conclusion concerning the size of the exceptional set; for any given $A > 1$ the number of $n \leq N$ satisfying the congruence condition and are not representable in one of specific shapes, is $O(N(\log N)^{-A})$.

A closely related problem is the determination of the smallest s such that the equation

$$\sum_{k=1}^{s} x_k^{k+1} = n \tag{1.8}$$

has solutions for all large natural numbers n. This has attracted many writers since it was first treated by Roth [25] with $s = 50$. The current record $s = 14$ is due to Ford [8]. Early work on the problem was based on diminishing ranges techniques, and has immediate applications to solutions of (1.8) in primes. This is explicitly mentioned in Thanigasalam [27] where it is shown that when $s = 23$ there are prime solutions for all large odd n. An improvement of this result may well be within reach, and we intend to return to this topic elsewhere.

When one seeks for solutions in primes, one may also add a linear term in (1.8), and still faces a non-trivial problem. In this direction, Prachar [23] showed that

$$p_1 + p_2^2 + p_3^3 + p_4^4 + p_5^5 = n$$

is soluble in primes p_i for all large odd n. Although we are unable to sharpen this result by removing a term from the equation, conclusions of this type are possible with some variables as almost primes. For example, it follows easily from the proof of Theorem 2 (ii) that for all large even n the equation

$$p_1 + p_2^2 + y^3 + p_4^4 = n$$

has solutions in primes p_i and a P_4-number y. We may also obtain conclusions which are sharper than those stemming directly from the above results.

Theorem 4. (i) *For all sufficiently large even n, the equation*

$$p_1 + x^2 + p_2^3 + p_3^4 = n$$

has solutions in primes p_i and a P_3-number x.

(ii) *For all sufficiently large even* n, *the equation*

$$p_1 + x^2 + p_2^3 + p_3^5 = n$$

has solutions in primes p_i *and a* P_4-*number* x.

Further we have a result when the linear term is allowed to be an almost prime.

Theorem 5. *For each integer* k *with* $3 \leq k \leq 5$, *and for all sufficiently large even* n, *the equation*

$$x + p_1^2 + p_2^3 + p_3^k = n$$

has solutions in primes p_i *and a* P_2-*number* x.

All results in this paper are based on a common principle. One first solves the diophantine equation at hand with the prospective almost prime variable an ordinary integer. Then the linear sieve is applied to the set of solutions. The sieve input is supplied by various applications of the circle method. This idea was first used by Heath-Brown [15], and for problems of Waring's type, by the first author [3].

A simplicistic application of this circle of ideas suffices to prove Theorem 5. For the other theorems we proceed by adding in refined machinery from sieve theory such as the bilinear structure of the error term due to Iwaniec [19], and the switching principle of Iwaniec [18] and Chen [5]. The latter was already used in problems cognate to those in this paper by the second author [22]. Another novel feature occurs in the proof of Theorem 2 (ii) where the factoriability of the sieving weights is used to perform an efficient differencing in a cubic exponential sum. We refer the reader to §6 and Lemma 4.5 below for details; it is hoped that such ideas prove profitable elsewhere.

2.   NOTATION AND PRELIMINARY RESULTS

We use the following notation throughout. We write $e(\alpha) = \exp(2\pi i \alpha)$, and denote the divisor function and Euler's totient function by $\tau(q)$ and $\varphi(q)$, respectively. The symbol $x \sim X$ is utilized as a shorthand for $X < x < 5X$, and $N \asymp M$ is a shorthand for $M \ll N \ll M$. The letter p, with or without subscript, always stands for prime numbers. We also adopt the familiar convention concerning the letter ε: whenever ε appears in a statement, we assert that the statement holds for each $\varepsilon > 0$, and implicit constants may depend on ε.

We suppose that N is a sufficiently large parameter, and for a natural number k, we put

$$X_k = \frac{1}{5}N^{\frac{1}{k}}.$$

We define

$$f_k(\alpha; d) = \sum_{\substack{x \sim X_k \\ x \equiv 0 \,(\mathrm{mod}\, d)}} e(x^k \alpha), \qquad g_k(\alpha; Q) = \sum_{p \sim Q} e(p^k \alpha),$$

$$S_k(q, a) = \sum_{r=1}^{q} e\left(\frac{a}{q}r^k\right), \qquad S_k^*(q, a) = \sum_{\substack{r=1 \\ (r,q)=1}}^{q} e\left(\frac{a}{q}r^k\right),$$

$$u_k(\beta) = \int_{X_k}^{5X_k} e(t^k \beta)\,dt, \qquad v_k(\beta; Q) = \int_{Q}^{5Q} \frac{e(t^k \beta)}{\log t}\,dt,$$

and write

$$g_k(\alpha) = g_k(\alpha; X_k), \qquad v_k(\beta) = v_k(\beta; X_k).$$

By a well-known theorem of van der Corput, there exists a constant A such that the following inequalities are valid for all $X \geq 2$ and for all integers k with $1 \leq k \leq 5$;

$$\sum_{\substack{x,\,y \sim X \\ x \neq y}} \tau(|x^k - y^k|)^5 \ll X^2 (\log X)^A,$$

$$\sum_{\substack{x_j \sim X \,(1 \leq j \leq 4) \\ x_1^k + x_2^k \neq x_3^k + x_4^k}} \tau(|x_1^k + x_2^k - x_3^k - x_4^k|)^{10} \ll X^4 (\log X)^A. \tag{2.1}$$

We fix such a number $A > 500$, and put

$$L = (\log N)^{500A}, \qquad \mathfrak{M}(q, a) = \{\alpha \in [0,1]; \, |\alpha - a/q| \leq L/N \,\}.$$

Then denote by $\mathfrak{M}$ the union of all $\mathfrak{M}(q, a)$ with $0 \leq a \leq q \leq L$ and $(q, a) = 1$, and write $\mathfrak{m} = [0,1] \setminus \mathfrak{M}$. It is straightforward, for the most part, to handle the various integrals over the major arcs $\mathfrak{M}$ that we encounter later. In order to dispose of such routines simultaneously, we prepare the scene with an exotic lemma.

Lemma 2.1. *Let s be either 1 or 2, and let k and k_j $(0 \leq j \leq s)$ be natural numbers less than 6. Suppose that $w(\beta)$ is a function satisfying $w(\beta) = Cv_{k_0}(\beta) + O(X_{k_0}(\log N)^{-2})$ with a constant C, and that the function $h(\alpha)$ has the property*

$$h(\alpha) = \varphi(q)^{-1}S_{k_0}^*(q, a)w(\alpha - a/q) + O(X_{k_0}L^{-5}),$$

for $\alpha \in \mathfrak{M}(q, a) \subset \mathfrak{M}$. Suppose also that $\sqrt{X_{k_j}} < Q_j \leq X_{k_j}$ for $1 \leq j \leq s$, and write

$$A_d(q, n) = q^{-1}\varphi(q)^{-1-s} \sum_{\substack{a=1 \\ (a,q)=1}}^{q} S_k(q, ad^k) \prod_{j=0}^{s} S_{k_j}^*(q, a)\, e(-an/q), \quad (2.2)$$

$$\mathfrak{S}_d(n, L) = \sum_{q \leq L} A_d(q, n), \quad (2.3)$$

$$I(n) = \int_{-\infty}^{\infty} v_k(\beta)v_{k_0}(\beta) \prod_{j=1}^{s} v_{k_j}(\beta; Q_j)\, e(-n\beta)d\beta, \quad (2.4)$$

$$J(n) = \int_{-L/N}^{L/N} u_k(\beta)w(\beta) \prod_{j=1}^{s} v_{k_j}(\beta; Q_j)\, e(-n\beta)d\beta, \quad (2.5)$$

and put $Q = \prod_{j=1}^{s} Q_j$. Then for $N \leq n \leq (6/5)N$ and $1 \leq d \leq X_k L^{-6}$, one has

$$\int_{\mathfrak{M}} f_k(\alpha; d)h(\alpha) \prod_{j=1}^{s} g_{k_j}(\alpha; Q_j)\, e(-n\alpha)d\alpha$$

$$= \frac{1}{d}\mathfrak{S}_d(n, L)J(n) + O\left(\frac{X_k X_{k_0} Q}{dNL}\right), \quad (2.6)$$

$$J(n) = (C \log X_k + O(\log L))I(n), \quad (2.7)$$

and

$$I(n) \asymp X_k X_{k_0} Q N^{-1}(\log N)^{-2-s}. \quad (2.8)$$

Proof. When $\alpha = a/q + \beta$, $(q, a) = 1$, $q \leq L$ and $|\beta| \leq L/N$, it is known that

$$f_k(\alpha; d) = \frac{S_k(q, ad^k)}{qd}u_k(\beta) + O(L),$$

$$g_{k_j}(\alpha; Q_j) = \frac{S_{k_j}^*(q, a)}{\varphi(q)}v_{k_j}(\beta; Q_j) + O\left(\frac{Q_j}{L^5}\right), \quad (2.9)$$

for $1 \leq j \leq s$. These are in fact immediate from Theorem 4.1 of Vaughan [30] and a minor modification of Lemma 7.15 of Hua [17]. Thus straightforward computation leads to (2.6).

It is also known, by a combination of a trivial estimate together with a partial integration, that

$$u_k(\beta) \ll X_k(1+N|\beta|)^{-1}, \quad v_{k_0}(\beta) \ll X_{k_0}(\log N)^{-1}(1+N|\beta|)^{-1}. \quad (2.10)$$

By these bounds and the trivial bounds $v_{k_j}(\beta; Q_j) \ll Q_j(\log N)^{-1}$ for $1 \leq j \leq s$, we swiftly obtain the upper bound for $I(n)$ contained in (2.8). To show the lower bound for $I(n)$, we appeal to Fourier's inversion formula, and observe that

$$I(n) = \int \left(n - \sum_{j=0}^{s} t_j^{k_j}\right)^{\frac{1}{k}-1} \left(\log\left(n - \sum_{j=0}^{s} t_j^{k_j}\right)\right)^{-1} \prod_{j=0}^{s} \frac{dt_j}{\log t_j},$$

where the region of integration is given by the inequalities $X_{k_0} \leq t_0 \leq 5X_{k_0}$, $Q_j \leq t_j \leq 5Q_j$ $(1 \leq j \leq s)$ and $n - (5X_k)^k \leq \sum_{j=0}^{s} t_j^{k_j} \leq n - X_k^k$. Noticing that all of these inequalities are satisfied when $(N/5)^{1/k_0} \leq t_0 \leq 1.01(N/5)^{1/k_0}$, $Q_j \leq t_j \leq 1.01Q_j$ $(1 \leq j \leq s)$ and $N \leq n \leq (6/5)N$, we obtain the required lower bound for $I(n)$.

It remains to confirm (2.7). By the assumption on $w(\beta)$, together with (2.10) and the trivial bounds for $v_{k_j}(\beta; Q_j)$, we see

$$J(n) = C \int_{-L/N}^{L/N} u_k(\beta) v_{k_0}(\beta) \prod_{j=1}^{s} v_{k_j}(\beta; Q_j) \, e(-n\beta) d\beta$$

$$+ O\left(X_k X_{k_0} Q(\log N)^{-2-s} \int_{-L/N}^{L/N} (1 + N|\beta|)^{-1} d\beta\right).$$

Then we use the relation

$$u_k(\beta) = \int_{X_k}^{5X_k} \frac{\log X_k + O(1)}{\log t} e(t^k \beta) dt = v_k(\beta) \log X_k + O(X_k(\log N)^{-1}),$$

and appeal to (2.10) once again. We consequently obtain

$$J(n) = CI(n) \log X_k + O(X_k X_{k_0} Q N^{-1} \log L(\log N)^{-2-s}),$$

which yields (2.7), in view of (2.8), and the proof of the lemma is completed.

When we appeal to the switching principle in our sieve procedure, we require some information on the generating functions associated with almost primes. We write

$$\Pi(z) = \prod_{p < z} p, \tag{2.11}$$

and denote by $\Omega(x)$ the number of prime factors of x, counted according to multiplicity. Then define

$$g_k(\alpha; X, r, z) = \sum_{\substack{x \sim X \\ (x, \Pi(z)) = 1 \\ \Omega(x) = r}} e(x^k \alpha) = \sum_{\substack{z \leq p_1 \leq \cdots \leq p_r \\ p_1 \cdots p_r \sim X}} e((p_1 \dots p_r)^k \alpha). \tag{2.12}$$

Also, let $C_1(u) = 1$ or 0 according as $u \geq 1$ or $u < 1$, and then define $C_r(u)$ inductively for $r \geq 2$ by

$$C_r(u) = \int_r^{\max\{u,r\}} \frac{C_{r-1}(\xi - 1)}{\xi - 1} d\xi.$$

Lemma 2.2. *Let B and δ be fixed positive numbers, k be a natural number, and X be sufficiently large number. Suppose that $z \geq X^\delta$, and that $\alpha = a/q + \beta$, $(q,a) = 1$, $q \leq (\log X)^B$ and $|\beta| \leq (\log X)^B X^{-k}$. Then for $r \geq 2$, one has*

$$g_k(\alpha; X, r, z) = \varphi(q)^{-1} S_k^*(q,a) w_k(\beta; X, r, z) + O(X(\log X)^{-5B}), \quad (2.13)$$

where the function $w_k(\beta; X, r, z)$ satisfies

$$w_k(\beta; X, r, z) = C_r\left(\frac{\log X}{\log z}\right) v_k(\beta; X) + O(X(\log X)^{-2}). \quad (2.14)$$

Here the implicit constants may depend only on k, B and δ.

Proof. We begin with the expression on the rightmost side of (2.12), and write $b = p_1 \ldots p_{r-1}$ for concision. The innermost sum over p_r becomes, by the corresponding analogue to the latter formula in (2.9) ([17], Lemma 7.15),

$$\varphi(q)^{-1} S_k^*(q, ab^k) \int_{\max\{X/b, p_{r-1}\}}^{5X/b} \frac{c(u^k b^k \beta)}{\log u} du + O(X_k b^{-1}(\log X)^{-5B})$$

$$= \varphi(q)^{-1} S_k^*(q, a) \int_{\max\{X, bp_{r-1}\}}^{5X} \frac{c(t^k \beta)}{b \log(t/b)} dt + O(X_k b^{-1}(\log X)^{-5B}),$$

since $(q, b) = 1$. Thus, on putting

$$c(t; r, z) = \sum_{\substack{z \leq p_1 \leq \cdots \leq p_{r-1} \\ bp_{r-1} \leq t}} \frac{1}{b} \frac{\log t}{\log(t/b)},$$

$$w_k(\beta; X, r, z) = \int_X^{5X} c(t; r, z) \frac{e(t^k \beta)}{\log t} dt, \quad (2.15)$$

we obtain the formula (2.13).

We shall next establish the formula

$$c(t; r, z) = C_r\left(\frac{\log t}{\log z}\right) + O((\log z)^{-1}), \quad (2.16)$$

for $r \geq 2$. Proving this is an exercise in elementary prime number theory, and we indicate only an outline here. It is enough to consider the case

$z^r \leq t$, because $c(t; r, z) = C_r(\log t / \log z) = 0$ otherwise. When $r = 2$, the formula (2.16) follows from Mertens' formula. Then for $r \geq 3$, one may prove (2.16) by induction on r, based on Mertens' formula and the recursive formula

$$c(t; r, z) = \sum_{z \leq p_1 \leq t^{1/r}} \frac{\log t}{p_1 \log(t/p_1)} c(t/p_1; r - 1, p_1).$$

From (2.16) and the definition of $w_k(\beta; X, r, z)$ in (2.15), we can immediately deduce (2.14), completing the proof of the lemma.

3.   SINGULAR SERIES

In this section we handle the partial singular series $\mathfrak{S}_d(n, L)$ defined in (2.3), keeping the conventions in Lemma 2.1 in mind. Namely, s is either 1 or 2, and the natural numbers k and k_j ($0 \leq j \leq s$) are less than 6. In addition we introduce the following notation which are related to (2.2) and (2.3);

$$A(q, n) = \varphi(q)^{-2-s} \sum_{\substack{a=1 \\ (a,q)=1}}^{q} S_k^*(q, a) \prod_{j=0}^{s} S_{k_j}^*(q, a)\, e(-an/q), \qquad (3.1)$$

$$B_d(p, n) = \sum_{h \geq 0} A_d(p^h, n), \qquad B(p, n) = \sum_{h \geq 0} A(p^h, n). \qquad (3.2)$$

The series defining $B_d(p, n)$ and $B(p, n)$ are finite sums in practice, because of the following lemma.

Lemma 3.1. *Let $\theta(p, k)$ be the number such that $p^{\theta(p,k)}$ is the highest power of p dividing k, and let*

$$\gamma(p, k) = \begin{cases} \theta(p, k) + 2, & \text{when } p = 2 \text{ and } k \text{ is even,} \\ \theta(p, k) + 1, & \text{when } p > 2 \text{ or } k \text{ is odd.} \end{cases} \qquad (3.3)$$

Then one has $S_k^*(p^h, a) = 0$ when $p \nmid a$ and $h > \gamma(p, k)$.

Proof. See Lemma 8.3 of Hua [17].

Next we assort basic properties of $A_d(q, n)$ et al.

Lemma 3.2. *Under the above convention, one has the following.*

(i) *$A_d(q, n)$ and $A(q, n)$ are multplicative functions with respect to q.*

(ii) *$B_d(p, n)$ and $B(p, n)$ are always non-negative rational numbers.*

(iii) $A_d(p, n) = A(p, n) = 0$, when $p \geq 7$ and $h \geq 2$, or when $p \leq 5$ and $h \geq 5$.

Proof. The first two assertions are proved via standard arguments (refer to the proofs of Lemmata 2.10-2.12 of Vaughan [30] and Lemmata 8.1 and 8.6 of Hua [17]). The part (iii) is immediate from Lemma 3.1, since $k_j < 6$ $(0 \leq j \leq s)$.

Lemma 3.3. *Assume that*

$$\min_{0 \leq j \leq s} \{\gamma(p, k_j)\} \leq k.$$

Then one has $B_d(p, n) = B_{(p,d)}(p, n)$. *One also has*

$$B_1(p, n) - \frac{1}{p}B_p(p, n) = \left(1 - \frac{1}{p}\right)B(p, n).$$

Proof. The assumption and Lemma 3.1 imply that $A_d(p^h, n) = A(p^h, n) = 0$ for $h > k$, thus

$$B_d(p, n) = \sum_{h=0}^{k} A_d(p^h, n), \quad B(p, n) = \sum_{h=0}^{k} A(p^h, n). \tag{3.4}$$

For $h \leq k$, we may observe that $S_k(p^h, ad^k) = S_k(p^h, a(p, d)^k)$, which gives $A_d(p^h, n) = A_{(p,d)}(p^h, n)$. So the former assertion of the lemma follows from (3.4).

Next we have

$$A_1(p^h, n) - \frac{1}{p}A_p(p^h, n)$$

$$= p^{-h}\varphi(p^h)^{-1-s} \sum_{\substack{a=1 \\ (a,p)=1}}^{p^h} \left(S_k(p^h, a) - \frac{1}{p}S_k(p^h, ap^k)\right) \prod_{j=0}^{s} S_{k_j}^*(p^h, a) \, e\left(-\frac{a}{p^h}n\right).$$

But, when $1 \leq h \leq k$, we see

$$S_k(p^h, a) = S_k^*(p^h, a) + p^{h-1} = S_k^*(p^h, a) + \frac{1}{p}S_k(p^h, ap^k),$$

whence

$$A_1(p^h, n) - \frac{1}{p}A_p(p^h, n) = \left(1 - \frac{1}{p}\right)A(p^h, n).$$

Obviously the last formula holds for $h = 0$ as well. Hence the latter assertion of the lemma follows from (3.4).

Now we commence our treatment of the singular series appearing in our ternary problems, where we set $s = 1$.

Lemma 3.4. *Let $A_d(q, n)$ be defined by (2.2) with $s = 1$, and with natural numbers k, k_0 and k_1 less than 6. Then for any prime p with $p \nmid d$, one has*

$$|A_d(p, n)| < 4kk_0k_1 p^{-1}(p, n)^{1/2}.$$

When $p|d$ one has

$$|A_d(p, n)| \le ((k_0, k_1) - 1)p^{-1/2}(p, n)^{1/2} + O(p^{-1/2}).$$

Proof. For a natural number l, let $\mathcal{A}_l$ be the set of all the non-principal Dirichlet characters χ modulo p such that χ^l is principal. Note that

$$\text{card}(\mathcal{A}_l) \le (l, p - 1) - 1 \le l - 1. \tag{3.5}$$

For a character χ modulo p and an integer m, we write

$$\tau(\chi, m) = \sum_{r=1}^{p-1} \chi(r)e(rm/p).$$

As for the Gauss sum $\tau(\chi, 1)$, we know that $|\tau(\chi, 1)| = p^{1/2}$, when χ is non-principal. It is also easy to observe that when χ is non-principal, we have $\tau(\chi, m) = \overline{\chi}(m)\tau(\chi, 1)$. When χ is principal, on the other hand, we see that $\tau(\chi, m) = p - 1$ or -1 depending on whether $p|m$ or not. In particular, we have

$$|\tau(\chi, m)| \le p^{1/2}(p, m)^{1/2}, \tag{3.6}$$

for any character χ modulo p and any integer m.

By Lemma 4.3 of Vaughan [30], we know that

$$S_l(p, a) = \sum_{\chi \in \mathcal{A}_l} \tau(\overline{\chi}, 1)\chi(a), \tag{3.7}$$

whenever $p \nmid a$, and obviously $S_l^*(p, a) = S_l(p, a) - 1$. So when $p \nmid d$, we have $S_k(p, ad^k) = S_k(p, a)$ and

$A_d(p, n)$

$$= \frac{1}{p(p-1)^2} \sum_{a=1}^{p-1} \left(\sum_{\chi \in \mathcal{A}_k} \tau(\overline{\chi}, 1)\chi(a) \right) \prod_{j=0}^{1} \left(\sum_{\psi_j \in \mathcal{A}_{k_j}} \tau(\overline{\psi}_j, 1)\psi_j(a) - 1 \right) e\left(-\frac{a}{p}n \right)$$

$$= \frac{1}{p(p-1)^2} \bigg(\sum_{\substack{\chi \in \mathcal{A}_k \\ \psi_0 \in \mathcal{A}_{k_0} \\ \psi_1 \in \mathcal{A}_{k_1}}} \tau(\overline{\chi}, 1) \tau(\overline{\psi}_0, 1) \tau(\overline{\psi}_1, 1) \tau(\chi \psi_0 \psi_1, -n)$$

$$- \sum_{j=0}^{1} \sum_{\substack{\chi \in \mathcal{A}_k \\ \psi_j \in \mathcal{A}_{k_j}}} \tau(\overline{\chi}, 1) \tau(\overline{\psi}_j, 1) \tau(\chi \psi_j, -n) + \sum_{\chi \in \mathcal{A}_k} \tau(\overline{\chi}, 1) \tau(\chi, -n) \bigg).$$

By appealing to (3.5) and (3.6), a straightforward estimation yields

$$|A_d(p, n)| \leq \frac{(k-1)p(p,n)^{1/2}}{p(p-1)^2} \prod_{j=0}^{1} ((k_j - 1)p^{1/2} + 1)$$

$$< 4kk_0 k_1 p^{-1}(p,n)^{1/2}.$$

When $p | d$, we have $S_k(p, ad^k) = p$, and the proof proceeds similarly. By using (3.5), (3.6) and (3.7), we have

$$A_d(p, n) = \frac{1}{(p-1)^2} \sum_{\substack{\psi_0 \in \mathcal{A}_{k_0} \\ \psi_1 \in \mathcal{A}_{k_1}}} \tau(\overline{\psi}_0, 1) \tau(\overline{\psi}_1, 1) \tau(\psi_0 \psi_1, -n) + O(p^{-1/2}).$$

Thus when $p | d$ but $p \nmid n$, we have $A_d(p, n) = O(p^{-1/2})$ by (3.6). When $p | n$, we know $\tau(\psi_0 \psi_1, -n) = 0$ unless $\psi_0 \psi_1$ is principal, in which case we have $\tau(\psi_0 \psi_1, -n) = p - 1$ and $\psi_1 = \overline{\psi}_0$, and then notice that $\psi_0 \in \mathcal{A}_{(k_0, k_1)}$, because both of $\psi_0^{k_0}$ and $\psi_0^{k_1}$ are principal. Therefore when $p | d$ and $p | n$, we have

$$|A_d(p, n)| \leq \frac{1}{(p-1)^2} \sum_{\psi_0 \in \mathcal{A}_{(k_0, k_1)}} p(p-1) + O(p^{-1/2})$$

$$\leq ((k_0, k_1) - 1) + O(p^{-1/2}),$$

by (3.5), and the proof of the lemma is complete.

Lemma 3.5. *Let $A_d(q, n)$, $\mathfrak{S}_d(n, L)$ and $B_d(p, n)$ be defined by (2.2), (2.3) and (3.2), respectively, with $s = 1$, and with natural numbers k, k_0 and k_1 less than 6. Moreover put $Y = \exp(\sqrt{\log N})$ and write*

$$\mathcal{P}_d(n, Y) = \prod_{p \leq Y} B_d(p, n). \tag{3.8}$$

Then one has

$$\sum_{n \sim N} \sum_{d \leq N} \frac{\tau(d)}{d} |\mathfrak{S}_d(n, L) - \mathcal{P}_d(n, Y)| \ll NL^{-1/3}.$$

Proof. We define $\mathcal{Q}$ to be the set of all natural numbers q such that every prime divisor of q does not exceed Y, so that we may write

$$\mathcal{P}_d(n, Y) - \mathfrak{S}_d(n, L) = \sum_{\substack{q > L \\ q \in \mathcal{Q}}} A_d(q, n), \tag{3.9}$$

in view of Lemma 3.2 (i). We begin by considering the contribution of integers q greater than $N^{1/5}$ to the latter sum. Put $\eta = 10(\log N)^{-1/2}$. Then, for $q > N^{1/5}$, we see $1 < (q/N^{1/5})^\eta = q^\eta Y^{-2}$, and

$$\sum_{\substack{q > N^{1/5} \\ q \in \mathcal{Q}}} |A_d(q, n)| < Y^{-2} \sum_{q \in \mathcal{Q}} q^\eta |A_d(q, n)| = Y^{-2} \prod_{p \leq Y} \left(\sum_{h=0}^\infty p^{\eta h} |A_d(p^h, n)| \right).$$

Since $p^\eta \leq Y^\eta = e^{10}$, it follows from Lemmata 3.4 and 3.2 (iii) that

$$\sum_{h=0}^\infty p^{\eta h} |A_d(p^h, n)| = 1 + O\left(p^{-1}(p, d)^{\frac{1}{2}}(p, n)^{\frac{1}{2}}\right),$$

which means that there is an absolute constant $C > 0$ such that

$$\sum_{\substack{q > N^{1/5} \\ q \in \mathcal{Q}}} |A_d(q, n)| < Y^{-2} \prod_{p \leq Y}(1 + C/p) \prod_{p | d} C \prod_{p | n} C.$$

Therefore a simple calculation reveals that

$$\sum_{n \sim N} \sum_{d \leq N} \frac{\tau(d)}{d} \left| \sum_{\substack{q > N^{1/5} \\ q \in \mathcal{Q}}} A_d(q, n) \right|$$

$$\ll Y^{-2}(\log Y)^C \left(\sum_{n \sim N} \prod_{p | n} C \right) \left(\sum_{d \leq N} \frac{\tau(d)}{d} \prod_{p | d} C \right)$$

$$\ll NY^{-1}. \tag{3.10}$$

Next we consider the sum

$$\Phi = \sum_{n \sim N} \sum_{d \leq N} \frac{1}{d} \left| \sum_{\substack{L < q \leq N^{1/5} \\ q \in \mathcal{Q}}} A_d(q, n) \right|^2. \tag{3.11}$$

Now write $T_d(q, a) = q^{-1}\varphi(q)^{-2} S_k(q, ad^k) S_{k_0}^*(q, a) S_{k_1}^*(q, a)$ for short. By (3.7) and (3.6) we have

$$S_k(p, ad^k) \ll p^{1/2}(p, d)^{1/2}, \quad S_{k_j}^*(p, a) \ll p^{1/2}, \tag{3.12}$$

whence $T_d(p, a) \ll p^{-3/2}(p, d)^{1/2}$ for all primes p with $p \nmid a$. From the latter result we may plainly deduce the bound

$$T_d(q, a) \ll q^{\varepsilon - 3/2}(q, d)^{1/2}, \tag{3.13}$$

for all natural numbers q with $(q, a) = 1$, in view of Lemma 2.10 of [30], Lemma 8.1 of [17], as well as our Lemma 3.2 (iii). In the meantime we observe that

$$\sum_{n \sim N} \left| \sum_{\substack{L < q \leq N^{1/5} \\ q \in \mathcal{Q}}} A_d(q, n) \right|^2$$

$$= \sum_{\substack{L < q_1, q_2 \leq N^{1/5} \\ q_1, q_2 \in \mathcal{Q}}} \sum_{\substack{1 \leq a_j \leq q_j \ (j=1,2) \\ (a_1, q_1) = (a_2, q_2) = 1}} T_d(q_1, a_1) \overline{T_d(q_2, a_2)} \sum_{n \sim N} e\left(\left(\frac{a_2}{q_2} - \frac{a_1}{q_1} \right) n \right).$$

Unless $q_1 = q_2$ and $a_1 = a_2$, we have $\|a_2/q_2 - a_1/q_1\| \geq 1/(q_1 q_2) \geq N^{-2/5}$, where $\|\beta\| = \min_{m \in \mathbb{Z}} |\beta - m|$, so the last expression is

$$\ll N \sum_{L < q \leq N^{1/5}} \sum_{\substack{a=1 \\ (a,q)=1}}^{q} |T_d(q, a)|^2 + N^{\frac{2}{5}} \left(\sum_{q \leq N^{1/5}} \sum_{\substack{a=1 \\ (a,q)=1}}^{q} |T_d(q, a)| \right)^2$$

$$\ll N \sum_{q > L} q^{\varepsilon - 2}(q, d) + N^{\frac{4}{5} + \varepsilon},$$

by using (3.13). Consequently we obtain the estimate

$$\Phi \ll N \sum_{q > L} q^{\varepsilon - 2} \sum_{d \leq N} \frac{(q, d)}{d} + N^{\frac{5}{6}} \ll N L^{\varepsilon - 1} \log N \log L,$$

which yields

$$\sum_{n \sim N} \sum_{d \leq N} \frac{\tau(d)}{d} \left| \sum_{\substack{L < q \leq N^{1/5} \\ q \in \mathcal{Q}}} A_d(q, n) \right| \ll \left(N \sum_{d \leq N} \frac{\tau(d)^2}{d} \right)^{\frac{1}{2}} \cdot \Phi^{\frac{1}{2}} \ll N L^{-\frac{1}{3}}.$$

The lemma follows from (3.9), (3.10) and the last estimate.

Lemma 3.6. *Let* $B(p, n)$ *be defined by (3.2) with* $s = 1$, $k = 2$, $k_0 = 3$ *and* $3 \leq k_1 \leq 5$.

(i) *When* $k_1 = 5$ *and* n *is odd, one has* $B(p, n) > p^{-2}$ *for all primes* p.

(ii) *When $k_1 = 4$, n is odd and $n \not\equiv 2 \pmod 3$, one has $B(p, n) > p^{-2}$ for all primes p.*

(iii) *When $k_1 = 3$, n is odd and $n \not\equiv 5 \pmod 7$, one has $B(p, n) > p^{-2}$ for all primes p.*

Proof. Since $\min\{\gamma(p, 2), \gamma(p, 3)\} = 1$ for every p, Lemma 3.1 yields that

$$B(p, n) = 1 + A(p, n)$$

$$= (p - 1)^{-3} \sum_{a=1}^{p} S_2^*(p, a) S_3^*(p, a) S_{k_1}^*(p, a) e\left(-\frac{a}{p} n\right)$$

$$= \frac{p}{(p - 1)^3} M(p, n),$$

where $M(p, n)$ denotes the number of solutions of the congruence $x_1^2 + x_2^3 + x_3^{k_1} \equiv n \pmod p$ with $1 \leq x_j < p$ $(1 \leq j \leq 3)$. Thus in order to show $B(p, n) > p^{-2}$, it suffices to confirm that either $|A(p, n)| < 1$ or $M(p, n) > 0$.

It is fairly easy to check directly that $M(p, n) > 0$ in the following cases; (i) $p = 2$ and n is odd, (ii) $p = 3$ and $k_1 = 3$ or 5, (iii) $p = 3$, $k_1 = 4$ and $n \not\equiv 2 \pmod 3$.

Next we note that for each l coprime to p, the number of the integers m with $1 \leq m < p$ such that $l^k \equiv m^k \pmod p$ is exactly $(p - 1, k)$. Thus it follows that

$$\sum_{a=1}^{p-1} |S_k^*(p, a)|^2 = \sum_{a=1}^{p} |S_k^*(p, a)|^2 - (p - 1)^2$$

$$= p(p - 1)(p - 1, k) - (p - 1)^2 = (\nu(p, k)p + 1)(p - 1),$$

where we put $\nu(p, k) = (p - 1, k) - 1$. When $p \nmid a$, meanwhile, we know that $|S_2(p, a)| = \sqrt{p}$ by (3.7), whence $|S_2^*(p, a)| \leq \sqrt{p} + 1$. Consequently we have

$$|A(p, n)| \leq \frac{\sqrt{p} + 1}{(p - 1)^3} \left(\sum_{a=1}^{p-1} |S_3^*(p, a)|^2\right)^{\frac{1}{2}} \left(\sum_{a=1}^{p-1} |S_{k_1}^*(p, a)|^2\right)^{\frac{1}{2}}$$

$$= (\sqrt{p} + 1)(p - 1)^{-2}(\nu(p, 3)p + 1)^{\frac{1}{2}}(\nu(p, k_1)p + 1)^{\frac{1}{2}}. \quad (3.14)$$

When $p \equiv 3 \pmod 4$, moreover, we know that $S_2(p, a)$ is pure imaginary unless $p|a$, which gives the sharper bound $|S_2^*(p, a)| \leq \sqrt{p + 1}$. For such primes, therefore, we may substitute $\sqrt{p + 1}$ for the factor $\sqrt{p} + 1$ appearing in (3.14).

Since $k_1 \leq 5$, we derive from (3.14) that

$$|A(p,n)| \leq (\sqrt{p}+1)(p-1)^{-2}(2p+1)^{\frac{1}{2}}(4p+1)^{\frac{1}{2}} < 1,$$

for $p \geq 17$. If $p \not\equiv 1 \pmod 3$, then we deduce from (3.14) that

$$|A(p,n)| \leq (\sqrt{p}+1)(p-1)^{-2}(4p+1)^{\frac{1}{2}} < 1,$$

for $p \geq 5$. Thus it remains to consider only the primes $p = 7$ and 13.

When $p = 13$, it follows from (3.14) that $|A(13,n)| < 1$ for $k_1 = 3$ and 5. When $p = 13$ and $k_1 = 4$, we can check that $M(13,n) > 0$ for each n with $0 \leq n \leq 12$ by finding a solution of the relevant congruence.

When $p = 7$, replacing the factor $\sqrt{p}+1$ by $\sqrt{p+1}$ in (3.14) according to the remark following (3.14), we have

$$|A(7,n)| \leq \sqrt{8} \cdot 6^{-2} \cdot \sqrt{15}\sqrt{((6,k_1)-1) \cdot 7 + 1} < 1,$$

for $k_1 = 4$ and 5. When $p = 7$ and $k_1 = 3$, we can check by hand again that $M(7,n) > 0$ unless $n \equiv 5 \pmod 7$.

Collecting all the conclusions, we obtain the lemma.

We next turn to the singular series which occur in our quaternary problems. In such circumstances, we set $s = 2$.

Lemma 3.7. *Let $A_d(q,a)$ and $\mathfrak{S}_d(n,L)$ be defined by (2.2) and (2.3) with $s = 2$ and natural numbers k and k_j $(0 \leq j \leq 2)$ which are less than 6, and suppose that $\min\{k,k_0\} = 1$. Then the infinite series $\mathfrak{S}_d(n) = \sum_{q=1}^{\infty} A_d(q,n)$ converges absolutely, and one has*

$$\mathfrak{S}_d(n) = \sum_{q=1}^{\infty} A_d(q,n) = \prod_p B_d(p,n), \tag{3.15}$$

as well as

$$\sum_{d \leq N} \frac{\tau(d)}{d} |\mathfrak{S}_d(n,L) - \mathfrak{S}_d(n)| \ll L^{-1/3}.$$

Proof. Suppose first that $k_0 = 1$. Since $S_1^*(p,a) = -1$ when $p \nmid a$, we derive from (3.12) that

$$|A_d(p,n)| \ll p^{-4} \cdot p \cdot p^{\frac{1}{2}}(p,d)^{\frac{1}{2}}(p^{\frac{1}{2}})^2 \ll p^{-\frac{3}{2}}(p,d)^{\frac{1}{2}}.$$

Thus, by Lemma 3.2 (i) and (iii), we have

$$|A_d(q,n)| \ll q^{\varepsilon-3/2}(q,d), \tag{3.16}$$

for all natural numbers q.

When $k = 1$, alternatively, it is obvious that $S_1(q, ad) \leq (q, d)$ whenever $(q, a) = 1$. So the estimate (3.16) is valid again by (3.12) and Lemma 3.2.

Hence we have (3.16) in all cases. Then the absolute convergence of $\mathfrak{S}_d(n)$ is obvious, and the latter equality sign in (3.15) is assured by Lemma 3.2 (i). Moreover, a simple estimation gives

$$\sum_{d \leq N} \frac{\tau(d)}{d} \left| \mathfrak{S}_d(n, L) - \mathfrak{S}_d(n) \right| \ll \sum_{q > L} q^{\varepsilon - \frac{3}{2}} \sum_{d \leq N} \frac{\tau(d)}{d}(q, d) \ll L^{-1/3}.$$

Lemma 3.8. *Let $B(p, n)$ be defined by (3.1) and (3.2) with $s = 2$, $k = 1$, $k_0 = 2$, $k_1 = 3$ and any k_2. Then one has $B(p, n) > p^{-3}$ for all even n and primes p.*

Proof. It is readily confirmed that the congruence $x_1 + x_2^2 + x_3^3 + x_4^{k_2} \equiv n$ (mod p) has a solution with $1 \leq x_j < p$ $(1 \leq j \leq 4)$ for every even n and every prime p. The desired conclusion follows from this, as in the proof of Lemma 3.6.

4.   ESTIMATION OF INTEGRALS

In this section we provide various estimates for integrals required later, mainly for integrals over the minor arcs $\mathfrak{m}$ defined in §2. We begin with a technical lemma, which generalizes an idea occurring in the proof of Lemma 6 of Brüdern [3].

Lemma 4.1. *Let X and D be real numbers ≥ 2 satisfying $\log D \ll \log X$, k be a fixed natural number, t be a fixed non-negative real number, and let $r = r(d)$ and $b = b(d)$ be integers with $r > 0$ and $(r, b) = 1$ for each natural number $d \leq D$. Also suppose that q and a are coprime integers satisfying $|q\alpha - a| \leq X^{-k/2}$ and $1 \leq q \leq X^{k/2}$. Then one has*

$$\sum_{d \leq D} \frac{X}{d} \tau(r)^t \left(r + \left(\frac{X}{d} \right)^k |rd^k\alpha - b| \right)^{-\frac{1}{k}} \ll \frac{\tau(q)^{t+1} X \log X}{(q + X^k |q\alpha - a|)^{1/k}} + X^{\frac{1}{2}+\varepsilon} D.$$

Proof. Let $\mathcal{D}$ be the set of all the natural numbers $d \leq D$ such that

$$r \leq X^{k/2}/(3D^k) \quad \text{and} \quad |rd^k\alpha - b| \leq 1/(3X^{k/2}).$$

When $d \leq D$ but $d \notin \mathcal{D}$, we have

$$\tau(r)^t \left(r + (X/d)^k |rd^k\alpha - b| \right)^{-1/k} \ll \left(r + (X/d)^k |rd^k\alpha - b| \right)^{(\varepsilon-1)/k}$$
$$\ll (\sqrt{X}/D)^{\varepsilon-1},$$

whence

$$\sum_{\substack{d \le D \\ d \notin \mathcal{D}}} \frac{X}{d} \tau(r)^t \left(r + \left(\frac{X}{d}\right)^k |rd^k\alpha - b| \right)^{-\frac{1}{k}} \ll X^{\frac{1}{2}+\varepsilon} D. \qquad (4.1)$$

When $d \in \mathcal{D}$, we have

$$|qb - rd^k a| \le q|rd^k\alpha - b| + rd^k |q\alpha - a|$$
$$\le X^{k/2} \cdot (3X^{k/2})^{-1} + X^{k/2}(3D^k)^{-1} \cdot D^k \cdot X^{-k/2} < 1,$$

which means $rd^k a = qb$. Since $(q, a) = (r, b) = 1$, it follows that $r = q/(q, d^k)$, and therefore

$$\sum_{d \in \mathcal{D}} \frac{X}{d} \tau(r)^t \left(r + \left(\frac{X}{d}\right)^k |rd^k\alpha - b| \right)^{-\frac{1}{k}}$$

$$\le \frac{X\tau(q)^t}{(q + X^k|q\alpha - a|)^{1/k}} \sum_{d \le D} \frac{(q, d^k)^{1/k}}{d}. \qquad (4.2)$$

Moreover we have easily

$$\sum_{d \le D} d^{-1}(q, d^k)^{1/k} \le \sum_{d \le D} d^{-1}(q, d) \ll \tau(q) \log X.$$

The lemma follows from (4.1), (4.2) and the last inequality.

We proceed to the main objective of this section, and particularly recall the notation $f_k(\alpha; d)$, $g_k(\alpha)$, L, $\mathfrak{M}$ and $\mathfrak{m}$ defined in §2. In addition to these, we introduce some extra notation which is used throughout this section.

We define the intervals

$$\mathfrak{N}(q, a; Q) = \{\alpha \in [0, 1] \,;\, |q\alpha - a| \le Q/N\},$$

denote by $\mathfrak{N}(Q)$ the union of all $\mathfrak{N}(q, a; Q)$ with $0 \le a \le q \le Q$ and $(q, a) = 1$, and write $\mathfrak{n}(Q) = [0, 1] \setminus \mathfrak{N}(Q)$, for a positive number Q. Note that the intervals $\mathfrak{N}(q, a; Q)$ composing $\mathfrak{N}(Q)$ are pairwise disjoint provided that $Q \le X_2$.

For the interest of saving space, we also introduce the notation

$$h_k(\alpha) = h_k(\alpha; (a_x)) = \sum_{x \sim X_k} a_x e(x^k\alpha),$$

where (a_x) is an arbitrary sequence satisfying

$$0 \le a_x \ll 1,$$

for all $x \sim X_k$. In our later application, we shall regard $h_k(\alpha)$ as $g_k(\alpha)$ or some generating functions associated with certain almost primes, which are represented by sums of the functions $g_k(\alpha; X, r, z)$ introduced in (2.12).

Lemma 4.2. *For given sequences (λ_u) and (μ_v) satisfying*

$$|\lambda_u| \le 1, \qquad |\mu_v| \le 1, \tag{4.3}$$

define

$$F_2(\alpha) = F_2(\alpha; D, (\lambda_u), (\mu_v)) = \sum_{u \le D^{2/3}} \lambda_u \sum_{v \le D^{1/3}} \mu_v f_2(\alpha; uv). \tag{4.4}$$

Let $k = 4$ or 5, and $D = X_2^\theta$ with $0 < \theta < (6 - k)/(2k)$. Then one has

$$\int_{\mathfrak{m}} |F_2(\alpha) h_3(\alpha) g_k(\alpha)|^2 d\alpha \ll N^{\frac{2}{3} + \frac{2}{k}} (\log N)^{-3A},$$

and

$$\int_{\mathfrak{m}} |g_1(\alpha) F_2(\alpha) g_3(\alpha) g_k(\alpha)| d\alpha \ll N^{\frac{5}{6} + \frac{1}{k}} (\log N)^{-A}.$$

Proof. The latter estimate follows from the former, since we see, by Schwarz's inequality and orthogonality,

$$\int_{\mathfrak{m}} |g_1 F_2 g_3 g_k| d\alpha \ll \left(\int_0^1 |g_1|^2 d\alpha \right)^{\frac{1}{2}} \left(\int_{\mathfrak{m}} |F_2 g_3 g_k|^2 d\alpha \right)^{\frac{1}{2}}$$

$$\ll N^{\frac{1}{2}} \left(\int_{\mathfrak{m}} |F_2 g_3 g_k|^2 d\alpha \right)^{\frac{1}{2}}.$$

So it suffices to prove the former inequality.

We begin with estimating $F_2(\alpha)$. For each pair of u and v, we can find coprime integers r and b satisfying $|ru^2 v^2 \alpha - b| \le uv/X_2$ and $1 \le r \le X_2/(uv)$, by Dirichlet's theorem (Lemma 2.1 of Vaughan [30]). Then using Theorem 4.1, (4.13), Theorem 4.2 and Lemma 2.8 of Vaughan [30], we have the bound

$$f_2(\alpha; uv) \ll \frac{X_2(uv)^{-1}}{(r + (X_2/(uv))^2 |ru^2 v^2 \alpha - b|)^{1/2}} + (X_2(uv)^{-1})^{\frac{1}{2} + \epsilon}.$$

We next take integers s, c, q and a such that

$$|su^2 \alpha - c| \le u/X_2, \quad s \le X_2/u, \quad (s, c) = 1,$$
$$|q\alpha - a| \le 1/X_2, \quad q \le X_2, \quad (q, a) = 1.$$

Then by (4.3) and repeated application of Lemma 4.1, we have

$$F_2(\alpha)$$

$$\ll \sum_{u \le D^{2/3}} \sum_{v \le D^{1/3}} \frac{X_2(uv)^{-1}}{\left(r + (X_2/(uv))^2|ru^2v^2\alpha - b|\right)^{1/2}} + X_2^{\frac{1}{2}+\epsilon}D^{\frac{1}{2}}$$

$$\ll \sum_{u \le D^{2/3}} \left(\frac{X_2 u^{-1}\tau(s)\log X_2}{\left(s + (X_2/u)^2|su^2\alpha - c|\right)^{1/2}} + \left(\frac{X_2}{u}\right)^{\frac{1}{2}+\epsilon}D^{1/3} \right) + X_2^{\frac{1}{2}+\epsilon}D^{\frac{1}{2}}$$

$$\ll \frac{X_2\tau(q)^2(\log N)^2}{(q + N|q\alpha - a|)^{1/2}} + X_2^{\frac{1}{2}+\epsilon}D^{\frac{2}{3}}. \tag{4.5}$$

We now define, for $\alpha \in \mathfrak{N}(q, a; X_2) \subset \mathfrak{N}(X_2)$,

$$G_2(\alpha) = X_2\tau(q)^2(\log N)^2(q + N|q\alpha - a|)^{-1/2},$$

and $G_2(\alpha) = 0$ for $\alpha \in \mathfrak{n}(X_2)$, so that we may express the estimate (4.5) as

$$F_2(\alpha) \ll G_2(\alpha) + X_2^{\frac{1}{2}+\epsilon}D^{\frac{2}{3}}, \tag{4.6}$$

for $\alpha \in [0, 1]$.

We next set

$$I = \int_{\mathfrak{m}} |F_2 h_3 g_k|^2 d\alpha, \quad I_1 = \int_0^1 |F_2 h_3^2 g_k^2| d\alpha, \quad I_2 = \int_{\mathfrak{m}} |G_2 h_3 g_k|^2 d\alpha.$$

The inequality (4.6) yields

$$I \ll X_2^{\frac{1}{2}+\epsilon}D^{\frac{2}{3}}I_1 + \int_{\mathfrak{m}} |G_2 F_2 h_3^2 g_k^2| d\alpha,$$

but the last integral is $\ll I^{1/2}I_2^{1/2}$ by Schwarz's inequaity. Thus we have

$$I \ll X_2^{\frac{1}{2}+\epsilon}D^{\frac{2}{3}}I_1 + I_2. \tag{4.7}$$

To estimate I_1, we first note that the first inequality in (2.1) swiftly implies the bound

$$\int_0^1 |h_l|^4 d\alpha \ll X_l^2(\log N)^A, \tag{4.8}$$

for each integer l with $1 \le l \le 5$, by considering the underlying diophantine equation. Secondly we estimate the number, say S, of the solutions of the equation $x_1^2 + y_1^k + y_2^k = x_2^2 + y_3^k + y_4^k$ subject to $x_1, x_2 \sim X_2$ and $y_j \sim X_k$ ($1 \le j \le 4$), not only for the immediate use. Estimating the number of such solutions separately according as $x_1 \neq x_2$ or $x_1 = x_2$, we have

$$S \ll \sum_{\substack{y_j \sim X \ (1 \leq j \leq 4) \\ y_1^k + y_2^k \neq y_3^k + y_4^k}} \tau(|y_1^k + y_2^k - y_3^k - y_4^k|) + X_2 \int_0^1 |f_k(\alpha; 1)|^4 d\alpha$$

$$\ll (X_k^4 + X_2 X_k^2)(\log N)^A, \tag{4.9}$$

by (4.8) and our convention (2.1). Thirdly we note that

$$F_2(\alpha) = \sum_{x \sim X_2} b_x e(x^2 \alpha), \quad \text{where} \quad b_x = \sum_{\substack{uv|x \\ u \leq D^{2/3}, \, v \leq D^{1/3}}} \lambda_u \mu_v, \tag{4.10}$$

by (4.4), and that $b_x \ll X_2^\varepsilon$ for $x \sim X_2$ by (4.3) and the well-known estimate for the divisor function. Consequently, by orthogonality, we have

$$\int_0^1 |F_2^2 g_k^4| d\alpha \ll X_2^\varepsilon S \ll N^{\frac{1}{2} + \frac{2}{k} + \varepsilon},$$

for $k = 4$ and 5. Finally it follows at once from (4.8) and the last inequality that

$$I_1 \ll \left(\int_0^1 |h_3|^4 d\alpha \right)^{\frac{1}{2}} \left(\int_0^1 |F_2^2 g_k^4| d\alpha \right)^{\frac{1}{2}} \ll N^{\frac{7}{12} + \frac{1}{k} + \varepsilon}. \tag{4.11}$$

We turn to I_2. For later use, we note that the following deliberation on I_2 is valid for $3 \leq k \leq 5$. Our treatment of integrals involving $G_2(\alpha)$ or its kin is motivated by the proof of Lemma 2 of Brüdern [2]. Putting $L' = (\log N)^{12A}$, we denote by $\mathfrak{N}_1$ the union of all $\mathfrak{N}(q, a; X_2)$ with $N^{1/3} L' < q \leq X_2$, $1 \leq a \leq q$ and $(a, q) = 1$, and by $\mathfrak{N}_2$ the union of all $\mathfrak{N}(q, a; X_2)$ with $0 \leq a \leq q \leq N^{1/3} L'$ and $(a, q) = 1$. We remark that $\mathfrak{N}(X_2) = \mathfrak{N}_1 \cup \mathfrak{N}_2$.

By the above definitions we have

$$\int_{\mathfrak{N}_1} |G_2 g_k|^4 d\alpha$$

$$\ll X_2^4 (\log N)^8 \sum_{N^{1/3} L' < q \leq X_2} \frac{\tau(q)^8}{q^2} \sum_{a=1}^q \int_0^{X_2/(qN)} \frac{|g_k(a/q + \beta)|^4}{(1 + N\beta)^2} d\beta. \tag{4.12}$$

We define ψ_l to be the number of solutions of the equation $x_1^k + x_2^k - x_3^k - x_4^k = l$ in primes x_j subject to $x_j \sim X_k$ ($1 \leq j \leq 4$), so that

$|g_k(\alpha)|^4 = \sum_l \psi_l e(l\alpha)$. As we know that $\psi_0 \ll X_k^{2+\varepsilon}$, we see

$$\sum_{a=1}^{q} |g_k(a/q + \beta)|^4 = q \sum_{\substack{l \equiv 0 \ (\mathrm{mod}\ q)}} \psi_l e(l\beta) \ll q \sum_{\substack{l \equiv 0 \ (\mathrm{mod}\ q) \\ l \neq 0}} \psi_l + q X_k^{2+\varepsilon}.$$

We substitute this into (4.12). Then, after modest operation we easily arrive at the estimation

$$\int_{\mathfrak{N}_1} |G_2 g_k|^4 d\alpha$$

$$\ll X_2^4 N^{-1} (\log N)^8 \left(\sum_{l \neq 0} \psi_l \sum_{\substack{N^{1/3} L' < q \leq X_2 \\ q | l}} \frac{\tau(q)^8}{q} + X_k^{2+\varepsilon} \sum_{q \leq X_2} \frac{\tau(q)^8}{q} \right). \quad (4.13)$$

Since we are assuming that A is so large that (2.1) holds, we have

$$\sum_{l \neq 0} \psi_l \sum_{\substack{N^{1/3} L' < q \leq X_2 \\ q | l}} \tau(q)^8 q^{-1} \leq (N^{1/3} L')^{-1} \sum_{l \neq 0} \psi_l \tau(l)^9$$

$$\ll X_k^4 N^{-1/3} (\log N)^{-11A}.$$

Thus, recalling that k is at most 5, we deduce from (4.13) that

$$\int_{\mathfrak{N}_1} |G_2 g_k|^4 d\alpha \ll X_2^4 X_k^4 N^{-4/3} (\log N)^{8-11A}.$$

By this and (4.8), we have

$$\int_{\mathfrak{N}_1} |G_2 h_3 g_k|^2 d\alpha \ll \left(\int_0^1 |h_3|^4 d\alpha \right)^{\frac{1}{2}} \left(\int_{\mathfrak{N}_1} |G_2 g_k|^4 d\alpha \right)^{\frac{1}{2}}$$

$$\ll N^{\frac{2}{3}+\frac{2}{k}} (\log N)^{4-5A}. \quad (4.14)$$

Next define ψ_l' by means of the formula $|h_3(\alpha)|^2 = \sum_l \psi_l' e(l\alpha)$, and write ψ_l'' for the number of solutions of $x^3 - y^3 = l$ subject to $x, y \sim X_3$. It follows from our convention on $h_3(\alpha)$ that $\psi_l' \ll \psi_l''$. Then, proceeding as above, we have

$$\int_{\mathfrak{N}_2} |G_2 h_3|^2 d\alpha \ll X_2^2 (\log N)^4 \sum_{q \leq N^{1/3} L'} \frac{\tau(q)^4}{q} \sum_{a=1}^{q} \int_0^1 \frac{|h_3(a/q + \beta)|^2}{1 + N\beta} d\beta$$

$$\ll (\log N)^5 \left(\sum_{l \neq 0} \psi_l'' \sum_{q|l} \tau(q)^4 + \psi_0'' \sum_{q \leq N^{1/3} L'} \tau(q)^4 \right).$$

The sums appearing in the last expression can be estimated by (2.1) and a well-known result on the divisor function. Thus we may conclude that

$$\int_{\mathfrak{N}_2} |G_2 h_3|^2 d\alpha \ll N^{2/3} (\log N)^{12A+20}.$$

Moreover, when $\alpha \in \mathfrak{N}(q, a; X_2)$ but $\alpha \notin \mathfrak{M}$, we must have

$$G_2(\alpha) \ll X_2 (\log N)^4 (q + N|q\alpha - a|)^{\varepsilon - 1/2} \ll X_2 L^{-1/3}. \qquad (4.15)$$

Therefore, using the trivial bound $h_3 \ll X_3$ also, we obtain

$$\int_{\mathfrak{m} \cap \mathfrak{N}_2} |G_2 h_3|^{\frac{17}{8}} d\alpha \ll \left(X_2 L^{-\frac{1}{3}} X_3 \right)^{\frac{1}{8}} \int_{\mathfrak{N}_2} |G_2 h_3|^2 d\alpha \ll N^{\frac{37}{48}} (\log N)^{-7A}.$$
$$(4.16)$$

On the other hand, it is known within the classical theory of the circle method that

$$\int_0^1 |g_k(\alpha)|^{34} d\alpha \ll \int_0^1 |f_k(\alpha; 1)|^{34} d\alpha \ll X_k^{34-k}, \qquad (4.17)$$

for $k \le 5$ (see Vaughan [30], Ch. 2). Thus we have

$$\int_{\mathfrak{m} \cap \mathfrak{N}_2} |G_2 h_3 g_k|^2 d\alpha \ll \left(\int_{\mathfrak{m} \cap \mathfrak{N}_2} |G_2 h_3|^{\frac{17}{8}} d\alpha \right)^{\frac{16}{17}} \left(\int_0^1 |g_k|^{34} d\alpha \right)^{\frac{1}{17}}$$
$$\ll N^{\frac{2}{3} + \frac{2}{k}} (\log N)^{-6A}, \qquad (4.18)$$

and it follows from this and (4.14) that

$$I_2 \ll N^{\frac{2}{3} + \frac{2}{k}} (\log N)^{-3A}. \qquad (4.19)$$

By (4.7), (4.11) and (4.19), we obtain

$$I \ll N^{\frac{5}{6} + \frac{1}{k} + \frac{1}{3}\theta + \varepsilon} + N^{\frac{2}{3} + \frac{2}{k}} (\log N)^{-3A},$$

which implies the conclusion of the lemma immediately.

Lemma 4.3. *Let* $F_2(\alpha) = F_2(\alpha; D, (\lambda_u), (\mu_v))$ *be as in Lemma 4.2, and let* $D = X_2^\theta$ *with* $0 < \theta < 5/12$. *Then one has*

$$\int_{\mathfrak{m}} |F_2(\alpha) h_3(\alpha) g_3(\alpha; X_3^{5/6})|^2 d\alpha \ll N^{\frac{11}{9}} (\log N)^{-10A}.$$

Proof. Write $\tilde{g}_3 = g_3(\alpha; X_3^{5/6})$ for short, and set

$$J = \int_{\mathfrak{m}} |F_2 h_3 \tilde{g}_3|^2 d\alpha, \quad J_1 = \int_0^1 |F_2^{\frac{3}{2}} h_3^2 \tilde{g}_3^2| d\alpha,$$

$$J_2 = \int_{\mathfrak{m} \cap \mathfrak{N}(X_3^{5/6})} |G_2 h_3 \tilde{g}_3|^2 d\alpha.$$

Since we have $G_2(\alpha) \ll X_2^{13/18+\varepsilon}$ whenever $\alpha \in \mathfrak{n}(X_3^{5/6})$, we deduce from (4.6) that

$$J \ll (X_2^{\frac{13}{18}+\varepsilon} + X_2^{\frac{1}{2}+\varepsilon} D^{\frac{2}{3}})^{\frac{1}{2}} J_1 + \int_{\mathfrak{m} \cap \mathfrak{N}(X_3^{5/6})} |G_2^{\frac{1}{2}} F_2^{\frac{3}{2}} h_3^2 \tilde{g}_3^2| d\alpha.$$

The last integral is $\ll J_2^{1/4} J^{3/4}$ by Hölder's inequality, whence

$$J \ll (X_2^{\frac{13}{36}+\varepsilon} + X_2^{\frac{1}{4}+\varepsilon} D^{\frac{1}{3}}) J_1 + J_2. \tag{4.20}$$

As for J_1, we appeal to the inequalities

$$\int_0^1 |F_2|^4 d\alpha \ll X_2^{2+\varepsilon}, \quad \int_0^1 |h_3|^8 d\alpha \ll X_3^{5+\varepsilon}, \quad \int_0^1 |h_3^2 \tilde{g}_3^4| d\alpha \ll X_3^{\frac{8}{3}+\varepsilon}. \tag{4.21}$$

The first one is plainly obtained in view of (4.10) and (4.8), the second one comes from Hua's inequality (Lemma 2.5 of Vaughan [30]), and the last one is due to Theorem of Vaughan [29]. Thus we have

$$J_1 \ll \left(\int_0^1 |F_2|^4 d\alpha \right)^{\frac{3}{8}} \left(\int_0^1 |h_3|^8 d\alpha \right)^{\frac{1}{8}} \left(\int_0^1 |h_3^2 \tilde{g}_3^4| d\alpha \right)^{\frac{1}{2}} \ll N^{\frac{37}{36}+\varepsilon}. \tag{4.22}$$

To estimate J_2, we may first follow the proof of (4.16) to confirm that

$$\int_{\mathfrak{N}(X_3^{5/6})} |G_2 \tilde{g}_3|^2 d\alpha \ll X_3^{5/3} (\log N)^{A+5} \tag{4.23}$$

holds, and then, by this together with (4.15) and (4.17), we infer the bound

$$J_2 \ll \left(\left(\sup_{\alpha \in \mathfrak{m}} |G_2| X_3^{\frac{5}{6}} \right)^{\frac{1}{8}} \int_{\mathfrak{N}(X_3^{5/6})} |G_2 \tilde{g}_3|^2 d\alpha \right)^{\frac{16}{17}} \left(\int_0^1 |h_3|^{34} d\alpha \right)^{\frac{1}{17}}$$

$$\ll N^{\frac{11}{9}} (\log N)^{-15A}.$$

Now it follows from (4.20), (4.22) and the last inequality that

$$J \ll N^{\frac{83}{72}+\frac{1}{6}\theta+\varepsilon} + N^{\frac{11}{9}} (\log N)^{-15A},$$

which gives the lemma.

Lemma 4.4. *For a given sequence (λ_d) satisfying $|\lambda_d| \leq 1$, define*

$$F_3(\alpha) = F_3(\alpha; D, (\lambda_d)) = \sum_{d \leq D} \lambda_d f_3(\alpha; d),$$

and let $D = X_3^\theta$ with $0 < \theta < 1/3$. Then one has

$$\int_m |h_2(\alpha) F_3(\alpha) g_3(\alpha; X_3^{5/6})|^2 d\alpha \ll N^{\frac{11}{9}} (\log N)^{-10A}.$$

Proof. Define

$$G_3(\alpha) = X_3 \tau(q)(\log N)(q + N|q\alpha - a|)^{-1/3},$$

for $\alpha \in \mathfrak{N}(q, a; X_2) \subset \mathfrak{N}(X_2)$, and $G_3(\alpha) = 0$ for $\alpha \in \mathfrak{n}(X_2)$. Then Lemma 6 of Brüdern [3] gives the bound

$$F_3(\alpha) \ll G_3(\alpha) + X_3^{3/4+\varepsilon} D^{1/4}, \tag{4.24}$$

for $\alpha \in [0, 1]$. (It is apparent from the proof that the factor q^ε appearing in Lemma 6 of [3] may be replaced by $\tau(q)$.)

We write $\tilde{g}_3 = g_3(\alpha; X_3^{5/6})$ again, and put

$$K = \int_m |h_2 F_3 \tilde{g}_3|^2 d\alpha, \quad K_1 = \int_0^1 |h_2^2 F_3 \tilde{g}_3^2| d\alpha,$$

$$K_2 = \int_{m \cap \mathfrak{N}(X_3^{3/4})} |h_2 G_3 \tilde{g}_3|^2 d\alpha.$$

Since $G_3(\alpha) \ll X_3^{3/4+\varepsilon}$ for $\alpha \in \mathfrak{n}(X_3^{3/4})$, we have

$$K \ll X_3^{3/4+\varepsilon} D^{1/4} K_1 + K^{1/2} K_2^{1/2},$$

by virtue of (4.24), whence

$$K \ll X_3^{3/4+\varepsilon} D^{1/4} K_1 + K_2. \tag{4.25}$$

We may express $F_3(\alpha)$ as $\sum_{x \sim X_3} b_x e(x^3 \alpha)$ with $b_x \ll X_3^\varepsilon$, and therefore, by orthogonality and the last inequality in (4.21), we see

$$\int_0^1 |F_3^2 \tilde{g}_3^4| d\alpha \ll X_3^\varepsilon \cdot X_3^{8/3+\varepsilon} \ll N^{8/9+\varepsilon}.$$

By this and (4.8), we have

$$K_1 \ll \left(\int_0^1 |h_2|^4 d\alpha \right)^{\frac{1}{2}} \left(\int_0^1 |F_3^2 \tilde{g}_3^4| d\alpha \right)^{\frac{1}{2}} \ll N^{\frac{17}{18}+\varepsilon}. \tag{4.26}$$

Next we note that $G_3(\alpha) \ll G_2(\alpha)^{2/3}$ for all $\alpha \in [0,1]$. So we see

$$\int_{\mathfrak{m}\cap\mathfrak{N}(X_3^{3/4})} |G_3\tilde{g}_3|^4 d\alpha \ll \int_{\mathfrak{m}\cap\mathfrak{N}(X_3^{3/4})} |G_2^{8/3}\tilde{g}_3^4| d\alpha$$

$$\ll (\sup_{\alpha\in\mathfrak{m}} |G_2|)^{\frac{2}{3}} (X_3^{\frac{5}{6}})^2 \int_{\mathfrak{N}(X_3^{3/4})} |G_2\tilde{g}_3|^2 d\alpha \ll N^{\frac{13}{9}} (\log N)^{-100A},$$

in view of (4.15) and (4.23). Thus, using (4.8) again, we obtain

$$K_2 \ll \left(\int_0^1 |h_2|^4 d\alpha\right)^{\frac{1}{2}} \left(\int_{\mathfrak{m}\cap\mathfrak{N}(X_3^{3/4})} |G_3\tilde{g}_3|^4 d\alpha\right)^{\frac{1}{2}} \ll N^{\frac{11}{9}} (\log N)^{-10A}.$$

(4.27)

The lemma follows from (4.25), (4.26) and (4.27), via a modicum of computation.

Lemma 4.5. *For a given sequence* (λ_d) *satisfying* $|\lambda_d| \leq 1$*, define*

$$\tilde{F}_3(\alpha) = \tilde{F}_3(\alpha; D, (\lambda_d)) = \sum_{p\sim X_{21}} \sum_{d\leq D} \lambda_d f_3(\alpha; pd),$$

and let $D = (X_3^{6/7})^\theta$ *with* $0 < \theta < 1/3$*. Then one has*

$$\int_{\mathfrak{m}} |g_2(\alpha)\tilde{F}_3(\alpha)g_4(\alpha)|^2 d\alpha \ll N^{\frac{7}{6}} (\log N)^{-10A}.$$

For each $p \sim X_{21}$, take co-prime integers $r = r(p)$ and $b = b(p)$ such that $|rp^3\alpha - b| \leq (X_3/p)^{-3/2}$ and $1 \leq r \leq (X_3/p)^{3/2}$. Then applying Lemma 6 of Brüdern [3] again, we have

$$\sum_{d\leq D} \lambda_d f_3(\alpha; pd) \ll \frac{X_3 p^{-1}\tau(r)(\log N)}{(r + (X_3/p)^3|rp^3\alpha - b|)^{1/3}} + (X_3/p)^{3/4+\varepsilon} D^{1/4}.$$

Accordingly Lemma 4.1 yields

$$\tilde{F}_3(\alpha) \ll \tilde{G}_3(\alpha) + X_3^{3/4+\varepsilon}(X_{21}D)^{1/4}, \tag{4.28}$$

where $\tilde{G}_3(\alpha)$ is the function on $[0,1]$ defined by

$$\tilde{G}_3(\alpha) = X_3\tau(q)^2(\log N)^2(q + N|q\alpha - a|)^{-1/3}$$

for $\alpha \in \mathfrak{N}(q, a; X_2) \subset \mathfrak{N}(X_2)$, and $\tilde{G}_3(\alpha) = 0$ for $\alpha \in \mathfrak{n}(X_2)$.

We now put

$$U_0 = \int_{\mathfrak{m}} |\tilde{F}_3|^8 d\alpha, \quad U_1 = \int_0^1 |\tilde{F}_3|^6 d\alpha, \quad U_2 = \int_{\mathfrak{m}} |\tilde{G}_3|^8 d\alpha.$$

By (4.28), we have

$$U_0 \ll (X_3^{3/4+\varepsilon} X_{21}^{1/4} D^{1/4})^2 U_1 + \int_{\mathfrak{m}} |\tilde{G}_3^2 \tilde{F}_3^6| d\alpha$$
$$\ll N^{11/21+\varepsilon} D^{1/2} U_1 + U_0^{3/4} U_2^{1/4},$$

whence

$$U_0 \ll N^{11/21+\varepsilon} D^{1/2} U_1 + U_2. \tag{4.29}$$

As regards U_1, considering the number of solutions of the underlying diophantine equation with a well-known estimate for the divisor function, we may observe that Lemma 6 of Brüdern and Watt [4] swiftly implies the bound

$$U_1 \ll X_3^{7/2+\varepsilon} X_{21}^{-3/2} \ll N^{23/21+\varepsilon}.$$

Also, we have the straightforward estimate

$$U_2 \ll \left(\sup_{\alpha \in \mathfrak{m}} |\tilde{G}_3(\alpha)| \right) \sum_{q \leq X_2} q \int_0^1 \frac{X_3^7 \tau(q)^{14} (\log N)^{14}}{(q + Nq\beta)^{7/3}} d\beta \ll X_3^5 L^{-1/4}.$$

Therefore we derive from (4.29) that

$$U_0 \ll N^{\frac{34}{21}+\frac{1}{7}\theta+\varepsilon} + X_3^5 L^{-\frac{1}{4}} \ll N^{\frac{5}{3}} (\log N)^{-100A}.$$

Making use of (4.8), (4.9) and the last result on U_0, we conclude that

$$\int_{\mathfrak{m}} |g_2 \tilde{F}_3 g_4|^2 d\alpha \ll \left(\int_0^1 |g_2|^4 d\alpha \right)^{\frac{1}{4}} \left(\int_0^1 |g_2^2 g_4^4| d\alpha \right)^{\frac{1}{2}} U_0^{\frac{1}{4}}$$
$$\ll N^{\frac{7}{6}} (\log N)^{-10A},$$

as required.

Lemma 4.6. *For a given sequence (λ_d) satisfying $|\lambda_d| \leq 1$, define*

$$F_1(\alpha) = F_1(\alpha; D, (\lambda_d)) = \sum_{d \leq D} \lambda_d f_1(\alpha; d),$$

and let $D = X_1^\theta$ with $0 < \theta < 17/30$. Then for each integer k with $3 \leq k \leq 5$, one has

$$\int_{\mathfrak{m}} |F_1(\alpha) h_2(\alpha) g_3(\alpha) g_k(\alpha)| d\alpha \ll N^{\frac{5}{6}+\frac{1}{k}} (\log N)^{-A}.$$

We remark that when $k \leq 4$, the restriction on θ in this lemma can be relaxed to $0 < \theta < 2/3$, as the following proof shows.

Proof. We know $f_1(\alpha; d) \ll \min\{N/d, \|\alpha d\|^{-1}\}$, where $\|\beta\|$ denotes the distance from β to the nearest integer. So Lemma 2.2 of Vaughan [30] gives

$$F_1(\alpha) \ll \sum_{d \leq D} \min\{N/d, \|\alpha d\|^{-1}\} \ll (N/q + D + q) \log(qN), \quad (4.30)$$

whenever $|q\alpha - a| \leq q^{-1}$ and $(q, a) = 1$.

We can find coprime integers q and a satisfying $|q\alpha - a| \leq N^{-1/2}$ and $q \leq N^{1/2}$ by Dirichlet's theorem. If $|q\alpha - a| \leq q/N$, then we obtain the bound

$$F_1(\alpha) \ll \left(N(q + N|q\alpha - a|)^{-1} + D + \sqrt{N}\right) \log N, \quad (4.31)$$

from (4.30). In the opposite case $|q\alpha - a| > q/N$ we take coprime integers r and b such that $|r\alpha - b| \leq |q\alpha - a|/2$ and $r \leq 2/|q\alpha - a|$, according to Dirichlet's theorem again. Since b/r cannot be identical with a/q now, we see

$$1 \leq |qb - ra| \leq q|r\alpha - b| + r|q\alpha - a| \leq 1/2 + r|q\alpha - a|,$$

which implies that $r \geq (2|q\alpha - a|)^{-1} \gg N^{1/2}$. Thus it follows from (4.30) that

$$F_1(\alpha) \ll (N/r + D + r) \log N \ll (\sqrt{N} + D + |q\alpha - a|^{-1}) \log N.$$

Hence the estimate (4.31) holds whenever $|q\alpha - a| \leq N^{-1/2}$, $q \leq N^{1/2}$ and $(q, a) = 1$.

Recycling the function $G_2(\alpha)$ defined in the proof of Lemma 4.2, we may express (4.31) as $F_1(\alpha) \ll G_2(\alpha)^2 + (D + X_2) \log N$. Now put

$$V = \int_{\mathfrak{m}} |F_1 h_2 g_3 g_k| d\alpha, \quad V_1 = \int_0^1 |F_1^{\frac{3}{4}} h_2 g_3 g_k| d\alpha, \quad V_2 = \int_{\mathfrak{m}} |G_2^2 h_2 g_3 g_k| d\alpha,$$

and observe that $V \ll (D + X_2)^{1/4+\varepsilon} V_1 + V^{3/4} V_2^{1/4}$, whence

$$V \ll (D + X_2)^{1/4+\varepsilon} V_1 + V_2. \quad (4.32)$$

By an application of Hölder's inequality, we see

$$V_1 \ll \left(\int_0^1 |F_1|^2 d\alpha\right)^{\frac{3}{8}} \left(\int_0^1 |h_2|^4 d\alpha\right)^{\frac{1}{8}} \left(\int_0^1 |g_3|^4 d\alpha\right)^{\frac{1}{4}} \left(\int_0^1 |h_2^2 g_k^4| d\alpha\right)^{\frac{1}{4}}.$$

The first integral appearing on the right hand side is obviously $O(N^{1+\varepsilon})$ by orthogonality, and we refer to (4.8) and (4.9) for the other integrals. Consequently we obtain

$$V_1 \ll N^{\frac{2}{3}+\varepsilon}(N^{\frac{1}{k}} + N^{\frac{1}{8} + \frac{1}{2k}}).$$

On the other hand, making use of (4.8), (4.19) and the simple estimate

$$\int_0^1 |G_2|^4 d\alpha \ll \sum_{q \leq X_2} q \int_0^1 \frac{X_2^4 \tau(q)^8 (\log N)^8}{(q + qN\beta)^2} d\beta \ll N(\log N)^{264},$$

we observe that

$$V_2 \ll \left(\int_0^1 |h_2|^4 d\alpha \right)^{\frac{1}{4}} \left(\int_0^1 |G_2|^4 d\alpha \right)^{\frac{1}{4}} \left(\int_m |G_2 g_3 g_k|^2 d\alpha \right)^{\frac{1}{2}}$$

$$\ll N^{\frac{5}{6}+\frac{1}{k}} (\log N)^{-A}.$$

Hence it follows from (4.32) that

$$V \ll N^{\frac{2}{3}+\frac{1}{k}+\frac{1}{4}\theta+\varepsilon} + N^{\frac{19}{24}+\frac{1}{2k}+\frac{1}{4}\theta+\varepsilon} + N^{\frac{5}{6}+\frac{1}{k}} (\log N)^{-A}.$$

This establishes the lemma.

We close this section by showing an easy lemma, which allows us to ignore the integers divisible by a square of a larger prime, in our application of weighted sieves.

Lemma 4.7. *Let* δ *be an arbitrary fixed positive number, and* k *be an integer with* $3 \leq k \leq 5$. *Then one has*

$$\sum_{n \sim N} \sum_{p > X_2^\delta} \int_0^1 f_2(\alpha; p^2) g_3(\alpha) g_k(\alpha) e(-n\alpha) d\alpha \ll N^{\frac{5}{6}+\frac{1}{k}-\frac{1}{2}\delta},$$

and

$$\sum_{p > X_1^\delta} \int_0^1 f_1(\alpha; p^2) g_2(\alpha) g_3(\alpha) g_k(\alpha) e(-n\alpha) d\alpha \ll N^{\frac{5}{6}+\frac{1}{k}-\frac{1}{3}\delta}.$$

Proof. Since $\sum_{n \sim N} e(-n\alpha) \ll \min\{N, \|\alpha\|^{-1}\}$, the expression on the left hand side of the first inequality is dominated by

$$\sum_{p > X_2^\delta} \frac{X_2 X_3 X_k}{p^2 (\log N)^2} \int_0^1 \min\{N, \|\alpha\|^{-1}\} d\alpha \ll X_2^{1-\delta} X_3 X_k.$$

Also it is indeed an easy exercise to establish the estimates

$$\int_0^1 \left| \sum_{p > X_1^\delta} f_1(\alpha; p^2) \right|^2 d\alpha \ll N^{1-\delta+\varepsilon} \quad \text{and} \quad \int_0^1 |g_2 g_3 g_k|^2 d\alpha \ll N^{\frac{2}{3}+\frac{2}{k}+\varepsilon}.$$

Therefore the latter conclusion of the lemma follows via Schwarz's inequality.

5. WEIGHTED SIEVES

Having finished the preparation concerning the Hardy-Littlewood circle method, we may proceed to application of sieve theory, and in this section we appeal to weighted linear sieves. The uninitiated reader may consult Brüdern [3], §2, for a general outline of the interaction of sieves and the circle method in problems of the kind considered in this paper.

The simplest of our proofs is that of Theorem 5.

The proof of Theorem 5. Let n be an even integer satisfying $N \leq n \leq (6/5)N$ with a sufficiently large number N, k_2 be an integer with $3 \leq k_2 \leq 5$, and let $R_d(n)$ be the number of representations of n in the form

$$n = x + p_1^2 + p_2^3 + p_3^{k_2}, \tag{5.1}$$

with integers x and primes p_j satisfying $x \equiv 0 \pmod{d}$ and

$$x \sim X_1, \quad p_1 \sim X_2, \quad p_2 \sim X_3, \quad p_3 \sim X_{k_2}. \tag{5.2}$$

For any measurable set $\mathfrak{B} \subset [0, 1]$, we write

$$R_d(n; \mathfrak{B}) = \int_{\mathfrak{B}} f_1(\alpha; d) g_2(\alpha) g_3(\alpha) g_{k_2}(\alpha) e(-n\alpha) d\alpha,$$

so that

$$R_d(n) = R_d(n; [0, 1]) = R_d(n; \mathfrak{M}) + R_d(n; \mathfrak{m}). \tag{5.3}$$

We can estimate $R_d(n; \mathfrak{M})$ by applying Lemma 2.1 in a trivial way. Namely this time we set

$$s = 2, \quad k = 1, \quad k_0 = 2, \quad k_1 = 3, \quad 3 \leq k_2 \leq 5, \tag{5.4}$$

and $h(\alpha) = g_2(\alpha)$, $w(\beta) = v_2(\beta)$, $Q_1 = X_3$ and $Q_2 = X_{k_2}$. Moreover we adopt the notation introduced in Lemma 2.1 with these specializations. Then Lemma 2.1 implies that

$$R_d(n; \mathfrak{M}) = d^{-1} \mathfrak{S}_d(n, L) J(n) + O(N^{\frac{5}{6} + \frac{1}{k_2}}/(dL)), \tag{5.5}$$

and

$$J(n) \asymp N^{\frac{5}{6} + \frac{1}{k_2}} (\log N)^{-3}. \tag{5.6}$$

Also we recall the definitions (3.1), (3.2) and (3.15) with the current arrangement (5.4). Lemmata 3.2 (ii), 3.3 and 3.8 assure that $B_1(p, n) > p^{-3} - p^{-4}$ for all primes p, while the estimate $B_1(p, n) = 1 + O(p^{\varepsilon - 3/2})$ follows from (3.16). Thus we see that

$$\mathfrak{S}_1(n) \asymp 1, \tag{5.7}$$

and we may define the multiplicative function $\omega_n(d)$ by

$$\omega_n(d) = \mathfrak{S}_d(n)/\mathfrak{S}_1(n) = \prod_{p|d}(B_p(p,n)/B_1(p,n)). \qquad (5.8)$$

Then by (5.3) and (5.5) we have

$$R_d(n) = \frac{\omega_n(d)}{d}\mathfrak{S}_1(n)J(n) + E_d(n), \qquad (5.9)$$

where

$$E_d(n) = \frac{1}{d}\big(\mathfrak{S}_d(n,L) - \mathfrak{S}_d(n)\big)J(n) + R_d(n;\mathfrak{m}) + O(N^{\frac{5}{6}+\frac{1}{k_2}}/(dL)).$$

We know $\omega_n(d)$ is nonnegative, and Lemmata 3.3 and 3.8 yield that

$$0 < B_1(p,n) - B_p(p,n)/p = B_1(p,n)(1 - \omega_n(p)/p),$$

whence

$$0 \leq \omega_n(p) < p, \qquad (5.10)$$

for all primes p. In the meantime we deduce from (3.16) that

$$\omega_n(p) = B_p(p,n)/B_1(p,n) = 1 + O(p^{\varepsilon-1/2}). \qquad (5.11)$$

Hence our situation belongs to the linear sieve problems.

Next we put $D = X_1^{5/9}$. Then, for any sequence (λ_d) with $|\lambda_d| \leq 1$, Lemma 4.6 ensures that

$$\sum_{d \leq D} \lambda_d R_d(n;\mathfrak{m}) = \int_{\mathfrak{m}} F_1(\alpha)g_2(\alpha)g_3(\alpha)g_{k_2}(\alpha)e(-n\alpha)d\alpha$$

$$\ll N^{\frac{5}{6}+\frac{1}{k_2}}(\log N)^{-A},$$

since $5/9 < 17/30$. Therefore we have

$$\sum_{d \leq D} \lambda_d E_d(n) \ll N^{\frac{5}{6}+\frac{1}{k_2}}(\log N)^{-A} \ll \mathfrak{S}_1(n)J(n)(\log N)^{-2}; \qquad (5.12)$$

here we have used Lemma 3.7, (5.6) and (5.7) to confirm (5.12). Further, for any fixed $\delta > 0$, it follows from Lemma 4.7, (5.6) and (5.7) that

$$\sum_{p > X_1^\delta} R_{p^2}(n) = \sum_{p > X_1^\delta} R_{p^2}(n;[0,1]) \ll N^{\frac{5}{6}+\frac{1}{k_2}-\frac{\delta}{3}} \ll \mathfrak{S}_1(n)J(n)N^{-\delta/4}.$$

$$(5.13)$$

Now Theorem 5 is a direct consequence of a weighted linear sieve. Here we refer to Richert's linear sieve (Theorem 9.3 of Halberstam and Richert [12]). Although the estimate (5.13) is weaker than the constraint (Ω_3) of Halberstam and Richert [12], but it is clear from the proof of Theorem 9.3 of [12] that (5.13) is an adequate and sufficient substitute. All other requirements of the latter theorem are satisfied (with $r = 2$ and $\alpha = 5/9$) in view of (5.9)-(5.12). In particular, we note that

$$\frac{5}{9} \cdot \left(3 - \frac{\log(18/5)}{\log 3} - 10^{-2}\right) > 1.$$

Then, as regards the number $R(n)$ of representations of n in the form (5.1) with P_2-numbers x and primes p_j satisfying (5.2), Theorem 9.3 of [12] gives the lower bound $R(n) \gg \mathfrak{S}_1(n)J(n)(\log N)^{-1}$. This completes the proof of Theorem 5.

We next turn to Theorem 1 and Theorem 2 (i). At this stage we benefit from the bilinear error term in Iwaniec's linear sieve. Without it, we can prove Theorems 1 and 2 (i) only with P_{16} and P_7, respectively, at present. Thus we require Iwaniec's bilinear error terms within a weighted sieve. This has been made available by Halberstam and Richert [13]. Rather than stating here their result in its general form, we just mention its effect on our particular problem within the proof below. Indeed the inequality (5.27) below is derived from Theorems A and B of Halberstam and Richert [13] (see also the comment on (8.4) in [13], following Theorem B), by taking $U = T = 2/3$, $V = 1/4$ and $E = 1/9$, for example. Here we should make a minor change in the error term in Theorem A of [13], where the bilinear error term is expressed by using the supremum over all the sequences (λ_u) and (μ_v) satisfying $|\lambda_u| \leq 1$ and $|\mu_v| \leq 1$, instead of the form appearing in (5.27). This change is negligible in the argument of Halberstam and Richert [13], and as the following proof shows, it is convenient for our aim to keep the error term in the original form given by Iwaniec [19].

In order to establish (5.27) below, we may alternatively combine Richert's weighted sieve with Iwaniec's linear sieve, while Halberstam and Richert [13] appealed to Greaves' weighted sieve. Although Greaves' weights give stronger conclusions, Richert's weights are simpler, and much easier to combine with Iwaniec's sieve. It is actually a straightforward task to utilize Iwaniec's sieve within the proof of Theorem 9.3 of Halberstam and Richert [12], and such a topic is discussed in §6.2 of the unpublished lecture note [20] of Iwaniec. The latter device is still adequate to our purpose, proving (5.27). We may leave the details of the verification of (5.27) to the reader.

Before launching the proofs of Theorems 1 and 2 (i), we record a simple fact as a lemma.

Lemma 5.1. *Let $\nu(n)$ be the number of distinct prime divisors of n, and A be any constant exceeding 3. Then one has $\nu(n) \leq 2A \log \log N$ for all but $O(N(\log N)^{-A})$ natural numbers $n \leq N$.*

Proof. In view of the well-known estimate

$$\sum_{n \leq N} \tau(n) \ll N \log N,$$

the number of $n \leq N$ such that $\tau(n) \geq (\log N)^{A+1}$ is $O(N(\log N)^{-A})$. Since $\tau(n) \geq 2^{\nu(n)}$, the lemma follows.

The proofs of Theorems 1 and 2 (i). Let $k_1 = 4$ or 5, and let $R_d(n)$ be the number of representations of n in the form

$$n = x^2 + p_1^3 + p_2^{k_1}, \tag{5.14}$$

with integers $x \equiv 0 \pmod d$ and primes p_1, p_2 satisfying

$$x \sim X_2, \quad p_1 \sim X_3, \quad p_2 \sim X_{k_1}. \tag{5.15}$$

For any measurable set $\mathfrak{B} \subset [0,1]$, we write

$$R_d(n; \mathfrak{B}) = \int_{\mathfrak{B}} f_2(\alpha; d) g_3(\alpha) g_{k_1}(\alpha) e(-n\alpha) d\alpha,$$

so that $R_d(n) = R_d(n; [0,1]) = R_d(n; \mathfrak{M}) + R_d(n; \mathfrak{m})$.
   This time we set

$$s = 1, \quad k = 2, \quad k_0 = 3 \quad \text{and} \quad k_1 = 4 \text{ or } 5,$$

and, under these specializations, we recall the definitions of $\mathfrak{S}_d(n, L)$, $J(n)$, Y, $\mathcal{P}_d(n, Y)$ et al from the statements of Lemmata 2.1 and 3.5, together with the definitions (3.1) and (3.2). Then Lemma 2.1 implies that

$$R_d(n; \mathfrak{M}) = d^{-1}\mathfrak{S}_d(n, L)J(n) + O(N^{\frac{1}{k_1} - \frac{1}{6}}/(dL)), \tag{5.16}$$

and that

$$J(n) \asymp N^{1/k_1 - 1/6}(\log N)^{-2}, \tag{5.17}$$

for every integer n with $N \leq n \leq (6/5)N$.
   To facilitate our subsequent description, we denote by $\mathbb{N}(5)$ the set of all the odd integers in the interval $[N, (6/5)N]$, and put $\mathbb{N}(4) = \mathbb{N}_1 \cap$

$[N, (6/5)N]$ where $\mathbb{N}_1$ is the set defined in the statement of Theorem 2. Also we say simply *"for almost all n"* instead of "for all $n \in \mathbb{N}(k_1)$ with at most $O(N(\log N)^{-A})$ possible exceptions", within the current section. Since $\min\{\gamma(p, 3), \gamma(p, k_1)\} = 1$ for all primes p, it follows from Lemmata 3.2 (ii), 3.3, 3.5 (i) and (ii) that

$$B_1(p, n) > (2p^2)^{-1}, \tag{5.18}$$

for all $n \in \mathbb{N}(k_1)$ and primes p. We can therefore define the multiplicative function $\omega_n(d)$ by

$$\omega_n(d) = \mathcal{P}_d(n, Y)/\mathcal{P}_1(n, Y) = \prod_{\substack{p \le Y \\ p \mid d}} \left(B_p(p, n)/B_1(p, n)\right),$$

for $n \in \mathbb{N}(k_1)$, so that, by (5.16), we may write

$$R_d(n) = \frac{\omega_n(d)}{d} \mathcal{P}_1(n, Y) J(n) + E_d(n), \tag{5.19}$$

where

$$E_d(n) = d^{-1}\left(\mathfrak{S}_d(n, L) - \mathcal{P}_d(n, Y)\right) J(n) + R_d(n; \mathfrak{m}) + O(N^{\frac{1}{k_1} - \frac{1}{6}}/(dL)).$$

By the definition, we see that $\omega_n(p) = B_p(p, n)/B_1(p, n)$ or 1, according to $p \le Y$ or $p > Y$, and also that $\omega_n(p^l) = \omega_n(p)$ for all $l \ge 1$. Then we may confirm that

$$0 \le \omega_n(p) < p \quad \text{and} \quad \omega_n(p^l) = \omega_n(p) = 1 + O(p^{-1/2}), \tag{5.20}$$

for all primes p and integers $l \ge 1$. In fact, we can show the former by Lemmata 3.3 and 3.6 (i) and (ii), following the verification of (5.10). As regards the latter, it suffices to note that, in the current situation, we always have

$$B_d(p, n) = 1 + A_d(p, n) \quad \text{and} \quad A_d(p, n) \ll p^{-1/2} \tag{5.21}$$

by Lemmata 3.1 and 3.4, respectively.

We next discuss a lower bound for $\mathcal{P}_1(n, Y)$. By using (5.18), and by combining the former formula in (5.21) with Lemma 3.4, we have

$$\mathcal{P}_1(n, Y) \ge \prod_{p \le 10^5} (2p^2)^{-1} \prod_{10^5 < p \le Y} (1 - 120p^{-1}) \prod_{\substack{p \mid n \\ p > 10^5}} (1 - 120p^{-1/2}),$$

$$\tag{5.22}$$

for $n \in \mathbb{N}(k_1)$. If we denote by $\tilde{p}_n$ the $\nu(n)$-th prime exceeding 10^5 where $\nu(n)$ is defined in Lemma 5.1, then we have

$$\prod_{\substack{p|n \\ p>10^5}} (1 - 120p^{-1/2}) \geq \prod_{10^5 < p \leq \tilde{p}_n} (1 - 120p^{-1/2}) > \exp(-500\tilde{p}_n^{1/2}), \quad (5.23)$$

as well as $\tilde{p}_n \ll \nu(n)^{3/2}$. Now Lemma 5.1 implies that, for almost all n, we have $\tilde{p}_n^{1/2} \ll (\log \log N)^{3/4}$ and, by (5.22) and (5.23),

$$\mathcal{P}_1(n, Y) \gg (\log Y)^{-120}(\log N)^{-\varepsilon} \gg (\log N)^{-61}. \quad (5.24)$$

Further, as an immediate consequence of Lemma 4.7, (5.17) and (5.24), we note that for any fixed $\delta > 0$, the chain of inequalities

$$\sum_{p>X^\delta} R_{p^2}(n) \ll N^{\frac{1}{k_1} - \frac{1}{6} - \frac{1}{3}\delta} \ll \mathcal{P}_1(n, Y)J(n)N^{-\frac{1}{4}\delta} \quad (5.25)$$

holds for almost all n.

Now let $R(n, r)$ be the number of representations of n in the form (5.14) subject to P_r-numbers x and primes p_1, p_2 satisfying (5.15). Then, based on the above conclusions (5.19), (5.20) and (5.25), we can apply a weighted linear sieve, for almost all n, to obtain a lower bound for $R(n, r)$. Indeed, as we announced in the account prior to Lemma 5.1, we conclude from Halberstam and Richert [13] (or Iwaniec [20], §6.2) that for every integer r satisfying

$$r > \frac{3}{2}\frac{\log X_2}{\log D} - \frac{1}{6}, \quad (5.26)$$

there exists a positive absolute constant η such that

$$R(n, r) \geq \eta \mathcal{P}_1(n, Y)J(n)(\log D)^{-1} - \sum_{h \leq H} \sum_{u \leq D^{2/3}} \sum_{v \leq D^{1/3}} \lambda_{u,h}\mu_{v,h}E_{uv}(n), \quad (5.27)$$

for almost all n, where the sequences $(\lambda_{u,h})$ and $(\mu_{v,h})$ are independent of n, and satisfy $|\lambda_{u,h}| \leq 1$ and $|\mu_{v,h}| \leq 1$, and where $1 \leq H \ll \log N$.

We put

$$D = X_2^{\theta(k_1)}, \quad \theta(k_1) = \frac{6 - k_1}{2k_1 + 10^{-1}}, \quad r(k_1) = \frac{3k_1}{6 - k_1},$$

getting $r(5) = 15$ and $r(4) = 6$. Note that (5.26) is satisfied with $r = r(k_1)$. We denote by $\mathcal{E}(n)$ the error term in (5.27), namely,

$$\mathcal{E}(n) = \sum_{h \leq H} \sum_{u \leq D^{2/3}} \sum_{v \leq D^{1/3}} \lambda_{u,h}\mu_{v,h}E_{uv}(n).$$

We shall estimate $\mathcal{E}(n)$ on average over n. To this end, we first estimate the expression

$$\Upsilon = \sum_{h \leq H} \sum_{n \sim N} \Big| \sum_{u \leq D^{2/3}} \sum_{v \leq D^{1/3}} \lambda_{u,h} \mu_{v,h} R_{uv}(n,m) \Big|^2.$$

By Bessel's inequality and Lemma 4.2, recalling the notation (4.4), we have

$$\Upsilon \ll \sum_{h \leq H} \int_{\mathfrak{m}} |F_2(\alpha; D, (\lambda_{u,h}), (\mu_{v,h})) g_3(\alpha) g_{k_1}(\alpha)|^2 \, d\alpha$$

$$\ll N^{\frac{2}{3} + \frac{2}{k_1}} (\log N)^{1-3A}.$$

Then we deduce from Lemma 3.5, (5.17) and the definition of $E_d(n)$, with an application of Cauchy's inequality, that

$$\sum_{n \sim N} |\mathcal{E}(n)| \ll N^{\frac{5}{6} + \frac{1}{k_1}} H L^{-\frac{1}{3}} + (NH\Upsilon)^{\frac{1}{2}} \ll N^{\frac{5}{6} + \frac{1}{k_1}} (\log N)^{1 - \frac{3}{2}A}. \quad (5.28)$$

This implies that for almost all n we have

$$|\mathcal{E}(n)| \leq N^{\frac{1}{k_1} - \frac{1}{6}} (\log N)^{1 - \frac{1}{2}A} \ll \mathcal{P}_1(n, Y) J(n) (\log N)^{64 - \frac{1}{2}A}, \quad (5.29)$$

by (5.17) and (5.24). In view of (5.27) and (5.29) with the constraint (5.26), we conclude that $R(n, r(k_1)) \gg \mathcal{P}_1(n, Y) J(n) (\log D)^{-1}$ for almost all n, from which Theorems 1 and 2 (i) follow immediately.

6. SWITCHING PRINCIPLE

The remaining theorems are proved by the switching principle in sieve theory.

Throughout this section we denote Euler's constant by γ. We begin with introducing the basic functions $\phi_0(u)$ and $\phi_1(u)$ concerning the linear sieve. These functions are defined by

$$\phi_0(u) = 0, \qquad \phi_1(u) = 2e^\gamma / u,$$

for $0 < u \leq 2$, and then by the difference-differential equations

$$(u\phi_0(u))' = \phi_1(u - 1), \qquad (u\phi_1(u))' = \phi_0(u - 1),$$

for $u \geq 2$. It is known that $0 \leq \phi_0(u) \leq 1 \leq \phi_1(u)$ for $u > 0$, and that

$$\phi_0(u) = \frac{2e^\gamma}{u} \log(u - 1) \quad \text{for } 2 \leq u \leq 4, \quad \phi_1(u) = \frac{2e^\gamma}{u} \quad \text{for } 0 < u \leq 3.$$

$$(6.1)$$

Then we refer to Iwaniec's linear sieve in the following form (see Iwaniec [19], or [20], for a proof).

Lemma 6.1. *Let Q, U, V, X be real numbers ≥ 1, and suppose that $D = UV$ is sufficiently large. Let $\omega(d)$ be a multiplicative function such that $0 \leq \omega(p) < p$ and $\omega(p^l) \ll 1$ for all primes p and integers $l \geq 1$, and suppose that*

$$\prod_{w \leq p < z} \left(1 - \frac{\omega(p)}{p}\right)^{-1} \leq \frac{\log z}{\log w}\left(1 + O\left(\frac{(\log \log \log D)^3}{\log w}\right)\right), \qquad (6.2)$$

whenever $z > w \geq 2$. Further, let $r(x)$ be a non-negative arithmetical function, z be a real number with $2 \leq z \leq D^{1/2}$, $\Pi(z)$ be as in (2.11), and write

$$E_d = \sum_{\substack{x \sim Q \\ x \equiv 0 \,(\mathrm{mod}\ d)}} r(x) \; -\frac{\omega(d)}{d} X, \qquad W(z) = \prod_{p < z}\left(1 - \frac{\omega(p)}{p}\right), \qquad s = \frac{\log D}{\log z}.$$

Then there exist sequences $(\lambda_{u,h}^{(j)})$ and $(\mu_{v,h}^{(j)})$ for $j = 0$, 1, satisfying $|\lambda_{u,h}^{(j)}| \leq 1$, $|\mu_{v,h}^{(j)}| \leq 1$, such that

$$\sum_{\substack{x \sim Q \\ (x,\Pi(z))=1}} r(x) > XW(z)(\phi_0(s) + O((\log \log D)^{-1/50})) - \mathcal{E}^{(0)},$$

as well as

$$\sum_{\substack{x \sim Q \\ (x,\Pi(z))=1}} r(x) < XW(z)(\phi_1(s) + O((\log \log D)^{-1/50})) + \mathcal{E}^{(1)},$$

where

$$\mathcal{E}^{(j)} = \sum_{h < \log D} \sum_{u \leq U} \sum_{v \leq V} \lambda_{u,h}^{(j)} \mu_{v,h}^{(j)} E_{uv},$$

for $j = 0$, 1. In addition, these sequences $(\lambda_{u,h}^{(j)})$ and $(\mu_{v,h}^{(j)})$ depend at most on U and V.

The following proof of Theorem 2 (ii) can serve as a model of our strategy when using the switching principle.

The proof of Theorem 2 (ii). Let $R_d(n)$ be the number of representations of n in the form

$$n = p_1^2 + (p_2 x)^3 + p_3^4, \qquad (6.3)$$

subject to primes p_1, p_2, p_3 and integers x satisfying

$$p_1 \sim X_2, \quad p_2 \sim X_{21}, \quad x \sim X_3/p_2, \quad p_3 \sim X_4, \tag{6.4}$$

and $x \equiv 0 \pmod d$. For any measurable set $\mathfrak{B} \subset [0, 1]$, we write

$$R_d(n; \mathfrak{B}) = \sum_{p_2 \sim X_{21}} \int_{\mathfrak{B}} g_2(\alpha) f_3(\alpha; p_2 d) g_4(\alpha) e(-n\alpha) d\alpha,$$

so that $R_d(n) = R_d(n; [0, 1]) = R_d(n; \mathfrak{M}) + R_d(n; \mathfrak{m})$. We now apply Lemma 2.1 with

$$s = 1, \quad k = 3, \quad k_0 = 2, \quad k_1 = 4, \tag{6.5}$$

and $h(\alpha) = g_2(\alpha)$, $w(\beta) = v_2(\beta)$, $C = 1$. We accordingly specify the relevant notation $\mathfrak{S}_d(n, L)$, $J(n)$, $I(n)$, Y, $\mathcal{P}_d(n, Y)$ et al, introduced in Lemmata 2.1 and 3.5. Note here that $\mathfrak{S}_{p_2 d}(n, L) = \mathfrak{S}_d(n, L)$ for $p_2 \geq X_{21} > L$, in view of the definitions (2.2) and (2.3). Then Lemma 2.1 implies that

$$R_d(n; \mathfrak{M}) = \sum_{p_2 \sim X_{21}} \left((p_2 d)^{-1} \mathfrak{S}_d(n, L) J(n) + O(N^{1/12}(p_2 d L)^{-1}) \right),$$

and

$$J(n) = (\log X_3 + O(\log L)) I(n), \qquad I(n) \asymp N^{1/12}(\log N)^{-3}, \tag{6.6}$$

for every integer n with $N \leq n \leq (6/5)N$.

Confirming swiftly the prerequisite of Lemma 3.3 in the current case, we note that $\mathcal{P}_1(n, Y) > 0$ whenever $n \in \mathbb{N}_1$, by Lemmata 3.2 (ii), 3.3 and 3.6 (ii), where $\mathbb{N}_1$ is the set introduced in the statement of Theorem 2. Thus we may define

$$\omega_n(d) = \mathcal{P}_d(n, Y)/\mathcal{P}_1(n, Y) = \prod_{\substack{p \leq Y \\ p \mid d}} \left(B_p(p, n)/B_1(p, n) \right),$$

for $n \in \mathbb{N}_1$, and then, we may write

$$R_d(n) = \frac{\omega_n(d)}{d} \varrho \mathcal{P}_1(n, Y) J(n) + E_d,$$

where

$$\varrho = \sum_{p_2 \sim X_{21}} \frac{1}{p_2},$$

$$E_d = \frac{\varrho}{d}(\mathfrak{S}_d(n, L) - \mathcal{P}_d(n, Y)) + R_d(n; \mathfrak{m}) + O(N^{1/12}(dL)^{-1}). \tag{6.7}$$

Note that Mertens' formula gives the estimate

$$\varrho \asymp (\log N)^{-1}. \tag{6.8}$$

Hereafter, until the end of the proof of Theorem 2 (ii), we say simply *"for almost all n"*, meaning "for all but $O(N(\log N)^{-A})$ values of $n \in \mathbb{N}_1 \cap [N, (6/5)N]$". Following the verification about (5.24), we derive from Lemmata 3.3, 3.4 and 3.6 (ii) that

$$\mathcal{P}_1(n, Y) \gg (\log N)^{-49}, \tag{6.9}$$

for almost all n.

We confirm that $0 \le \omega_n(p) < p$ and $\omega_n(p^l) \ll 1$ for all primes p, integers $l \ge 1$ and $n \in \mathbb{N}_1$, by using Lemmata 3.2 (ii), 3.3, 3.4, 3.6 (ii) and the definition of $\omega_n(d)$, following the allied argument in the proof of Theorem 2 (i) in the preceding section. Next we have to discuss the property of $\omega_n(d)$ corresponding to (6.2).

When $10^5 < p \le Y$, we derive from Lemma 3.4 that

$$\omega_n(p) = \frac{B_p(p, n)}{B_1(p, n)} < \frac{1 + p^{-1/2}(p, n)^{1/2} + O(p^{-1/2})}{1 - 96p^{-1}(p, n)^{1/2}}$$

$$\le 1 + \frac{(p, n)^{1/2}}{p^{1/2}} + O(p^{-1/2}), \tag{6.10}$$

while we have $\omega_n(p) = 1$ for $p > Y$. On putting $w_0 = \log \log N$, when $z > w \ge w_0$ we have

$$\sum_{w \le p < z} \frac{(p, n)^{1/2}}{p^{3/2}} \ll w^{-1/2} + \frac{1}{\log w} \sum_{\substack{w \le p < z \\ p \mid n}} \frac{\log p}{p}$$

$$\ll \frac{1}{\log w} \sum_{p \le \tilde{p}_n} \frac{\log p}{p} \ll \frac{\log \tilde{p}_n}{\log w},$$

recalling $\tilde{p}_n$ introduced in connection with (5.23). Appealing to Lemma 5.1 and Mertens' formula, we consequently see that for almost all n,

$$\sum_{w \le p < z} \omega_n(p)/p \le \log \log z - \log \log w + O(\log \log \log N / \log w),$$

whence

$$\prod_{w \le p < z} (1 - \omega_n(p)/p)^{-1} \le \frac{\log z}{\log w}(1 + O(\log \log \log N / \log w)),$$

when $z > w \geq w_0$. To consider the case where $2 \leq w \leq w_0$, we note that $1 - \omega_n(p)/p \gg 1$ for $p \leq 10^5$ and $n \in \mathbb{N}_1$, by Lemmata 3.3, 3.4 and 3.6 (ii). Then we have by (6.10)

$$\prod_{w \leq p < z} (1 - \omega_n(p)/p)^{-1} \ll \prod_{w \leq p < z} (1 + p^{-1}) \prod_{w \leq p < z} (1 + p^{-3/2}(p,n)^{1/2})$$

$$\ll \frac{\log z}{\log w}(\log \tilde{p}_n) \ll \frac{\log z}{\log w} \frac{(\log \log \log N)^2}{\log w},$$

when $2 \leq w \leq w_0$, for almost all n, by virtue of Lemma 5.1. Hence, for almost all n, we conclude that

$$\prod_{w \leq p < z} (1 - \omega_n(p)/p)^{-1} \leq \frac{\log z}{\log w}(1 + O((\log \log \log N)^2/\log w)), \quad (6.11)$$

whenever $z > w \geq 2$.

Now we put

$$D = (X_3^{6/7})^\theta, \quad \theta = 0.333, \quad z = (X_3^{6/7})^{1/7},$$

and let $R(n)$ be the number of solutions of (6.3) subject to primes p_j $(1 \leq j \leq 3)$ and integer x satisfying (6.4) and

$$(x, \Pi(z)) = 1. \quad (6.12)$$

By the above argument, we know that we can apply the linear sieve, Lemma 6.1, to obtain a lower bound for $R(n)$, for almost all n. Indeed, taking $U = D$ and $V = 3/2$ in the latter lemma, and writing

$$W(n, z) = \prod_{p < z}(1 - \omega_n(p)/p),$$

we have the bound

$$R(n) > \left(\phi_0(7\theta) + O((\log \log N)^{-1/50})\right)\varrho \mathcal{P}_1(n, Y)J(n)W(n, z) - \mathcal{E}_0(n), \quad (6.13)$$

for almost all n, where

$$\mathcal{E}_0(n) = \sum_{h < \log D} \sum_{d \leq D} \lambda_{d,h} E_d,$$

with a certain sequence $(\lambda_{d,h})$ satisfying $|\lambda_{d,h}| \leq 1$. In much the same way as the deliberation on (5.28), we deduce from Lemmata 3.5 and 4.5 that

$$\sum_{n \sim N} |\mathcal{E}_0(n)| \ll N^{\frac{13}{12}}(\log N)^{-3A},$$

whence

$$|\mathcal{E}_0(n)| \ll N^{1/12}(\log N)^{-2A} \ll \varrho P_1(n, Y)J(n)W(n, z)(\log N)^{-A}, \quad (6.14)$$

for almost all n, by (6.6), (6.8), (6.9) and (6.11).

We next write

$$\mathcal{P}(n, Y) = \prod_{p \leq Y} B(p, n), \quad (6.15)$$

where $B(p, n)$ is defined by (3.2) with (3.1), under the current specialization (6.5). Then Lemma 3.3 gives the formula

$$(1 - \omega_n(p)/p)B_1(p, n) = (1 - 1/p)B(p, n),$$

for $p \leq Y$. Recalling that $\omega_n(p) = 1$ for $p > Y$, we see that

$$P_1(n, Y)W(n, z) = \mathcal{P}(n, Y) \prod_{p < z}(1 - p^{-1})$$

$$= \mathcal{P}(n, Y)\frac{e^{-\gamma}}{\log z}(1 + O((\log z)^{-1})), \quad (6.16)$$

by Mertens' formula. In view of (6.9), (6.11) and (6.16), we remark that

$$\mathcal{P}(n, Y) \gg (\log N)^{-50}$$

holds for almost all n. Also it follows from (6.1), (6.6), (6.8), (6.13), (6.14) and (6.16) that for almost all n,

$$R(n) > \frac{7}{30}(\log(7\theta - 1) + O((\log \log N)^{-1/50}))\varrho\mathcal{P}(n, Y)I(n)$$

$$> 2\varrho\mathcal{P}(n, Y)I(n). \quad (6.17)$$

We denote by $\tilde{R}(n)$ the number of solutions of (6.3) subject to primes p_j ($1 \leq j \leq 3$) and integer x satisfying (6.4), (6.12) and $\Omega(x) \geq 4$, where $\Omega(x)$ is defined in §2, prior to (2.12). We aim to show that $R(n) - \tilde{R}(n) > 0$ for almost all n. The latter conclusion implies that for almost all n, the equation (6.3) has a solution in primes p_j ($1 \leq j \leq 3$) and an integer x satisfying (6.4), (6.12) and $\Omega(x) \leq 3$, whence Theorem 2 (ii) follows. To this end we need an upper bound for $\tilde{R}(n)$ valid for almost all n, and such a bound is obtained by the switching principle.

Now let $R'_d(n)$ be the number of representations of n in the form

$$n = y^2 + (p_2 x)^3 + p_3^4, \quad (6.18)$$

subject to primes p_2, p_3 and integers x, y satisfying

$$y \sim X_2, \quad p_2 \sim X_{21}, \quad x \sim X_3/p_2, \quad (x, \Pi(z)) = 1, \quad \Omega(x) \geq 4, \quad p_3 \sim X_4, \quad (6.19)$$

and $y \equiv 0 \pmod{d}$. Note that $\Omega(x) \leq 7$ when x satisfies the conditions in (6.19). Recalling the notation (2.12), we write

$$R'_d(n; \mathcal{B}) = \sum_{p_2 \sim X_{21}} \sum_{r=4}^{7} \int_{\mathcal{B}} f_2(\alpha; d) g_3(p_2^3\alpha; X_3/p_2, r, z) g_4(\alpha) e(-n\alpha) d\alpha,$$

and then observe that $R'_d(n) = R'_d(n; [0, 1]) = R'_d(n; \mathfrak{M}) + R'_d(n; \mathfrak{m})$.

When $\alpha = a/q + \beta$, $q \leq L$ and $|\beta| \leq L/N$, we note that $S_3^*(q, ap_2^3) = S_3^*(q, a)$ for $p_2 > X_{21} > L$, and deduce from Lemma 2.2 that

$$g_3(p_2^3\alpha; X_3/p_2, r, z) = \varphi(q)^{-1} S_3^*(q, a) w_3(p_2^3\beta; X_3/p_2, r, z)$$
$$+ O(X_3 p_2^{-1} L^{-5}), \qquad (6.20)$$

for $4 \leq r \leq 7$. When $p \sim X_{21}$, we also note that

$$C_r\left(\frac{\log(X_3/p_2)}{\log z}\right) = C_r(7) + O((\log N)^{-1}),$$

and that, by change of variable,

$$v_3(p_2^3\beta; X_3/p_2) = \frac{1}{p_2} \int_{X_3}^{5X_3} \frac{e(t^3\beta)}{\log(t/p_2)} dt = \frac{1}{p_2}\left(\frac{7}{6} v_3(\beta) + O(X_3(\log N)^{-2})\right).$$

Thus by Lemma 2.2 we have

$$w_3(p_2^3\beta; X_3/p_2, r, z) = \frac{7}{6p_2} C_r(7) v_3(\beta) + O(X_3 p_2^{-1}(\log N)^{-2}). \quad (6.21)$$

Then in order to estimate $R'_d(n; \mathfrak{M})$, we apply Lemma 2.1 for each $p_2 \sim X_{21}$, with taking

$$h(\alpha) = p_2 \sum_{r=4}^{7} g_3(p_2^3\alpha; X_3/p_2, r, z), \quad w(\beta) = p_2 \sum_{r=4}^{7} w_3(p_2^3\beta; X_3/p_2, r, z),$$

and

$$C = \frac{7}{6} \sum_{r=4}^{7} C_r(7).$$

But we require some notation here. We set

$$s = 1, \quad k = 2, \quad k_0 = 3, \quad k_1 = 4, \qquad (6.22)$$

and recall the definitions (2.2)-(2.5), (3.1), (3.2) and (3.8), but we basically attach primes, for distinction, to the respective symbols. Namely we define

$$A'_d(q, n) = q^{-1}\varphi(q)^{-2} \sum_{\substack{a=1 \\ (a,q)=1}}^{q} S_2(q, ad^2) S_3^*(q, a) S_4^*(q, a) e(-an/q),$$

$$\mathfrak{S}'_d(n, L) = \sum_{q \leq L} A'_d(q, n), \quad B'_d(p, n) = \sum_{h=0}^{\infty} A'_d(p^h, n),$$

$$\mathcal{P}'_d(n, Y) = \prod_{p \leq Y} B'_d(p, n),$$

where $Y = \exp(\sqrt{\log N})$, and

$$J'(n) = J'(n, p_2)$$

$$= \int_{-L/N}^{L/N} u_2(\beta) \Big(p_2 \sum_{r=4}^{7} w_3(p_2^3 \beta; X_3/p_2, r, z) \Big) v_4(\beta) \, e(-n\beta) d\beta.$$

As regards $I(n)$, $A(q, n)$, $B(p, n)$ and $P(n, Y)$, we note that the settings (6.5) and (6.22) cause no difference in their definitions (2.4), (3.1), (3.2) and (6.15). Indeed we define

$$A(q, n) = \varphi(q)^{-3} \sum_{\substack{a=1 \\ (a,q)=1}}^{q} S_2^*(q, a) S_3^*(q, a) S_4^*(q, a) e(-an/q),$$

$$B(p, n) = \sum_{h=0}^{\infty} A(p^h, n),$$

$$P(n, Y) = \prod_{p \leq Y} B(p, n), \quad I(n) = \int_{-\infty}^{\infty} v_2(\beta) v_3(\beta) v_4(\beta) e(-n\beta) d\beta,$$

and, in particular, it should be stressed that $I(n)$ and $P(n, Y)$ are exactly the same as those appearing in (6.6), (6.15) and (6.17). We then derive from Lemma 2.1 that

$$R'_d(n; \mathfrak{M}) = \sum_{p_2 \sim X_{21}} \frac{1}{p_2} \Big(\frac{1}{d} \mathfrak{S}'_d(n, L) J'(n) + O(N^{1/12}/(dL)) \Big), \quad (6.23)$$

and that for each $p_2 \sim X_{21}$,

$$J'(n) = J'(n, p_2) = \Big(\frac{7}{6} \sum_{r=4}^{7} C_r(7) \log X_2 + O(\log L) \Big) I(n). \quad (6.24)$$

We now remark that $\mathcal{P}'_d(n, Y)$ is identical with the function $\mathcal{P}_d(n, Y)$ that appeared in the proof of Theorem 2 (i), with $k_1 = 4$, in §5. According to the deliberation there, we remember that

$$\mathcal{P}'_1(n, Y) \gg (\log N)^{-61}, \quad (6.25)$$

for almost all n, and we can define $w'_n(d) = \mathcal{P}'_d(n, Y)/\mathcal{P}'_1(n, Y)$ for $n \in \mathbb{N}_1$. Then we may write by (6.23)

$$R'_d(n) = \frac{w'_n(d)}{d} \mathcal{P}'_1(n, Y)J''(n) + E'_d(n),$$

where

$$J''(n) = \sum_{p_2 \sim X_{21}} p_2^{-1} J'(n),$$

and

$$E'_d(n) = \frac{1}{d}(\mathfrak{S}'_d(n, L) - \mathcal{P}'_d(n, Y))J''(n) + R'_d(n, \mathfrak{m}) + O(N^{1/12}(dL)^{-1}).$$

Recalling ϱ defined at (6.7), we derive from (6.24) that

$$J''(n) = \varrho\Big(\frac{7}{6} \sum_{r=4}^{7} C_r(7) \log X_2 + O(\log L)\Big)I(n), \qquad (6.26)$$

Put

$$D' = X_2^{\theta'}, \quad \theta' = 0.249, \quad z' = (D')^{1/3},$$

and let $R'(n)$ be the number of representations of n in the form (6.18) subject to primes p_2, p_3 and integers x, y satisfying (6.19) and $(y, \Pi(z')) = 1$. The last condition is satisfied when y is a prime with $y \sim X_2$, we see that $\tilde{R}(n) \le R'(n)$. Since we know that the property (5.20) holds with w'_n in place of w_n, we can apply Iwaniec's linear sieve in the form of Lemma 6.1, to obtain the upper bound

$$R'(n) < (\phi_1(3) + O((\log \log N)^{-1/50}))\mathcal{P}'_1(n, Y)J''(n)W'(n, z') + \mathcal{E}_1(n), \qquad (6.27)$$

where

$$W'(n, z') = \prod_{p<z'}(1 - w'_n(p)/p), \quad \mathcal{E}_1(n) = \sum_{h<\log D'} \sum_{u \le U} \sum_{v \le V} \lambda_{u,h} \mu_{v,h} E'_{uv}(n),$$

with $U = (D')^{2/3}$, $V = (D')^{1/3}$, and certain sequences $(\lambda_{u,h})$ and $(\mu_{v,h})$ satisfying $|\lambda_{u,h}| \le 1$ and $|\mu_{v,h}| \le 1$. Note that

$$W'(n, z') \gg (\log z')^{-1}, \qquad (6.28)$$

for $n \in \mathbb{N}_1$, by the result corresponding to (5.20).

Via the same way as the verification of (5.29), we deduce from Lemmata 3.5 and 4.2, together with (6.6), (6.8), (6.25), (6.26) and (6.28), that

$$|\mathcal{E}_1(n)| \ll \mathcal{P}'_1(n)J''(n)W'(n, z')(\log N)^{-2},$$

for almost all n. Further, Lemma 3.3 gives the formula

$$(1 - \omega'_n(p)/p)B'_1(p,n) = (1 - 1/p)B(p,n),$$

for $p \leq Y$, and we know $\omega'_n(p) = 1$ for $p > Y$, thus

$$P'_1(n,Y)W'(n,z') = P(n,Y)\prod_{p<z'}(1 - p^{-1})$$

$$= P(n,Y)\frac{e^{-\gamma}}{\log z'}(1 + O((\log z')^{-1})).$$

Hence it follows from (6.27), (6.26) and (6.1) that

$$R'(n) < \left(\frac{7}{30'}\sum_{r=4}^{7}C_r(7) + O((\log\log N)^{-1/50})\right)\varrho P(n,Y)I(n), \quad (6.29)$$

for almost all n. Here we can confirm the numerical estimates

$$C_4(7) < 0.1734, \quad C_5(7) < 0.00821, \quad C_6(7) < 3.4 \times 10^{-5}, \quad (6.30)$$

and $C_7(7) = 0$, as is recorded in (12) of Kawada [22]. (To confirm these bounds, we can appeal to Theorem 1 of Grupp and Richert [9]. We remark that the function $I_r(u)$ in [9] and our $C_r(u)$ satisfy the relation $C_r(u) = uI_r(u)$.)

Finally, by (6.29) and (6.30), we obtain the estimate

$$\tilde{R}(n) \leq R'(n) < 1.8\varrho P(n,Y)I(n),$$

for almost all n. We conclude that $R(n) - \tilde{R}(n) > 0$ for almost all n, by (6.17) and the last inequality, and the proof of Theorem 2 (ii) is now completed.

We proceed to the proofs of Theorems 3 and 4, following the methods introduced thus far. These proofs are somewhat simpler than that of Theorem 2 (ii) above. Concerning Theorem 3, moreover, the limits of "level of distribution" D assured by Lemmata 4.3 and 4.4 are not worse than those appearing in Lemmata 4.5 and 4.2, and the latter constraints were still adequate for getting a "P_3" as we saw in the above proof of Theorem 2 (ii), whence the conclusions in Theorem 3 are essentially obvious to the experienced reader. For these reasons, we shall be brief in the following proofs.

The proof of Theorem 3. Within this proof, we use the terminology "*for almost all n*", for short, to mean that "for all but $O(N(\log N)^{-A})$ values of $n \in \mathbb{N}_2 \cap [N, (6/5)N]$", where $\mathbb{N}_2$ is the set introduced in the statement of Theorem 3.

When $\mathcal{X}$ and $\mathcal{Y}$ are sets of integers, we denote by $R(n, \mathcal{X}, \mathcal{Y})$ the number of representations of n in the form $n = x^2 + y^3 + p^3$ subject to $x \in \mathcal{X}$, $y \in \mathcal{Y}$ and primes $p \sim X_3^{5/6}$. We denote by $\mathcal{X}(d)$ and $\mathcal{Y}(d)$, respectively, the set of the multiples of d in the intervals $(X_2, 5X_2)$ and $(X_3, 5X_3)$, and by $\mathcal{X}_0$ and $\mathcal{Y}_0$, respectively, the sets of primes in the intervals $(X_2, 5X_2)$ and $(X_3, 5X_3)$. We further put

$$z = X_2^{1/7}, \qquad z' = X_3^{1/7},$$

and define the sets

$$\begin{aligned}
\mathcal{X}_1 &= \{x \sim X_2 \, ; \, (x, \Pi(z)) = 1\}, \\
\mathcal{X}_2 &= \{x \in \mathcal{X}_1 \, ; \, \Omega(x) \geq 4\}, \\
\mathcal{X}_3 &= \mathcal{X}_1 \setminus \mathcal{X}_2, \\
\mathcal{Y}_1 &= \{y \sim X_3 \, ; \, (y, \Pi(z')) = 1\}, \\
\mathcal{Y}_2 &= \{y \in \mathcal{Y}_1 \, ; \, \Omega(y) \geq 4\}, \\
\mathcal{Y}_3 &= \mathcal{Y}_1 \setminus \mathcal{Y}_2.
\end{aligned}$$

Trivially, all the numbers in $\mathcal{X}_3$ and $\mathcal{Y}_3$ are P_3-numbers.

By orthogonality, we have the formulae

$$R(n, \mathcal{X}(d), \mathcal{Y}_0) = \int_0^1 f_2(\alpha; d) g_3(\alpha) g_3(\alpha; X_3^{5/6}) e(-n\alpha) d\alpha,$$

$$R(n, \mathcal{X}_2, \mathcal{Y}(d)) = \int_0^1 \left(\sum_{r=4}^7 g_2(\alpha; X_2, r, z) \right) f_3(\alpha; d) g_3(\alpha; X_3^{5/6}) e(-n\alpha) d\alpha.$$

The contributions from the major arcs $\mathfrak{M}$ to these integrals are immediately estimated by Lemma 2.1, with the aid of Lemma 2.2 in the latter case. As in the proof of Theorem 2 (ii), then we can apply Iwaniec's linear sieve, Lemma 6.1, to obtain a lower estimate for $R(n, \mathcal{X}_1, \mathcal{Y}_0)$ and an upper estimate for $R(n, \mathcal{X}_2, \mathcal{Y}_1)$, both valid for almost all n.

Now let $I(n)$ and $\mathcal{P}(n, Y)$ be defined by (2.4), (3.1), (3.2) and (6.15), with $s = 1$, $k = 2$ and $k_0 = k_1 = 3$, and put

$$D = X_2^\theta, \quad \theta = 0.416, \quad D' = X_3^{\theta'}, \quad \theta' = 0.333.$$

By Lemmata 3.1, 3.3, 3.4, 3.6 and 5.1, we may show that

$$\mathcal{P}(n, Y) \gg (\log N)^{-50},$$

for almost all n, and Lemma 2.1 asserts that $I(n) \asymp N^{1/9} (\log N)^{-3}$ for $N \leq n \leq (6/5)N$.

When we apply Lemma 6.1 to $R(n, \mathcal{X}_1, \mathcal{Y}_0)$, we are concerned with a remainder term corresponding to $\mathcal{E}^{(0)}$ in Lemma 6.1, with $U = D^{2/3}$ and $V = D^{1/3}$. We can regard this remainder term as negligible for almost all n by Lemmata 3.5 and 4.3. Using Lemma 3.3 in addition, we can establish the lower bound

$$R(n, \mathcal{X}_1, \mathcal{Y}_0)$$
$$> \left(\phi_0 \left(\frac{\log D}{\log z} \right) + O((\log \log N)^{-1/50}) \right) \cdot P(n, Y) \frac{e^{-\gamma}}{\log z} \cdot (\log X_2) I(n)$$
$$= \left(\frac{2}{\theta} \log(7\theta - 1) + O((\log \log N)^{-1/50}) \right) P(n, Y) I(n), \qquad (6.31)$$

for almost all n. On the other hand, we apply Lemma 6.1 to $R(n, \mathcal{X}_2, \mathcal{Y}_1)$, with taking $U = D'$ and $V = 3/2$, and dispose of the error term corresponding to $\mathcal{E}^{(1)}$ in Lemma 6.1 by Lemmata 3.5 and 4.4. In this way we arrive at the upper bound

$$R(n, \mathcal{X}_2, \mathcal{Y}_1) < \left(\phi_1 \left(\frac{\log D'}{\log z'} \right) + O((\log \log N)^{-1/50}) \right) \cdot P(n, Y) \frac{e^{-\gamma}}{\log z'} \times$$
$$\left(\sum_{r=4}^{7} C_r(7) \log X_3 \right) I(n)$$
$$= \left(\frac{2}{\theta'} \sum_{r=4}^{7} C_r(7) + O((\log \log N)^{-1/50}) \right) P(n, Y) I(n), \quad (6.32)$$

which is valid for almost all n.

Since $\mathcal{Y}_0 \subset \mathcal{Y}_1$, we see $R(n, \mathcal{X}_2, \mathcal{Y}_0) \leq R(n, \mathcal{X}_2, \mathcal{Y}_1)$, and

$$R(n, \mathcal{X}_3, \mathcal{Y}_0) \geq R(n, \mathcal{X}_1, \mathcal{Y}_0) - R(n, \mathcal{X}_2, \mathcal{Y}_1) > 0,$$

for almost all n, by (6.31), (6.32) and (6.30) with modest numerical computation. This proves Theorem 3 (i).

Similarly we can show that

$$R(n, \mathcal{X}_0, \mathcal{Y}_1) > \left(\frac{2}{\theta'} \log(7\theta' - 1) + O((\log \log N)^{-1/50}) \right) P(n, Y) I(n),$$
$$R(n, \mathcal{X}_1, \mathcal{Y}_2) < \left(\frac{2}{\theta} \sum_{r=4}^{7} C_r(7) + O((\log \log N)^{-1/50}) \right) P(n, Y) I(n),$$

for almost all n, and then that

$$R(n, \mathcal{X}_0, \mathcal{Y}_3) \geq R(n, \mathcal{X}_0, \mathcal{Y}_1) - R(n, \mathcal{X}_1, \mathcal{Y}_2) > 0,$$

for almost all n, by (6.30). This establishes Theorem 3 (ii).

The proof of Theorem 4. Let $k_2 = 4$ or 5, and put

$$D = X_2^{\theta(k_2)}, \quad \theta(k_2) = \frac{6 - k_2}{2k_2} - 10^{-3}, \quad z = D^{1/3},$$

and

$$r(4) = 3, \qquad r(5) = 4.$$

We denote by $R(n)$ the number of representations of n in the form

$$n = p_1 + x^2 + p_2^3 + p_3^{k_2},$$

subject to primes p_1, p_2, p_3 and integers x satisfying

$$p_1 \sim X_1, \quad x \sim X_2, \quad p_2 \sim X_3, \quad p_3 \sim X_{k_2}, \quad \text{and} \quad (x, \Pi(z)) = 1.$$

Also we denote by $\tilde{R}(n)$ the number of representations counted by $R(n)$ with the additional constraint $\Omega(x) > r(k_2)$. We aim to prove Theorem 4 by showing that $R(n) - \tilde{R}(n) > 0$ for every even integer $n \in [N, (6/5)N]$.

We first fix some notation. Let $I(n)$ and $B(p, n)$ be defined by (2.4), (3.1) and (3.2) with $s = 2$, $k = 1$, $k_0 = 2$, $k_1 = 3$ and $k_2 = 4$ or 5, and put

$$\mathfrak{S}(n) = \prod_p B(p, n).$$

It is easily confirmed that $B(p, n) = 1 + O(p^{-3/2})$, which assures the absolute convergence of the last infinite product, as well as the lower bound $\mathfrak{S}(n) \gg 1$ for even n, in combination with Lemma 3.8. Meanwhile, Lemma 2.1 gives the estimate $I(n) \asymp N^{5/6+1/k_2}(\log N)^{-4}$ valid for $N \leq n \leq (6/5)N$.

Proceeding along with the lines in the proof of Theorem 5 in §5 (see, in particular, the argument from (5.5) through to (5.11)), with trivial adjustment of notation, we see that Lemma 6.1 can be applied to $R(n)$ for every even $n \in [N, (6/5)N]$. In the latter application we take $U = D^{2/3}$ and $V = D^{1/3}$, and then Lemmata 3.7 and 4.2 show that the error term $\mathcal{E}^{(0)}$ in Lemma 6.1 is negligible in this instance. Using Lemma 3.3 also, we can conclude that, for all even $n \in [N, (6/5)N]$,

$$R(n) > (\phi_0(3) + O((\log \log N)^{-1/50}))\mathfrak{S}(n)\frac{e^{-\gamma}}{\log z}(\log X_2)I(n)$$

$$= ((2 \log 2)/\theta(k_2) + O((\log \log N)^{-1/50}))\mathfrak{S}(n)I(n). \qquad (6.33)$$

Next we put

$$D' = X_1^{\theta'}, \quad \theta' = \frac{5}{9}, \quad z' = (D')^{1/3},$$

and denote by $R'(n)$ the number of representations of n in the form

$$n = y + x^2 + p_2^3 + p_3^{k_2},$$

subject to primes $p_2 \sim X_3$, $p_3 \sim X_{k_2}$ and integers x and y satisfying

$$x \sim X_2, \quad (x, \Pi(z)) = 1, \quad \Omega(x) > r(k_2), \quad y \sim X_1, \quad (y, \Pi(z')) = 1.$$

Obviously we see $\tilde{R}(n) \leq R'(n)$. Again we can apply Lemma 6.1 to obtain an upper bound for $R'(n)$, this time taking $U = D'$ and $V = 3/2$, and then it is negligible the contribution of the remainder term corresponding to $\mathcal{E}^{(1)}$ in Lemma 6.1. To see this, we just follow the argument from (5.5) through to (5.12), replacing $g_2(\alpha)$ with

$$\sum_{r(k_2) < r \leq 3/\theta(k_2)} g_2(\alpha; X_2, r, z).$$

Hence, by using mainly Lemmata 2.1, 2.2, 3.3, 3.7, 4.6 and 6.1, we can establish the upper bound

$$\tilde{R}(n) \leq R'(n)$$

$$< (\phi_1(3) + O((\log \log N)^{-1/50}))\mathfrak{S}(n)\frac{e^{-\gamma}}{\log z'}(K(k_2)\log X_1)I(n)$$

$$= \big(2K(k_2)/\theta' + O((\log \log N)^{-1/50})\big)\mathfrak{S}(n)I(n), \qquad (6.34)$$

valid for all even $n \in [N, (6/5)N]$, where

$$K(k_2) = \sum_{r(k_2) < r \leq 3/\theta(k_2)} C_r(3/\theta(k_2)).$$

To examine $K(k_2)$, the following observation is useful. For a fixed real number u exceeding 1 and a large real number X, let $\pi(X, u)$ be the number of integers x with $x \sim X$ and $(x, \Pi(X^{1/u})) = 1$. Then we have

$$\pi(X, u) = \sum_{1 \leq r \leq u} g_1(0; X, r, X^{1/u}) = \Big(\sum_{r \geq 1} C_r(u) + O((\log X)^{-1})\Big)\frac{4X}{\log X},$$

by Lemma 2.2 (with the latter formula in (2.9) for $r = 1$). On the other hand, to estimate $\pi(X, u)$ is the simplest linear sieve problem, and it is easy to prove that

$$\pi(X, u) < (\phi_1(u) + o(1)) \cdot 4X \prod_{p < X^{1/u}} (1 - p^{-1}) = (\phi_1(u) + o(1))\frac{4ue^{-\gamma}X}{\log X},$$

as $X \to \infty$. Moreover, Grupp and Richert [9] showed that $\phi_1(u) < 1 + 3 \cdot 10^{-10}$ for $u \geq 10$ (see [9], p.212). Consequently we must have

$$\sum_{r \geq 1} C_r(u) \leq ue^{-\gamma}\phi_1(u) < ue^{-\gamma}(1 + 10^{-9}), \qquad (6.35)$$

for $u \geq 10$.

Note now that $C_1(u) = 1$ and $C_2(u) = \log(u - 1)$ for $u \geq 2$, by the definition. In the case $k_2 = 4$, we have $C_3(3/\theta(4)) > 2$, so we see by (6.35) that

$$K(4) \leq \sum_{r \geq 1} C_r(3/\theta(4)) - 1 - \log(3/\theta(4) - 1) - 2 < 1.37,$$

whence $2K(4)/\theta' < 5$. Since $(2\log 2)/\theta(4) > 5.5$, we conclude by (6.33) and (6.34) that $R(n) - \tilde{R}(n) \geq R(n) - R'(n) > 0$ for every even $n \in [N, (6/5)N]$, which establishes Theorem 4 (i).

In the case $k_2 = 5$, we have $C_3(3/\theta(5)) > 4.9$ and $C_4(3/\theta(5)) > 4$, thus we see by (6.35)

$$K(5) \leq \sum_{r \geq 1} C_r(3/\theta(5)) - 1 - \log(3/\theta(5) - 1) - 4.9 - 4 < 3.75,$$

whence $2K(5)/\theta' < 13.5$. Since $(2\log 2)/\theta(5) > 14$, we conclude by (6.33) and (6.34) that $R(n) - \tilde{R}(n) > 0$ again for every even $n \in [N, (6/5)N]$, which completes the proof of Theorem 4 (ii).

References

[1] J. Brüdern, *Iterationsmethoden in der additiven Zahlentheorie.* Thesis, Göttingen 1988.

[2] J. Brüdern, *A problem in additive number theory.* Math. Proc. Cambridge Philos. Soc. **103** (1988), 27–33.

[3] J. Brüdern, *A sieve approach to the Waring-Goldbach problem I: Sums of four cubes.* Ann. Scient. Éc. Norm. Sup. (4) **28** (1995), 461–476.

[4] J. Brüdern and N. Watt, *On Waring's problem for four cubes.* Duke Math. J. **77** (1995), 583–606.

[5] J.-R. Chen, *On the representation of a large even integer as the sum of a prime and the product of at most two primes.* Sci. Sinica **16** (1973), 157–176.

[6] H. Davenport and H. Heilbronn, *On Waring's problem: two cubes and one square.* Proc. London Math. Soc. (2) **43** (1937), 73–104.

[7] H. Davenport and H. Heilbronn, *Note on a result in additive theory of numbers.* Proc. London Math. Soc. (2) **43** (1937), 142–151.

[8] K. B. Ford, *The representation of numbers as sums of unlike powers, II.* J. Amer. Math. Soc. **9** (1996), 919–940.

[9] F. Grupp and H.-E. Richert, *The functions of the linear sieve.* J. Number Th. **22** (1986), 208–239.

[10] H. Halberstam, *Representations of integers as sums of a square, a positive cube, and a fourth power of a prime.* J. London Math. Soc. **25** (1950), 158–168.

[11] H. Halberstam, *Representations of integers as sums of a square of a prime, a cube of a prime, and a cube.* Proc. London Math. Soc. (2) **52** (1951), 455–466.

[12] H. Halberstam and H.-E. Richert, Sieve Methods. Academic Press, London, 1974.

[13] H. Halberstam and H.-E. Richert, *A weighted sieve of Greaves' type II.* Banach Center Publ. Vol. 17 (1985), 183–215.

[14] G. H. Hardy and J. E. Littlewood, *Some problems of "Partitio Numerorum": VI Further researches in Waring's problem.* Math. Z. **23** (1925), 1–37.

[15] D. R. Heath-Brown, *Three primes and an almost prime in arithmetic progression.* J. London Math. Soc. (2) **23** (1981), 396–414.

[16] C. Hooley, *On a new approach to various problems of Waring's type.* Recent progress in analytic number theory (Durham, 1979), Vol. 1. Academic Press, London-New York, 1981, 127–191.

[17] L. K. Hua, Additive Theory of Prime Numbers. Amer. Math. Soc., Providence, Rhode Island, 1965.

[18] H. Iwaniec, *Primes of the type $\phi(x, y) + A$, where ϕ is a quadratic form.* Acta Arith. **21** (1972), 203–224.

[19] H. Iwaniec, *A new form of the error term in the linear sieve.* Acta Arith. **37** (1980), 307–320.

[20] H. Iwaniec, Sieve Methods. Unpublished lecture note, 1996.

[21] W. Jagy and I. Kaplansky, *Sums of squares, cubes and higher powers.* Experiment. Math. **4** (1995), 169–173.

[22] K. Kawada, *Note on the sum of cubes of primes and an almost prime.* Arch. Math. **69** (1997), 13–19.

[23] K. Prachar, *Über ein Problem vom Waring-Goldbach'schen Typ II.* Monatsh. Math. **57** (1953) 113–116.

[24] K. F. Roth, *Proof that almost all positive integers are sums of a square, a cube and a fourth power.* J. London Math. Soc. **24** (1949), 4–13.

[25] K. F. Roth, *A problem in additive number theory.* Proc. London Math. Soc. (2) **53** (1951), 381–395.

[26] W. Schwarz, Zur Darstellung von Zahlen durch Summen von Primzahlpotenzen, II. J. Reine Angew. Math. **206** (1961), 78–112.

[27] K. Thanigasalam, *On sums of powers and a related problem.* Acta Arith. **36** (1980), 125–141.

[28] R. C. Vaughan, *A ternary additive problem.* Proc. London Math. Soc. (3) **41** (1980), 516–532.

[29] R. C. Vaughan, *Sums of three cubes.* Bull. London Math. Soc. **17** (1985), 17–20.

[30] R. C. Vaughan, The Hardy-Littlewood Method, 2nd ed., Cambridge Univ. Press, 1997.

A GENERALIZATION OF E. LEHMER'S CONGRUENCE AND ITS APPLICATIONS

Tianxin CAI

Department of Mathematics, Zhejiang University, Hangzhou, 310028, P.R.China

Keywords: quotients of Euler, Bernoulli polynomials, binomial coefficients

Abstract   In this paper, we announce the result that for any odd $n > 1$,

$$\sum_{\substack{i=1 \\ (i,n)=1}}^{(n-1)/2} \frac{1}{i} \equiv -2q_2(n) + nq_2^2(n) \pmod{n^2},$$

where $q_r(n) = (r^{\phi(n)} - 1)/n$, $(r, n) = 1$ is Euler's quotient of n with base r, which is a generalization of E. Lehmer's congruence. As applications, we mention some generalizations of Morley's congruence and Jacobstahl's Theorem to modulo arbitary positive integers. The details of the proof will partly appear in Acta Arithmetica.

2000 Mathematics Subject Classification: 11A25, 11B65, 11B68.

1.   INTRODUCTION

In 1938 E. Lehmer [5] established the following congruence:

$$\sum_{i=1}^{(p-1)/2} \frac{1}{i} \equiv -2q_2(p) + pq_2^2(p) \pmod{p^2} \tag{1}$$

for any odd prime p, which is an improvement of Eisenstein's famous congruence (1850):

$$q_2(p) \equiv -\frac{1}{2}\left(1 + \frac{1}{2} + \frac{1}{3} + \dots + \frac{1}{\frac{p-1}{2}}\right) \pmod{p},$$

Partly supported by the project NNSFC..

C. Jia and K. Matsumoto (eds.), Analytic Number Theory, 93–98.
© 2002 *Kluwer Academic Publishers. Printed in the Netherlands.*

where

$$q_2(p) = \frac{2^{p-1} - 1}{p}$$

is Fermat's quotient, using (1) and other similar congruences, he obtained various criteria for the first case of Fermat's Last Theorem (Cf. [8]). The proof of (1) followed the method of Glaisher [2], which depends on Bernoulli polynomials of fractional arguments. In this paper, we follow the same way to generalize (1) to modulo arbitrary positive integers, however, we need establish special congruences concerning the quotients of Euler. The main theorem we obtain is the following,

Theorem 1. *If* $n > 1$ *is odd, then*

$$\sum_{\substack{i=1 \\ (i,n)=1}}^{(n-1)/2} \frac{1}{i} \equiv -2q_2(n) + nq_2^2(n) \pmod{n^2},$$

where

$$q_r(n) = \frac{r^{\phi(n)} - 1}{n}, \ (r,n) = 1$$

is Euler's quotient of n *with base* r.

Corollary 1. *If* n *is odd, then*

$$\sum_{\substack{i=1 \\ (i,2n)=1}}^{n} \frac{1}{i} \equiv q_2(n) - nq_2^2(n)/2 \begin{cases} \pmod{n^2} & \text{for} \quad 3 \nmid n \\ \pmod{n^2/3} & \text{for} \quad 3 \mid n. \end{cases}$$

Similarly as Theorem 1, we can generalize other congruences by Lehmer to modulo arbitary positive integers. Among those, the most interesting one might be the following,

Theorem 2. *If* n *is odd, then*

$$\sum_{\substack{i=1 \\ (i,n)=1}}^{(n-1)/2} i^{\phi(n)-1} \equiv -2q_2(n)(1 \pm nw_n) + 2nq_2^2(n) \pmod{n^2},$$

where

$$w_n = \left(\prod_{\substack{r=1 \\ (r,n)=1}}^{n} \pm 1 \right) \Big/ n$$

is generalized Wilson's quotient or Gaussian quotient, the negative signs are to be chosen only when n *is not a prime power.*

Corollary 2. *If n is odd, then*

$$\sum_{\substack{i=1 \\ (i,n)=1}}^{(n-1)/2} \frac{q_i(n)}{i} \equiv \pm 2q_2(n)w_n + q_2^2(n) \pmod{n},$$

where the negative sign is to be chosen only when n is not a prime power.

In 1895, Morley [7] showed that

$$(-1)^{\frac{p-1}{2}} \binom{p-1}{(p-1)/2} \equiv 4^{p-1} \pmod{p^3} \tag{2}$$

for any prime $p \geq 5$, this is one of the most beautiful congruences concerning binomial coefficients. However, his ingenious proof, which is based on an explicit of De Moivre's Theorem, cannot be modified to investigate other binomial coefficients, we use Theorem 1 to present a generalization of (2), i.e.,

Theorem 3. *If n is odd, then*

$$(-1)^{\frac{\phi(n)}{2}} \prod_{d|n} \binom{d-1}{(d-1)/2}^{\mu(n/d)} \equiv 4^{\phi(n)} \begin{cases} \pmod{n^3} & \text{for} \quad 3 \nmid n \\ \pmod{n^3/3} & \text{for} \quad 3 \mid n, \end{cases} \tag{3}$$

where $\mu(n)$ is Möbius' function, and $\phi(n)$ is Euler's function. In particular, if $n \geq 5$ is prime, (3) becomes (2).

Corollary 3. *If $p \geq 5$ is prime, then*

$$(-1)^{\frac{p-1}{2}} \binom{p^l-1}{(p^l-1)/2} \Big/ \binom{p^{l-1}-1}{(p^{l-1}-1)/2} \equiv 4^{\phi(p^l)} \pmod{p^{3l}},$$

and

$$(-1)^{\frac{(p-1)l}{2}} \binom{p^l-1}{(p^l-1)/2} \equiv 4^{p^l-1} \pmod{p^3}$$

for any $l \geq 1$.

Corollary 4. *For each $l \geq 1$, there are exactly two primes up to 4×10^{12} such that*

$$\binom{p^l-1}{(p^l-1)/2} \equiv \pm 1 \pmod{p^2},$$

the positive sign is to be chosen when $p = 1093$ and the negative sign is to be chosen when $p = 3511$.

Corollary 5. *If $p, q \geq 5$ are distinct odd primes, then*

$$\binom{pq-1}{(pq-1)/2} \equiv 4^{(p-1)(q-1)} \binom{p-1}{(p-1)/2} \binom{q-1}{(q-1)/2} \pmod{p^3 q^3}.$$

Moreover, we have the following,

Theorem 4. *Let $n \geq 1$ be an integer, then*

$$\prod_{d|n} \binom{ud}{vd}^{\mu(n/d)} \equiv 1 \quad \begin{cases} \pmod{n^3} & if \quad 3 \nmid n, \ n \neq 2^a \\ \pmod{n^3/3} & if \quad 3 \mid n \\ \pmod{n^3/2} & if \quad n = 2^a, \ a \geq 2 \\ \pmod{n^3/4} & if \quad n = 2 \end{cases} \tag{4}$$

for any integers $u > v > 0$. In particular, if $p \geq 5$ is prime, then (4) becomes

$$\binom{up}{vp} \Big/ \binom{u}{v} \equiv 1 \pmod{p^3},$$

this is Jacobstahl's Theorem.

Corollary 6. *If $p, q \geq 5$ are distinct primes, then*

$$\binom{up}{vp} \binom{uq}{vq} \equiv \binom{u}{v} \binom{upq}{vpq} \pmod{p^3 q^3}$$

for any integers $u > v > 0$.

Corollary 7. *If $p \geq 5$ is prime, then*

$$\binom{2p^l - 1}{p^l - 1} \Big/ \binom{2p^{l-1} - 1}{p^{l-1} - 1} \equiv 1 \pmod{p^{3l}},$$

and

$$\binom{2p^l - 1}{p^l - 1} \equiv 1 \pmod{p^3}$$

for any $l \geq 1$.

Corollary 8. *If $p, q \geq 5$ are distinct primes, then*

$$\binom{2pq-1}{pq-1} \equiv \binom{2p-1}{p-1} \binom{2q-1}{q-1} \pmod{p^3 q^3}.$$

In 1862, Wolstenholme showed that

$$\binom{2p-1}{p-1} \equiv 1 \pmod{p^3} \tag{5}$$

for any prime $p \geq 5$. This is a consequence of Jacobstahl's Theorem, and therefore a consequence of Theorem 4. The exponent 3 in (5) can be increased only if $p|B_{p-3}$, here B_{p-3} is the $p-3$th Bernoulli number. Jones (Cf. [3]) has asked for years that whether the converse for (5) is true. As direct consequences of Corollary 7 and Corollary 8, we present two equivalences for Jones' problem, i.e.,

Theorem 5. *If the congruence*

$$\binom{2n-1}{n-1} \equiv 1 \pmod{n^3} \tag{6}$$

has a solution of prime power p^l $(l > 1)$, then p must satisfy

$$\binom{2p-1}{p-1} \equiv 1 \pmod{p^6}.$$

The converse is also true. Meanwhile, if the congruence (6) has a solution of product of distinct odd primes p and q, then

$$\binom{2q-1}{q-1} \equiv 1 \pmod{p^3}, \quad \binom{2p-1}{p-1} \equiv 1 \pmod{q^3}.$$

The converse is also true.

In particular, if $l = 2$, the first part of Theorem 5 was obtained by R. J. McIntosh [6] in 1995.

Acknowledgments

The author is very grateful to Prof. Andrew Granville for his constructive comments and valuable suggestions.

References

[1] T. Cai and A. Granville, *On the residue of binomial coefficients and their products modulo prime powers*, preprint.

[2] J. W. L. Glaisher, Quart. J. Math., **32** (1901), 271–305.

[3] R. Guy, *Unsolved problems in number theory*, Springer-Verlag, Second Edition, 1994.

[4] G. H. Hardy and E. M. Wright, *An introduction to the theory of numbers*, Oxford, Fourth Edition, 1971.

[5] E. Lehmer, *On congruences involving Bernoulli numbers and the quotients of Fermat and Wilson*, Ann. of Math., **39** (1938), 350–359.

[6] R. J. McIntosh, *On the converse of Wolstenholmes theorem*, Acta Arith., **71** (1995), 381–389.

[7] F. Morley, *Note on the congruence $2^{4n} \equiv (-1)^n (2n)!/(n!)^2$, where $2n + 1$ is prime*, Ann. of Math., **9** (1895), 168–170.

[8] P. Ribenboim, *The new book of prime number records*, Springer-Verlag, Third Edition, 1996.

ON CHEN'S THEOREM

CAI Yingchun
and LU Minggao
Department of Mathematics, Shanghai University, Shanghai 200436, P. R. China

Keywords: sieve, application of sieve method, Goldbach problem

Abstract   Let N be a sufficiently large even integer and $S(N)$ denote the number of solutions of the equation

$$N = p + P_2,$$

where p denotes a prime and P_2 denotes an almost-prime with at most two prime factors. In this paper we obtain

$$S(N) \geq \frac{0.8285 C(N) N}{\log^2 N},$$

where

$$C(N) = \prod_{p>2} \left(1 - \frac{1}{(p-1)^2}\right) \prod_{p \mid N, p>2} \frac{p-1}{p-2}.$$

2000 Mathematics Subject Classification: 11N05, 11N36.

1.   INTRODUCTION

In 1966 Chen Jingrun[1] made a considerable progress in the research of the binary Goldbach conjecture, he[2] proved the remarkable Chen's Theorem: let N be a sufficiently large even integer and $S(N)$ denote the number of solutions of the equation

$$N = p + P_2,$$

where p is a prime and P_2 is an almost-prime with at most two prime factors, then

$$S(N) \geq \frac{0.67 C(N) N}{\log^2 N},$$

Project supported by The National Natural Science Fundation of China (grant no.19531010, 19801021)

C. Jia and K. Matsumoto (eds.), Analytic Number Theory, 99–119.

where

$$C(N) = \prod_{p>2} \left(1 - \frac{1}{(p-1)^2}\right) \prod_{p|N,p>2} \frac{p-1}{p-2}.$$

The oringinal proof of Chen's was simplified by Pan Chengdong, Ding Xiaqi, Wang Yuan[3], Halberstam-Richert[4], Halberstam[5], Ross[6]. In [4] Halberstam and Richert announced that they obtained the constant 0.689 and a detail proof was given in [5]. In page 338 of [4] it says: "It would be interesting to know whether the more elaborate weighting procedure could be adapted to the numerical improvements and could be important". In 1978 Chen Jingrun[7] introduced a new sieve procedure to show

$$S(N) \geq \frac{0.81C(N)N}{\log^2 N}.$$

In this paper we shall prove

Theorem.

$$S(N) \geq \frac{0.8285C(N)N}{\log^2 N}.$$

The constant 0.8285 is rather near to the limit obtained by the method employed in this paper.

2.   SOME LEMMAS

Let $\mathcal{A}$ denote a finite set of integers, $\mathcal{P}$ denote an infinite set of primes, $\overline{\mathcal{P}}$ denote the set of primes that do not belong to $\mathcal{P}$. Let $z \geq 2$, put

$$P(z) = \prod_{p<z,p\in\mathcal{P}} p, \qquad \mathcal{P}(q) = \{p|p \in \mathcal{P}, (p,q) = 1\},$$

$$S(\mathcal{A}, z) = S(\mathcal{A}; \mathcal{P}, z) = \sum_{a\in\mathcal{A},(a,P(z))=1} 1,$$

$$\mathcal{A}_d = \{a|a \in \mathcal{A}, a \equiv 0(\mathrm{mod} d)\}.$$

Lemma 1[8]. *If*

$A_1)$ $|\mathcal{A}_d| = \dfrac{\omega(d)}{d}X + r_d,$ $\qquad \mu(d) \neq 0,$ $\qquad (d,\overline{\mathcal{P}}) = 1;$

$A_2)$ $\displaystyle\sum_{z_1\leq p<z_2} \frac{\omega(p)}{p} = \log\frac{\log z_2}{\log z_1} + O\left(\frac{1}{\log z_1}\right),$ $\qquad z_2 > z_1 \geq 2,$

where $\omega(d)$ is a multiplicative function, $0 \leq \omega(p) < p$, X is independent of d. Then

$$S(\mathcal{A}, \mathcal{P}, z) \geq XV(z)\left\{ f(s) + O\left(\frac{1}{\log^{\frac{1}{3}} D}\right)\right\} - R_D,$$

$$S(\mathcal{A}, \mathcal{P}, z) \leq XV(z)\left\{ F(s) + O\left(\frac{1}{\log^{\frac{1}{3}} D}\right)\right\} + R_D,$$

where

$$s = \frac{\log D}{\log z}, \qquad R_D = \sum_{d < D, d \mid P(z)} |r_d|,$$

$$V(z) = C(\omega)\frac{e^{-\gamma}}{\log z}\left(1 + O\left(\frac{1}{\log z}\right)\right),$$

$$C(\omega) = \prod_p \left(1 - \frac{\omega(p)}{p}\right)\left(1 - \frac{1}{p}\right)^{-1},$$

where γ denotes the Euler's constant, $f(s)$ and $F(s)$ are determined by the following differential-difference equations

$$\begin{cases} F(s) = \dfrac{2e^\gamma}{s}, & f(s) = 0, & 0 < s \leq 2, \\ (sF(s))' = f(s-1), & (sf(s))' = F(s-1), & s \geq 2. \end{cases}$$

Lemma 2[9].

$$F(s) = \frac{2e^\gamma}{s}, \qquad 0 < s \leq 3;$$

$$F(s) = \frac{2e^\gamma}{s}\left(1 + \int_2^{s-1} \frac{\log(t-1)}{t}dt\right), \qquad 3 \leq s \leq 5;$$

$$f(s) = \frac{2e^\gamma \log(s-1)}{s}, \qquad 2 \leq s \leq 4;$$

$$f(s) = \frac{2e^\gamma}{s}\left(\log(s-1) + \int_3^{s-1} \frac{dt}{t}\int_2^{t-1} \frac{\log(u-1)}{u}du\right), \qquad 4 \leq s \leq 6.$$

Lemma 3[9]. *For any given constant $A > 0$, there exists a constant $B = B(A) > 0$ such that*

$$\sum_{d \leq D} \max_{(l,d)=1} \max_{y \leq x} \left| \sum_{\substack{a \leq E(x), (a,d)=1}} g(x,a)H(y; a, d, l)\right| \ll \frac{x}{\log^A x},$$

where

$$H(y; a, d, l) = \sum_{\substack{ap \leq y \\ ap \equiv l (\bmod d)}} 1 - \frac{1}{\varphi(d)} \sum_{ap \leq y} 1,$$

$$\frac{1}{2} \leq E(x) \ll x^{1-\alpha}, \qquad 0 < \alpha \leq 1,$$

$$g(x, a) \ll d_r(a), \qquad D = x^{\frac{1}{2}} \log^{-B} x.$$

Remark 1[9]. Let $r_1(y)$ be a positive function depending on x and satisfying $r_1(y) \ll x^\alpha$ for $y \leq x$. Then under the conditions in Lemma 3, we have

$$\sum_{d \leq D} \max_{(l,d)=1} \max_{y \leq x} \left| \sum_{a \leq E(x),(a,d)=1} g(x,a) H(ar_1(y); a, d, l) \right| \ll \frac{x}{\log^A x}.$$

Remark 2[9]. Let $r_2(a)$ be a positive function depending on x and y such that $ar_2(a) \ll x$ for $a \leq E(x), y \leq x$. Then under the conditions in Lemma 3, we have

$$\sum_{d \leq D} \max_{(l,d)=1} \max_{y \leq x} \left| \sum_{a \leq E(x),(a,d)=1} g(x,a) H(ar_2(a); a, d, l) \right| \ll \frac{x}{\log^A x}.$$

Lemma 4[9,10]. *Let*

$$x > 1, \qquad z = x^{\frac{1}{u}}, \qquad Q(z) = \prod_{p < z} p.$$

Then for $u \geq u_0 > 1$, we have

$$\sum_{\substack{n \leq x \\ (n, Q(z))=1}} 1 = w(u) \frac{x}{\log z} + O\left(\frac{x}{\log^2 z}\right),$$

where $w(u)$ is determined by the following differential-difference equation

$$\begin{cases} w(u) = \dfrac{1}{u}, & 1 \leq u \leq 2, \\ (uw(u))' = w(u-1), & u \geq 2. \end{cases}$$

Moreover, we have

$$w(u) < \frac{1}{1.7803}, \qquad u \geq 4.$$

3.   WEIGHTED SIEVE METHOD

Let N be a sufficiently large even integer and put

$$\mathcal{A} = \{a \,|\, a = N - p, p \le N\}, \tag{3.1}$$
$$\mathcal{P} = \{p|\ (p, N) = 1\}. \tag{3.2}$$

Lemma 5[7]. *Let* $0 < \alpha < \beta < \frac{1}{3}$ *and* $\alpha + 3\beta > 1$. *Then*

$$S(N) \ge \sum_{\substack{a \in \mathcal{A} \\ (a, NP(N^\alpha))=1}} \left(1 - \frac{1}{2}\rho_1(a) - \frac{1}{2}\rho_2(a) - \rho_3(a) + \frac{1}{2}\rho_4(a)\right) + O(N^{1-\alpha})$$

$$\ge S(\mathcal{A}, N^\alpha) - \frac{1}{2} \sum_{N^\alpha \le p < N^\beta, (p,N)=1} S(\mathcal{A}_p, N^\alpha)$$

$$-\frac{1}{2} \sum_{\substack{N^\alpha \le p_1 < N^\beta \le p_2 < \left(\frac{N}{p_1}\right)^{\frac{1}{2}} \\ (p_1 p_2, N)=1}} S(\mathcal{A}_{p_1 p_2}, p_2)$$

$$- \sum_{\substack{N^\beta \le p_1 < p_2 < \left(\frac{N}{p_1}\right)^{\frac{1}{2}} \\ (p_1 p_2, N)=1}} S(\mathcal{A}_{p_1 p_2}, p_2)$$

$$+\frac{1}{2} \sum_{\substack{N^\alpha \le p_1 < p_2 < p_3 < N^\beta \\ (p_1 p_2 p_3, N)=1}} S(\mathcal{A}_{p_1 p_2 p_3}; \mathcal{P}(p_1), p_2) + O(N^{1-\alpha}),$$

where

$$\rho_1(a) = \sum_{\substack{p|a,(p,N)=1 \\ N^\alpha \le p < N^\beta}} 1,$$

$$\rho_2(a) = \begin{cases} 1, & a = p_1 p_2 p_3,\ N^\alpha \le p_1 < N^\beta \le p_2 < p_3,\ (a, N) = 1; \\ 0, & otherwise. \end{cases}$$

$$\rho_3(a) = \begin{cases} 1, & a = p_1 p_2 p_3,\ N^\beta \le p_1 < p_2 < p_3,\ (a, N) = 1; \\ 0, & otherwise. \end{cases}$$

$$\rho_4(a) = \begin{cases} 1, & a = p_1 p_2 p_3 n,\ N^\alpha \le p_1 < p_2 < p_3 < N^\beta,\ (a, Np_1^{-1}P(p_2)) = 1; \\ 0, & otherwise. \end{cases}$$

Proof. Since the second inequlity can be deduced from the first one easily, so it suffice to prove the first inequality. Let

$$v_2(a) = \sum_{p^m | a} 1, \qquad \lambda(a) = \begin{cases} 1, & v_2(a) \le 2, \\ 0, & v_2(a) > 2. \end{cases}$$

Then

$$S(N) \ge \sum_{\substack{a \in \mathcal{A} \\ (a, P(N^\alpha)) = 1}} \lambda(a)$$

$$= \sum_{\substack{a \in \mathcal{A}, (a, N) = 1 \\ (a, P(N^\alpha)) = 1}} \mu^2(a) \lambda(a) + O(N^{1-\alpha}).$$

On the other hand,

$$\sum_{\substack{a \in \mathcal{A} \\ (a, P(N^\alpha)) = 1}} \left(1 - \frac{1}{2} \rho_1(a) - \frac{1}{2} \rho_2(a) - \rho_3(a) + \frac{1}{2} \rho_4(a) \right)$$

$$= \sum_{\substack{a \in \mathcal{A}, (a, N) = 1 \\ (a, P(N^\alpha)) = 1}} \mu^2(a) \left(1 - \frac{1}{2} \rho_1(a) - \frac{1}{2} \rho_2(a) - \rho_3(a) + \frac{1}{2} \rho_4(a) \right) + O(N^{1-\alpha}).$$

For

$$\mu^2(a) = 1, \qquad (a, P(N^\alpha)) = 1, \qquad (a, N) = 1$$

we have

1) $v_2(a) \le 2$. Then $\rho_4(a) = 0$.

$$\lambda(a) = 1 \ge 1 - \frac{1}{2} \rho_1(a) - \frac{1}{2} \rho_2(a) - \rho_3(a) + \frac{1}{2} \rho_4(a);$$

2) $v_2(a) = 3$. If $\rho_1(a) = 0$, then $\rho_3(a) = 1$, $\rho_2(a) = \rho_4(a) = 0$,

$$\lambda(a) = 0 \ge 1 - \frac{1}{2} \rho_1(a) - \frac{1}{2} \rho_2(a) - \rho_3(a) + \frac{1}{2} \rho_4(a);$$

If $\rho_1(a) = 1$, then $\rho_3(a) = \rho_4(a) = 0$, and $\rho_2(a) = 1$, hence

$$\lambda(a) = 0 = 1 - \frac{1}{2} \rho_1(a) - \frac{1}{2} \rho_2(a) - \rho_3(a) + \frac{1}{2} \rho_4(a);$$

If $\rho_1(a) = 2$, then $\rho_2(a) = \rho_3(a) = \rho_4(a) = 0$, and

$$\lambda(a) = 0 = 1 - \frac{1}{2} \rho_1(a) - \frac{1}{2} \rho_2(a) - \rho_3(a) + \frac{1}{2} \rho_4(a);$$

If $\rho_1(a) = 3$, then $\rho_2(a) = \rho_3(a) = 0$, $\rho_4(a) = 1$, and

$$\lambda(a) = 0 = 1 - \frac{1}{2}\rho_1(a) - \frac{1}{2}\rho_2(a) - \rho_3(a) + \frac{1}{2}\rho_4(a);$$

3) $v_2(a) \geq 4$. Then $\rho_1(a) \geq 2$. If $\rho_1(a) = 2$, then $\rho_2(a) = \rho_3(a) = \rho_4(a) = 0$,

$$\lambda(a) = 0 = 1 - \frac{1}{2}\rho_1(a) - \frac{1}{2}\rho_2(a) - \rho_3(a) + \frac{1}{2}\rho_4(a);$$

If $\rho_1(a) \geq 3$, then $\rho_2(a) = \rho_3(a) = 0$, $\rho_4(a) = 1$,

$$\lambda(a) = 0 \geq 1 - \frac{1}{2}\rho_1(a) - \frac{1}{2}\rho_2(a) - \rho_3(a) + \frac{1}{2}\rho_4(a).$$

Combining the above arguments we complete the proof of Lemma 5.

Lemma 6.

$$2S(N) \geq 2S(\mathcal{A}, N^{\frac{1}{12}}) - \frac{1}{2} \sum_{N^{\frac{1}{12}} \leq p < N^{\frac{1}{3.047}}, (p,N)=1} S(\mathcal{A}_p, N^{\frac{1}{12}})$$

$$- \frac{1}{2} \sum_{\substack{N^{\frac{1}{12}} \leq p_1 < N^{\frac{1}{3.047}} \leq p_2 < \left(\frac{N}{p_1}\right)^{\frac{1}{2}} \\ (p_1 p_2, N)=1}} S(\mathcal{A}_{p_1 p_2}, p_2)$$

$$- \sum_{\substack{N^{\frac{1}{3.047}} \leq p_1 < p_2 < \left(\frac{N}{p_1}\right)^{\frac{1}{2}} \\ (p_1 p_2, N)=1}} S(\mathcal{A}_{p_1 p_2}, p_2)$$

$$- \frac{1}{2} \sum_{N^{\frac{1}{12}} \leq p < N^{\frac{1}{8.813}}, (p,N)=1} S(\mathcal{A}_p, p)$$

$$- \frac{1}{2} \sum_{N^{\frac{1}{12}} \leq p < N^{\frac{1}{3.383}}, (p,N)=1} S(\mathcal{A}_p, N^{\frac{1}{12}})$$

$$+ \frac{1}{2} \sum_{\substack{N^{\frac{1}{12}} \leq p_1 < p_2 < N^{\frac{1}{8.813}} \\ (p_1 p_2, N)=1}} S(\mathcal{A}_{p_1 p_2}, N^{\frac{1}{12}})$$

$$+ \frac{1}{2} \sum_{\substack{N^{\frac{1}{12}} \leq p_1 < N^{\frac{1}{8.813}} \leq p_2 < N^{\frac{1}{3}} p_1^{-1} \\ (p_1 p_2, N)=1}} S(\mathcal{A}_{p_1 p_2}, N^{\frac{1}{12}})$$

$$-\frac{1}{2} \sum_{\substack{N^{\frac{1}{8.813}} \le p_1 < N^{\frac{1}{3.383}} \le p_2 < \left(\frac{N}{p_1}\right)^{\frac{1}{2}} \\ (p_1 p_2, N)=1}} S(\mathcal{A}_{p_1 p_2}, p_2)$$

$$- \sum_{\substack{N^{\frac{1}{3.383}} \le p_1 < p_2 < \left(\frac{N}{p_1}\right)^{\frac{1}{2}} \\ (p_1 p_2, N)=1}} S(\mathcal{A}_{p_1 p_2}, p_2)$$

$$-\frac{1}{2} \sum_{\substack{N^{\frac{1}{12}} \le p_1 < p_4 < p_2 < p_3 < N^{\frac{1}{8.813}} \\ (p_1 p_2 p_3 p_4, N)=1}} S(\mathcal{A}_{p_1 p_2 p_3 p_4}; \mathcal{P}(p_1), p_4)$$

$$-\frac{1}{2} \sum_{\substack{N^{\frac{1}{12}} \le p_1 < p_4 < p_2 < N^{\frac{1}{8.813}} \le p_3 < N^{\frac{1}{3}} p_2^{-1} \\ (p_1 p_2 p_3 p_4, N)=1}} S(\mathcal{A}_{p_1 p_2 p_3 p_4}; \mathcal{P}(p_1), p_4)$$

$$+ O(N^{\frac{11}{12}})$$

$$= 2\Sigma - \frac{1}{2}\Sigma_1 - \frac{1}{2}\Sigma_2 - \Sigma_3 - \frac{1}{2}\Sigma_4 - \frac{1}{2}\Sigma_5 + \frac{1}{2}\Sigma_6 + \frac{1}{2}\Sigma_7$$

$$- \frac{1}{2}\Sigma_8 - \Sigma_9 - \frac{1}{2}\Sigma_{10} - \frac{1}{2}\Sigma_{11} + O(N^{\frac{11}{12}}).$$

Proof. By Buchstab's identity

$$S(\mathcal{A}, z_2) = S(\mathcal{A}, z_1) - \sum_{z_1 \le p < z_2} S(\mathcal{A}_p, p), \qquad 2 \le z_1 \le z_2,$$

we have

$$S(\mathcal{A}, N^{\frac{1}{8.813}}) = S(\mathcal{A}, N^{\frac{1}{12}}) - \frac{1}{2} \sum_{N^{\frac{1}{12}} \le p < N^{\frac{1}{8.813}}, (p,N)=1} S(\mathcal{A}_p, p)$$

$$- \frac{1}{2} \sum_{N^{\frac{1}{12}} \le p < N^{\frac{1}{8.813}}, (p,N)=1} S(\mathcal{A}_p, N^{\frac{1}{12}})$$

$$+ \frac{1}{2} \sum_{\substack{N^{\frac{1}{12}} \le p_1 < p_2 < N^{\frac{1}{8.813}} \\ (p_1 p_2, N)=1}} S(\mathcal{A}_{p_1 p_2}, N^{\frac{1}{12}})$$

$$- \frac{1}{2} \sum_{\substack{N^{\frac{1}{12}} \le p_1 < p_2 < p_3 < N^{\frac{1}{8.813}} \\ (p_1 p_2 p_3, N)=1}} S(\mathcal{A}_{p_1 p_2 p_3}, p_1), \qquad (3.3)$$

$$\sum_{N^{\frac{1}{8.813}} \leq p < N^{\frac{1}{3.383}}, (p,N)=1} S(\mathcal{A}_p, N^{\frac{1}{8.813}})$$

$$\leq \sum_{N^{\frac{1}{8.813}} \leq p < N^{\frac{1}{4}}, (p,N)=1} S(\mathcal{A}_p, N^{\frac{1}{12}})$$

$$- \sum_{\substack{N^{\frac{1}{12}} \leq p_1 < N^{\frac{1}{8.813}} \leq p_2 < N^{\frac{1}{4}} \\ (p_1 p_2, N)=1}} S(\mathcal{A}_{p_1 p_2}, p_1)$$

$$+ \sum_{N^{\frac{1}{4}} \leq p < N^{\frac{1}{3.383}}, (p,N)=1} S(\mathcal{A}_p, N^{\frac{1}{12}})$$

$$\leq \sum_{N^{\frac{1}{8.813}} \leq p < N^{\frac{1}{3.383}}, (p,N)=1} S(\mathcal{A}_p, N^{\frac{1}{12}})$$

$$- \sum_{\substack{N^{\frac{1}{12}} \leq p_1 < N^{\frac{1}{8.813}} \leq p_2 < N^{\frac{1}{3}} p_1^{-1} \\ (p_1 p_2, N)=1}} S(\mathcal{A}_{p_1 p_2}, p_1)$$

$$= \sum_{N^{\frac{1}{8.813}} \leq p < N^{\frac{1}{3.383}}, (p,N)=1} S(\mathcal{A}_p, N^{\frac{1}{12}})$$

$$- \sum_{\substack{N^{\frac{1}{12}} \leq p_1 < N^{\frac{1}{8.813}} \leq p_2 < N^{\frac{1}{3}} p_1^{-1} \\ (p_1 p_2, N)=1}} S(\mathcal{A}_{p_1 p_2}, N^{\frac{1}{12}})$$

$$+ \sum_{\substack{N^{\frac{1}{12}} \leq p_1 < p_2 < N^{\frac{1}{8.813}} \leq p_3 < N^{\frac{1}{3}} p_2^{-1} \\ (p_1 p_2 p_3, N)=1}} S(\mathcal{A}_{p_1 p_2 p_3}, p_1), \qquad (3.4)$$

$$\sum_{\substack{N^{\frac{1}{12}} \leq p_1 < p_2 < p_3 < N^{\frac{1}{3.047}} \\ (p_1 p_2 p_3, N)=1}} S(\mathcal{A}_{p_1 p_2 p_3}; \mathcal{P}(p_1), p_2)$$

$$- \sum_{\substack{N^{\frac{1}{12}} \leq p_1 < p_2 < p_3 < N^{\frac{1}{8.813}} \\ (p_1 p_2 p_3, N)=1}} S(\mathcal{A}_{p_1 p_2 p_3}, p_1)$$

$$- \sum_{\substack{N^{\frac{1}{12}} \leq p_1 < p_2 < N^{\frac{1}{8.813}} \leq p_3 < N^{\frac{1}{3}} p_2^{-1} \\ (p_1 p_2 p_3, N)=1}} S(\mathcal{A}_{p_1 p_2 p_3}, p_1)$$

$$\geq \sum_{\substack{N^{\frac{1}{12}} \leq p_1 < p_2 < p_3 < N^{\frac{1}{8.813}} \\ (p_1 p_2 p_3, N)=1}} (S(\mathcal{A}_{p_1 p_2 p_3}; \mathcal{P}(p_1), p_2) - S(\mathcal{A}_{p_1 p_2 p_3}, p_1))$$

$$+ \sum_{\substack{N^{\frac{1}{12}} \leq p_1 < p_2 < N^{\frac{1}{8.813}} \leq p_3 < N^{\frac{1}{3}} p_2^{-1} \\ (p_1 p_2 p_3, N)=1}} (S(\mathcal{A}_{p_1 p_2 p_3}; \mathcal{P}(p_1), p_2) - S(\mathcal{A}_{p_1 p_2 p_3}, p_1))$$

$$\geq - \sum_{\substack{N^{\frac{1}{12}} \leq p_1 < p_4 < p_2 < p_3 < N^{\frac{1}{8.813}} \\ (p_1 p_2 p_3 p_4, N)=1}} S(\mathcal{A}_{p_1 p_2 p_3}; \mathcal{P}(p_1), p_4)$$

$$- \sum_{\substack{N^{\frac{1}{12}} \leq p_1 < p_4 < p_2 < N^{\frac{1}{8.813}} \leq p_3 < N^{\frac{1}{3}} p_2^{-1} \\ (p_1 p_2 p_3 p_4, N)=1}} S(\mathcal{A}_{p_1 p_2 p_3}; \mathcal{P}(p_1), p_4). \tag{3.5}$$

By Lemma 5 with $(\alpha, \beta) = (\frac{1}{12}, \frac{1}{3.047})$ and $(\alpha, \beta) = (\frac{1}{8.813}, \frac{1}{3.383})$ and (3.3)—(3.5), we complete the proof of Lemma 6.

4.   PROOF OF THE THEOREM

In this section, sets $\mathcal{A}$ and $\mathcal{P}$ are defined by (3.1) and (3.2) respectively. Let

$$X = \mathrm{Li}N \sim \frac{N}{\log N}, \quad (d, N) = 1,$$

$$r_d = \pi(N; d, N) - \frac{\mathrm{Li}N}{\varphi(d)},$$

$$\omega(d) = \frac{d}{\varphi(d)}, \qquad \mu(d) \neq 0.$$

1) Evaluation of $\Sigma, \Sigma_1, \Sigma_4, \Sigma_5, \Sigma_6, \Sigma_7$.

Let $D = \frac{N^{\frac{1}{2}}}{\log^B N}$ with $B = B(5) > 0$, by Lemma 3 we get

$$R_D = \sum_{d \leq D} \left| \pi(N; d, N) - \frac{\mathrm{Li}N}{\varphi(d)} \right|$$

$$\leq \sum_{d \leq D} \max_{y \leq N} \max_{(l,d)=1} \left| \pi(y; d, l) - \frac{\mathrm{Li}y}{\varphi(d)} \right|$$

$$\ll \frac{N}{\log^5 N}, \tag{4.1}$$

and

$$C(\omega) = \prod_p \left(1 - \frac{\omega(p)}{p}\right)\left(1 - \frac{1}{p}\right)^{-1}$$

$$= \prod_{p|N} \left(1 - \frac{1}{p}\right)^{-1} \prod_{(p,N)=1} \left(1 - \frac{1}{p-1}\right)\left(1 - \frac{1}{p}\right)^{-1}$$

$$= \prod_{p|N} \frac{p}{p-1} \prod_{(p,N)=1} \frac{p(p-2)}{(p-1)^2}$$

$$= 2 \prod_{p|N,p>2} \left(\frac{p}{p-1} \cdot \frac{(p-1)^2}{p(p-2)}\right) \prod_{p>2} \frac{p(p-2)}{(p-1)^2}$$

$$= 2 \prod_{p>2} \left(1 - \frac{1}{(p-1)^2}\right) \prod_{p|N,p>2} \frac{p-1}{p-2}$$

$$= 2C(N). \tag{4.2}$$

By Lemma 1, Lemma 2, (4.1), (4.2) and some routine arguments we get that

$$\Sigma \ge 8(1 + o(1))\frac{C(N)N}{\log^2 N}\left(\log 5 + \int_2^4 \frac{\log(s-1)}{s} \log\frac{5}{s+1}ds\right)$$

$$\ge 13.4735\frac{C(N)N}{\log^2 N}. \tag{4.3}$$

For $N^{\frac{1}{12}} \le p < N^{\frac{1}{3.047}}$, by Lemma 1 and (4.2) we get that

$$S(A_p, \mathcal{P}, N^{\frac{1}{12}}) \le 24(1 + o(1))C(N)e^{-\gamma}$$

$$\times \frac{N}{p\log^2 N}F\left(6 - 12\frac{\log p}{\log N}\right) + R_D(p), \tag{4.4}$$

where

$$R_D(p) = \sum_{d < \frac{D}{p}, d|P(N^{\frac{1}{12}})} |r_{dp}|.$$

By Lemma 3 we have

$$\sum_{\substack{N^{\frac{1}{12}} \le p < N^{\frac{1}{3.047}} \\ (p,N)=1}} R_D(p) = \sum_{\substack{N^{\frac{1}{12}} \le p < N^{\frac{1}{3.047}} \\ (p,N)=1}} \sum_{d < \frac{D}{p}, d|P(N^{\frac{1}{12}})} \left|\pi(N; dp, N) - \frac{\text{Li}N}{\varphi(dp)}\right|$$

$$\le \sum_{d \le D} \max_{(l,d)=1} \max_{y \le N} \left|\pi(y; d, l) - \frac{\text{Li}y}{\varphi(d)}\right|$$

$$\ll \frac{N}{\log^5 N}. \tag{4.5}$$

By (4.4), (4.5), the prime number theorem and summation by parts we get that

$$\Sigma_1 \le 24(1 + o(1))C(N)e^{-\gamma}\frac{N}{\log^2 N}$$

$$\times \sum_{\substack{N^{\frac{1}{12}} \le p < N^{\frac{1}{3.047}} \\ (p,N)=1}} \frac{1}{p}F\left(6 - 12\frac{\log p}{\log N}\right)$$

$$= 24(1 + o(1))C(N)e^{-\gamma}\frac{N}{\log^2 N}$$

$$\times \int_{N^{\frac{1}{12}}}^{N^{\frac{1}{3.047}}} \frac{1}{u\log u}F\left(6 - 12\frac{\log u}{\log N}\right)du$$

$$\le 8(1 + o(1))\frac{C(N)N}{\log^2 N}\left(\log\frac{10}{1.047} + \int_2^4 \frac{\log(s-1)}{s}\log\frac{5(5-s)}{s+1}ds\right)$$

$$\le 20.4456\frac{C(N)N}{\log^2 N}. \tag{4.6}$$

Similarly,

$$\Sigma_4 \le 8(1 + o(1))\frac{C(N)N}{\log^2 N}\left(\log\frac{10}{6.813}\left(1 + \int_2^{2.4065}\frac{\log(s-1)}{s}ds\right)\right.$$

$$\left. + \int_{2.4065}^4 \frac{\log(s-1)}{s}\log\frac{5}{s+1}ds\right)$$

$$\le 3.6569\frac{C(N)N}{\log^2 N}, \tag{4.7}$$

$$\Sigma_5 \le 8(1 + o(1))\frac{C(N)N}{\log^2 N}\left(\log\frac{10}{1.383} + \int_2^4 \frac{\log(s-1)}{s}\log\frac{5(5-s)}{s+1}ds\right)$$

$$\le 18.21899\frac{C(N)N}{\log^2 N}, \tag{4.8}$$

$$\Sigma_6 \ge 8(1 + o(1))\frac{C(N)N}{\log^2 N}\int_{\frac{1}{12}}^{\frac{1}{8.813}}\int_{t_1}^{\frac{1}{8.813}}\frac{\log(5 - 12(t_1 + t_2))}{t_1 t_2(1 - 2(t_1 + t_2))}dt_1 dt_2,$$

$$\Sigma_7 \ge 8(1 + o(1))\frac{C(N)N}{\log^2 N}\int_{\frac{1}{12}}^{\frac{1}{8.813}}\int_{\frac{1}{8.813}}^{\frac{1}{3}-t_1}\frac{\log(5 - 12(t_1 + t_2))}{t_1 t_2(1 - 2(t_1 + t_2))}dt_1 dt_2,$$

$$\Sigma_6 + \Sigma_7 \ge 2.6942\frac{C(N)N}{\log^2 N}. \tag{4.9}$$

2) Evaluation of Σ_{10}, Σ_{11}.

We have

$$\Sigma_{10} = \sum_{\substack{N^{\frac{1}{12}} \leq p_1 < p_4 < p_2 < p_3 < N^{\frac{1}{8.813}} \\ (p_1 p_2 p_3 p_4, N)=1}} \sum_{\substack{a \in \mathcal{A}, p_1 p_2 p_3 p_4 | a \\ (a, p_1^{-1} N P(p_4))=1}} 1 + O(N^{\frac{11}{12}})$$

$$= \sum_{\substack{N^{\frac{1}{12}} \leq p_1 < p_4 < p_2 < p_3 < N^{\frac{1}{8.813}} \\ (p_1 p_2 p_3 p_4, N)=1}} \sum_{\substack{p=N-p_1 p_2 p_3 p_4 n \\ 1 \leq n \leq \frac{N}{p_1 p_2 p_3 p_4}, (n, p_1^{-1} N P(p_4))=1}} 1 + O(N^{\frac{11}{12}})$$

$$= \Sigma'_{10} + O(N^{\frac{11}{12}}), \tag{4.10}$$

where

$$\Sigma'_{10} = \sum_{\substack{N^{\frac{1}{12}} \leq p_1 < p_4 < p_2 < N^{\frac{1}{8.813}} \\ (p_1 p_2 p_4, N)=1}} \sum_{\substack{1 \leq n \leq \frac{N}{p_1 p_2^2 p_4} \\ (n, p_1^{-1} N P(p_4))=1}} \sum_{\substack{p=N-(p_1 p_2 p_4 n) p_3 \\ p_2 < p_3 < \min\left(N^{\frac{1}{8.813}}, \frac{N}{n p_1 p_2 p_4}\right)}} 1. \tag{4.11}$$

Now we consider the set

$$\mathcal{E} = \left\{ e \,\middle|\, e = n p_1 p_2 p_4, N^{\frac{1}{12}} \leq p_1 < p_4 < p_2 < N^{\frac{1}{8.813}}, (p_1 p_2 p_4, N) = 1, \right.$$
$$\left. 1 \leq n \leq \frac{N}{p_1 p_2^2 p_4}, (n, p_1^{-1} N P(p_4)) = 1 \right\}.$$

By the definition of the set $\mathcal{E}$, it is easy to see that for every $e \in \mathcal{E}, p_1, p_2, p_4$ are determined by e uniquely. Let $p_2 = r(e)$, then we have

$$N^{\frac{1}{12}} < r(e) < N^{\frac{1}{8.813}}, \qquad er(e) < N.$$

Let

$$\mathcal{L} = \left\{ l \,\middle|\, l = N - e p_3, e \in \mathcal{E}, r(e) < p_3 < \min\left(N^{\frac{1}{8.813}}, \frac{N}{e}\right) \right\}.$$

We have

$$N^{\frac{1}{4}} < e < N^{\frac{11}{12}}, \qquad e \in \mathcal{E},$$
$$|\mathcal{E}| \ll N^{\frac{11}{12}}, \qquad \sum_{l \in \mathcal{L}, l \leq N^{\frac{1}{4}}} 1 \ll N^{\frac{11}{12}}.$$

Σ'_{10} does not exceeding the number of primes in $\mathcal{L}$, hence

$$\Sigma'_{10} \leq S(\mathcal{L}, z) + O(N^{\frac{11}{12}}), \qquad z \leq N^{\frac{1}{4}}. \tag{4.12}$$

By Lemma 1 we get

$$S(\mathcal{L}, D^{\frac{1}{2}}) \leq 8(1 + o(1))\frac{C(N)|\mathcal{L}|}{\log N} + R_1 + R_2, \qquad (4.13)$$

where

$$D = N^{\frac{1}{2}} \log^{-B} N \qquad (B = B(5) > 0),$$

$$R_1 = \sum_{d \leq D,(d,N)=1} \left| \sum_{\substack{e \in \mathcal{E} \\ (e,d)=1}} \left(\sum_{\substack{r(e)<p_3<\min\left(N^{\frac{1}{8.813}},\frac{N}{e}\right) \\ ep_3 \equiv N(d)}} 1 \right. \right.$$

$$\left. \left. - \frac{1}{\varphi(d)} \sum_{r(e)<p_3<\min\left(N^{\frac{1}{8.813}},\frac{N}{e}\right)} 1 \right) \right|,$$

$$R_2 = \sum_{d \leq D,(d,N)=1} \frac{1}{\varphi(d)} \sum_{\substack{e \in \mathcal{E} \\ (e,d)=1}} \sum_{r(e)<p_3<\min\left(N^{\frac{1}{8.813}},\frac{N}{e}\right)} 1.$$

Let

$$g(a) = \sum_{e=a,e \in \mathcal{E}} 1,$$

then

$$R_1 = \sum_{d \leq D} \left| \sum_{\substack{N^{\frac{1}{4}}<a<N^{\frac{11}{12}} \\ (a,d)=1}} g(a) \left(\sum_{\substack{r(e)<p_3<\min\left(N^{\frac{1}{8.813}},\frac{N}{a}\right) \\ ap_3 \equiv N(d)}} 1 \right. \right.$$

$$\left. \left. - \frac{1}{\varphi(d)} \sum_{r(e)<p_3<\min\left(N^{\frac{1}{8.813}},\frac{N}{a}\right)} 1 \right) \right|,$$

$$R_2 = \sum_{d \leq D} \frac{1}{\varphi(d)} \sum_{\substack{N^{\frac{1}{4}}<a<N^{\frac{11}{12}} \\ (a,d) \geq N^{\frac{1}{12}}}} \sum_{r(e)<p_3<\min\left(N^{\frac{1}{8.813}},\frac{N}{a}\right)} 1.$$

It is easy to show that

$$g(a) \leq 1.$$

Now

$$R_2 \ll \sum_{d \leq D} \frac{1}{\varphi(d)} \sum_{\substack{N^{\frac{1}{4}} < a < N^{\frac{11}{12}} \\ (a,d) \geq N^{\frac{1}{12}}}} \frac{N}{a}$$

$$\ll \sum_{d \leq D} \frac{1}{\varphi(d)} \sum_{\substack{m \mid d, m \geq N^{\frac{1}{12}}}} \sum_{\substack{a < N^{\frac{11}{12}} \\ (a,d)=m}} \frac{N}{a}$$

$$\ll N \log N \sum_{d \leq D} \frac{1}{\varphi(d)} \sum_{\substack{m \mid d, m \geq N^{\frac{1}{12}}}} \frac{1}{m}$$

$$\ll N \log N \sum_{N^{\frac{1}{12}} \leq m \leq D} \frac{1}{m\varphi(m)} \sum_{d \leq \frac{D}{m}} \frac{1}{\varphi(d)}$$

$$\ll N^{\frac{11}{12}} \log^2 N, \tag{4.14}$$

$$R_1 \leq R_3 + R_4 + R_5, \tag{4.15}$$

where

$$R_3 = \sum_{d \leq D, (d,N)=1} \left| \sum_{\substack{N^{\frac{1}{4}} < a < N^{\frac{7.813}{8.813}} \\ (a,d)=1}} g(a) \left(\sum_{\substack{p_3 < N^{\frac{1}{8.813}} \\ ap_3 \equiv N(d)}} 1 - \frac{1}{\varphi(d)} \sum_{p_3 < N^{\frac{1}{8.813}}} 1 \right) \right|,$$

$$R_4 = \sum_{d \leq D, (d,N)=1} \left| \sum_{\substack{N^{\frac{7.813}{8.813}} < a < N^{\frac{11}{12}} \\ (a,d)=1}} g(a) \left(\sum_{\substack{ap_3 < N \\ ap_3 \equiv N(d)}} 1 - \frac{1}{\varphi(d)} \sum_{ap_3 < N} 1 \right) \right|,$$

$$R_5 = \sum_{d \leq D, (d,N)=1} \left| \sum_{\substack{N^{\frac{1}{4}} < a < N^{\frac{11}{12}} \\ (a,d)=1}} g(a) \left(\sum_{\substack{p_3 < r(a) \\ ap_3 \equiv N(d)}} 1 - \frac{1}{\varphi(d)} \sum_{p_3 < r(a)} 1 \right) \right|.$$

By Lemma 3, Remark 1, Remark 2 we get

$$R_j \ll \frac{N}{\log^4 N}, \qquad j = 3, 4, 5. \tag{4.16}$$

Now by Lemma 4 we have

$$|\mathcal{L}| = \sum_{e \in \mathcal{E}} \sum_{r(e) < p_3 < \min\left(N^{\frac{1}{8.813}}, \frac{N}{e}\right)} 1$$

$$= \sum_{\substack{N^{\frac{1}{12}} \le p_1 < p_4 < p_2 < p_3 < N^{\frac{1}{8.813}} \\ (p_1 p_2 p_3 p_4, N) = 1}} \sum_{\substack{1 \le n \le \frac{N}{p_1 p_2 p_3 p_4} \\ (n, p_1^{-1} N P(p_4)) = 1}} 1 + O(N^{\frac{11}{12}})$$

$$< \frac{1}{1.7803}\left(1 + O\left(\frac{1}{\log N}\right)\right)$$

$$\times \sum_{N^{\frac{1}{12}} \le p_1 < p_4 < p_2 < p_3 < N^{\frac{1}{8.813}}} \frac{N}{p_1 p_2 p_3 p_4 \log p_4} + O(N^{\frac{11}{12}})$$

$$= \frac{N}{1.7803 \log N}(1 + o(1))$$

$$\times \int_{\frac{1}{12}}^{\frac{1}{8.813}} \frac{dt_1}{t_1} \int_{t_1}^{\frac{1}{8.813}} \frac{1}{t_2}\left(\frac{1}{t_1} - \frac{1}{t_2}\right) \log \frac{1}{8.813 t_2} dt_2. \qquad (4.17)$$

By (4.10)–(4.17) we get

$$\Sigma_{10} \le \frac{8}{1.7803}(1 + o(1))\frac{C(N)N}{\log^2 N}$$

$$\times \int_{\frac{1}{12}}^{\frac{1}{8.813}} \frac{dt_1}{t_1} \int_{t_1}^{\frac{1}{8.813}} \frac{1}{t_2}\left(\frac{1}{t_1} - \frac{1}{t_2}\right) \log \frac{1}{8.813 t_2} dt_2. \qquad (4.18)$$

By a similar method we get

$$\Sigma_{11} \le \frac{8}{1.7803}(1 + o(1))\frac{C(N)N}{\log^2 N}$$

$$\times \int_{\frac{1}{12}}^{\frac{1}{8.813}} \frac{dt_1}{t_1} \int_{t_1}^{\frac{1}{8.813}} \frac{1}{t_2}\left(\frac{1}{t_1} - \frac{1}{t_2}\right) \log \left(8.813\left(\frac{1}{3} - t_2\right)\right) dt_2.$$

$$\qquad (4.19)$$

By (4.18) and (4.19) we obtain

$$\Sigma_{10} + \Sigma_{11} \le \frac{8}{1.7803}(1 + o(1))\frac{C(N)N}{\log^2 N}$$

$$\times \int_{\frac{1}{12}}^{\frac{1}{8.813}} \frac{dt_1}{t_1} \int_{t_1}^{\frac{1}{8.813}} \frac{1}{t_2}\left(\frac{1}{t_1} - \frac{1}{t_2}\right) \log \left(\frac{1}{3 t_2} - 1\right) dt_2$$

$$\le 0.1768 \frac{C(N)N}{\log^2 N}. \qquad (4.20)$$

3) Evaluation of $\Sigma_2, \Sigma_3, \Sigma_8, \Sigma_9$.

We have

$$\Sigma_2 = \sum_{\substack{N^{\frac{1}{12}} \leq p_1 < N^{\frac{1}{3.047}} \leq p_2 < \left(\frac{N}{p_1}\right)^{\frac{1}{2}} \\ (p_1 p_2, N)=1}} \sum_{\substack{a \in \mathcal{A}, a = p_1 p_2 p_3 \\ p_2 < p_3, (p_3, N)=1}} 1$$

$$= \sum_{\substack{N^{\frac{1}{12}} \leq p_1 < N^{\frac{1}{3.047}} \leq p_2 < \left(\frac{N}{p_1}\right)^{\frac{1}{2}} \\ (p_1 p_2, N)=1}} \sum_{\substack{p = N - p_1 p_2 p_3 \\ p_2 < p_3 < \frac{N}{p_1 p_2}, (p_3, N)=1}} 1$$

$$= S. \tag{4.21}$$

Consider the sets

$$\mathcal{E} = \left\{ e \,\middle|\, e = p_1 p_2, N^{\frac{1}{12}} \leq p_1 < N^{\frac{1}{3.047}} \leq p_2 < \left(\frac{N}{p_1}\right)^{\frac{1}{2}}, (p_1 p_2, N) = 1 \right\},$$

$$\mathcal{L} = \{ l \,|\, l = N - ep, e \in \mathcal{E}, ep < N, (p, N) = 1 \}.$$

We have

$$|\mathcal{E}| \leq \sum_{N^{\frac{1}{12}} \leq p_1 < N^{\frac{1}{3.047}}} \left(\frac{N}{p_1}\right)^{\frac{1}{2}} < N^{\frac{2}{3}},$$

$$N^{\frac{1}{3}} < e \leq N^{\frac{2}{3}}, \qquad e \in \mathcal{E}.$$

The number of elements in the set $\mathcal{L}$ which are less than $N^{\frac{1}{3}}$ does not exceed $N^{\frac{2}{3}}$, S does not exceed the number of primes in $\mathcal{L}$. Therefore

$$S \leq S(\mathcal{L}, \mathcal{P}, z) + O(N^{\frac{2}{3}}), \qquad z \leq N^{\frac{1}{3}}. \tag{4.22}$$

Now we apply Lemma 1 to estimate the upper bound of $S(\mathcal{L}, \mathcal{P}, z)$. For the set $\mathcal{L}$,

$$X = \sum_{e \in \mathcal{E}} \text{Li}\left(\frac{N}{e}\right),$$

$$\omega(d) = \frac{d}{\varphi(d)}, \qquad \mu(d) \neq 0, \qquad (d, N) = 1,$$

$$D = \frac{N^{\frac{1}{2}}}{\log^B N}, \qquad B = B(5) > 0, \qquad z = D^{\frac{1}{2}}.$$

Then by Lemma 1 we get that

$$S(\mathcal{L}, \mathcal{P}, z) \leq 8(1 + o(1))C(N)\frac{X}{\log N} + R_1 + R_2, \qquad (4.23)$$

where

$$R_1 = \sum_{d \leq D, (d,N)=1} \left| \sum_{\substack{e \in \mathcal{E} \\ (e,d)=1}} \left(\sum_{\substack{ep < N \\ ep \equiv N(d)}} 1 - \frac{1}{\varphi(d)} \text{Li}\left(\frac{N}{e}\right) \right) \right|,$$

$$R_2 = \sum_{d \leq D, (d,N)=1} \frac{1}{\varphi(d)} \sum_{e \in \mathcal{E}, (e,d)>1} \text{Li}\left(\frac{N}{e}\right).$$

In view of $N^{\frac{1}{3}} \leq e < N^{\frac{2}{3}}$ for $e \in \mathcal{E}$, let

$$g(a) = \sum_{e=a, e \in \mathcal{E}} 1,$$

then we have

$$R_1 = \sum_{d \leq D, (d,N)=1} \left| \sum_{\substack{N^{\frac{1}{3}} < a < N^{\frac{2}{3}} \\ (a,d)=1}} g(a) \left(\sum_{\substack{ap < N \\ ap \equiv N(d)}} 1 - \frac{1}{\varphi(d)} \text{Li}\left(\frac{N}{a}\right) \right) \right|,$$

$$R_2 = \sum_{d \leq D, (d,N)=1} \frac{1}{\varphi(d)} \sum_{\substack{N^{\frac{1}{3}} < a < N^{\frac{2}{3}} \\ (a,d) \geq N^{\frac{1}{12}}}} g(a) \text{Li}\left(\frac{N}{a}\right).$$

It is easy to show that $g(a) \leq 1$. Now

$$R_2 \ll \sum_{d \leq D} \frac{1}{\varphi(d)} \sum_{\substack{N^{\frac{1}{3}} < a < N^{\frac{2}{3}} \\ (a,d) \geq N^{\frac{1}{12}}}} \frac{N}{a}$$

$$\ll \sum_{d \leq D} \frac{1}{\varphi(d)} \sum_{m|d, m \geq N^{\frac{1}{12}}} \sum_{\substack{a < N^{\frac{2}{3}} \\ (a,d)=m}} \frac{N}{a}$$

$$\ll N \log N \sum_{d \leq D} \frac{1}{\varphi(d)} \sum_{m|d, m \geq N^{\frac{1}{12}}} \frac{1}{m}$$

$$\ll N \log N \sum_{N^{\frac{1}{12}} \leq m \leq D} \frac{1}{m\varphi(m)} \sum_{d \leq \frac{D}{m}} \frac{1}{\varphi(d)}$$

$$\ll N^{\frac{11}{12}} \log^2 N. \tag{4.24}$$

By Lemma 3 we get that

$$R_1 \ll \frac{N}{\log^5 N}. \tag{4.25}$$

By the prime number theorem and integeration by parts we have

$$X = (1 + o(1)) \sum_{e \in \mathcal{E}} \mathrm{Li}\left(\frac{N}{e}\right)$$

$$= (1 + o(1))N \sum_{N^{\frac{1}{12}} \leq p_1 < N^{\frac{1}{3.047}} \leq p_2 < \left(\frac{N}{p_1}\right)^{\frac{1}{2}}} \frac{1}{p_1 p_2 \log \frac{N}{p_1 p_2}}$$

$$= (1 + o(1))N \int_{N^{\frac{1}{12}}}^{N^{\frac{1}{3.047}}} \frac{dt}{t \log t} \int_{N^{\frac{1}{3.047}}}^{\left(\frac{N}{t}\right)^{\frac{1}{2}}} \frac{du}{u \log u \log \frac{N}{ut}}$$

$$= (1 + o(1)) \frac{N}{\log N} \int_{2.047}^{11} \frac{\log\left(2.047 - \frac{3.047}{s+1}\right)}{s} ds. \tag{4.26}$$

By (4.21)—(4.26) we get that

$$\Sigma_2 \leq 8(1 + o(1)) \frac{C(N)N}{\log N} \sum_{N^{\frac{1}{12}} \leq p_1 < N^{\frac{1}{3.047}} \leq p_2 < \left(\frac{N}{p_1}\right)^{\frac{1}{2}}} \frac{1}{p_1 p_2 \log \frac{N}{p_1 p_2}}$$

$$\leq 8(1 + o(1)) \frac{C(N)N}{\log^2 N} \int_{2.047}^{11} \frac{\log\left(2.047 - \frac{3.047}{t+1}\right)}{t} dt$$

$$\leq 5.1577 \frac{C(N)N}{\log^2 N}. \tag{4.27}$$

By similar methods, we have

$$\Sigma_3 \leq 8(1 + o(1)) \frac{C(N)N}{\log N} \sum_{N^{\frac{1}{3.047}} \leq p_1 < p_2 < \left(\frac{N}{p_1}\right)^{\frac{1}{2}}} \frac{1}{p_1 p_2 \log \frac{N}{p_1 p_2}}$$

$$\leq 8(1 + o(1)) \frac{C(N)N}{\log^2 N} \int_2^{2.047} \frac{\log(t - 1)}{t} dt$$

$$\leq 0.0043 \frac{C(N)N}{\log^2 N}, \tag{4.28}$$

$$\Sigma_8 \leq 8(1 + o(1))\frac{C(N)N}{\log N} \sum_{N^{\frac{1}{8.813}} \leq p_1 < N^{\frac{1}{3.383}} \leq p_2 < \left(\frac{N}{p_1}\right)^{\frac{1}{2}}} \frac{1}{p_1 p_2 \log \frac{N}{p_1 p_2}}$$

$$\leq 8(1 + o(1))\frac{C(N)N}{\log^2 N} \int_{2.383}^{7.813} \frac{\log\left(2.383 - \frac{3.383}{t+1}\right)}{t} dt$$

$$\leq 5.1429 \frac{C(N)N}{\log^2 N}, \tag{4.29}$$

$$\Sigma_9 \leq 8(1 + o(1))\frac{C(N)N}{\log N} \sum_{N^{\frac{1}{3.383}} \leq p_1 < p_2 < \left(\frac{N}{p_1}\right)^{\frac{1}{2}}} \frac{1}{p_1 p_2 \log \frac{N}{p_1 p_2}}$$

$$\leq 8(1 + o(1))\frac{C(N)N}{\log^2 N} \int_2^{2.383} \frac{\log(t-1)}{t} dt$$

$$\leq 0.2329 \frac{C(N)N}{\log^2 N}. \tag{4.30}$$

4) Proof of the Theorem.

By (4.3),(4.6)–(4.9), (4.20), (4.27)–(4.30) we get

$$\begin{aligned} 2S(N) \geq &\left(2 \times 13.4735 - \frac{20.4456}{2} - \frac{5.1577}{2} - 0.0043 \right. \\ &- \frac{3.6569}{2} - \frac{18.21899}{2} + \frac{2.6942}{2} - \frac{5.1429}{2} \\ &\left. - 0.2329 - \frac{0.1768}{2} \right) \frac{C(N)N}{\log^2 N} \\ &> \frac{1.657 C(N)N}{\log^2 N}, \\ S(N) > &\frac{0.8285 C(N)N}{\log^2 N}. \end{aligned}$$

The Theorem is proved.

References

[1] Chen Jingrun, On the representation of a large even integer as the sum of a prime and the product of at most two primes, Kexue Tongbao **17** (1966), 385–386. (in Chinese)

[2] Chen Jingrun, On the representation of a large even integer as the sum of a prime and the product of at most two primes, Sci. Sin., **16** (1973), 157–176.

[3] Pan Chengdong, Ding Xiaqi, Wang Yuan, On the representation of a large even integer as the sum of a prime and an almost prime, Sci. Sin., **18** (1975), 599–610.

[4] H. Halberstam, H. E. Richert, Sieve Methods, Academic Press, London, 1974.

[5] H. Halberstam, A proof of Chen's Theorem, Asterisque, **24-25** (1975), 281–293.

[6] P. M. Ross, On Chen's theorem that each large even number has the form $p_1 + p_2$ or $p_1 + p_2 p_3$, J. London. Math. Soc. (2) **10** (1975), 500–506.

[7] Chen Jingrun, On the representation of a large even integer as the sum of a prime and the product of at most two primes, Sci. Sin., **21** (1978), 477–494. (in Chinese)

[8] H. Iwaniec, Rosser's sieve, Recent Progress in Analytic Number Theory II, 203–230, Academic Press, 1981.

[9] Pan Chengdong, Pan Chengbiao, Goldbach Conjecture, Science Press, Beijing, China, 1992.

[10] Chaohua Jia, Almost all short intervals containing prime numbers, Acta Arith. **76** (1996), 21–84.

ON A TWISTED POWER MEAN OF $L(1,\chi)$

Shigeki EGAMI
Faculty of Engineering, Toyama University,
Gofuku 3190, Toyama City, Toyama 930, Japan
megami@eng.toyama-u.ac.jp

Keywords: $L(1,\chi)$, Kloosterman sum

Abstract   The k-th power moment of $L(1,\chi)$ over the odd characters modulo p (a prime number) twisted by the argument of the Gaussian sum is estimated.

2000 Mathematics Subject Classification: 11M20.

Let p be a prime number and k be a natural number. For a Dirichlet character χ modulo p we denote by $L(s,\chi)$ the Dirichlet L-function for χ as usual. The power moment of $L(1,\chi)$ over odd characters were investigated after the results of Ankeny and Chowla [1], Walum [2] on mean square, and now we have general asymptotic formula by Zhang [5], [6]:

$$\sum_{\chi(-1)=-1} |L(1,\chi)|^{2k} = C_k p + O\left(\exp\left(\frac{2k \log p}{\log\log p}\right)\right)$$

and

$$\sum_{\chi(-1)=-1} L(1,\chi)^k = D_k p + O\left(\exp\left(\frac{k \log p}{\log\log p}\right)\right),$$

where C_k and D_k are certain positive constants depending only on k. We note that Zhang's results are actually given for any composite modulus and the constants C_k and D_k are given explicitly.

In this paper we report a curious "vanishing" phenomenon on the similar power moment twisted by the arguments of the Gaussian sum. Let $G(\chi)$ denotes the Gaussian sum i.e.

$$G(\chi) = \sum_{a=1}^{p-1} \chi(a) e_p(a),$$

121

C. Jia and K. Matsumoto (eds.), Analytic Number Theory, 121–125.
© 2002 *Kluwer Academic Publishers. Printed in the Netherlands.*

where $e_p(x) = e^{2\pi i \frac{x}{p}}$. Define

$$\theta_\chi = \frac{\overline{G(\chi)}}{\sqrt{p}}.$$

Note that the absolute value of θ_χ is 1. Then our result is the following:

Theorem.

$$\sum_{\chi(-1)=-1} (\theta_\chi L(1,\chi))^k = O(\sqrt{p}\log^k p). \tag{1}$$

The author expresses his sincere gratitude to Professor Kohji Matsumoto for valuable information and discussions on Zhang's works, and to the referee for several important comments for correction and improvement of the manuscript.

Hereafter we use the following conventions:

$$Z_p = \{1,\dots,p\}$$
$$Z_p^* = \{1,\dots,p-1\}.$$

Lemma 1.

$$\sum_{\chi(-1)=-1} (\theta_\chi L(1,\chi))^k = \pi^k i^k p^{-\frac{3}{2}k}\left((p-1)S - \left(\frac{p(p-1)}{2}\right)^k\right), \tag{2}$$

where

$$S = \sum_{\substack{(a_1,\dots,a_k)\in(Z_p^*)^k \\ a_1\cdots a_k \equiv 1 \pmod{p}}} a_1\cdots a_k.$$

Proof. The class number formula says if $\chi(-1)=-1$ (see e.g. [3] p.37, Theorem 4.9)

$$L(1,\chi) = \frac{\pi i G(\chi)}{p^2}h(\chi),$$

where $h(\chi) = \sum_{a=1}^{p-1}\overline{\chi(a)}a$. Then we have

$$\sum_{\chi(-1)=-1} (\theta_\chi L(1,\chi))^k = \sum_{\chi(-1)=-1}\left(\pi i p^{-\frac{3}{2}}h(\chi)\right)^k$$

$$= \pi^k i^k p^{-\frac{3}{2}}\left(\sum_{\chi \pmod{p}} h(\chi)^k - \left(\frac{p(p-1)}{2}\right)^k\right).$$

Here we have used the facts $h(\chi) = 0$ for even non-principal χ and $h(\chi) = \frac{p(p-1)}{2}$ for principal χ. The first term of the last expression is calculated as

$$\sum_{\chi \pmod p} h(\chi)^k = \sum_{\chi \pmod p} \sum_{a_1=1}^{p-1} \cdots \sum_{a_k=1}^{p-1} \overline{\chi(a_1) \cdots \chi(a_k)} a_1 \cdots a_k$$

$$= \sum_{(a_1,\ldots,a_k) \in (Z_p^{*k})} a_1 \cdots a_k \sum_{\chi \pmod p} \overline{\chi(a_1 \cdots a_k)}$$

$$= (p-1)S,$$

which shows the conclusion.   □

Lemma 2.

$$S = \sum_{(n_1,\ldots,n_k) \in Z_p^{*k}} K(n_1,\ldots,n_k)\Phi(n_1) \cdots \Phi(n_k), \qquad (3)$$

where

$$K(n_1,\ldots,n_k) = p^{-k} \sum_{\substack{(a_1,\ldots,a_k) \in Z_p^{*k} \\ a_1 \cdots a_k \equiv 1 \pmod p}} e_p(-a_1 n_1 - \cdots - a_k n_k)$$

and

$$\Phi(n) = \sum_{a=1}^{p-1} a e_p(-an).$$

Proof. Let $I(a_1,\ldots,a_k)$ denote the characteristic function of the set defined by $a_1 \cdots a_k \equiv 1 \pmod p$. Then by Fourier analysis on Z_p^k (considerd as an additive group) we have

$$I(a_1,\ldots,a_k) = \sum_{(n_1,\ldots,n_k) \in Z_p^k} K(n_1,\ldots,n_k)e_p(a_1 n_1 + \cdots + a_k n_k).$$

Hence

$$S = \sum_{(a_1,\ldots,a_k) \in Z_p^k} a_1 \cdots a_k I(a_1,\ldots,a_k)$$

$$= \sum_{(n_1,\ldots,n_k) \in Z_p^k} K(n_1,\ldots,n_k)$$

$$\times \sum_{a_1=1}^{p-1} \cdots \sum_{a_k=1}^{p-1} a_1 \cdots a_k e_p(-a_1 n_1 - \cdots - a_k n_k)$$

$$= \sum_{(n_1,\ldots,n_k)\in Z_p^{*k}} K(n_1,\ldots,n_k)\Phi(n_1)\cdots\Phi(n_k).$$

□

For $(n_1,\ldots,n_k) \in Z_p^k$ we denote by $r = r(n_1,\ldots,n_k)$ the number of zeros among the numbers $n_1,\ldots,n_k$.

Lemma 3.

$$K(n_1,\ldots,n_k) = \begin{cases} (p-1)^{r-1}(-1)^{k-r}p^{-k} & r \geq 1 \\ O(p^{-\frac{k+1}{2}}) & r = 0. \end{cases} \tag{4}$$

Proof. First we assume $r \geq 1$. Without loss of generality we may assume $(n_1,\ldots,n_k) = (n_1,\ldots,n_{k-r},0,\ldots,0)$, and all of $n_1,\ldots,n_{k-r}$ are not zero. Then we have

$$K(n_1,\ldots,n_k) = \frac{1}{p^k} \sum_{(a_1,\ldots,a_{k-r})\in Z_p^{*k-r}} e_p(-a_1 n_1 - \cdots - a_{k-r}n_{k-r})$$

$$\times \sum_{\substack{(a_{k-r+1},\ldots,a_k)\in Z_p^{*r} \\ a_{k-r+1}\cdots a_k \equiv (a_1\cdots a_{k-r})^{-1} \pmod{p}}} 1$$

$$= p^{-k}(p-1)^{r-1}(-1)^{k-r}.$$

The latter half is a direct consequence of the Deligne estimate of the hyper-Kloosterman sums (see [4]). □

Proof of Theorem. From **Lemma 2** and **Lemma 3** we have

$$S = \sum_{\nu=0}^{k} \sum_{r(n_1,\ldots,n_k)=\nu} K(n_1,\ldots,n_k)\Phi(n_1)\cdots\Phi(n_k)$$

$$= \sum_{\nu=1}^{k} \binom{k}{\nu} p^{-k}(-1)^{k-\nu}(p-1)^{\nu-1}\Phi(0)^{\nu} \left(\sum_{n=1}^{p-1}\Phi(n)\right)^{k-\nu}$$

$$+ \sum_{(n_1,\ldots,n_k)\in Z_p^{*k}} K(n_1,\ldots,n_k)\Phi(n_1)\cdots\Phi(n_k).$$

Since $\Phi(0) = \frac{p(p-1)}{2}$ and $\sum_{n=1}^{p-1}\Phi(n) = -\frac{p(p-1)}{2}$ (note that $\sum_{n=0}^{p-1}\Phi(n) = 0$) we observe the first term to be

$$\frac{(p-1)^{k-1}}{2^k}(p^k - 1).$$

While the second term can be estimated to $O(p^{\frac{3}{2}k-\frac{1}{2}} \log^k p)$ since

$$\sum_{n=1}^{p-1} |\Phi(n)| = O(p^2 \log p).$$

Thus **Theorem** follows immediately from **Lemma 1**.   □

References

[1] Ankeny, N. C., Chowla, S., The class number of cyclotomic field., Canad. J. Math., **3** (1951), 486–494.

[2] Walum, H., An exact formula for an average of L series, Illinois J. Math., **26** (1982), 1–3.

[3] Washington, L. C., Introduction to Cyclotomic Fields, Graduate Texts in Mathematics Vol. 83, Springer Verlag, 1982.

[4] Weinstein, L., The hyper-Kloosterman sum, Enseignement Math., **27** (1981), 29–40.

[5] Zhang, W., On the forth power moment of Dirichlet L-functions. (Chinese) Acta Math. Sinica **34** (1991), 138–142.

[6] Zhang, W., The kth power mean formula for L-functions. (Chinese) J. Math. (Wuhan) **12** (1992), 162–166.

ON THE PAIR CORRELATION OF THE ZEROS OF THE RIEMANN ZETA FUNCTION

Akio FUJII

Department of mathematics, Rikkyo University, Nishi-ikebukuro, Toshimaku, Tokyo, Japan

Keywords: Riemann zeta function, Dirichlet L-function, the pair correlation, zeros

Abstract   An estimate on the Montgomery's conjecture on the pair correlation of the zeros of the Riemann zeta function is given. An extension is given also to the zeros of Dirichlet L-functions .

2000 Mathematics Subject Classification: 11M26.

1.   INTRODUCTION

We shall give a remark concerning the Montgomery's conjecture on the pair correlation of the zeros of the Riemann zeta function $\zeta(s)$.

We start by recalling Montgomery's conjecture [10]. We shall state it in the following form, where γ and γ' run over the imaginary parts of the zeros of $\zeta(s)$.

Montgomery's pair correlation conjecture.
For all $T > T_0$ and for any $\alpha > 0$,

$$\sum_{\substack{0 < \gamma, \gamma' \leq T \\ 0 < \gamma - \gamma' \leq \frac{2\pi\alpha}{\log \frac{T}{2\pi}}}} 1 = \frac{T}{2\pi} \log T \cdot \left\{ \int_0^\alpha \left(1 - \left(\frac{\sin \pi t}{\pi t} \right)^2 \right) dt + o(1) \right\}.$$

In this article, we shall prove that the above conjecture is correct as far as the asymptotic behavior as $\alpha \to \infty$ is concerned. We [4] have already proved this under the Riemann Hypothesis. Here we shall give a proof without assuming any unproved hypothesis.

C. Jia and K. Matsumoto (eds.), Analytic Number Theory, 127–142.
© 2002 *Kluwer Academic Publishers. Printed in the Netherlands.*

An importnat point of the above conjecture is, as noticed by Dyson, that the density function

$$1 - \left(\frac{\sin \pi t}{\pi t}\right)^2$$

is exactly the density function of the pair correlation of the eigenvalues of Gaussian Unitary Ensembles. This observation has stimulated many researches since then. One of the latest works along this line can be seen in the works of Katz-Sarnak [9] and Connes [1].

We start by recalling the Riemann-von Mangoldt formula for the number $N(T)$ of the zeros $\beta + i\gamma$ of $\zeta(s)$ in $0 < \gamma \le T$ and $0 \le \beta \le 1$, the zeros on $\gamma = T$ being counted one half only. Let

$$S(T) = \frac{1}{\pi}\arg \zeta(\frac{1}{2} + iT) \quad \text{for} \quad T \ne \gamma,$$

where the argument is obtained by the continuous variation along the straight lines joining 2, $2 + iT$, and $\frac{1}{2} + iT$, starting with the value zero. When $T = \gamma$, we put

$$S(T) = \frac{1}{2}\big(S(T+0) + S(T-0)\big).$$

Then the well known Riemann-von Mangoldt formula (cf. p.212 of Titchmarsh [12]) states that

$$N(T) = \frac{1}{\pi}\vartheta(T) + 1 + S(T),$$

where $\vartheta(T)$ is the continuous function defined by

$$\vartheta(T) = \Im\big(\log \Gamma(\frac{1}{4} + \frac{iT}{2})\big) - \frac{1}{2}T \log \pi$$

with

$$\vartheta(0) = 0$$

and has the following expansion

$$\vartheta(T) = \frac{T}{2}\log \frac{T}{2\pi} - \frac{T}{2} - \frac{\pi}{8} + \frac{1}{48T} + \frac{7}{5760T^3} + \cdots$$

and we have

$$S(T) \ll \log T,$$

$\Gamma(s)$ being the Γ−function.

Now using the Riemann-von Mangoldt formula, we see that for any $\alpha > 0$ and for all $T > T_0$,

$$\sum_{\substack{0<\gamma,\gamma'\leq T \\ 0<\gamma-\gamma'\leq \frac{2\pi\alpha}{\log \frac{T}{2\pi}}}} 1 = \sum_{0<\gamma\leq T} \sum_{\gamma-\frac{2\pi\alpha}{\log \frac{T}{2\pi}}\leq\gamma'<\gamma} 1$$

$$= \frac{T}{2\pi}\log\frac{T}{2\pi}\int_0^\alpha dt - \sum_{0<\gamma\leq T} S(\gamma - \frac{2\pi\alpha}{\log\frac{T}{2\pi}}) + O(T).$$

Thus it is clear that the following is equivalent to Montgomery's pair correlation conjecture.

Conjecture. *For all $T > T_0$ and for any $\alpha > 0$, we have*

$$\sum_{0<\gamma\leq T} S(\gamma - \frac{2\pi\alpha}{\log\frac{T}{2\pi}}) = \frac{T}{2\pi}\log\frac{T}{2\pi}\left\{\int_0^\alpha \left(\frac{\sin \pi t}{\pi t}\right)^2 dt + o(1)\right\}.$$

The above argument has been noticed in pp. 242–243 of Fujii [5] with slightly more details. Thus it may be realized that to study the sum

$$\sum_{0<\gamma\leq T} S(\gamma - \frac{2\pi\alpha}{\log\frac{T}{2\pi}})$$

is very important. Under the assumption of the Riemann Hypothesis, we have shown in Fujii [4] that *for all $T > T_0$ and for any positive $\alpha \ll T^A$ with some positive constant A, we have*

$$\sum_{0<\gamma\leq T} S(\gamma - \frac{2\pi\alpha}{\log\frac{T}{2\pi}}) \ll T\log T.$$

The first purpose of the present article is to eliminate the Riemann Hypothesis from the above result and prove the following result.

Theorem I. *Suppose that $T > T_0$ and $0 \leq |a| \ll T$. Then we have*

$$\sum_{0<\gamma\leq T, \gamma-a>0} S(\gamma - a) \ll T\log T.$$

Thus we have the following consequence without assuming any unproved hypothesis.

Corollary I. *For all $T > T_0$ and for any $\alpha > 0$, we have*

$$\sum_{\substack{0<\gamma,\gamma'\leq T \\ 0<\gamma-\gamma'\leq \frac{2\pi\alpha}{\log \frac{T}{2\pi}}}} 1 = \frac{T}{2\pi}\log\frac{T}{2\pi}\cdot\left\{\int_0^\alpha dt + O(1)\right\}.$$

The second purpose of the present article is to consider a more general situation. Let χ and ψ be primitive Dirichlet characters with the conductor $d \geq 1$ and $k \geq 1$, respectively. Let $L(s,\chi)$ and $L(s,\psi)$ be the corresponding Dirichlet L-functions. Let $\gamma(\chi)$ (and $\gamma'(\psi)$) run over the imaginary parts of the zeros of $L(s,\chi)$ (and $L(s,\psi)$, respectively). One general problem is to study the following quantity.

$$\sum_{\substack{0<\gamma(\chi),\gamma'(\psi)\leq T \\ 0<\gamma(\chi)-\gamma'(\psi)\leq \frac{2\pi\alpha}{\log \frac{T}{2\pi}}}} 1,$$

where $T > T_0$ and α is any positive number. To state our conjecture and the results on this, we have to start by stating the following Riemann-von Mangoldt formula for $N(T,\chi)$, the number of the zeros of $L(s,\chi)$ in $0 \leq \Re(s) \leq 1$ and $0 \leq \Im(s) = t \leq T$, possible zeros with $t = 0$ or $t = T$ counting one half only. Then it is well-known (cf. p.283 of Selberg [11]) that for all $T > 0$, we have

$$N(T,\chi) = \frac{1}{2\pi}T\log\frac{dT}{2\pi e} - \frac{\chi(-1)}{8} + S(T,\chi) - S(0,\chi) + O\left(\frac{1}{T+1}\right),$$

where $S(T,\chi)$ is defined as in p.283 of Selberg [11].

Now if $\chi = \psi$, then the asymptotic behavior of the above sum must be similar to the one conjectured in Montgomery's conjecture. On the other hand, if $\chi \neq \psi$, then it may be natural to suppose that there may be no strong tendency to separate $\gamma(\chi)$ and $\gamma'(\psi)$. Thus we may have the following.

Conjecture. *Let χ and ψ be primitive Dirichlet characters with the conductor $d \geq 1$ and $k \geq 1$, respectively. Then for all $T > T_0$ and for any $\alpha > 0$, we have*

$$\sum_{\substack{0<\gamma(\chi),\gamma'(\psi)\leq T \\ 0<\gamma(\chi)-\gamma'(\psi)\leq \frac{2\pi\alpha}{\log \frac{T}{2\pi}}}} 1 = \frac{T}{2\pi}\log T\cdot\left\{\int_0^\alpha\left(1-\delta(\chi,\psi)\cdot\left(\frac{\sin \pi t}{\pi t}\right)^2\right)dt + o(1)\right\},$$

where we put

$$\delta(\chi, \psi) = \begin{cases} 1 & if \quad \chi = \psi \\ 0 & if \quad \chi \neq \psi. \end{cases}$$

This has been already stated in p.250 of Fujii [5].

. If we use the Riemann-von Mangoldt formula for $N(T, \chi)$ as stated above, we see that our conjecture is equivalent to the following.

Conjecture. *Let χ and ψ be primitive Dirichlet characters with the conductor $d \geq 1$ and $k \geq 1$, respectively. Then for all $T > T_0$ and for any $\alpha > 0$, we have*

$$\sum_{0 < \gamma(\chi) \leq T} \left(S(\gamma(\chi), \psi) - S(\gamma(\chi) - \frac{2\pi\alpha}{\log \frac{T}{2\pi}}, \psi) \right)$$

$$= \frac{T}{2\pi} \log \frac{T}{2\pi} \cdot \delta(\chi, \psi) \cdot \left\{ \int_0^\alpha \left(\frac{\sin \pi t}{\pi t} \right)^2 dt + o(1) \right\}.$$

In the present article, we shall prove the following result which is a generalization of Theorem I.

Theorem II. *Let χ and ψ be primitive Dirichlet characters with the conductor $d \geq 1$ and $k \geq 1$, respectively. Suppose that $T > T_0$ and $0 \leq |a| \ll T$. Then we have*

$$\sum_{0 < \gamma(\chi) \leq T, \gamma(\chi) - a > 0} S(\gamma(\chi) - a, \psi) \ll T \log T.$$

This implies, in particular, the following.

Corollary II. *Let χ and ψ be primitive Dirichlet characters with the conductor $d \geq 1$ and $k \geq 1$, respectively. Then for all $T > T_0$ and for any $\alpha > 0$, we have*

$$\sum_{\substack{0 < \gamma(\chi), \gamma'(\psi) \leq T \\ 0 < \gamma(\chi) - \gamma'(\psi) \leq \frac{2\pi\alpha}{\log \frac{T}{2\pi}}}} 1 = \frac{T}{2\pi} \log T \cdot \left\{ \int_0^\alpha 1 \, dt + O(1) \right\}.$$

We understand that our results do not reach to the point where either the factor

$$\left(\frac{\sin \pi t}{\pi t}\right)^2$$

or the factor

$$\delta(\chi, \psi)$$

plays an essential role.

Finally, we notice that Theorem I is announced in Fujii [7]. Here we shall give the details of the proof of Theorem II in the next section. We shall not trace the dependence on d or k , for simplicity.

2.   PROOF OF THEOREM II

We shall use the following explicit formula for $S(T, \chi)$ due to Selberg (cf. p.330–334 of Selberg [11]).

Lemma. *For* $2 \leq X \leq t^2$, $t \geq 2$, *we have*

$$S(t, \chi) = \frac{1}{\pi}\Im\left\{\sum_{p<X^3} \frac{\chi(p)}{p^{\frac{1}{2}+it}}\right\} + O(\Delta(t, X, \chi)),$$

where we put

$$\Delta_1(t, X, \chi) = \sum_{p<X^3} \frac{\Lambda(p) - \Lambda_X(p)}{\sqrt{p}\log p \cdot p^{it}} \chi(p),$$

$$\Delta_2(t, X, \chi) = \sum_{p<X^{\frac{3}{2}}} \frac{\Lambda_X(p^2)\chi(p^2)}{p\log p \cdot p^{2it}},$$

$$\Delta_3(t, X, \chi) = (\sigma_{\chi,t} - \frac{1}{2})\log dt,$$

$$\Delta_4(t, X, \chi) = (\sigma_{\chi,t} - \frac{1}{2})X^{(\sigma_{\chi,t}-\frac{1}{2})}\int_{\frac{1}{2}}^{\infty} X^{\frac{1}{2}-\sigma}\left|\sum_{p<X^3} \frac{\Lambda_X(p)\log(Xp)\chi(p)}{p^{\sigma+it}}\right| d\sigma,$$

$$\Delta(t, X, \chi) = |\Delta_1(t, X, \chi)| + |\Delta_2(t, X, \chi)| + \Delta_3(t, X, \chi) + \Delta_4(t, X, \chi),$$

$$\Lambda_X(n) = \begin{cases} \Lambda(n) & \text{for } 1 \leq n \leq X \\ \Lambda(n)\dfrac{(\log \frac{X^3}{n})^2 - 2(\log \frac{X^2}{n})^2}{2(\log X)^2} & \text{for } X \leq n \leq X^2 \\ \Lambda(n)\dfrac{(\log \frac{X^3}{n})^2}{2(\log X)^2} & \text{for } X^2 \leq n \leq X^3 \end{cases}$$

and

$$\sigma_{\chi,t} = \frac{1}{2} + 2 \max_{\varrho(\chi)} \left(|\beta(\chi) - \frac{1}{2}|, \frac{2}{\log X} \right),$$

$\varrho(\chi)$ *running here through all zeros* $\beta(\chi) + i\gamma(\chi)$ *of* $L(s, \chi)$ *for which*

$$|t - \gamma(\chi)| \le \frac{X^{3|\beta(\chi) - \frac{1}{2}|}}{\log X}.$$

We shall divide our proof into two cases.

We suppose first that $a > 0$.
We put $X = T^b$ with a sufficiently small positive b and $T_1 = \sqrt{X} + a$.
We may suppose that

$$T_1 = \sqrt{X} + a \le T.$$

In this case, we have first

$$\sum_{0 < \gamma(\chi) \le T,\, \gamma(\chi) - a > 0} S(\gamma(\chi) - a, \psi)$$

$$= \sum_{a < \gamma(\chi) \le T_1} S(\gamma(\chi) - a, \psi) + \sum_{T_1 < \gamma(\chi) \le T} S(\gamma(\chi) - a, \psi)$$

$$= S_1 + S_2, \quad \text{say.}$$

Using the estimate $S(\gamma(\chi) - a, \psi) \ll \log T$, we get, simply,

$$S_1 \ll \log T \sum_{a < \gamma(\chi) \le T_1} 1 \ll (T_1 - a) \log^2 T \ll \sqrt{X} \log^2 T \ll T \log T.$$

For S_2, we can use the above lemma and get

$$S_2 = \frac{1}{\pi} \Im \left\{ \sum_{T_1 < \gamma(\chi) \le T} \sum_{p < X^3} \frac{\psi(p)}{\sqrt{p} p^{i(\gamma(\chi) - a)}} \right\}$$

$$+ \sum_{T_1 < \gamma(\chi) \le T} O(\Delta(\gamma(\chi) - a, X, \psi))$$

$$= S_3 + S_4, \quad \text{say.}$$

Using the Riemann-von Mangoldt formula, we get first

$$S_3 = \frac{1}{\pi} \sum_{p < X^3} \frac{1}{\sqrt{p}} \Im \left\{ \int_{T_1}^{T} p^{-i(t-a)} \psi(p) \frac{1}{2\pi} \log \frac{dt}{2\pi} dt \right\}$$

$$+ \frac{1}{\pi} \sum_{p<X^3} \frac{1}{\sqrt{p}} \Im\left\{ \int_{T_1}^{T} p^{-i(t-a)} \psi(p) \, dS(t,\chi) \right\} + O\left(\sum_{p<X^3} \frac{1}{\sqrt{p}} \right)$$

$$= S_5 + S_6 + S_7, \text{ say.}$$

By the integration by parts, we get

$$S_5 \ll \sum_{p<X^3} \frac{1}{\sqrt{p}} \frac{\log T}{\log p} \ll \log T \frac{X^{\frac{3}{2}}}{\log^2 X} \ll T.$$

By the integration by parts, we get also

$$S_6 = \frac{1}{\pi} \sum_{p<X^3} \frac{1}{\sqrt{p}} \Im\left\{ \left[p^{-i(t-a)} \psi(p) S(t,\chi) \right]_{T_1}^{T} \right\}$$

$$- \frac{1}{\pi} \sum_{p<X^3} \Im\left\{ \frac{-i\log p}{\sqrt{p}} \int_{T_1}^{T} p^{-i(t-a)} \psi(p) S(t,\chi) \, dt \right\}$$

$$= S_8 - S_9, \text{ say.}$$

Since $S(T,\chi) \ll \log T$, we get immediately,

$$S_8 \ll \log T \frac{X^{\frac{3}{2}}}{\log X} \ll T.$$

To treat S_9, we apply the above lemma again. Since $\sqrt{X} \le T_1$, we have

$$S_9 = \frac{1}{\pi} \sum_{p<X^3} \Im\left\{ \frac{-i\log p}{\sqrt{p}} \int_{T_1}^{T} p^{-i(t-a)} \psi(p) \left(\frac{1}{\pi} \Im\left\{ \sum_{q<X^3} \frac{\chi(q)}{q^{\frac{1}{2}+it}} \right\} \right) dt \right\}$$

$$+ \frac{1}{\pi} \sum_{p<X^3} \Im\left\{ \frac{-i\log p}{\sqrt{p}} \int_{T_1}^{T} p^{-i(t-a)} \psi(p) O\left(\Delta(t,X,\chi) \right) dt \right\}$$

$$= S_{10} + S_{11}, \quad \text{say.}$$

We get, easily,

$$S_{10} \ll T \sum_{p<X^3} \frac{\log p}{p} + \sum_{p<X^3, q<X^3, p\neq q} \frac{\log q}{\sqrt{pq} \, |\log \frac{p}{q}|} \ll T \log X \ll T \log T.$$

By Cauchy's inequality, we get

$$S_{11} \ll \left(\int_{T_1}^{T} \left| \frac{1}{\pi} \sum_{p<X^3} \frac{\log p}{\sqrt{p}} p^{i(t-a)} \psi(p) \right|^2 dt \right)^{\frac{1}{2}} \cdot \left(\int_{T_1}^{T} |\Delta(t,X,\chi)|^2 \, dt \right)^{\frac{1}{2}}.$$

Since

$$\int_{T_1}^{T} \left| \frac{1}{\pi} \sum_{p<X^3} \frac{\log p}{\sqrt{p}} p^{i(t-a)} \psi(p) \right|^2 dt \ll \sum_{p<X^3} \frac{\log^2 p}{p} (T+p) \ll T \log^2 X,$$

we get

$$S_{11} \ll \sqrt{T} \log X \cdot \left(\sum_{j=1}^{4} \int_{T_1}^{T} |\Delta_j(t, X, \chi)|^2 \, dt \right)^{\frac{1}{2}}$$

$$= \sqrt{T} \log X \cdot (S_{12} + S_{13} + S_{14} + S_{15})^{\frac{1}{2}}, \quad \text{say.}$$

Using Fujii [2] (as in pp.246–251 of Selberg [11]), we get

$$S_{12} + S_{13} + S_{14} + S_{15} \ll T,$$

$$S_{11} \ll T \log T,$$

$$S_9 \ll T \log T$$

and

$$S_6 \ll T \log T.$$

Since

$$S_7 \ll \frac{X^{\frac{3}{2}}}{\log X} \ll T,$$

we get

$$S_3 \ll T \log T.$$

We shall next estimate S_4.

$$S_4 \ll \sum_{j=1}^{4} \sum_{T_1 < \gamma(\chi) \le T} \Delta_j(\gamma(\chi) - a, X, \psi)$$

$$\ll \sum_{j=1}^{3} \sqrt{T \log T} \sqrt{\sum_{T_1 < \gamma(\chi) \le T} |\Delta_j(\gamma(\chi) - a, X, \psi)|^2}$$

$$+ \sum_{T_1 < \gamma(\chi) \le T} \Delta_4(\gamma(\chi) - a, X, \psi).$$

For $j = 1, 2$, we use the following inequality

$$\sum_{T_1 < \gamma(\chi) \leq T} |\Delta_j(\gamma(\chi) - a, X, \psi)|^2$$

$$\ll \log T \int_{T_1}^T |\Delta_j(t - a, X, \psi)|^2 \, dt$$

$$+ \log T \max_{T_1 \leq t \leq T} |\Delta_j(t - a, X, \psi)|^2$$

$$+ \left| \int_{T_1}^T \left(|\Delta_j(t - a, X, \psi)|^2 \right)' S(t, \chi) \, dt \right|$$

$$= U_j(1) + U_j(2) + U_j(3), \quad \text{say.}$$

We see easily that

$$U_1(1) \ll \sum_{p < X^3} \frac{|\Lambda(p) - \Lambda_X(p)|^2}{p \log^2 p} (T + p)$$

$$\ll T \sum_{X < p < X^3} \frac{1}{p} + \frac{X^3}{\log^3 X} \ll T,$$

$$U_2(1) \ll \sum_{p < X^{\frac{3}{2}}} \frac{\Lambda_X^2(p^2)}{p^2 \log^2 p} (T + p) \ll T,$$

and

$$U_1(2), \ U_2(2) \ll T.$$

To treat $U_j(3)$ for $j = 1, 2$, we use the above lemma again. Then we have

$$U_j(3) \ll \left| \int_{T_1}^T \Delta_j(t - a, X, \psi)' \overline{\Delta}_j(t - a, X, \psi) \sum_{r < X^3} \frac{1}{\sqrt{r}} r^{it} \chi(r) dt \right|$$

$$+ \left| \int_{T_1}^T \Delta_j(t - a, X, \psi)' \overline{\Delta}_j(t - a, X, \psi) \sum_{r < X^3} \frac{1}{\sqrt{r}} r^{-it} \chi(r) dt \right|$$

$$+ \left(\int_{T_1}^T |\Delta_j(t - a, X, \psi)'|^4 \, dt \right)^{\frac{1}{4}} \left(\int_{T_1}^T |\Delta_j(t - a, X, \psi)|^4 dt \right)^{\frac{1}{4}}$$

$$\cdot \sqrt{\int_{T_1}^T |\Delta(t, X, \chi)|^2 \, dt}$$

$$= |U_j(4)| + |U_j(5)| + U_j^{\frac{1}{4}}(6) U_j^{\frac{1}{4}}(7) \sqrt{U(8)}, \quad \text{say.}$$

$$U_1(4) = -\int_{T_1}^{T} \sum_{p<X^3} \frac{\left(\Lambda(p) - \Lambda_X(p)\right) \log p \cdot \psi(p)}{\sqrt{p} \log p \cdot p^{i(t-a)}}$$

$$\cdot \sum_{q<X^3} \frac{\left(\Lambda(q) - \Lambda_X(q)\right)\overline{\psi}(q)}{\sqrt{q} \log q} q^{i(t-a)} \sum_{r<X^3} \frac{1}{\sqrt{r}} r^{it} \chi(r) \, dt$$

$$\ll \sum_{p,q<X^3, r<X^3} \frac{\left(\Lambda(p) - \Lambda_X(p)\right) \log p \left(\Lambda(q) - \Lambda_X(q)\right)}{\sqrt{pq} \log p \cdot |\log \frac{qr}{p}|}$$

$$\ll T.$$

Similarly, we get

$$U_j(4), \; U_j(5) \ll T.$$

As above, we get

$$U(8) \ll T\left(\frac{\log T}{\log X}\right)^2 \ll T.$$

Next, we have

$$U_1(6)$$

$$= \int_{T_1}^{T} \left| \sum_{X<p,q<X^3} \frac{\left(\Lambda(p) - \Lambda_X(p)\right)\psi(p)\left(\Lambda(q) - \Lambda_X(q)\right)\psi(q)}{\sqrt{pq}(pq)^{ia}} (pq)^{-it} \right|^2 dt$$

$$\ll \int_{T_1}^{T} \left| \sum_{X^2<n<X^6} \frac{a(n)}{\sqrt{n}} n^{-it} \right|^2 dt$$

$$\ll \sum_{X^2<n<X^6} \frac{|a(n)|^2}{n}(T + n)$$

$$\ll T \log^4 X \left(\sum_{X^2<p<X^6} \frac{1}{p} \right)^2$$

$$\ll T \log^4 X,$$

where

$$a(n) = \sum_{pq=n, X<p,q<X^3} \frac{\left(\Lambda(p) - \Lambda_X(p)\right)\psi(p)\left(\Lambda(q) - \Lambda_X(q)\right)\psi(q)}{(pq)^{ia}}$$

$$\ll \log^2 X \sum_{pq=n, X<p,q<X^3} 1.$$

In the same manner, we get

$$U_1(7) \ll T.$$

Similarly, we get

$$U_2(6) \ll T \log^4 X$$

and

$$U_2(7) \ll T.$$

Thus we get

$$U_j^{\frac{1}{4}}(6) U_j^{\frac{1}{4}}(7) \sqrt{U(8)} \ll T \log T.$$

Consequently,

$$U_j(3) \ll T \log T.$$

Hence, we get

$$\sum_{T_1 < \gamma(\chi) \leq T} |\Delta_1(\gamma(\chi) - a, X, \psi)|^2 \ll T \log T$$

and

$$\sum_{T_1 < \gamma(\chi) \leq T} |\Delta_2(\gamma(\chi) - a, X, \psi)|^2 \ll T \log T.$$

Extending the argument in pp. 69–71 of Fujii [6], we get

$$\sum_{T_1 < \gamma(\chi) \leq T} |\Delta_3(\gamma(\chi) - a, X, \psi)|^2 \ll T \log T \left(\frac{\log T}{\log X}\right)^2.$$

Finally, extending again the argument in pp. 69–71 of Fujii [6], we get first

$$\sum_{T_1 < \gamma(\chi) \leq T} \Delta_4(\gamma(\chi) - a, X, \psi)$$

$$\ll \left(\sum_{T_1 < \gamma(\chi) \leq T} \left(\sigma_{\psi,(\gamma(\chi)-a)} - \frac{1}{2}\right)^2 X^{2(\sigma_{\psi,(\gamma(\chi)-a)} - \frac{1}{2})} \right)^{\frac{1}{2}}$$

$$\cdot \left(\int_{\frac{1}{2}}^{\infty} X^{\frac{1}{2}-\sigma} d\sigma \int_{\frac{1}{2}}^{\infty} X^{\frac{1}{2}-\sigma} \sum_{T_1 < \gamma(\chi) \leq T} \left| \sum_{p < X^3} \frac{\Lambda_X(p) \log(Xp) \psi(p)}{p^{\sigma+i(\gamma(\chi)-a)}} \right|^2 d\sigma \right)^{\frac{1}{2}}$$

$$\ll \sqrt{\frac{T \log T}{\log^2 X} \left(\frac{1}{\log X} \cdot \int_{\frac{1}{2}}^{\infty} X^{\frac{1}{2}-\sigma} \sum_{T_1 < \gamma(\chi) \leq T} |\Xi(\gamma(\chi))|^2 d\sigma \right)^{\frac{1}{2}}},$$

where we put

$$\Xi(t) = \sum_{p < X^3} \frac{g(p)}{p^\sigma} p^{-it}$$

with

$$g(p) = \frac{\Lambda_X(p)\log(Xp)\psi(p)}{p^{ia}} \ll \log X \cdot \log p.$$

Hence we are reduced to the estimate of the sum

$$\sum_{T_1 < \gamma(\chi) \leq T} |\Xi(\gamma(\chi))|^2.$$

Now as before, we have first

$$\sum_{T_1 < \gamma(\chi) \leq T} |\Xi(\gamma(\chi))|^2 \ll \log T \int_{T_1}^{T} |\Xi(t)|^2 \, dt$$

$$+ X^3 \log T \cdot \log^2 X + \int_{T_1}^{T} \left(|\Xi(t)|^2\right)' S(t, \chi) \, dt.$$

The first term in the right hand side is

$$\ll \log T \sum_{p < X^3} \frac{g^2(p)}{p^{2\sigma}} (T + p)$$

$$\ll T \log T \cdot \log^2 X \sum_{p < X^3} \frac{\log^2 p}{p} + X^3 \log^3 X \cdot \log T$$

$$\ll T \log T \cdot \log^4 X.$$

The last integral is

$$\ll \left| \int_{T_1}^{T} \sum_{p < X^3} \frac{g(p)\log p}{p^\sigma} p^{-it} \sum_{q < X^3} \frac{g(q)}{q^\sigma} q^{it} \sum_{r < X^3} \frac{1}{r^{\frac{1}{2}}} r^{it} \chi(r) dt \right|$$

$$+ \left| \int_{T_1}^{T} \sum_{p < X^3} \frac{g(p)\log p}{p^\sigma} p^{-it} \sum_{q < X^3} \frac{g(q)}{q^\sigma} q^{it} \sum_{r < X^3} \frac{1}{r^{\frac{1}{2}}} r^{-it} \chi(r) dt \right|$$

$$+ \left(\int_{T_1}^{T} |\Xi(t)|^4 dt \right)^{\frac{1}{4}} \cdot \left(\int_{T_1}^{T} |\Xi'(t)|^4 dt \right)^{\frac{1}{4}} \sqrt{\int_{T_1}^{T} |\Delta(t, X, \chi)|^2 dt}$$

$$= |W_1| + |W_2| + (W_3)^{\frac{1}{4}} \cdot (W_4)^{\frac{1}{4}} \sqrt{W_5}.$$

We see easily that

$$W_1 \ll \sum_{p<X^3} \frac{g(p)\log p}{p^\sigma} \sum_{q<X^3} \frac{g(q)}{q^\sigma} \sum_{r<X^3} \frac{1}{r^{\frac{1}{2}}} \int_{T_1}^T \left(\frac{qr}{p}\right)^{it} dt$$

$$\ll \sum_{p<X^3,q<X^3,r<X^3} \frac{g(p)\log p \cdot g(q)}{(pqr)^{\frac{1}{2}}|\log \frac{qr}{p}|}$$

$$\ll T.$$

W_2 has the same upper bound.

$$W_4 \ll \int_{T_1}^T \left| \sum_{p,q<X^3} \frac{g(p)\log p \cdot g(q)\log q}{(pq)^\sigma}(pq)^{-it} \right|^2 dt$$

$$\ll \sum_{n<X^6} \frac{|b(n)|^2}{n}(T+n),$$

where

$$b(n) = \sum_{\substack{n=pq \\ p,q<X^3}} g(p)\log p \cdot g(q)\log q$$

$$\ll \log^2 X \sum_{\substack{n=pq \\ p,q<X^3}} \log^2 p \cdot \log^2 q.$$

Now since we have

$$\sum_{n<X^6} \frac{|b(n)|^2}{n} \ll \log^4 X \sum_{n<X^6} \frac{\left(\sum_{\substack{n=pq \\ p,q<X^3}} \log^2 p \cdot \log^2 q\right)^2}{n}$$

$$\ll \log^4 X \sum_{p,q<X^3} \frac{\log^4 p \cdot \log^4 q}{pq}$$

$$\ll \log^4 X \cdot \left(\sum_{p<X^3} \frac{\log^4 p}{p} \right)^2$$

$$\ll \log^{12} X,$$

we get

$$W_4 \ll T\log^{12} X.$$

Similarly, we get

$$W_3 \ll T\log^8 X.$$

Since
$$W_5 \ll T,$$

we get
$$\int_{T_1}^{T} \left(|\Xi(t)|^2\right)' S(t,\chi)\, dt \ll T \log T \log^4 X.$$

Consequently, we get

$$\sum_{T_1 < \gamma(\chi) \le T} \Delta_4(\gamma(\chi) - a, X, \psi)$$

$$\ll \sqrt{\frac{T \log T}{\log^2 X}} \left(\frac{1}{\log X} \cdot \int_{\frac{1}{2}}^{\infty} X^{\frac{1}{2} - \sigma} T \log T \cdot \log^4 X\, d\sigma\right)^{\frac{1}{2}}$$

$$\ll \sqrt{\frac{T \log T}{\log^2 X}} (T \log T \cdot \log^2 X)^{\frac{1}{2}}$$

$$\ll T \log T.$$

Hence, we get
$$S_4 \ll T \log T$$

and
$$\sum_{a < \gamma(\chi) \le T} S(\gamma(\chi) - a, \psi) \ll T \log T.$$

We suppose next that
$$a = -a' \le 0.$$

We put $T_1 = \max(\sqrt{X} - a', 0)$. We may suppose that $T_1 \le T$. We suppose first that
$$0 < a' < \sqrt{X}.$$

In this case, we have first

$$\sum_{0 < \gamma(\chi) \le T} S(\gamma(\chi) - a) = \sum_{0 < \gamma(\chi) \le T} S(\gamma(\chi) + a')$$

$$= \sum_{0 < \gamma(\chi) \le T_1} S(\gamma(\chi) + a') + \sum_{T_1 < \gamma(\chi) \le T} S(\gamma(\chi) + a').$$

We can modify the proof given above and get our conclusion as stated in Theorem II.

When
$$a' \geq \sqrt{X},$$

then $T_1 = 0$ and we can follow the same argument as above and get our conclusion as stated in Theorem II.

Thus in any case, we have the same conclusion as stated in Theorem II.

References

[1] A. Connes, Trace formula on the adele class space and Weil positivity, Current developments in mathematics, 1997, International Press, 1999, 5–64.

[2] A. Fujii, On the zeros of Dirichlet L-functions I, Trans. A. M. S. **196**, (1974), 225–235.

[3] A. Fujii, On the distribution of the zeros of the Riemann zeta function in short intervals, Proc. of Japan Academy **66**, (1990), 75–79.

[4] A. Fujii, On the gaps between the consecutive zeros of the Riemann zeta function, Proc. of Japan Academy **66**, (1990), 97–100.

[5] A. Fujii, Some observations concerning the distribution of the zeros of the zeta functions I, Advanced Studies in Pure Math. **21**, (1992), 237–280.

[6] A. Fujii, An additive theory of the zeros of the Riemann zeta function, Comment. Math. Univ. Sancti Pauli, **45**, (1996), 49–116.

[7] A. Fujii, Explicit formulas and oscillations, Emerging applications of number theory, IMA Vol. **109**, Springer, 1999, 219–267.

[8] A. Fujii and A. Ivić, On the distribution of the sums of the zeros of the Riemann zeta function, Comment. Math. Univ. Sancti Pauli, **49**, (2000), 43–60.

[9] N. Katz and P. Sarnak, Random matrices, Frobenius eigenvalues, and monodromy, AMS Colloq. Publ., Vol. **45**, 1999.

[10] H. L. Montgomery, The pair correlation of the zeros of the zeta function, Proc. Symp. Pure Math. **24**, (1973), 181–193.

[11] A. Selberg, Collected Works, Vol. **I**, 1989, Springer.

[12] E. C. Titchmarsh, The theory of the Riemann zeta function (2nd ed. rev. by D. R. Heath-Brown), Oxford Univ. Press, 1951 (1988).

DISCREPANCY OF
SOME SPECIAL SEQUENCES

Kazuo GOTO

Faculty of Education and Regional Sciences, Tottori University, Tottori-shi, 680-0945, Japan

goto@mail.fed.tottori-u.ac.jp

Yukio OHKUBO

Faculty of Economics, The International University of Kagoshima, Shimofukumoto-cho, Kagoshima-shi, 891-0191, Japan

ohkubo@eco.iuk.ac.jp

Keywords: Discrepancy, uniform distribution mod 1, irrational number, finite approximation type

Abstract In this paper, we show that the discrepancy of some special sequence $(f(n))_{n=1}^{N}$ satisfies $D_N(f(n)) \ll N^{-\frac{1}{\eta+1/2}+\varepsilon}$, where ε is any positive number and twice differentiable function $f(x)$ satisfies that $(f'(x) - \alpha)f''(x) < 0$ for $x \geq 1$ and $f'(x) = \alpha + O(|f''(x)|^{1/2})$ for an irrational number α of finite type η. We show further that if α is an irrational number of constant type, then the discrepancy of the sequence $(f(n))_{n=1}^{N}$ satisfies $D_N(f(n)) \ll N^{-2/3} \log N$. We extend the results much more by van der Corput's inequality.

1991 Mathematics Subject Classification: Primary 11K38; Secondary 11K06.

1.   INTRODUCTION AND RESULTS

For a real number x, let $\{x\}$ denote the fractional part of x. A sequence (x_n), $n = 1, 2, \ldots$, of real numbers is said to be uniformly distributed modulo 1 if

$$\lim_{N \to \infty} \frac{1}{N} \sum_{n=1}^{N} \chi_{[a,b)}(x_n) = b - a$$

143

C. Jia and K. Matsumoto (eds.), Analytic Number Theory, 143–155.
© 2002 *Kluwer Academic Publishers. Printed in the Netherlands.*

for every pair a, b of real numbers with $0 \le a < b \le 1$, where $\chi_{[a,b)}(x)$ is the characteristic function of $[a,b) + \mathbb{Z}$, that is, $\chi_{[a,b)}(x) = 1$ for $a \le \{x\} < b$ and $\chi_{[a,b)}(x) = 0$ otherwise. It is known that the sequence (x_n) is uniformly distributed modulo 1 if and only if $\lim_{N\to\infty} D_N(x_n) = 0$, where

$$D_N(x_n) = \sup_{0 \le a < b \le 1} \left| \frac{1}{N} \sum_{n=1}^{N} \chi_{[a,b)}(x_n) - (b-a) \right|$$

is the discrepancy of the sequence (x_n) (see [2, p.88]).

Let α be an irrational number. Let ψ be a non-decreasing positive function that is defined at least for all positive integers. We shall say that α is of type $< \psi$ if $h\|h\alpha\| > 1/\psi(h)$ holds for each positive integer h, where $\|x\| = \min\{\{x\}, 1 - \{x\}\}$ for $x \in \mathbb{R}$. If ψ is a constant function, then α of type $< \psi$ is called of constant type (see [2, p.121, Definition 3.3]). Let η be a positive real number. The irrational number α is said to be of finite type η if η is infimum of all real numbers τ for which there exists a positive constant $c = c(\tau, \alpha)$ such that α is of type $< \psi$, where $\psi(q) = cq^{\tau-1}$ (see [2, p.121, Definition 3.4 and Lemma 3.1]).

By Dirichlet's theorem, if α is of finite type η, then $\eta \ge 1$ holds, and if α is of constant type, then α is of finite type $\eta = 1$.

In [4], Tichy and Turnwald investigated the distribution behaviour modulo 1 of the sequence $(\alpha n + \beta \log n)$, where α and β are real numbers with $\beta \ne 0$. They showed that for any $\varepsilon > 0$

$$D_N(\alpha n + \beta \log n) \ll_{\alpha,\beta,\varepsilon} N^{-\frac{1}{\eta+1}+\varepsilon}, \tag{1.1}$$

provided that α is the irrational number of finite type η and $\beta \ne 0$.

In [3], Ohkubo improved the estimate (1.1) as follows: for any $\varepsilon > 0$

$$D_N(\alpha n + \beta \log n) \ll_{\beta,\varepsilon} N^{-\frac{1}{\eta+1/2}+\varepsilon} \tag{1.2}$$

holds whenever α is the irrational number of finite type η and $\beta \ne 0$.

The following theorem includes as a special case the above result (1.2).

Theorem 1. *Let $f(x)$ be a twice differentiable function defined for $x \ge 1$. Suppose that there exists an irrational number α of finite type η such that for $x \ge 1$ either*

$$f'(x) > \alpha, \ f''(x) < 0 \quad \text{or} \quad f'(x) < \alpha, \ f''(x) > 0$$

and $f'(x) = \alpha + O(|f''(x)|^{1/2})$. Then for any $\varepsilon > 0$

$$D_N(f(n)) \ll N^{-\frac{1}{\eta+1/2}+\varepsilon}.$$

Remark 1. *The following statement was shown by van der Corput : If* $f(x)$, $x \geq 1$, *is differentiable for sufficiently large* x *and* $\lim_{x\to\infty} f'(x) = \alpha$ *(irrational), then the sequence* $(f(n))$ *is uniformly distributed* mod 1 *(see* [2, p.28, Theorem 3.3 *and* p.31, Exercise 3.5]*). If the function* $f(x)$ *in Theorem 1 also satisfies the condition* $\lim_{x\to\infty} f''(x) = 0$, *then* $\lim_{x\to\infty} f'(x) = \alpha$. *Therefore, Theorem 1 gives a quantitative aspect of van der Corput's result.*

For the irrational number α of constant type, we obtain the following:

Theorem 2. *Let* α *be an irrational of constant type and let* $f(x)$ *be as in Theorem 1. Then*

$$D_N(f(n)) \ll N^{-2/3} \log N.$$

Example 1. $f(x) = \alpha x + \beta \log\log x$ *and* $f(x) = \alpha x + \beta \log x$ *satisfy the conditions of Theorem 1 and 2.*

Furthermore, we extend Theorem 1 by van der Corput's inequality (Lemma 5).

Theorem 3. *Let* q *be a non-negative integer and let* $Q = 2^q$. *Suppose that* f *is* $q + 2$ *times continuously differentiable on* $[1, \infty)$. *Suppose also that there exists an irrational number* α *of finite type* η *such that for* $x \geq 1$ *either*

$$f^{(q+1)}(x) > \alpha, \quad f^{(q+2)}(x) < 0 \quad \text{or} \quad f^{(q+1)}(x) < \alpha, \quad f^{(q+2)}(x) > 0$$

and $f^{(q+1)}(x) = \alpha + O(|f^{(q+2)}(x)|^{1/2})$. *Then for any* $\varepsilon > 0$

$$D_N(f(n)) \ll N^{-\frac{1}{2Q^2 + 2(\eta-1)Q - \eta + 1/2} + \varepsilon}.$$

Example 2. *Let* q *be a non-negative integer, let* $f(x) = \alpha_0 x^{q+1} + \alpha_1 x^q + \cdots + \alpha_q x + \beta x^q \log x$, *where* $\alpha_i \in \mathbb{R}$ $(i = 0, \ldots, q)$, $0 < \beta \in \mathbb{R}$ *and* α_0 *is the irrational number of finite type* η. *Then* $f(x)$ *satisfies the conditions of Theorem 3:*

$$f^{(q+1)}(x) = (q+1)!\,\alpha_0 + \beta c x^{-1} \quad \text{for some } c > 0,$$
$$f^{(q+1)}(x) > (q+1)!\,\alpha_0 \quad \text{for } x \geq 1,$$
$$f^{(q+2)}(x) = -\beta c x^{-2} < 0 \quad \text{for } x \geq 1,$$
$$f^{(q+1)}(x) = (q+1)!\,\alpha_0 + O\left(|f^{(q+2)}(x)|^{1/2}\right).$$

Theorem 4. *Let* α *be an irrational number of constant type and let* q, Q, *and* $f(x)$ *be as in Theorem 3. Then*

$$D_N(f(n)) \ll N^{-\frac{1}{2Q^2 - 1/2}} \log N.$$

2.   SOME LEMMAS

First, Erdös-Turán's inequality is stated (see [2, p.114]).

Lemma 1 (Erdös-Turán's inequality). *For any finite sequence* (x_n) *of real numbers and any positive integer* m, *it follows*

$$D_N(x_n) \ll \frac{1}{m} + \sum_{h=1}^{m} \frac{1}{h} \left| \frac{1}{N} \sum_{n=1}^{N} e^{2\pi i h x_n} \right|.$$

Subsequently, two known results are stated (see [5, p.74, Lemma 4.7] for the first one and [6, p.226, Lemma 10.5] for the second one).

Lemma 2 (van der Corput). *Let* a *and* b *be real numbers with* $a < b$. *Let* $f(x)$ *be a real-valued function with a continuous and steadily decreasing* $f'(x)$ *in* (a, b), *and let* $f'(b) = \alpha, f'(a) = \beta$. *Then*

$$\sum_{a<n\leq b} e^{2\pi i f(n)} = \sum_{\alpha-\eta<\nu<\beta+\eta} \int_a^b e^{2\pi i (f(x)-\nu x)} dx + O\left(\log(\beta - \alpha + 2)\right),$$

where η *is any positive constant less than* 1.

Lemma 3 (Salem). *Let* a *and* b *be real numbers with* $a < b$. *Let* $r(x)$ *be a positive decreasing and differentiable function. Suppose that* $f(x)$ *is a real-valued function such that* $f(x) \in C^2[a, b]$, $f''(x)$ *is of constant sign and* $r'(x)/f''(x)$ *is monotone for* $a \leq x \leq b$. *Then*

$$\left| \int_a^b r(x) e^{2\pi i f(x)} dx \right| \leq 8 \max_{a\leq x\leq b} \left(\frac{r(x)}{|f''(x)|^{1/2}} \right) + \max_{a\leq x\leq b} \left| \frac{r'(x)}{f''(x)} \right|.$$

We need the following inequality.

Lemma 4. *Let* q *be a non-negative integer, and let* $Q = 2^q$. *If* α *be an irrational number of type* $< \psi$, *then for any positive integer* m

$$\sum_{h=1}^{m} h^{1/(2Q)-1} \|h\alpha\|^{-1/Q}$$

$$\ll \psi(2m)^{1/Q} m^{1/(2Q)} g_0(m) + \sum_{h=1}^{m} \psi(2h)^{1/Q} h^{1/(2Q)-1} g_0(h),$$

where $g_0(m) = \log m$ *if* $q = 0$, *and* $g_0(m) = 1$ *if* $q \geq 1$.

Proof. The proof is almost as in the proof of [2, p.123, Lemma 3.3].

By Abel's summation formula, we have

$$\sum_{h=1}^{m} h^{1/(2Q)-1} \|h\alpha\|^{-1/Q}$$

$$\leq \left(1 - \frac{1}{2Q}\right) \sum_{h=1}^{m} h^{1/(2Q)-2} s_h + (m+1)^{1/(2Q)-1} s_m, \qquad (2.1)$$

where $s_h = \sum_{j=1}^{h} \|j\alpha\|^{-1/Q}$. For $0 \leq p < q \leq h$, we obtain

$$\|q\alpha \pm p\alpha\| \geq \frac{1}{(q \pm p)\psi(q \pm p)} \geq \frac{1}{2h\psi(2h)}.$$

It follows that

$$\left| \|q\alpha\| - \|p\alpha\| \right| \geq \frac{1}{2h\psi(2h)} \qquad \text{for} \quad 0 \leq p < q \leq h. \qquad (2.2)$$

From (2.2), it follows that in each of the intervals

$$\left[\frac{j}{2h\psi(2h)}, \frac{j+1}{2h\psi(2h)} \right), \; j = 0, \ldots, h,$$

there exists at most one number of the form $\|j\alpha\|$, $1 \leq j \leq h$, with no such number lying in the first interval. Therefore, we have

$$s_h = \sum_{j=1}^{h} \|j\alpha\|^{-1/Q} \leq \sum_{j=1}^{h} \left(\frac{2h\psi(2h)}{j} \right)^{1/Q} \ll (2h\psi(2h))^{1/Q} g_1(h),$$

where $g_1(h) = \log h$ if $q = 0$, and $g_1(h) = h^{1-1/Q}$ if $q \geq 1$. Using this with (2.1), we have

$$\sum_{h=1}^{m} h^{1/(2Q)-1} \|h\alpha\|^{-1/Q}$$

$$\ll \sum_{h=1}^{m} h^{3/(2Q)-2} \psi(2h)^{1/Q} g_1(h) + m^{3/(2Q)-1} \psi(2m)^{1/Q} g_1(m).$$

This completes the proof. $\qquad \qquad \qquad \square$

We quote the following lemma from [1, p.14, Lemma 2.7].

Lemma 5 (van der Corput's inequality). *Let q be a positive integer, and let $Q = 2^q$. If $0 < H \le b - a, H_q = H, H_{i-1} = H_i^{1/2}$ ($i = q, q - 1, \ldots, 2$) then*

$$\left| \sum_{a < n \le b} e^{2\pi i f(n)} \right|^Q \ll \frac{(b-a)^Q}{H} + \frac{(b-a)^{Q-1}}{H_1 \cdots H_q} \sum_{1 \le h_1 \le H_1} \cdots \sum_{1 \le h_q \le H_q} |S_q(h)|,$$

where $f(x) \in C^q[a, b]$, $h = (h_1, \ldots, h_q)$, $S_q(h) = \sum_{n \in I(h)} e^{2\pi i f_q(n; h)}$, $f_q(n; h) = \int_0^1 \cdots \int_0^1 \frac{\partial^q}{\partial t_1 \cdots \partial t_q} f(n + h \cdot t) dt_1 \cdots dt_q$, $t = (t_1, t_2, \ldots, t_q)$, $h \cdot t$ is the inner product, and $I(h) = (a, b - h_1 - h_2 - \cdots - h_q]$.

3.   PROOF OF THEOREMS

Proof of Theorem 1. Let h be a positive integer. Applying Lemma 2, we get

$$\left| \sum_{n=1}^N e^{2\pi i h f(n)} \right| \ll \sum_{A - 1/2 < \nu < B + 1/2} \left| \int_1^N e^{2\pi i (h f(x) - \nu x)} dx \right| + \log(B - A + 2),$$

where $A = h f'(N)$ and $B = h f'(1)$. We set $g(x) = h(f(x) - \alpha x)$. Using integration by parts, we have

$$\int_1^N e^{2\pi i (h f(x) - \nu x)} dx = \int_1^N e^{2\pi i ((h\alpha - \nu)x + g(x))} dx$$

$$= \int_1^N e^{2\pi i (h\alpha - \nu)x} e^{2\pi i g(x)} dx$$

$$= \left[\frac{e^{2\pi i (h\alpha - \nu)x}}{2\pi i (h\alpha - \nu)} e^{2\pi i g(x)} \right]_1^N - \frac{1}{h\alpha - \nu} \int_1^N g'(x) e^{2\pi i ((h\alpha - \nu)x + g(x))} dx.$$

Hence,

$$\int_1^N e^{2\pi i (h f(x) - \nu x)} dx$$

$$\ll \frac{1}{|h\alpha - \nu|} + \frac{1}{|h\alpha - \nu|} \left| \int_1^N g'(x) e^{2\pi i ((h\alpha - \nu)x + g(x))} dx \right|.$$

We suppose that

$$f'(x) > \alpha \text{ and } f''(x) < 0 \text{ for } x \ge 1.$$

From Lemma 3 and the hypothesis, it follows that

$$\left| \int_1^N g'(x) e^{2\pi i ((h\alpha - \nu)x + g(x))} dx \right| \ll h^{1/2} \max_{1 \le x \le N} \left(\frac{f'(x) - \alpha}{|f''(x)|^{1/2}} \right) + 1 \ll h^{1/2}.$$

Hence we have

$$\left| \sum_{n=1}^{N} e^{2\pi i h f(n)} \right| \ll h^{1/2} \sum_{A-1/2 < \nu < B+1/2} \frac{1}{|h\alpha - \nu|} + \log(B - A + 2) \quad (3.1)$$

$$\ll h^{1/2} \left(\frac{1}{\|h\alpha\|} + \int_{\|h\alpha\|}^{h(f'(1)-\alpha)+1/2} \frac{1}{x} dx \right)$$

$$+ \log(h(f'(1) - f'(N)) + 2)$$

$$\ll h^{1/2} \left(\frac{1}{\|h\alpha\|} + \log\left(h(f'(1) - \alpha) + 2 \right) \right).$$

Since α is of type η, for any $\delta > 0$, α is of type $< \psi$ with $\psi(q) = cq^{\eta-1+\delta/2}$ for some $c > 0$. Then, Lemma 4 with $q = 0$ implies

$$\sum_{h=1}^{m} \frac{1}{h^{1/2}\|h\alpha\|} \ll m^{\eta-1/2+\delta}. \quad (3.2)$$

Applying Lemma 1, by (3.1) and (3.2), we obtain

$$D_N(f(n)) \ll \frac{1}{m} + \frac{1}{N} \left(\sum_{h=1}^{m} \frac{1}{h^{1/2}\|h\alpha\|} + \sum_{h=1}^{m} \frac{\log(h(f'(1) - \alpha) + 2)}{h^{1/2}} \right)$$

$$\ll \frac{1}{m} + \frac{1}{N} \left(m^{\eta-1/2+\delta} + m^{1/2} \log m \right)$$

$$\ll \frac{1}{m} + \frac{1}{N} m^{\eta-1/2+\delta}, \quad (3.3)$$

for any $\delta > 0$.

Choosing $m = \left[N^{\frac{1}{\eta+1/2}} \right]$, we have

$$D_N(f(n)) \ll N^{-\frac{1}{\eta+1/2}} + N^{-\frac{1}{\eta+1/2}+\frac{\delta}{\eta+1/2}} \ll N^{-\frac{1}{\eta+1/2}+\varepsilon}.$$

In the case $f'(x) < \alpha$, $f''(x) > 0$ for $x \geq 1$, the proof runs along the same lines as above. $\qquad\square$

Proof of Theorem 2. The proof runs the same lines as in the proof of Theorem 1. Since α is of constant type, by Lemma 4 with $q = 0$ and $\psi(x) = c$, we have

$$\sum_{h=1}^{m} \frac{1}{h^{1/2}\|h\alpha\|} \ll m^{1/2} \log m.$$

Then from the first inequality of (3.3)

$$D_N(f(n)) \ll \frac{1}{m} + \frac{1}{N} m^{1/2} \log m$$

for any positive integer m.

Choosing $m = \left[N^{2/3}\right]$, we have

$$D_N(f(n)) \ll N^{-2/3} \log N,$$

which completes the proof.   □

Proof of Theorem 3. The case $q = 0$ coincides with Theorem 1, and so, we may suppose $q \geq 1$. Let $1 \leq H \leq N$. Set $H_q = H$, $H_{i-1} = H_i^{1/2}$ $(i = q, q-1, \ldots, 2)$. Applying Lemma 5, for any positive integer k we have

$$\left|\sum_{n=1}^{N} e^{2\pi i k f(n)}\right|^Q \ll \frac{N^Q}{H} + \frac{N^{Q-1}}{H_1 \cdots H_q} \sum_{1 \leq h_1 \leq H_1} \cdots \sum_{1 \leq h_q \leq H_q} |S_q(k, h)|, \quad (3.4)$$

where $h = (h_1, \ldots, h_q)$, $S_q(k; h) = \sum_{n \in I(h)} e^{2\pi i k f_q(n;h)}$, $I(h) = (0, \ N - h_1 - h_2 - \cdots - h_q]$ and $f_q(x;h) = \int_0^1 \cdots \int_0^1 \frac{\partial^q}{\partial t_1 \cdots \partial t_q} f(x + h \cdot t) dt_1 \cdots dt_q = h_1 \cdots h_q \int_0^1 \cdots \int_0^1 f^{(q)}(x + h \cdot t) dt_1 \cdots dt_q$ with $t = (t_1, t_2, \ldots, t_q)$.

Suppose that $f^{(q+2)}(x) < 0$ for $x \geq 1$. Since $\int_0^1 \cdots \int_0^1 f^{(q+2)}(x + h \cdot t) dt_1 \cdots dt_q < 0$, $\frac{d}{dx} f_q(x;h)$ is steadily decreasing as x increases. Hence, by Lemma 2,

$$S_q(k; h) = \sum_{n=1}^{N-h_1-\cdots-h_q} e^{2\pi i k f_q(n;h)} \tag{3.5}$$

$$\ll \sum_{A-1/2<\nu<B+1/2} \left|\int_1^{N-h_1-\cdots-h_q} e^{2\pi i (k f_q(x;h)-\nu x)} dx\right| + \log(B - A + 2),$$

where $A = k h_1 \cdots h_q \int_0^1 \cdots \int_0^1 f^{(q+1)}(N - h_1 - \cdots - h_q + h \cdot t) dt_1 \cdots dt_q$, $B = k h_1 \cdots h_q \int_0^1 \cdots \int_0^1 f^{(q+1)}(1 + h \cdot t) dt_1 \cdots dt_q$.

Using integration by parts, we have

$$\int_1^{N-h_1-\cdots-h_q} e^{2\pi i (k f_q(x;h)-\nu x)} dx \tag{3.6}$$

$$= \int_1^{N-h_1-\cdots-h_q} e^{2\pi i (k h_1 \cdots h_q \alpha - \nu)x} e^{2\pi i g(x)} dx$$

$$\ll \frac{1}{|k h_1 \cdots h_q \alpha - \nu|} +$$

$$\frac{1}{|k h_1 \cdots h_q \alpha - \nu|} \left|\int_1^{N-h_1-\cdots h_q} g'(x) e^{2\pi i ((k h_1 \cdots h_q \alpha - \nu)x + g(x))} dx\right|,$$

where we set $g(x) = kh_1 \cdots h_q(\int_0^1 \cdots \int_0^1 f^{(q)}(x + \boldsymbol{h} \cdot \boldsymbol{t})dt_1 \cdots dt_q - \alpha x)$.

If $f^{(q+1)}(x) > \alpha$ for $x \geq 1$, then $\int_0^1 \cdots \int_0^1 f^{(q+1)}(x + \boldsymbol{h} \cdot \boldsymbol{t})dt_1 \cdots dt_q > \alpha$ $(x \geq 1)$. It follows from Lemma 3 that

$$\int_1^{N-h_1-\cdots-h_q} g'(x)e^{2\pi i((kh_1\cdots h_q\alpha-\nu)x+g(x))}dx$$

$$\ll (kh_1 \cdots h_q)^{\frac{1}{2}} \times$$

$$\max_{1\leq x\leq N-h_1-\cdots-h_q} \left(\frac{\int_0^1 \cdots \int_0^1 f^{(q+1)}(x + \boldsymbol{h} \cdot \boldsymbol{t})dt_1 \cdots dt_q - \alpha}{\left| \int_0^1 \cdots \int_0^1 f^{(q+2)}(x + \boldsymbol{h} \cdot \boldsymbol{t})dt_1 \cdots dt_q \right|^{\frac{1}{2}}} \right) + 1.$$

By the hypothesis $f^{(q+1)}(x) = \alpha + O(|f^{(q+2)}(x)|^{1/2})$ and Hölder's inequality

$$\int_0^1 \cdots \int_0^1 f^{(q+1)}(x + \boldsymbol{h} \cdot \boldsymbol{t})dt_1 \cdots dt_q - \alpha$$

$$= O\left(\left| \int_0^1 \cdots \int_0^1 f^{(q+2)}(x + \boldsymbol{h} \cdot \boldsymbol{t})dt_1 \cdots dt_q \right|^{\frac{1}{2}} \right).$$

Therefore, we have

$$\int_1^{N-h_1-\cdots-h_q} g'(x)e^{2\pi i((kh_1\cdots h_q\alpha-\nu)x+g(x))}dx \ll (kh_1 \cdots h_q)^{\frac{1}{2}}. \tag{3.7}$$

By combining (3.6) and (3.7), we get

$$\int_1^{N-h_1-\cdots-h_q} e^{2\pi i(kf_q(x;h)-\nu x)}dx \ll \frac{(kh_1 \cdots h_q)^{\frac{1}{2}}}{|kh_1 \cdots h_q\alpha - \nu|}. \tag{3.8}$$

Applying (3.8) to the right-hand side of (3.5), we have

$$S_q(k; \boldsymbol{h}) \ll \sum_{A-1/2<\nu<B+1/2} \frac{(kh_1 \cdots h_q)^{\frac{1}{2}}}{|kh_1 \cdots h_q\alpha - \nu|} + \log(B - A + 2)$$

$$\ll (kh_1 \cdots h_q)^{\frac{1}{2}} \left(\frac{1}{\|kh_1 \cdots h_q\alpha\|} + \int_{\|kh_1\cdots h_q\alpha\|}^{kh_1\cdots h_qc(h)+\frac{1}{2}} \frac{1}{x}dx \right)$$

$$+ \log(B - A + 2)$$

$$\ll (kh_1 \cdots h_q)^{\frac{1}{2}} \left(\frac{1}{\|kh_1 \cdots h_q\alpha\|} + \log(kh_1 \cdots h_qc(h) + \frac{1}{2}) \right)$$

$$+ \log(B - A + 2),$$

where we set $c(\boldsymbol{h}) = \int_0^1 \cdots \int_0^1 f^{(q+1)}(1 + \boldsymbol{h} \cdot \boldsymbol{t})dt_1 \cdots dt_q - \alpha$. Since $f^{(q+1)}(x)$ is non-increasing for $x \geq 1$, we have $0 < c(\boldsymbol{h}) \leq f^{(q+1)}(1) - \alpha$. By the same reason, we have

$$\log(B - A + 2) \ll \log(kh_1 \cdots h_q(f^{(q+1)}(1) - \alpha) + 2).$$

Therefore,

$$S_q(k; \boldsymbol{h}) \ll (kh_1 \cdots h_q)^{\frac{1}{2}} \times$$
$$\left(\frac{1}{\|kh_1 \cdots h_q \alpha\|} + \log(kh_1 \cdots h_q(f^{(q+1)}(1) - \alpha) + 2) \right). \quad (3.9)$$

Applying (3.9) to the right-hand side of (3.4), we have

$$\left| \sum_{n=1}^N e^{2\pi i k f(n)} \right|^Q$$
$$\ll \frac{N^Q}{H} + \frac{N^{Q-1}}{H_1 \cdots H_q} \sum_{1 \leq h_1 \leq H_1} \cdots \sum_{1 \leq h_q \leq H_q} (kh_1 \cdots h_q)^{\frac{1}{2}} \times$$
$$\left(\frac{1}{\|kh_1 \cdots h_q \alpha\|} + \log(kh_1 \cdots h_q(f^{(q+1)}(1) - \alpha) + 2) \right)$$
$$\ll \frac{N^Q}{H} + \frac{N^{Q-1}k^{\frac{1}{2}}}{H_1 \cdots H_q} \sum_{1 \leq h_1 \leq H_1} \cdots \sum_{1 \leq h_q \leq H_q} \frac{(h_1 \cdots h_q)^{\frac{1}{2}}}{\|kh_1 \cdots h_q \alpha\|} + \frac{N^{Q-1}k^{\frac{1}{2}}}{H_1 \cdots H_q} \times$$
$$\sum_{1 \leq h_1 \leq H_1} \cdots \sum_{1 \leq h_q \leq H_q} (h_1 \cdots h_q)^{\frac{1}{2}} \log(kh_1 \cdots h_q(f^{(q+1)}(1) - \alpha) + 2).$$

Thus, for any positive integer m, we obtain

$$\sum_{k=1}^m \frac{1}{k} \left| \sum_{n=1}^N e^{2\pi i k f(n)} \right|$$
$$\ll \sum_{k=1}^m \frac{1}{k} \left(\frac{N}{H^{\frac{1}{Q}}} + \frac{N^{1-\frac{1}{Q}}k^{\frac{1}{2Q}}}{(H_1 \cdots H_q)^{\frac{1}{Q}}} \sum_{1 \leq h_1 \leq H_1} \cdots \sum_{1 \leq h_q \leq H_q} \frac{(h_1 \cdots h_q)^{\frac{1}{2Q}}}{\|kh_1 \cdots h_q \alpha\|^{\frac{1}{Q}}} + \right.$$
$$\frac{N^{1-\frac{1}{Q}}k^{\frac{1}{2Q}}}{(H_1 \cdots H_q)^{\frac{1}{Q}}} \sum_{1 \leq h_1 \leq H_1} \cdots \sum_{1 \leq h_q \leq H_q} (h_1 \cdots h_q)^{\frac{1}{2Q}} \times$$
$$\left. \left(\log \left(kh_1 \cdots h_q(f^{(q+1)}(1) - \alpha) + 2 \right) \right)^{\frac{1}{Q}} \right). \quad (3.10)$$

Since α is of finite type η, for any $\varepsilon > 0$ there exists a positive constant $c = c(\varepsilon, \alpha)$ such that $k\|k\alpha\| \geq \dfrac{1}{ck^{\eta-1+\varepsilon}}$ for all positive integers k. Thus, for all positive integers $h_1, \ldots, h_q$

$$k\|k(h_1 \cdots h_q\alpha)\| = k\|(kh_1 \cdots h_q)\alpha\| \geq \frac{1}{c(h_1 \cdots h_q)^{\eta+\varepsilon}k^{\eta-1+\varepsilon}}$$

holds for all positive integers k, and so, $h_1 \cdots h_q\alpha$ is of type $< \psi$ with $\psi(x) = c(h_1 \cdots h_q)^{\eta+\varepsilon}x^{\eta-1+\varepsilon}$. Hence, by Lemma 4, we get

$$\sum_{k=1}^{m} k^{\frac{1}{2Q}-1}\|kh_1 \cdots h_q\alpha\|^{-\frac{1}{Q}}$$

$$\ll (h_1 \cdots h_q)^{\frac{\eta+\varepsilon}{Q}} m^{\frac{\eta-1+\varepsilon}{Q}} m^{\frac{1}{2Q}} + \sum_{k=1}^{m}(h_1 \cdots h_q)^{\frac{\eta+\varepsilon}{Q}} k^{\frac{\eta-1+\varepsilon}{Q}} k^{\frac{1}{2Q}-1}$$

$$\ll (h_1 \cdots h_q)^{\frac{\eta+\varepsilon}{Q}} m^{\frac{2\eta-1+2\varepsilon}{2Q}}.$$

Therefore,

$$\frac{N^{1-\frac{1}{Q}}}{(H_1 \cdots H_q)^{\frac{1}{Q}}} \sum_{k=1}^{m} k^{\frac{1}{2Q}-1} \sum_{1\leq h_1\leq H_1} \cdots \sum_{1\leq h_q\leq H_q} \frac{(h_1 \cdots h_q)^{\frac{1}{2Q}}}{\|kh_1 \cdots h_q\alpha\|^{\frac{1}{Q}}}$$

$$\ll \frac{N^{1-\frac{1}{Q}}}{(H_1 \cdots H_q)^{\frac{1}{Q}}} \sum_{1\leq h_1\leq H_1} \cdots \sum_{1\leq h_q\leq H_q} h_1^{\frac{1}{2Q}} \cdots h_q^{\frac{1}{2Q}} \times$$

$$\sum_{k=1}^{m} k^{\frac{1}{2Q}-1}\|kh_1 \cdots h_q\alpha\|^{-\frac{1}{Q}}$$

$$\ll \frac{N^{1-\frac{1}{Q}}}{(H_1 \cdots H_q)^{\frac{1}{Q}}} \sum_{1\leq h_1\leq H_1} \cdots \sum_{1\leq h_q\leq H_q} h_1^{\frac{1}{2Q}} \cdots h_q^{\frac{1}{2Q}} h_1^{\frac{\eta+\varepsilon}{Q}} \cdots h_q^{\frac{\eta+\varepsilon}{Q}} m^{\frac{2\eta-1+2\varepsilon}{2Q}}$$

$$= N^{1-\frac{1}{Q}} m^{\frac{2\eta-1+2\varepsilon}{2Q}} H^{(2+\frac{2(\eta+\varepsilon)-1}{Q})(1-\frac{1}{Q})}, \tag{3.11}$$

where we used $H_1 \cdots H_q = H^{2(1-1/Q)}$ in the last step.

On the other hand, since

$$\sum_{k=1}^{m} k^{\frac{1}{2Q}-1}(\log(kh_1 \cdots h_q + 2))^{\frac{1}{Q}} \ll (\log(mh_1 \cdots h_q + 2))^{\frac{1}{Q}} m^{\frac{1}{2Q}},$$

we have

$$\sum_{k=1}^{m} k^{\frac{1}{2Q}-1} \sum_{1\leq h_1\leq H_1} \cdots \sum_{1\leq h_q\leq H_q} (h_1 \cdots h_q)^{\frac{1}{2Q}} (\log(kh_1 \cdots h_q + 2))^{\frac{1}{Q}}$$

$$\ll m^{\frac{1}{2Q}} (H_1 \cdots H_q)^{1+\frac{1}{2Q}} (\log(mH_1 \cdots H_q + 2))^{\frac{1}{Q}}. \tag{3.12}$$

Combining (3.10), (3.11), and (3.12), we obtain

$$\sum_{k=1}^{m} \frac{1}{k} \left| \sum_{n=1}^{N} e^{2\pi i h f(n)} \right|$$

$$\ll \frac{N}{H^{\frac{1}{Q}}} \log m + N^{1-\frac{1}{Q}} m^{\frac{2\eta-1+2\varepsilon}{2Q}} H^{(2+\frac{2(\eta+\varepsilon)-1}{Q})(1-\frac{1}{Q})}$$

$$+ \frac{N^{1-\frac{1}{Q}} m^{\frac{1}{2Q}}}{(H_1 \cdots H_q)^{\frac{1}{Q}}} (H_1 \cdots H_q)^{1+\frac{1}{2Q}} (\log(mH_1 \cdots H_q + 2))^{\frac{1}{Q}}$$

$$\ll \frac{N}{H^{\frac{1}{Q}}} \log m + N^{1-\frac{1}{Q}} m^{\frac{2\eta-1+2\varepsilon}{2Q}} H^{(2+\frac{2(\eta+\varepsilon)-1}{Q})(1-\frac{1}{Q})}$$

$$+ N^{1-\frac{1}{Q}} m^{\frac{1}{2Q}} H^{2(1-\frac{1}{Q})(1-\frac{1}{2Q})} (\log(mH^2 + 2))^{\frac{1}{Q}}.$$

Since $\log(mH^2 + 2) \ll (mH^2)^{\varepsilon/2}$ and $\eta \ge 1$, we get

$$\sum_{k=1}^{m} \frac{1}{k} \left| \sum_{n=1}^{N} e^{2\pi i k f(n)} \right|$$

$$\ll N(\log m) H^{-\frac{1}{Q}} + N^{1-\frac{1}{Q}} m^{\frac{2\eta-1+2\varepsilon}{2Q}} H^{(2+\frac{2(\eta+\varepsilon)-1}{Q})(1-\frac{1}{Q})}$$

$$+ N^{1-\frac{1}{Q}} m^{\frac{1+\varepsilon}{2Q}} H^{2(1-\frac{1}{Q})(1-\frac{1}{2Q})+\frac{\varepsilon}{Q}}$$

$$\ll N(\log m) H^{-\frac{1}{Q}} + N^{1-\frac{1}{Q}} m^{\frac{2\eta-1+2\varepsilon}{2Q}} H^{(2+\frac{2\eta-1}{Q}+\frac{2\varepsilon}{Q})(1-\frac{1}{Q})}$$

$$= L(H), \text{ say.}$$

If

$$N \ge N^{1-\frac{1}{Q}} m^{\frac{2\eta-1+2\varepsilon}{2Q}}, \tag{3.13}$$

then the solution $H(= H_0)$ of the equation

$$N H^{-\frac{1}{Q}} = N^{1-\frac{1}{Q}} m^{\frac{2\eta-1+2\varepsilon}{2Q}} H^{(2+\frac{2\eta-1}{Q}+\frac{2\varepsilon}{Q})(1-\frac{1}{Q})}$$

satisfies $1 \le H_0 \le N$ with $H_0 = N^{uQ} m^{-vQ}$, where

$$u = \frac{1}{2Q^2 + 2(\eta - 1 + \varepsilon)Q - (2\eta - 1 + 2\varepsilon)},$$

$$v = \frac{2\eta - 1 + 2\varepsilon}{4Q^2 + 4(\eta - 1 + \varepsilon)Q - 2(2\eta - 1 + 2\varepsilon)}.$$

Hence we have

$$\sum_{k=1}^{m} \frac{1}{k} \left| \sum_{n=1}^{N} e^{2\pi i k f(n)} \right| \ll L(H_0) \ll N H_0^{-\frac{1}{Q}} \log m = N^{1-u} m^{v} \log m.$$

Applying Lemma 1, we conclude that

$$D_N(f(n)) \ll \frac{1}{m} + \frac{1}{N} \sum_{k=1}^{m} \frac{1}{k} \left| \sum_{n=1}^{N} e^{2\pi i k f(n)} \right| \ll \frac{1}{m} + N^{-u} m^v \log m.$$

By choosing $m = [N^w]$ with $w = 1/(2Q^2 + 2(\eta - 1)Q - \eta + 1/2)$, m satisfies the condition (3.13) for sufficiently large N, and so we have

$$D_N(f(n)) \ll N^{-w+\varepsilon'},$$

for any $\varepsilon' > 0$. $\qquad\square$

Proof of Theorem 4. Putting $\eta = 1$ and $\varepsilon = 0$ in the proof of Theorem 3, we obtain

$$D_N(f(n)) \ll \frac{1}{m} + N^{-\frac{1}{2Q^2-1}} m^{\frac{1}{4Q^2-2}} \log m.$$

Choosing $m = \left[N^{\frac{2}{4Q^2-1}}\right]$, we get the conclusion. $\qquad\square$

Acknowledgments

The authors thank the referee for comments and suggestions.

References

[1] S. W. Graham and G. Kolesnik, Van der Corput's method of exponential sums, Cambridge University Press, Cambridge, 1991.

[2] L. Kuipers and H. Niederreiter, Uniform Distribution of Sequences, John Wiley and Sons, New York, 1974.

[3] Y. Ohkubo, Notes on Erdös-Turán inequality, J. Austral. Math. Soc. (Series A), **67** (1999), 51-57.

[4] R. F. Tichy and G. Turnwald, Logarithmic uniform distribution of $(\alpha n + \beta \log n)$, Tsukuba J. Math., **10** (1986), 351-366.

[5] E. C. Titchmarsh, The Theory of the Riemann Zeta-function, 2nd ed. (revised by D. R. Heath-Brown), Clarendon Press, Oxford, 1986.

[6] A. Zygmund, Trigonometric Series Vol.I, Cambridge University Press, Cambridge, 1959.

PADÉ APPROXIMATION TO THE LOGARITHMIC DERIVATIVE OF THE GAUSS HYPERGEOMETRIC FUNCTION

Masayoshi HATA

Division of Mathematics, Faculty of Integrated Human Studies, Kyoto University, Kyoto 606-8501, Japan

Marc HUTTNER

U.F.R. de Mathématiques Pures et Appliquées, Université des Sciences et Technologies de Lille, F-59665 Villeneuve d'Ascq Cedex, France

Keywords: Padé approximation, Gauss hypergeometric function

Abstract   We construct explicitly $(n, n-1)$-Padé approximation to the logarithmic derivative of Gauss hypergeometric function for arbitrary parameters by the simple combinatorial method used by Maier and Chudnovsky.

1991 Mathematics Subject Classification: 41A21, 33C05.

1.   INTRODUCTION AND RESULT

For any $a \in \mathbb{C}$ and for any integer $n \in \mathbb{N}_0$ ($= \{0\} \cup \mathbb{N}$) we use the notation $(a)_n = a(a+1) \cdots (a+n-1)$, known as Pochhammer's symbol. (We adopt, of course, the rule $(a)_0 = 1$.) The Gauss hypergeometric function is defined by the power series

$$F(z) \equiv F(a, b, c\,;z) = \sum_{n=0}^{\infty} \frac{(a)_n (b)_n}{(c)_n\, n\,!}\, z^n,$$

where a, b, c are complex parameters satisfying $c \notin -\mathbb{N}_0$. As is well-known, it satisfies the following linear differential equation of the second order:

$$z(1 - z)F'' + \big(c - (a + b + 1)z\big)F' - abF = 0.$$

157

C. Jia and K. Matsumoto (eds.), Analytic Number Theory, 157–172.
© 2002 *Kluwer Academic Publishers. Printed in the Netherlands.*

If a or b is in $-\mathbb{N}_0$, then $F(z)$ becomes a polynomial; otherwise $F(z)$ has radius of convergence 1.

Let $m, n \in \mathbb{N}$. (m, n)-Padé approximation for a given formal power series $f(z) \in \mathbb{C}[[z]]$ means the functional relation

$$P_m(z) f(z) - Q_n(z) = O(z^{m+n+1})$$

with polynomials $P_m(z)$ and $Q_n(z)$, not all zero, of degrees less than or equal to m and n respectively. The existence of such polynomials follows easily from the theory of linear algebra, by comparing the number of coefficients of $P_m(z)$ and $Q_n(z)$ with the number of conditions imposed on them. To find the explicit construction of $P_m(z)$ and $Q_n(z)$, when m/n is comparatively close to 1, is very important in the study of arithmetical properties of the values of $f(z)$ at specific points z near to the origin. However this must be a difficult problem. For example, such explicit (m, n)-Padé approximations for $F(1/2, 1/2, 1; z)$ and for the polylogarithms of order greater than 1 are not known.

The purpose of the present paper is to give the explicit $(n, n-1)$-Padé approximation to the ratio

$$H(a, b, c; z) = \frac{F(a+1, b+1, c+1; z)}{F(a, b, c; z)}$$

for any $a, b, c \in \mathbb{C}$ satisfying $c \notin -\mathbb{N}_0$, which is equal to the logarithmic derivative of $F(a, b, c; z)$ multiplied by c/ab if $ab \neq 0$. It was the second author who first gave the explicit $(n, n-1)$-Padé approximation to $H(a, b, c; z)$ in the case $a + b = c = 1$ using the theory of monodromy. Especially he treated the case $a = b = 1/2$ and $c = 1$ in order to study arithmetical properties of the value $H(1/2, 1/2, 1; 1/q)$ for $q \in \mathbb{Z} \setminus \{0\}$, since it is related to the ratio $E(k)/K(k)$ where $qk^2 = 1$ and $E(k), K(k)$ are the first and the second kind elliptic integrals respectively [4, 5]. If $ab \neq 0$, then our function $H(a, b, c; z)$ satisfies the following *non-linear* differential equation of the first order:

$$z(1 - z)H' + \frac{ab}{c}z(1 - z)H^2 + \big(c - (a + b + 1)z\big)H - c = 0.$$

Remark. Gauss found the following continued fraction expansion:

$$G(a, b, c; z) \equiv \frac{F(a, b+1, c+1; z)}{F(a, b, c; z)}$$

$$= \frac{1|}{|1} + \frac{a_1 z|}{|1} + \frac{a_2 z|}{|1} + \frac{a_3 z|}{|1} + \cdots$$

where the coefficients a_j are determined by a, b, c and j explicitly. The numerator and the denominator of the convergents in this continued fraction expansion give the explicit $([(n+1)/2], [n/2])$-Padé approximation to $G(a, b, c; z)$. (See [2] for the details.) This expansion was generalized by Yu.V. Nesterenko [7] to generalized hypergeometric functions. The irrationality measures of the values $G(a, b, c; z)$ were discussed in [3] for small enough $x \in \mathbb{Q}$.

Our theorem can now be stated as follows:

Theorem. *Let a, b, c be any complex parameters satisfying $c \notin -\mathbb{N}_0$. We put*

$$A_{i,n} = \frac{\binom{a+n}{i}\binom{b+n}{i}}{\binom{c+2n-1}{i}},$$

$$B_{j,\epsilon} = \frac{\binom{a+j-\epsilon}{j}\binom{b+j-\epsilon}{j}}{\binom{c+j-\epsilon}{j}}.$$

for any $0 \le i, j \le n$, $n \in \mathbb{N}$ and $\epsilon \in \{0, 1\}$. Then the polynomials

$$P_n(z) = \sum_{\ell=0}^{n}\left(\sum_{i+j=\ell}(-1)^i A_{i,n} B_{j,1}\right)z^\ell,$$

$$Q_{n-1}(z) = \sum_{\ell=0}^{n-1}\left(\sum_{i+j=\ell}(-1)^i A_{i,n} B_{j,0}\right)z^\ell$$

satisfy the functional relation

$$P_n(z)F(a+1, b+1, c+1; z) - Q_{n-1}(z)F(a, b, c; z) = C_n\, z^{2n} R_n(z) \quad (1.1)$$

with

$$R_n(z) = F(a+n+1, b+n+1, c+2n+1; z)$$

and

$$C_n = \frac{\binom{a+n}{n}\binom{b+n}{n}\binom{c-a+n-1}{n}\binom{c-b+n-1}{n}}{\binom{2n}{n}^2\binom{c+2n-1}{2n}\binom{c+2n}{2n}}.$$

Therefore (1.1) gives the explicit $(n, n-1)$-Padé approximation to $H(a, b, c; z)$ since $P_n(0) = Q_{n-1}(0) = 1$. The existence of such a relation, without explicit expressions of P_n, Q_{n-1} and C_n, was first shown by Riemann (See [4, 5]). We will show this theorem by a marvelous application of the following simple combinatorial fact:

$$\sum_{i=0}^{n} (-1)^i \binom{n}{i} S(i) = 0$$

for all polynomial $S(z) \in \mathbb{C}[z]$ of degree less than n. More direct applications of this method have been made by W. Maier [6] and G.V. Chudnovsky [1] in order to obtain Padé-type approximations.

If $ab = 0$, then clearly $B_{0,1} = 1$ and $B_{j,1} = 0$ for all $j \in \mathbb{N}$. Thus the case $a = 0$ in the theorem (replacing b, c by $b - 1, c - 1$ respectively) implies the following corollary, giving the explicit $(n, n-1)$-Padé approximation to the Gauss hypergeometric function

$$F(1, b, c; z) = \sum_{n=0}^{\infty} \frac{\dbinom{b+n-1}{n}}{\dbinom{c+n-1}{n}} z^n.$$

Corollary. *Let b, c be any complex parameters satisfying $c \notin -\mathbb{N}_0$. Then the non-trivial polynomials*

$$P_n^*(z) = \sum_{\ell=0}^{n} (-1)^\ell \binom{n}{\ell} \frac{\dbinom{b+n-1}{\ell}}{\dbinom{c+2n-2}{\ell}} z^\ell,$$

$$Q_{n-1}^*(z) = \sum_{0 \le i+j < n} (-1)^i \binom{n}{i} \frac{\dbinom{b+n-1}{i} \dbinom{b+j-1}{j}}{\dbinom{c+2n-2}{i} \dbinom{c+j-1}{j}} z^{i+j}$$

satisfy the functional relation

$$P_n^*(z)F(1, b, c; z) - Q_{n-1}^*(z)$$
$$= C_n^* z^{2n} F(n+1, b+n, c+2n; z) \tag{1.2}$$

with

$$C_n^* = \frac{\binom{b+n-1}{n}\binom{c+n-2}{n}\binom{c-b+n-1}{n}}{\binom{2n}{n}^2\binom{c+2n-2}{2n}\binom{c+2n-1}{2n}}.$$

Note that C_n^* has a finite limit as $c \to 1$; so (1.2) is valid also for $c = 1$.

The first author is indebted to the Département de Mathématiques de l'U.S.T.L. for their kind hospitality in September, 1998.

2. PRELIMINARIES

It may be convenient to introduce Pochhammer's symbol with negative suffix

$$(x)_{-p} = \frac{1}{(x-p)(x-p+1)\cdots(x-1)}$$

for any $x \notin \mathbb{N}$ and $p \in \mathbb{N}$. We first show the following simple useful lemma concerning generalized Pochhammer's symbol:

Lemma 1. *Let $x \in \mathbb{C}$ and $p, q \in \mathbb{Z}$. Suppose that $x + p \notin \mathbb{N}$ if $q < 0$, $x \notin \mathbb{N}$ if $p + q < 0$, and that $x \notin -\mathbb{N}_0$ if $p > 0$. Then we have*

$$(x+p)_q = \frac{(x)_{p+q}}{(x)_p}.$$

Proof. First consider the case $q \geq 0$; $(x+p)_q = (x+p)(x+p+1)\cdots(x+p+q-1)$. If $p \geq 0$, then clearly

$$(x+p)_q = \frac{x(x+1)\cdots(x+p+q-1)}{x(x+1)\cdots(x+p-1)} = \frac{(x)_{p+q}}{(x)_p}.$$

Conversely, if $p < 0$, then

$$(x+p)_q = \begin{cases} (x+p)\cdots(x-1) \times x \cdots (x+p+q-1) = \dfrac{(x)_{p+q}}{(x)_p} \\ \qquad\qquad\qquad\qquad\qquad\qquad \text{for } p+q \geq 0, \\[2ex] \dfrac{(x+p)\cdots(x-1)}{(x+p+q)\cdots(x-1)} = \dfrac{(x)_{p+q}}{(x)_p} \quad \text{for } p+q < 0. \end{cases}$$

Next consider the case $q < 0$; $(x+p)_q = \big((x+p+q)(x+p+q+1)\cdots(x+p-1)\big)^{-1}$. If $p < 0$, then obviously

$$(x+p)_q = \frac{(x+p)(x+p+1)\cdots(x-1)}{(x+p+q)(x+p+q+1)\cdots(x-1)} = \frac{(x)_{p+q}}{(x)_p}.$$

Conversely, if $p \geq 0$, then

$$
(x+p)_q = \begin{cases}
\dfrac{x \cdots (x+p+q-1)}{x \cdots (x+p-1)} = \dfrac{(x)_{p+q}}{(x)_p} & \text{for } p+q \geq 0, \\[4mm]
\dfrac{1}{(x+p+q)\cdots(x-1)\times x \cdots (x+p-1)} = \dfrac{(x)_{p+q}}{(x)_p} & \\
& \text{for } p+q < 0,
\end{cases}
$$

as required. ∎

We put $B_j^* = B_{j,0}$ and $B_j = B_{j,1}$ for brevity. It is not difficult to see that B_j and B_j^* are the coefficients of z^j in the series $F(a,b,c;z)$ and $F(a+1,b+1,c+1;z)$ respectively. We also notice that the polynomial $Q_{n-1}(z)$ can be replaced by

$$
\sum_{\ell=0}^{n} \left(\sum_{i+j=\ell} (-1)^i A_{i,n} B_j^* \right) z^\ell.
$$

For we have

$$
\sum_{i+j=n} (-1)^i A_{i,n} B_j^* = \sum_{i+j=n} (-1)^i \frac{\dbinom{a+n}{i}\dbinom{b+n}{i}\dbinom{a+j}{j}\dbinom{b+j}{j}}{\dbinom{c+2n-1}{i}\dbinom{c+j}{j}}
$$

$$
= \frac{(a+1)_n (b+1)_n}{(c+1)_{2n-1}\, n!} \sum_{i=0}^{n} (-1)^i \binom{n}{i} S(i)
$$

$$
= 0,
$$

since $S(x) = (c+n+1-x)\cdots(c+2n-1-x)$ is a polynomial in x of degree $n-1$.

Let $\alpha_{\ell,n}$ be the coefficient of z^ℓ in the Taylor expansion of $P_n(z)F(a+1,b+1,c+1;z)$ at $z=0$; namely,

$$
\alpha_{\ell,n} = \sum_{\substack{i+j+k=\ell \\ i+j \leq n}} (-1)^i A_{i,n} B_j B_k^*.
$$

For arbitrarily fixed $i \in [0, \min(\ell,n)]$, the variable k varies in the range $[0, \ell-i]$ satisfying $k = \ell - i - j \geq \ell - n$. Hence, exchanging j and k, we have

$$
\alpha_{\ell,n} = \sum_{i=0}^{\min(\ell,n)} (-1)^i A_{i,n} \sum_{j=\max(0,\ell-n)}^{\ell-i} B_j^* B_{\ell-i-j}. \tag{2.1}
$$

Similarly let $\beta_{\ell,n}$ be the coefficient of z^ℓ in the Taylor expansion of $Q_{n-1}(z)F(a,b,c\,;z)$;

$$\beta_{\ell,n} = \sum_{\substack{i+j+k=\ell \\ i+j\leq n}} (-1)^i A_{i,n} B_j^* B_k.$$

For arbitrarily fixed $i \in [0, \min(\ell, n)]$, the variable j varies in the range $[0, n-i]$ satisfying $j = \ell - i - k \leq \ell - i$; hence

$$\beta_{\ell,n} = \sum_{i=0}^{\min(\ell,n)} (-1)^i A_{i,n} \sum_{j=0}^{\min(\ell,n)-i} B_j^* B_{\ell-i-j}. \qquad (2.2)$$

Put $\gamma_{\ell,n} = \alpha_{\ell,n} - \beta_{\ell,n}$ for $\ell \in \mathbb{N}_0$. We then must show that $\gamma_{\ell,n}$ vanishes for all $\ell \in [0, 2n)$. However this is trivial for $\ell \in [0, n]$, since $\max(0, \ell - n) = 0$ and $\min(\ell, n) = \ell$ in (2.1) and (2.2). Thus it remains to show that $\gamma_{\ell,n} = 0$ for $\ell \in (n, 2n)$, which will be treated in the next section.

3.   VANISHMENT OF $\gamma_{\ell,n}$ FOR $\ell \in (n, 2n)$

For a while we assume that $a, b \notin -\mathbb{N}_0$ and that $c \notin \mathbb{N}$. Let $m \in \mathbb{N}$ be fixed and put $\ell = m + n$. It then follows from (2.1) and (2.2) that

$$\gamma_{m+n,n} = \sum_{i=0}^{n} (-1)^i A_{i,n} \left\{ \sum_{j=m}^{m+n-i} B_j^* B_{m+n-i-j} - \sum_{j=0}^{n-i} B_j^* B_{m+n-i-j} \right\}.$$

The quantity in the above brace can be transformed into

$$\left(\sum_{j=0}^{m+n-i} - \sum_{j=0}^{m-1} \right) - \left(\sum_{j=0}^{m+n-i} - \sum_{j=n-i+1}^{m+n-i} \right) = \sum_{j=n-i+1}^{m+n-i} - \sum_{j=0}^{m-1}$$

$$= \sum_{j'=0}^{m-1} - \sum_{j=0}^{m-1}$$

by substituting $j' = m + n - i - j$ in the first sum of the right-hand side; therefore

$$\gamma_{m+n,n} = \sum_{i=0}^{n} (-1)^i A_{i,n} \sum_{j=0}^{m-1} \left(B_j B_{m+n-i-j}^* - B_j^* B_{m+n-i-j} \right).$$

We can write this in the following manner:

$$\gamma_{m+n,n} = \frac{(a+1)_n (b+1)_n}{(c+1)_{2n-1}\, n!} \sum_{i=0}^{n} (-1)^i \binom{n}{i} (\Delta_i^+ - \Delta_i^-), \qquad (3.1)$$

where

$$
\begin{cases}
\Delta_i^+ = i!(n-i)!\,A_{i,n}\dfrac{(c+1)_{2n-1}}{(a+1)_n(b+1)_n}\displaystyle\sum_{j=0}^{m-1} B_j\,B^*_{m+n-i-j}, \\[2ex]
\Delta_i^- = i!(n-i)!\,A_{i,n}\dfrac{(c+1)_{2n-1}}{(a+1)_n(b+1)_n}\displaystyle\sum_{j=0}^{m-1} B_j^*\,B_{m+n-i-j}.
\end{cases}
\tag{3.2}
$$

(We used here the assumption that $a, b \notin -\mathbb{N}_0$.)

We first deal with the quantity Δ_i^+ when $m < n$. Since

$$
A_{i,n} = \frac{(a+n-i+1)_i\,(b+n-i+1)_i}{(c+2n-i)_i\,i!},
$$

it follows that

$$
\Delta_i^+ = \sum_{j=0}^{m-1} B_j\,\frac{(\sigma+a)_{m-j}(\sigma+b)_{m-j}(\sigma+c+m-j)_{n-m+j-1}}{(\sigma)_{m-j}},
$$

where $\sigma \equiv n-i+1$. This expression shows that Δ_i^+ is a rational function in σ; so we write $\Delta^+(\sigma)$ instead of Δ_i^+, which can be written in the form (partial fraction expansion)

$$
\Delta^+(\sigma) = \sum_{k=0}^{m-1} \frac{d_k^+}{\sigma+k} + S^+(\sigma)
$$

where d_k^+ are constants and $S^+(\sigma)$ is a polynomial in σ of degree less than n. In fact each d_k^+ is the residue of $\Delta^+(\sigma)$ at the simple pole $\sigma = -k$. Since $(\sigma)_{m-j} = \sigma(\sigma+1)\cdots(\sigma+m-j-1)$ contains the factor $\sigma+k$ if and only if $0 \le j \le m-k-1$, we get

$$
\begin{aligned}
d_k^+ &= \lim_{\sigma \to -k} (\sigma+k)\,\Delta^+(\sigma) \\
&= \frac{(-1)^k}{k!}\sum_{j=0}^{m-k-1} B_j\,\frac{(a-k)_{m-j}(b-k)_{m-j}(c+m-j-k)_{n-m+j-1}}{(m-j-k-1)!}.
\end{aligned}
$$

Then it follows from Lemma 1 that

$$
(c+m-j-k)_{n-m+j-1} = \frac{(c)_{n-k-1}}{(c)_{m-j-k}}
$$

and

$$
(a-k)_{m-j} = \frac{(a)_{m-j-k}}{(a)_{-k}};
$$

hence we have

$$d_k^+ = (-1)^k \frac{(c)_{n-k-1}}{(a)_{-k}(b)_{-k}k!} \sum_{j=0}^{m-k-1} (m-j-k) B_j B_{m-j-k}. \tag{3.3}$$

Similarly the second quantity in (3.2) can be written as

$$\Delta^-(\sigma) = \sum_{j=0}^{m-1} (j+1) B_{j+1}$$

$$\times \frac{(\sigma+a)_{m-j-1}(\sigma+b)_{m-j-1}(\sigma+c+m-j-1)_{n-m+j}}{(\sigma)_{m-j}};$$

so we can put

$$\Delta^-(\sigma) = \sum_{k=0}^{m-1} \frac{d_k^-}{\sigma+k} + S^-(\sigma)$$

where d_k^- are constants and $S^-(\sigma)$ is a polynomial in σ of degree less than $n-1$. Then we have

$$d_k^- = \lim_{\sigma \to -k} (\sigma+k) \Delta^-(\sigma)$$

$$= \frac{(-1)^k}{k!} \sum_{j=0}^{m-k-1} (j+1) B_{j+1}$$

$$\times \frac{(a-k)_{m-j-1}(b-k)_{m-j-1}(c+m-j-k-1)_{n-m+j}}{(m-j-k-1)!};$$

hence, from Lemma 1,

$$d_k^- = (-1)^k \frac{(c)_{n-k-1}}{(a)_{-k}(b)_{-k}k!} \sum_{j=0}^{m-k-1} (j+1) B_{j+1} B_{m-j-k-1}. \tag{3.4}$$

Substituting $j' = m - k - 1 - j$, it is easily seen that $d_k^+ = d_k^-$ for all $k \in [0, m)$ from (3.3) and (3.4). Therefore

$$\Delta^+(\sigma) - \Delta^-(\sigma) = S^+(\sigma) - S^-(\sigma)$$

is a polynomial in σ, hence in i, of degree less than n. This implies that $\gamma_{m+n,n} = 0$ for all $m \in (0, n)$ from (3.1), as required.

4. CALCULATION OF THE REMAINDER TERM

On the remainder term of our approximation, as is mentioned in Section 1, Riemann determined its form; so it suffices to calculate the value of $C_n = \gamma_{2n,n}$ explicitly. However we will show the theorem directly by calculating $\gamma_{\ell,n}$ for all $\ell \geq 2n$.

We first consider $\Delta^+(\sigma)$ for $m \geq n$. Put $m = n + M, M \in \mathbb{N}_0$ for brevity. Then it can be seen that

$$\Delta^+(\sigma) = \sum_{j=0}^{n+M-1} B_j \frac{(\sigma + a)_{n+M-j}(\sigma + b)_{n+M-j}}{(\sigma)_{n+M-j}} \Omega_j^+(\sigma),$$

where

$$\Omega_j^+(\sigma) = \begin{cases} \dfrac{1}{(\sigma + c + n - 1)_{M-j+1}} & \text{for } 0 \leq j \leq M, \\ (\sigma + c + n + M - j)_{j-M-1} & \text{for } M < j < n + M. \end{cases}$$

Owing to the assumption that $c \notin \mathbb{N}$, all poles of $\Delta^+(\sigma)$ are simple; so, we can write

$$\Delta^+(\sigma) = \sum_{k=0}^{n+M-1} \frac{r_k^+}{\sigma + k} + \sum_{\ell=0}^{M} \frac{s_\ell^+}{\sigma + c + n - 1 + \ell} + T^+(\sigma), \qquad (4.1)$$

where r_k^+, s_ℓ^+ are constants and $T^+(\sigma)$ is a polynomial in σ of degree less than n. Since $(\sigma)_{n+M-j}$ contains the factor $\sigma + k$ if and only if $0 \leq j \leq n + M - k - 1$, we have

$$r_k^+ = \lim_{\sigma \to -k} (\sigma + k) \Delta^+(\sigma)$$

$$= \frac{(-1)^k}{k!} \sum_{j=0}^{n+M-k-1} B_j \frac{(a-k)_{n+M-j}(b-k)_{n+M-j}}{(n+M-j-k-1)!} \Omega_j^+(-k).$$

It then follows from Lemma 1 that

$$\Omega_j^+(-k) = (c + n + M - j - k)_{j-M-1} = \frac{(c)_{n-k-1}}{(c)_{n+M-j-k}}$$

and

$$(a - k)_{n+M-j} = \frac{(a)_{n+M-j-k}}{(a)_{-k}};$$

hence we obtain

$$r_k^+ = (-1)^k \frac{(c)_{n-k-1}}{(a)_{-k}(b)_{-k}\,k!} \sum_{j=0}^{n+M-k-1} (n + M - j - k)\, B_j\, B_{n+M-j-k}, \qquad (4.2)$$

which formally coincides with (3.3) when $m = n + M$.

The factor $\sigma + c + n - 1 + \ell$ appears in the denominator of $\Omega_j^+(\sigma)$ if and only if $0 \leq j \leq M - \ell$; therefore we get

$$
s_\ell^+ = \lim_{\sigma \to -c-n+1-\ell} (\sigma + c + n - 1 + \ell)\,\Delta^+(\sigma)
$$

$$
= \frac{(-1)^\ell}{\ell!} \sum_{j=0}^{M-\ell} B_j \frac{(1+a-c-n-\ell)_{n+M-j}(1+b-c-n-\ell)_{n+M-j}}{(1-c-n-\ell)_{n+M-j}\,(M-j-\ell)!}.
$$

Using Lemma 1 and the identity $(1-x)_k(x)_{-k} = (-1)^k$ for any $x \notin \mathbb{N}$ and $k \in \mathbb{N}_0$,

$$
(1+a-c-n-\ell)_{n+M-j} = \frac{(1+a-c)_{M-j-\ell}}{(1+a-c)_{-\ell-n}}
$$

$$
= (-1)^{\ell+n}(c-a)_{\ell+n}(1+a-c)_{M-j-\ell}.
$$

Thus

$$
s_\ell^+ = (-1)^n \frac{(c-a)_{\ell+n}(c-b)_{\ell+n}}{(c)_{\ell+n}\,\ell!} \sum_{j=0}^{M-\ell} B_j\, D_{M-j-\ell}, \tag{4.3}
$$

where D_k is the coefficient of z^k in the series $F(1+a-c, 1+b-c, 1-c\,; z)$.

We next consider

$$
\Delta^-(\sigma) = \sum_{j=0}^{n+M-1} (j+1)B_{j+1} \frac{(\sigma+a)_{n+M-j-1}(\sigma+b)_{n+M-j-1}}{(\sigma)_{n+M-j}} \Omega_j^-(\sigma),
$$

where

$$
\Omega_j^-(\sigma) = \begin{cases} \dfrac{1}{(\sigma+c+n-1)_{M-j}} & \text{for } 0 \leq j < M, \\[2mm] (\sigma+c+n+M-j-1)_{j-M} & \text{for } M \leq j < n+M. \end{cases}
$$

Then $\Delta^-(\sigma)$ is a rational function in σ and we can write

$$
\Delta^-(\sigma) = \sum_{k=0}^{n+M-1} \frac{r_k^-}{\sigma+k} + \sum_{\ell=0}^{M-1} \frac{s_\ell^-}{\sigma+c+n-1+\ell} + T^-(\sigma), \tag{4.4}
$$

where r_k^-, s_ℓ^- are constants and $T^-(\sigma)$ is a polynomial in σ of degree less than $n-1$. Since $(\sigma)_{n+M-j}$ contains the factor $\sigma+k$ if and only if $0 \leq j \leq n+M-k-1$, we get

$$
r_k^- = \lim_{\sigma \to -k} (\sigma+k)\,\Delta^-(\sigma)
$$

$$
= \frac{(-1)^k}{k!} \sum_{j=0}^{n+M-k-1} (j+1)B_{j+1} \frac{(a-k)_{n+M-j-1}(b-k)_{n+M-j-1}}{(n+M-j-k-1)!} \Omega_j^-(-k).
$$

It then follows from Lemma 1 that

$$\Omega_j^-(-k) = (c+n+M-j-k-1)_{j-M} = \frac{(c)_{n-k-1}}{(c)_{n+M-j-k-1}};$$

hence we obtain

$$r_k^- = (-1)^k \frac{(c)_{n-k-1}}{(a)_{-k}\,(b)_{-k}\,k\,!} \sum_{j=0}^{n+M-k-1} (j+1)\,B_{j+1}\,B_{n+M-j-k-1}, \quad (4.5)$$

which formally coincides with (3.4) when $m = n+M$. Substituting $j' = n+M-j-k-1$, it is easily seen that $r_k^+ = r_k^-$ for all $k \in [0, n+M)$ from (4.2) and (4.5).

Similarly we calculate s_ℓ^-. The factor $\sigma + c + n - 1 + \ell$ appears in the denominator of $\Omega_j^-(\sigma)$ if and only if $0 \le j \le M - \ell - 1$; hence

$$s_\ell^- = \lim_{\sigma \to -c-n+1-\ell} (\sigma + c + n - 1 + \ell)\,\Delta^-(\sigma)$$

$$= \frac{(-1)^\ell}{\ell!} \sum_{j=0}^{M-\ell-1} (j+1)B_{j+1}$$

$$\times \frac{(1+a-c-n-\ell)_{n+M-j-1}(1+b-c-n-\ell)_{n+M-j-1}}{(1-c-n-\ell)_{n+M-j}\,(M-j-\ell-1)!}.$$

It then follows from Lemma 1 that

$$s_\ell^- = \frac{(-1)^n ab}{c(1-c)} \cdot \frac{(c-a)_{\ell+n}(c-b)_{\ell+n}}{(c)_{\ell+n}\,\ell!} \sum_{j=0}^{M-\ell-1} B_j^*\,D_{M-j-\ell-1}^*, \quad (4.6)$$

where D_k^* is the coefficient of z^k in the series $F(1+a-c, 1+b-c, 2-c; z)$.

By noticing that $\sigma + c + n - 1 + \ell = c + 2n - i + \ell$ and that

$$\sum_{i=0}^{n} (-1)^i \binom{n}{i} \frac{1}{c+2n-i+\ell} = \int_0^1 x^{c+n-1+\ell}(x-1)^n\,dx$$

$$= (-1)^n\,n!\,\frac{(c)_{\ell+n}}{(c)_{\ell+2n+1}}$$

(this is true for $c > -n$; hence for $c \notin -\mathbb{N}_0$ by analytic continuation), it follows from (3.1), (4.1), (4.3), (4.4) and (4.6) that

$$\gamma_{2n+M,n} = (-1)^n \frac{(a+1)_n(b+1)_n}{(c+1)_{2n-1}} \left(\sum_{\ell=0}^{M} \frac{(c)_{\ell+n}}{(c)_{\ell+2n+1}} s_\ell^+ - \sum_{\ell=0}^{M-1} \frac{(c)_{\ell+n}}{(c)_{\ell+2n+1}} s_\ell^- \right);$$

in particular we have

$$C_n = \gamma_{2n,n} = \frac{(a+1)_n(b+1)_n(c-a)_n(c-b)_n}{(c+1)_{2n-1}(c)_{2n+1}},$$

as required. Moreover

$$\gamma_{2n+M,n} = \gamma_{2n,n}\left(\Phi_{n,M}^+ - \Phi_{n,M}^-\right)$$

with

$$\Phi_{n,M}^+ = \sum_{\ell=0}^{M} \frac{(c-a+n)_\ell(c-b+n)_\ell}{(c+2n+1)_\ell\,\ell!} \sum_{j=0}^{M-\ell} B_j D_{M-j-\ell}$$

$$= \sum_{j+k+\ell=M} B_j\, D_k\, E_\ell,$$

where E_ℓ is the coefficient of z^ℓ in the series $F(c-a+n, c-b+n, c+2n+1; z)$, and with

$$\Phi_{n,M}^- = \frac{ab}{c(1-c)} \sum_{\ell=0}^{M-1} \frac{(c-a+n)_\ell(c-b+n)_\ell}{(c+2n+1)_\ell\,\ell!} \sum_{j=0}^{M-\ell-1} B_j^*\, D_{M-j-\ell-1}^*$$

$$= \frac{ab}{c(1-c)} \sum_{j+k+\ell=M-1} B_j^*\, D_k^*\, E_\ell.$$

We thus obtain

$$R_n(z) \equiv \sum_{M=0}^{\infty} \left(\Phi_{n,M}^+ - \Phi_{n,M}^-\right) z^M$$

$$= \lambda(z)\, F(c-a+n, c-b+n, c+2n+1; z),$$

where

$$\lambda(z) = F(a, b, c; z)\, F(1+a-c, 1+b-c, 1-c; z)$$

$$- \frac{ab}{c(1-c)}\, z\, F(a+1, b+1, c+1; z)\, F(1+a-c, 1+b-c, 2-c; z).$$

Now, using the well-known formula:

$$F(\alpha, \beta, \gamma; z) = (1-z)^{\gamma-\alpha-\beta} F(\gamma-\alpha, \gamma-\beta, \gamma; z),$$

it is easily seen that

$$R_n(z) = \phi(z)\, F(a+n+1, b+n+1, c+2n+1; z)$$

where

$$\phi(z) = F(a, b, c; z) F(-a, -b, 1 - c; z)$$
$$- \frac{ab}{c(1-c)} z(1-z) F(a+1, b+1, c+1; z) F(1-a, 1-b, 2-c; z).$$

Put $f(z) = F(a, b, c; z)$ and $g(z) = F(-a, -b, 1 - c; z)$ for brevity. Then it follows that

$$\phi(z) = f(z)g(z) - \frac{z(1-z)}{ab} f'(z)g'(z),$$

from which we get

$$\phi' = f'g + fg' - \frac{1-2z}{ab} f'g' - \frac{z(1-z)}{ab} \left(f''g' + f'g'' \right).$$

We now use the differential equation satisfied by the Gauss hypergeometric function, stated in Section 1, that is,

$$z(1-z)f'' = \big((a+b+1)z - c \big) f' + abf$$

and

$$z(1-z)g'' = \big((1-a-b)z - 1 + c \big) g' + abg$$

in order to show that $\phi'(z) \equiv 0$; hence $\phi(z) \equiv 1$ by $\phi(0) = 1$. Therefore $R_n(z)$ is equal to $F(a+n+1, b+n+1, c+2n+1; z)$, as required.

Finally the assumption that $a, b \notin -\mathbb{N}_0$ and $c \notin \mathbb{N}$ can be easily removed by limit operation, since each $A_{i,n}, B_{j,\epsilon}$ and C_n are continuous at $a, b \in -\mathbb{N}_0$ and at $c \in \mathbb{N}$. This completes the proof of the theorem.

5.   REMARKS AND OPEN PROBREMS

It is easily seen from (1.1) that $H(a, b, c; z)$ is a rational function if and only if $C_n = 0$ for some $n \in \mathbb{N}$. Therefore the explicit expression of C_n implies that $H(a, b, c; z)$ is a rational function if and only if either $a \in -\mathbb{N}, b \in -\mathbb{N}, c - a \in -\mathbb{N}_0$ or $c - b \in -\mathbb{N}_0$.

Using the same method employed in Section 3, we can calculate the leading coefficient of the polynomial $P_n(z)$; namely,

$$\sum_{i+j=n} (-1)^i A_{i,n} B_j$$
$$= \frac{(a)_{n+1}(b)_{n+1}}{(c)_{2n}\, n!} \sum_{i=0}^{n} (-1)^i \binom{n}{i} \frac{(c+n-i)\cdots(c+2n-i-1)}{(a+n-i)(b+n-i)}$$

$$= (-1)^n \frac{(a)_{n+1}(c-b)_n - (b)_{n+1}(c-a)_n}{(a-b)(c)_{2n}}.$$

Similarly the leading coefficient of $Q_{n-1}(z)$ is

$$\sum_{i+j=n-1} (-1)^i A_{i,n} B_j^*$$

$$= \frac{(a+1)_n(b+1)_n}{(c)_{2n}\,(n-1)!} \sum_{i=0}^{n-1}(-1)^i \binom{n-1}{i} \frac{(c+n-i)\cdots(c+2n-i-1)}{(a+n-i)(b+n-i)}$$

$$= (-1)^{n-1} \frac{(a+1)_n(c-b)_n - (b+1)_n(c-a)_n}{(a-b)(c+1)_{2n-1}}.$$

Finally we propose the following interesting problems:

Problem 1. Find a continued fraction expansion of $H(a,b,c;z)$ analogous to Gauss' continued fraction expansion to $G(a,b,c;z)$.

Problem 2. Extend our theorem to the generalized hypergeometric function

$$F\left(\begin{matrix} a_1, a_2, \cdots, a_p \\ b_1, b_2, \cdots, b_q \end{matrix}; z\right) = \sum_{n=0}^{\infty} \frac{(a_1)_n(a_2)_n\cdots(a_p)_n}{(b_1)_n(b_2)_n\cdots(b_q)_n} \cdot \frac{z^n}{n!},$$

where $p \le q+1$ and $a_i \in \mathbb{C}, b_j \notin -\mathbb{N}_0$.

The simple combinatorial method employed in this paper will be certainly applied to construct Padé approximation of the second kind for the generalized hypergeometric function and its derivatives. Such approximations may have an interesting application to the irrationality problem of the ratio

$$\frac{L_n(1/q)}{L_m(1/q)}$$

for $q \in \mathbb{Z}\backslash\{0\}$, where $n \ne m \in \mathbb{N}$ and $L_n(z) = \sum_{k=1}^{\infty} z^k/k^n$ is the polylogarithm of order n.

References

[1] G. V. Chudnovsky, *Padé approximations to the generalized hypergeometric functions. I*, J. Math. Pure et Appl., **58** (1979), 445–476.

[2] N. I. Fel'dman and Yu. V. Nesterenko, Transcendental Numbers, Number Theory IV, Encyclopaedia of Mathematical Sciences, Vol. 44, Springer.

[3] M. Huttner, *Problème de Riemann et irrationalité d'un quotient de deux fonctions hypergéométriques de Gauss,* C. R. Acad. Sc. Paris Série I, **302** (1986), 603–606.

[4] M. Huttner, *Monodromie et approximation diophantienne d'une constante liée aux fonctions elliptiques,* C. R. Acad. Sc. Paris Série I, **304** (1987), 315-318.

[5] M. Huttner and T. Matala-aho, *Approximations diophantiennes d'une constante liée aux fonctions elliptiques,* Pub. IRMA, Lille, Vol. 38, XII, 1996, p.19.

[6] W. Maier, *Potenzreihen irrationalen grenzwertes,* J. Reine Angew. Math., **156** (1927), 93–148.

[7] Yu. V. Nesterenko, *Hermite-Padé approximations of generalized hypergeometric functions,* Russian Acad. Sci. Sb. Math., **83** (1995), No.1, 189–219.

THE EVALUATION OF THE SUM
OVER ARITHMETIC PROGRESSIONS
FOR THE COEFFICIENTS OF
THE RANKIN-SELBERG SERIES II

Yumiko ICHIHARA

Graduate School of Mathematics, Nagoya University, Chikusa-ku, Nagoya, 464-8602, Japan

m96046i@math.nagoya-u.ac.jp

Keywords: Rankin-Selberg series

Abstract We study $\sum_{\substack{n \le x \\ n \equiv a(p^r)}} c_n$, where c_ns are the coefficients of the Rankin-Selberg series, p is an odd prime, r is a natural number, and a is also a natural number satisfying $(a, p) = 1$. For any natural number d, we know the asymptotic formula for $\sum_{n \le x} c_n \chi(n)$, where χ is a primitive Dirichlet character mod d. This is obtained by using the Voronoï-formula of the Riesz-mean $\sum_{n \le x} c_n \chi(n)(x - n)^2$. In particular, in case $d = p^r$, the fourth power of the Gauss sum appears in that Voronoï-formula. We consider the sum over all characters mod p^r, then the fourth power of the Gauss sum produces the hyper-Kloosterman sum. Hence, applying the results of Deligne and Weinstein, we can estimate the error term in the asymptotic formula for $\sum_{\substack{n \le x \\ n \equiv a(p^r)}} c_n$.

1991 Mathematics Subject Classification: 11F30.

1. INTRODUCTION

Let a, d be integers, $(a, d) = 1$, $d \ge 1$. The author [3] studied the sum $\sum_{\substack{n \le x \\ n \equiv a(d)}} c_n$, in the case $d = p$ (p is an odd prime), where c_ns are the coefficients of the Rankin-Selberg series and $n \equiv a(d)$ means $n \equiv a$ mod d. In [3], the author showed the following results. If $x^2 \le p^3$, then

C. Jia and K. Matsumoto (eds.), Analytic Number Theory, 173–182.
© 2002 *Kluwer Academic Publishers. Printed in the Netherlands.*

we have

$$
\sum_{\substack{n \leq x \\ n \equiv a(p)}} c_n = \frac{1}{\phi(p)} \sum_{n \leq x} c_n \chi_0(n) + \frac{1}{\phi(p)} \sum_{\substack{\chi \bmod p \\ \chi \neq \chi_0}} \chi(a^{-1}) L_{f \otimes g}(0, \chi)
$$
$$
+ O\left(x^{3/5} p^{3/5} + x^{-\varepsilon} p^{3/2+4\varepsilon}\right) \tag{1.1}
$$

and

$$
\sum_{n \leq x} c_n \chi_0(n) = \delta_{f,g} \frac{\pi^2}{6} k\alpha (1 - \alpha_p^2 p^{-k})(1 - \overline{\alpha}_p^2 p^{-k})(1 - p^{-1})^2 x
$$
$$
+ O\left(x^{3/5} p^{4/5}\right) \tag{1.2}
$$

by using Deligne's estimate of the hyper-Kloosterman sum, where χ_0 is the principal character mod p. Some notations used in (1.1) and (1.2) are defined below. And if $p^3 > x^2$, we have

$$
\sum_{\substack{n \leq x \\ n \equiv a(p)}} c_n \ll x^{3/5} p^{3/5}.
$$

In this paper, we generalize the result of [3] to the case $d = p^r$ by using Weinstein's estimate and an induction argument. First of all, we introduce the Rankin-Selberg series.

Let $f(z)$ and $g(z)$ be normalized Hecke eigen cusp forms of weight k and l respectively for $SL_2(\mathbf{Z})$, and denote the Fourier expansion of them as

$$
f(z) = \sum_{n=1}^{\infty} a_n e^{2\pi i n z}
$$

and

$$
g(z) = \sum_{n=1}^{\infty} b_n e^{2\pi i n z},
$$

where $a_n, b_n \in \mathbf{R}$. The Rankin-Selberg series is defined by

$$
L_{f \otimes g}(s, \chi) = L(2s, \chi^2) \sum_{n=1}^{\infty} \frac{a_n b_n \chi(n)}{n^{s+(k+l)/2-1}}, \tag{1.3}
$$

in $\Re(s) > 1$. Here $L(s, \chi)$ is the Dirichlet L-function with a Dirichlet character χ. This Rankin-Selberg series has the Euler product

$$
\begin{aligned}
L_{f \otimes g}&(s, \chi) \\
&= \prod_p (1 - \alpha_p \beta_p \chi(p) p^{-s-(k+l)/2+1})^{-1} (1 - \alpha_p \overline{\beta}_p \chi(p) p^{-s-(k+l)/2+1})^{-1} \\
&\quad \times (1 - \overline{\alpha}_p \beta_p \chi(p) p^{-s-(k+l)/2+1})^{-1} (1 - \overline{\alpha_p \beta_p} \chi(p) p^{-s-(k+l)/2+1})^{-1},
\end{aligned}
$$

where α_p and β_p are complex numbers satisfying the following conditions;

$$
\alpha_p + \overline{\alpha}_p = a_p, \quad |\alpha_p| = p^{(k-1)/2}
$$

and

$$
\beta_p + \overline{\beta}_p = b_p, \quad |\beta_p| = p^{(l-1)/2}.
$$

Here bar means the complex conjugate. We find that

$$
L_{f \otimes g}(s, \chi) = \sum_{n=1}^{\infty} \frac{c_n \chi(n)}{n^s}
$$

and

$$
c_n = n^{1-(k+l)/2} \sum_{m^2 \mid n} a_{n/m^2} b_{n/m^2} m^{k+l-2} \in \mathbf{R}
$$

from (1.1). Deligne's estimate of a_n and b_n gives the estimate $c_n \ll n^{\varepsilon}$. We find $\sum_{n \le x} |c_n| \ll x$ easily by using the Cauchy-Schwarz inequality and the argument which is in the proof of Lemma 4 in Ivić-Matsumoto-Tanigawa [4].

On the sum of c_n over arithmetic progressions, we obtain the following theorem. Throughout this paper, ε is an arbitrarily small positive constant.

Theorem. *Let p be an odd prime, $r \ge 1$ a natural number and a an integer with $(a, p) = 1$. If $p^{3r} \ge x^2$, then we have*

$$
\sum_{\substack{n \le x \\ n \equiv a(p^r)}} c_n = O(x^{3/5} p^{3r/5}).
$$

If $p^{3r} \leq x^2$, then we have

$$\sum_{\substack{n \leq x \\ n \equiv a(p^r)}} c_n = \delta_{f,g} \frac{\pi^2 k \alpha}{6\phi(p^r)} (1 - \alpha_p^2 p^{-k})(1 - \overline{\alpha}_p^2 p^{-k})(1 - p^{-1})^2 x$$

$$+ \frac{1}{\phi(p^r)} \sum_{\substack{\chi \bmod p^r \\ \chi \neq \chi_0}} \chi(a^{-1}) L_{f \otimes g}(0, \chi)$$

$$+ O(x^{3/5} p^{3r/5} + x^{-\varepsilon} p^{3r/2 + 4\varepsilon}),$$

where ϕ is the Euler function, χ_0 is the principal character mod p, $k\alpha$ is the residue of $L_{f \otimes f}(s)$ at $s = 1$ and

$$\delta_{f,g} = \begin{cases} 1 & f = g \\ 0 & f \neq g. \end{cases}$$

In particular, the error term can be estimated as $O(x^{3/5} p^{3r/5})$ in the case $r \geq 3$. Here it is noted that $L_{f \otimes g}(0, \chi) = 0$, if $k = l$ and χ is not trivial.

The constant α is given by

$$\alpha = \frac{12(4\pi)^{k-1}}{\Gamma(k+1)} \int \int y^{k-2} |f(x + iy)|^2 dx dy,$$

where the integral runs over a fundamental domain for $SL_2(\mathbf{Z})$ in the upper half plane. This constant is given by Rankin [7].

The author would like to express her gratitude to Professor Shigeki Egami for his comment.

2.   THE PROOF OF THEOREM

We prepare several facts for the proof of Theorem. First, recall the following functional equation of $L_{f \otimes g}(s, \chi)$, which was got by Li [6]. Let

$$\Phi_{f \otimes g}(s, \chi) = \left(\frac{2\pi}{d}\right)^{-2s} \Gamma\left(s + \frac{k-l}{2}\right) \Gamma\left(s + \frac{k+l}{2} - 1\right) L_{f \otimes g}(s, \chi), (2.1)$$

then we have

$$\Phi_{f \otimes g}(s, \chi) = C_\chi \Phi_{f \otimes g}(1 - s, \overline{\chi}),$$

where C_χ is a constant depending on χ with $|C_\chi| = 1$. When χ is a non-real primitive character mod p^r (p is a prime), Li [6] shows $C_\chi =$

$W(\chi)^4/p^{2r}$ where $W(\chi)$ is the Gauss sum. There is no real primitive character mod p^r when p is an odd prime and $r \geq 2$. These facts yield the following lemma which is analogous to Lemma 2 of the author's article [3]. Deligne's estimate of the hyper-Kloosterman sum is the key of the proof of Lemma 2 in [3]. In order to prove the following lemma we need Weinstein's result [8] for the hyper-Kloosterman sum.

Lemma. *Let p be an odd prime, $r \geq 1$ a natural number and b a constant with $(b, p) = 1$. Then we have*

$$\sum_{\chi \bmod p^r}{}' C_\chi \overline{\chi}(b) \ll \phi(p^r) p^{-r/2},$$

where $\sum'$ means the sum over the primitive characters.

Proof of Lemma. The proof is complete in [3] when $r = 1$. We prove this Lemma when $r \geq 2$. First, we consider the case $r \geq 3$. Let b^{-1} be the integer satisfying $bb^{-1} \equiv 1 \bmod p^r$.

$$\sum_{\chi \bmod p^r}{}' C_\chi \overline{\chi}(b) = \frac{1}{p^{2r}} \sum_{\chi \bmod p^r}{}' W(\chi)^4 \overline{\chi}(b)$$

$$= \frac{1}{p^{2r}} \sum_{\chi \bmod p^r}{}' \chi(b^{-1}) \prod_{j=1}^{4} \sum_{y_j=1}^{p^r} \chi(y_j) e^{2\pi i(y_j)/p^r}$$

$$= \frac{1}{p^{2r}} \sum_{\chi \bmod p^r} \chi(b^{-1}) \prod_{j=1}^{4} \sum_{y_j=1}^{p^r} \chi(y_j) e^{2\pi i(y_j)/p^r}$$

$$- \frac{1}{p^{2r}} \sum_{\chi \bmod p^{r-1}} \chi(b^{-1}) \prod_{j=1}^{4} \sum_{y_j=1}^{p^r} \chi(y_j) e^{2\pi i(y_j)/p^r}$$

$$= \frac{\phi(p^r)}{p^{2r}} \sum_{[b]} e^{2\pi i(y_1+y_2+y_3+y_4)/p^r}$$

$$- \frac{\phi(p^{r-1})}{p^{2r}} \sum_{m=0}^{p-1} \sum_{[mp^{r-1}+b']} e^{2\pi i(y_1+y_2+y_3+y_4)/p^r}$$

$$= \frac{\phi(p^r)}{p^{2r}} \sum_{[1]} e^{2\pi i(y_1+y_2+y_3+by_4)/p^r}$$

$$- \frac{\phi(p^{r-1})}{p^{2r}} \sum_{m=0}^{p-1} \sum_{[1]} e^{2\pi i(y_1+y_2+y_3+(mp^{r-1}+b')y_4)/p^r}$$

where $\sum_{[b]}$ means the sum for y_1, y_2, y_3 and y_4 running over $1 \leq y_i \leq p^r$ ($1 \leq i \leq 4$) satisfying $\prod_{i=1}^{4} y_i \equiv b \bmod p^r$ and $(y_i, p) = 1$. And b' is an integer satisfying $b \equiv b' \bmod p^{r-1}$. By using Weinstein's estimate [8] we find

$$\sideset{}{'}\sum_{\chi \bmod p^r} C_\chi \overline{\chi}(b) \ll \phi(p^r) p^{-r/2}.$$

Secondly, we consider the case $r = 2$. We divide

$$\sideset{}{'}\sum_{\chi \bmod p^2} C_\chi \overline{\chi}(b) = \sum_{\chi \bmod p^2} C_\chi \overline{\chi}(b) - \sum_{\chi \bmod p} C_\chi \overline{\chi}(b),$$

and the first term can be treated as above. There is a little difference in the treatment of the second term, but it is estimated by Lemma 2 of [3]. Hence we obtain

$$\sideset{}{'}\sum_{\chi \bmod p^2} C_\chi \overline{\chi}(b) \ll \phi(p^2) p^{-1}$$

in this case.   ∎

The above lemma is the key for getting the following estimate.

Proposition. *Let p be an odd prime, $r > 1$ a natural number and a an integer coprime to p. If $p^{3r} > x^2$, then we have*

$$\sum_{\substack{n \leq x \\ n \equiv a(p^r)}} c_n = O(x^{3/5} p^{3r/5}).$$

If $p^{3r} \leq x^2$, then we have

$$\sum_{\substack{n \leq x \\ n \equiv a(p^r)}} c_n - \frac{1}{p} \sum_{\substack{n \leq x \\ n \equiv a(p^{r-1})}} c_n = \frac{1}{\phi(p^r)} \sideset{}{'}\sum_{\chi \bmod p^r} \chi(a^{-1}) L_{f \otimes g}(0, \chi)$$

$$+ O(x^{3/5} p^{3r/5} + x^{-\varepsilon} p^{3r/2 + 4\varepsilon}),$$

where ϕ is the Euler function and $\sum'$ means the sum over primitive characters. Here if $k = l$ and χ is not trivial, then $L_{f \otimes g}(0, \chi) = 0$.

    The basic structure of the proof of Proposition is same as the argument which is used in the proof of Proposition 1 of the author [3]. (This method is used in Golubeva-Fomenko [1] and the author [3], and the idea goes back to the works of Landau [5] and Walfisz [9].) Therefore we just give a sketch of the proof of Proposition in this paper.

Proof of Proposition. We use Lemma and Hafner's results [2] on Voronoï formulas. We apply his results to $D_\rho(x) = \Gamma(\rho+1)^{-1}\sum_{n\leq x} c_n\chi(n)(x-n)^\rho$, where χ is a primitive character mod d. Then we have

$$D_\rho(x) = Q_\rho(x) + \sum_{n=1}^{\infty} \frac{4\pi^2 C_\chi c_n \overline{\chi}(n)d^{-2}}{(16\pi^4 d^{-4}n)^{1+\rho}} f_\rho\left(\frac{16\pi^4 xn}{d^4}\right), \qquad (2.2)$$

$$Q_\rho(x) = \frac{1}{2\pi i}\int_C \frac{\Gamma(s)L_{f\otimes g}(s,\chi)x^{\rho+s}}{\Gamma(s+\rho+1)}ds$$

and

$$f_\rho(x) = \frac{1}{2\pi i}\int_{C_{a,b}} \frac{\Gamma(1-s)\Gamma(s+(k-l)/2)\Gamma(s+(k+l)/2-1)x^{1+\rho-s}}{\Gamma(2+\rho-s)\Gamma(-s+(k-l)/2+1)\Gamma(-s+(k+l)/2)}ds.$$

Here, the integral paths C and $C_{a,b}$ conform to Hafner's notation. Let R be a real number satisfying $R > (k+l)/2 - 1$. The path C is the rectangle with vertices $b\pm iR$ and $1-b\pm iR$ and has positive orientation. The path $C_{a,b}$ is the oriented polygonal path with vertices $a-i\infty$, $a-iR$, $b-iR$, $b+iR$, $a+iR$ and $a+i\infty$. In our case, $a = 0$ and $b > (k+l)/2-1$ (see Hafner [2]). From the definition, we see that $\frac{d}{dx}D_\rho(x) = D_{\rho-1}(x)$ and there are the analogous relations for $Q_\rho(x)$ and $f_\rho(x)$. Hafner [2] showed the asymptotic expansion of $f_\rho(x)$ for $x \geq 1$. (Actually Hafner stated that it holds for $x > 0$, but this is a slip.)

We start explaining the sketch of the proof of Proposition. The Voronoï formula (2.2) with $\rho = 2$ implies

$$\frac{1}{2}\sum_{n\leq x} c_n(x-n)^2 \sum_{\chi \bmod p^r}{}' \chi(a^{-1}n) = \sum_{\chi \bmod p^r}{}' Q_2(x)\chi(a^{-1})$$

$$+ \sum_{n=1}^{\infty} \frac{c_n p^{10r}}{(2\pi)^{10}n^3} f_2\left(\frac{16\pi^4 xn}{p^{4r}}\right) \sum_{\chi \bmod p^r}{}' c_n\overline{\chi}(na). \qquad (2.3)$$

The left-hand side of (2.3) is equal to

$$\frac{\phi(p^r)}{2} \sum_{\substack{n\leq x \\ n\equiv a\ (p^r)}} c_n(x-n)^2 - \frac{\phi(p^{r-1})}{2} \sum_{\substack{n\leq x \\ n\equiv a\ (p^{r-1})}} c_n(x-n)^2. \qquad (2.4)$$

Let τ be a real number satisfying $x^\varepsilon < \tau \leq x$, and we define the operator Δ_τ as

$$\Delta_\tau(h(x)) = h(x+2\tau) - 2h(x+\tau) + h(x), \qquad (2.5)$$

where $h(x)$ is a function. We consider the operation of Δ_τ to (2.3) and (2.4). Then we get the following result.

$$\Delta_\tau\left(\frac{\phi(p^r)}{2}\sum_{\substack{n\leq x \\ n\equiv a\,(p^r)}}c_n(x-n)^2\right) = \phi(p^r)\int_x^{x+\tau}\int_v^{v+\tau}\sum_{\substack{n\leq w \\ n\equiv a\,(p^r)}}c_n\,dwdv$$

$$= \phi(p^r)\int_x^{x+\tau}\int_v^{v+\tau}\left(\sum_{\substack{n\leq x \\ n\equiv a\,(p^r)}}c_n + \sum_{\substack{x<n\leq w \\ n\equiv a\,(p^r)}}c_n\right)dwdv$$

$$= \tau^2\phi(p^r)\sum_{\substack{n\leq x \\ n\equiv a\,(p^r)}}c_n + O(\phi(p^r)\tau^3). \tag{2.6}$$

In the same way, we have

$$\Delta_\tau\left(\frac{\phi(p^{r-1})}{2}\sum_{\substack{n\leq x \\ n\equiv a\,(p^{r-1})}}c_n(x-n)^2\right) = \tau^2\phi(p^{r-1})\sum_{\substack{n\leq x \\ n\equiv a\,(p^{r-1})}}c_n$$

$$+ O(\phi(p^{r-1})\tau^3). \tag{2.7}$$

From the definition of $Q_\rho(x)$, we get

$$\Delta_\tau\left(\sum_{\chi\bmod p^r}{}'Q_2(x)\chi(a^{-1})\right) = \tau^2\sum_{\chi\bmod p^r}{}'\chi(a^{-1})L_{f\otimes g}(0,\chi). \tag{2.8}$$

If $k = l$ and χ is a non-trivial character, from the functional equation (2.1) and the Euler product of $L_{f\otimes g}(s,\chi)$ we find that $L_{f\otimes g}(0,\chi) = 0$. We estimate the remaining part

$$\Delta_\tau\left(\sum_{n=1}^\infty\frac{c_np^{10r}}{(2\pi)^{10}n^3}f_2\left(\frac{16\pi^4xn}{p^{4r}}\right)\sum_{\chi\bmod p^r}{}'C_\chi\overline{\chi}(na)\right) \tag{2.9}$$

by using Lemma. And we have to study $\Delta_\tau(f_2(16\pi^4xn/p^{4r}))$ which is a part of (2.9). The relation $\frac{d}{dx}f_\rho(x) = f_{\rho-1}(x)$ implies

$$\Delta_\tau\left(f_2\left(\frac{16\pi^4xn}{p^{4r}}\right)\right) = \tau^2\left(\frac{2\pi}{p^r}\right)^8\int_x^{x+\tau}\int_v^{v+\tau}f_0\left(\frac{16\pi^4xn}{p^{4r}}\right)dwdv.$$

Using the mean value theorem, we find that there is a ξ in $[x, x+2\tau]$ satisfying

$$\Delta_\tau\left(f_2\left(\frac{16\pi^4xn}{p^{4r}}\right)\right) = \tau^2\left(\frac{2\pi}{p^r}\right)^8\tau^2 f_0\left(\frac{16\pi^4\xi n}{p^{4r}}\right). \tag{2.10}$$

We estimate the sum over $n > x^3 \tau^{-4} p^{4r}$ in (2.9) by using Hafner's estimate of $f_2(x)$. Then this part is estimated as $O(\tau^{1/2} x^{3/2} p^{3r/2} \phi(p^r))$. The estimate of the sum over $p^{4r}/16\pi^4 < n \le x^3 \tau^{-4} p^{4r}$ is obtained by using (2.10) and Hafner's estimate of $f_0(x)$. In fact, we can estimate it as $O(\tau^{1/2} x^{3/2} p^{3r/2} \phi(p^r))$. The reason of the restriction $p^{4r}/16\pi^4 < n$ is that Hafner's estimate of $f_\rho(x)$ can not use in $x < 1$. The estimate of the remaining part sum over $p^{4r}/16\pi^4 \ge n$ can be got by using (2.10) and moving the integral path of f_0 appropriately, similarly to the argument in [3]. We are able to estimate this part of (2.9) as $O(\tau^2 x^{-\epsilon} p^{3r/2+4\epsilon} \phi(p^r))$.

Collecting the above results, we obtain

$$\sum_{\substack{n \le x \\ n \equiv a \ (p^r)}} c_n - \frac{1}{p} \sum_{\substack{n \le x \\ n \equiv a \ (p^{r-1})}} c_n = \frac{1}{\phi(p^r)} \sum_{\chi \bmod p^r}{}' \chi(a^{-1}) L_{f \otimes g}(0, \chi)$$

$$+ O\left(\tau + x^{3/2} p^{3r/2} \tau^{-3/2} + p^{3r/2+4\epsilon} x^{-\epsilon}\right).$$

We put $\tau = x^{3/5} p^{3r/5}$, then we complete the proof of the latter half of Proposition. The first half is proved easily by using $\sum_{n \le x} |c_n| \ll x$. ∎

The claim of Proposition, (1.1) and (1.2) give the proof of Theorem by using induction.

References

[1] E. P. Golubeva and O. M. Fomenko, *Values of Dirichlet series associated with modular forms at the points $s = 1/2$, 1,* J. Soviet Math. **36** (1987) 79–93.

[2] J. L. Hafner, *On the representation of the summatory functions of a class of arithmetical functions,* Lec. Notes in Math. **899** (1981) 148–165.

[3] Y. Ichihara, *The evaluation of the sum over arithmetic progressions for the coefficients of the Rankin-Selberg series,* preprint.

[4] A. Ivić, K. Matsumoto and Y. Tanigawa, *On Riesz means of the coefficients of the Rankin-Selberg series,* Math. Proc. Camb. Phil. Soc. **127** (1999) 117–131.

[5] E. Landau, *Zur analytischen Zahlentheorie der definiten quadratischen Formen(Über die Gitterpunkte in einem mehrdimensionalen Ellipsoid),* Berliner Akademieberichte (1915) 458–476.

[6] W. Li, *L-series of Rankin-type and their functional equations,* Math. Ann. **244** (1979) 135–166.

[7] R. A. Rankin, *Contribution to the theory of Ramanujan's function $\tau(n)$ and similar arithmetical functions II*, Proc. Camb. Phil. Soc. **35** (1939) 357–372.

[8] L. Weinstein, *The hyper-Kloosterman sum*, Enseignement. Math. (2) **27** (1981) 29–40.

[9] A. Walfisz, *Über die Koeffizientensummen einiger Modulformen*, Math. Ann. **108**, No1 (1933) 75–90.

SUBSTITUTIONS, ATOMIC SURFACES, AND PERIODIC BETA EXPANSIONS

Shunji ITO
and Yuki SANO
Department of Mathematics and Computer Science, Tsuda College, 2-1-1, Tsuda-Machi,
Kodaira, Tokyo, 187-8577, Japan
ito@tsuda.ac.jp
sano@tsuda.ac.jp

Keywords: Beta-expansion, periodic points, substitution

Abstract   We study the pure periodicity of β-expansions where β is a Pisot number satisfing the following two conditions: the β-expansion of 1 is equal to $k_1 k_2 \ldots k_{d-1} 1$, $k_i \geq 0$, and the minimal polynomial of β is given by $x^d - k_1 x_{d-1} - \cdots - k_{d-1} x - 1$. From the substitution associated with the Pisot number β, a domain with a fractal boundary, called atomic surface, is constructed. The essential point of the proof is to define a natural extension of the β-transformation on a d-dimensional product space which consists of the unit interval and the atomic surface.

1991 Mathematics Subject Classification: 34C35; Secondary 58F22.

0.   INTRODUCTION

Let β be a real number and let T_β be the β-transformation on the unit interval $[0,1) : T_\beta : x \to \beta x \pmod 1$. Then any real number $x \in [0,1)$ is represented by $d_\beta(x) = (d_i)_{i \geq 1}$ where $d_i = \lfloor \beta T_\beta^{i-1}(x) \rfloor$. Here denote by $\lfloor y \rfloor$ the integer part of a number y. This representation of real numbers with a base β is called the β-expansion, which was introduced by Rényi [9]. Parry [7] characterized the set $\{d_\beta(x) | x \in [0,1)\}$.

A real number $x \in [0,1)$ is said to have an eventually periodic β-expansion if $d_\beta(x)$ is of the form uv^∞, where v^∞ will be denoted the sequence $vvv \ldots$. In particular, when $d_\beta(x)$ is of the form v^∞, x is said to have a purely periodic β-expansion.

For $x = 1$, we can also define the β-expansion of 1 in the same way: $d_\beta(1) = d_1 d_2 \ldots$, $d_i = \lfloor \beta T_\beta^{i-1}(1) \rfloor$. However the region of the definition

C. Jia and K. Matsumoto (eds.), Analytic Number Theory, 183–194.
© 2002 *Kluwer Academic Publishers. Printed in the Netherlands.*

of T_β is the right open interval $[0,1)$ which doesn't contain the element 1 unless we write $T_\beta(1)$ in this paper.

Bertrand [3] and K. Schmidt [11] studied eventually periodic β-expansions. A Pisot number is an algebraic integer (> 1) whose conjugates other than itself have modulus less than one. Let $\mathbb{Q}(\beta)$ be the smallest extension field of rational numbers $\mathbb{Q}$ containing β.

Theorem 0.1 (Bertrand, K. Schmidt). *Let β be a Pisot number and let x be a real number of $[0,1)$. Then x has an eventually β-expansion if and only if $x \in \mathbb{Q}(\beta)$.*

In [1], Akiyama investigated sufficient condition of pure periodicity where β belongs to a certain class of Pisot numbers. Authors [6] characterized numbers having purely periodic β-expansions where β is a Pisot number satisfying the polynomial $Irr(\beta) = x^3 - k_1 x^2 - k_2 x - 1, 0 \le k_2 \le k_1 \ne 0$. In [10], one of the authors gives necessary and sufficient condition of pure periodicity where β is a Pisot number whose minimal polynomial is given by

$$Irr(\beta) = x^d - k_1 x^{d-1} - k_2 x^{d-2} - \cdots - k_{d-1} x - 1,$$
$$k_i \in \mathbb{Z}, \text{ and } k_1 \ge k_2 \ge \cdots \ge k_{d-1} \ge 1.$$

And the main tool of the proof is a domain X with a fractal boundary. Many properties of this domain X from [2] are quoted in [10]. In this paper, instead of using these properties we define the domain X in another way, called atomic surface, using a graph associated with a substitution. Moreover, let β belong to a larger class of Pisot numbers. Hereafter, β is a Pisot number whose minimal polynomial is

$$Irr(\beta) = x^d - k_1 x^{d-1} - k_2 x^{d-2} - \cdots - k_{d-1} x - 1, \ k_i \ge 0 \qquad (1)$$

and which the β-expansion of 1 is

$$d_\beta(1) = k_1 k_2 \ldots k_{d-1} 1. \qquad (2)$$

From (2) we can see $k_1 = \lfloor \beta \rfloor \ge 1$ and $0 \le k_i \le k_1$ for any $2 \le i \le d-1$. Then we have the following result:

Theorem 0.2 (Main Theorem). *Let x be a real number in $\mathbb{Q}(\beta) \cap [0,1)$. Then x has a purely periodic β-expansion if and only if x is reduced.*

We define "reduced" in section 3. For our goal we introduce a d-dimensional domain $\widehat{Y}$ from the domain X and define a natural extension of T_β on $\widehat{Y}$. The main line of the proof is almost same as the way in [10]. In this paper, we would like to make a point of constructing the domain X.

1. ATOMIC SURFACES

Let $\mathcal{A}$ be an alphabet of d letters $\{1, 2, \ldots, d\}$. The free monoid on $\mathcal{A}$, that is to say, the set of finite words on $\mathcal{A}$, is denoted by $\mathcal{A}^* = \bigcup_{n=0}^{\infty} \mathcal{A}^n$. A substitution σ is a map from $\mathcal{A}$ to $\mathcal{A}^*$, such that, $\sigma(i)$ is a non-empty word for any i. The substitution σ extends in a natural way to an endomorphism on $\mathcal{A}^*$ by the rule $\sigma(UV) = \sigma(U)\sigma(V)$ for $U, V \in \mathcal{A}^*$.

For any $i \in \mathcal{A}$ we note $\sigma(i) = W^{(i)} = W_1^{(i)} W_2^{(i)} \ldots W_{l_i}^{(i)}$, $W_n^{(i)} \in \mathcal{A}$. We also write $\sigma(i) = W^{(i)} = P_n^{(i)} W_n^{(i)} S_n^{(i)}$ where $P_n^{(i)} = W_1^{(i)} \ldots W_{n-1}^{(i)}$ is the prefix of the length $n-1$ of the letter $W_n^{(i)}$ (this is an empty word for $n = 1$) and $S_n^{(i)} = W_{n+1}^{(i)} \ldots W_{l_i}^{(i)}$ is the suffix of the length $l_i - n$ of the letter $W_n^{(i)}$ (this is an empty word for $n = l_i$).

There is a natural homomorphism (abelianization) $f : \mathcal{A}^* \to \mathbb{Z}^d$ given by $f(i) = \mathbf{e}_i$ for any $i \in \mathcal{A}$ where $\{\mathbf{e}_1, \ldots, \mathbf{e}_d\}$ is the canonical basis of $\mathbb{R}^d$. For any finite word $W \in \mathcal{A}^*$, $f(W) = {}^t(x_1, \ldots, x_d)$. Here ^{t} indicates the transpose. We know that each x_i means the number of occurrences of the letter i in W. Then there exists a unique linear transformation ${}^0\sigma$ satisfying the following commutative diagram:

$$
\begin{array}{ccc}
\mathcal{A}^* & \xrightarrow{\sigma} & \mathcal{A}^* \\
f \downarrow & & \downarrow f \\
\mathbb{Z}^d & \xrightarrow[{}^0\sigma]{} & \mathbb{Z}^d.
\end{array}
$$

It is easily checked that the matrix ${}^0\sigma$ is given by ${}^0\sigma = (f(\sigma(1)), \ldots, f(\sigma(d)))$. Hence each $\left({}^0\sigma\right)_{ij}$, which means the (i, j)th entry of the matrix ${}^0\sigma$, represents the number of occurrences of the letter i in $\sigma(j)$.

Let σ be the substitution

$$\sigma(1) = \underbrace{1 \ldots 1}_{k_1} 2, \quad \sigma(2) = \underbrace{1 \ldots 1}_{k_2} 3, \quad \ldots, \quad \sigma(d-1) = \underbrace{1 \ldots 1}_{k_{d-1}} d, \quad \sigma(d) = 1. \quad (3)$$

Then the matrix ${}^0\sigma$ is

$$
{}^0\sigma = \begin{pmatrix}
k_1 & k_2 & \cdots & k_{d-1} & 1 \\
1 & 0 & \cdots & 0 & 0 \\
0 & 1 & \cdots & 0 & 0 \\
\vdots & \vdots & \ddots & \vdots & \vdots \\
0 & 0 & \cdots & 1 & 0
\end{pmatrix}.
$$

We see that the characteristic polynomial of ${}^0\sigma$ is $Irr(\beta)$. The eigenvector corresponding to the eigenvalue β of ${}^0\sigma$ and the eigenvector corresponding to β of the transpose of ${}^0\sigma$ are denoted by ${}^t(\alpha_1, \ldots, \alpha_d)$ and

$^{t}(\gamma_1,\ldots,\gamma_d)$. Since the matrix $^{0}\sigma$ is primitive, the Perron-Frobenius theory shows both eigenvectors are positive. A nonnegative matrix A is primitive if $A^{N} > 0$ for some $N \geq 1$. We put $\alpha_1 = \gamma_1 = 1$. The elements α_i and γ_i are given by

$$\alpha_1 = 1, \qquad \alpha_2 = \beta^{-1}, \qquad \alpha_3 = \beta^{-2}, \qquad \ldots, \qquad \alpha_d = \beta^{-(d-1)},$$

$$\gamma_1 = 1, \qquad \gamma_2 = T_\beta(1), \qquad \gamma_3 = T_\beta^2(1), \qquad \ldots, \qquad \gamma_d = T_\beta^{d-1}(1) = \frac{1}{\beta}.$$

Let $\mathcal{P}$ be the contractive invariant plane of $^{0}\sigma$, that is,

$$\mathcal{P} = \left\{ \mathbf{x} \in \mathbb{R}^d \mid \langle \mathbf{x}, {}^{t}(\gamma_1,\ldots,\gamma_d)\rangle = 0 \right\},$$

where $\langle\ ,\ \rangle$ indicates the standard inner product. Let $\pi : \mathbb{R}^d \to \mathcal{P}$ be the projection along the eigenvector $^{t}(\alpha_1,\ldots,\alpha_d)$.

Let us define the graph G with vertex set $\mathcal{V}(G) = \{1, 2, \ldots, d\}$ and edge set $\mathcal{E}(G)$. There is one edge $\begin{pmatrix} j \\ k \end{pmatrix}$ from the vertex i to the vertex j if $W_k^{(j)} = i$. Otherwise, there is no edge. This graph was introduced by Sh. Ito and Ei in [5].

Let X_G be the edge shift, that is,

$$X_G = \left\{ \xi = (\xi_n)_{n\in\mathbb{N}} \in \mathcal{E}(G)^{\mathbb{N}} \mid T(\xi_n) = I(\xi_{n+1}) \text{ for all } n \in \mathbb{N} \right\}.$$

Here each edge $e \in \mathcal{E}(G)$ starts at a vertex denoted by $I(e) \in \mathcal{V}(G)$ and terminates at a vertex $T(e) \in \mathcal{V}(G)$.

On the plane $\mathcal{P}$ we can define the sets X_i $(1 \leq i \leq d)$ and X using the edge shift X_G as follows:

$$X_i := \left\{ -\sum_{n=1}^{\infty} {}^{0}\sigma^{n-1}\pi\left(f\left(P_{k_n}^{(j_n)}\right)\right) \mid I\begin{pmatrix} j_1 \\ k_1 \end{pmatrix} = i \text{ and } \begin{pmatrix} j_n \\ k_n \end{pmatrix}_{n\in\mathbb{N}} \in X_G \right\},$$

$$X := \left\{ -\sum_{n=1}^{\infty} {}^{0}\sigma^{n-1}\pi\left(f\left(P_{k_n}^{(j_n)}\right)\right) \mid \begin{pmatrix} j_n \\ k_n \end{pmatrix}_{n\in\mathbb{N}} \in X_G \right\} = \bigcup_{i=1}^{d} X_i.$$

The set X is called the atomic surface associated with the substitution σ. (See Figure 1 in case $k_1 = k_2 = 1$.)

We see that X is bounded since the set

$$\left\{ f\left(P_k^{(j)}\right) \mid \begin{pmatrix} j \\ k \end{pmatrix} \in \mathcal{E}(G) \right\}$$

is a finite set and $^{0}\sigma$ is a contractive map on the plane $\mathcal{P}$. Moreover, $\begin{pmatrix} 1 \\ 1 \end{pmatrix}^{\infty} \in X_G$ and $I\begin{pmatrix} 1 \\ 1 \end{pmatrix} = 1$ show that the origin point $\mathbf{0}$ is in X_1.

Figure 1   The atomic surface ($k_1 = k_2 = 1$).

Proposition 1.1. *The sets X_i ($1 \leq i \leq d$) are closed.*

Proof. We take a sequence $\{x_l\}_{l=1}^{\infty}$ satisfing $x_l \in X_i$ and $x_l \to x$ ($l \to \infty$) for some x. For each l, we can put

$$x_l = - \sum_{n=1}^{\infty} {}^0\sigma^{n-1}\pi\left(f\left(P_{k_n(l)}^{(j_n(l))}\right)\right) \text{ for some } \begin{pmatrix} j_n(l) \\ k_n(l) \end{pmatrix}_{n \in \mathbb{N}} \in X_G.$$

Since the edge set $\mathcal{E}(G)$ is finite, there exists $\begin{pmatrix} j_1' \\ k_1' \end{pmatrix} \in \mathcal{E}(G)$ such that $\begin{pmatrix} j_1' \\ k_1' \end{pmatrix} = \begin{pmatrix} j_1(l) \\ k_1(l) \end{pmatrix}$ for infinitely many l. Thus we take the subsequence $\{x_{l_m}\}_{m=1}^{\infty}$ satisfying $\begin{pmatrix} j_1' \\ k_1' \end{pmatrix} = \begin{pmatrix} j_1(l_m) \\ k_1(l_m) \end{pmatrix}$. Similarly, we choose for all n $\begin{pmatrix} j_n' \\ k_n' \end{pmatrix} \in \mathcal{E}(G)$ and the subsequence $\{x_{l_s}\}_{s=1}^{\infty}$ such that $\begin{pmatrix} j_n' \\ k_n' \end{pmatrix} = \begin{pmatrix} j_n(l_s) \\ k_n(l_s) \end{pmatrix}$. If we put

$$x' = - \sum_{n=1}^{\infty} {}^0\sigma^{n-1}\pi\left(f\left(P_{k_n'}^{(j_n')}\right)\right) \in X_i,$$

then $x_{l_s} \to x'$ ($s \to \infty$). Therefore $x_l \to x'$ ($l \to \infty$). Hence we see $x = x' \in X_i$. This implies that the sets X_i are closed. $\qquad\square$

Moreover, the set X has the following property:

$$\bigcup_{z \in \mathcal{L}} (X + z) = \mathcal{P},$$

where $\mathcal{L}$ is the lattice given by $\mathcal{L} = \left\{ \sum_{i=2}^{d} n_i \pi (\mathbf{e}_1 - \mathbf{e}_i) \mid n_i \in \mathbb{Z} \right\}$. See the details in [4]. From the Baire-Hausdorff theory, we see that the set X has at least one inner point. Then $|X|$ is positive, where we note $|K|$ the measure of a set K. In [4], it is implied that X is the closure of the interior of X.

Proposition 1.2. *For any $1 \le i \le d$, the following set equations hold:*

$$
{}^0\sigma^{-1} X_i = \bigcup_{j=1}^{d} \bigcup_{P_k^{(j)} : W_k^{(j)} = i} \left(X_j - {}^0\sigma^{-1} \pi f(P_k^{(j)}) \right). \tag{4}
$$

Proof. The definitions of X_i imply that

$$
{}^0\sigma^{-1} X_i
$$

$$
= \left\{ - \sum_{n=1}^{\infty} {}^0\sigma^{n-2} \pi \left(f \left(P_{k_n}^{(j_n)} \right) \right) \; \middle| \; I \begin{pmatrix} j_1 \\ k_1 \end{pmatrix} = i \text{ and } \begin{pmatrix} j_n \\ k_n \end{pmatrix}_{n \in \mathbb{N}} \in X_G \right\},
$$

substituting $\begin{pmatrix} j \\ k \end{pmatrix}$ for $\begin{pmatrix} j_1 \\ k_1 \end{pmatrix}$ and $\begin{pmatrix} j_{n-1} \\ k_{n-1} \end{pmatrix}$ for $\begin{pmatrix} j_n \\ k_n \end{pmatrix}$ for all $n \ge 2$,

$$
= \left\{ - {}^0\sigma^{-1} \pi \left(f \left(P_k^{(j)} \right) \right) - \sum_{n=1}^{\infty} {}^0\sigma^{n-1} \pi \left(f \left(P_{k_n}^{(j_n)} \right) \right) \right.
$$

$$
\left. \middle| \; W_k^{(j)} = i, I \begin{pmatrix} j_1 \\ k_1 \end{pmatrix} = j, \text{ and } \begin{pmatrix} j_n \\ k_n \end{pmatrix}_{n \in \mathbb{N}} \in X_G \right\}
$$

$$
= \bigcup_{j=1}^{d} \bigcup_{P_k^{(j)} : W_k^{(j)} = i} \left(X_j - {}^0\sigma^{-1} \pi f(P_k^{(j)}) \right).
$$

We can get the set equations above.   □

Applying ${}^0\sigma$ to the equation (4), from the form of the substitution σ in the equation (3), we have

$$
X_1 = \bigcup_{i_1=0}^{k_1-1} ({}^0\sigma X_1 - i_1 \pi \mathbf{e}_1) \cup \cdots \cup \bigcup_{i_{d-1}=0}^{k_{d-1}-1} ({}^0\sigma X_{d-1} - i_{d-1} \pi \mathbf{e}_1) \cup {}^0\sigma X_d,
$$

$$
X_2 = {}^0\sigma X_1 - k_1 \pi \mathbf{e}_1,
$$

$$
\vdots
$$

$$
X_d = {}^0\sigma X_{d-1} - k_{d-1} \pi \mathbf{e}_1,
$$

$$X = \bigcup_{i=1}^{d} X_i$$

$$= \bigcup_{i_1=0}^{k_1} \left({}^0\sigma X_1 - i_1 \pi \mathbf{e}_1\right) \cup \cdots \cup \bigcup_{i_{d-1}=0}^{k_{d-1}} \left({}^0\sigma X_{d-1} - i_{d-1}\pi \mathbf{e}_1\right) \cup {}^0\sigma X_d.$$

$$(5)$$

In order to know X_i are disjoint each other up to a set of measure 0, we would prepare lemmas. The next result can be found in [2], originally in [8].

Lemma 1.3. *Let M be a primitive matrix with a maximal eigenvalue λ. Suppose that $\mathbf{v}$ is a positive vector such that $M\mathbf{v} \geq \lambda\mathbf{v}$. Then the inequality is an equality and $\mathbf{v}$ is the eigenvector with respect to λ.*

Lemma 1.4. *The vector of volumes ${}^t (|X_i|)_{1\leq i\leq d}$ satisfies the following inequality:*

$$
{}^0\sigma \begin{pmatrix} |X_1| \\ \vdots \\ |X_d| \end{pmatrix} \geq \beta \begin{pmatrix} |X_1| \\ \vdots \\ |X_d| \end{pmatrix}.
$$

Proof. From the form of X_i in the equation (4), we see

$$\left|{}^0\sigma^{-1} X_i\right| \leq \sum_{j=1}^{d} \left({}^0\sigma\right)_{ij} |X_j|.$$

Since the determinant of ${}^0\sigma^{-1}$ restricted to $\mathcal{P}$ is β, we know that $\left|{}^0\sigma^{-1} X_i\right| = \beta |X_i|$. Hence we arrive at the conclusion. $\qquad\square$

Two lemmas above imply the following result.

Corollary 1.5. *The sets X_i $(1 \leq i \leq d)$ are disjoint up to a set of measure 0. Therefore the atomic surface has the partition (5).*

Proof. Lemmas 1.3 and 1.4 show that the vector of volumes ${}^t (|X_i|)_{1\leq i\leq d}$ is the eigenvector corresponding to β of ${}^0\sigma$. Since the inequality become the equality in Lemma 1.4, for each i and j with $W_k^{(j)} = i$ the sets $\left(X_j - {}^0\sigma^{-1}\pi f(P_k^{(j)})\right)$ are disjoint up to a set of measure 0.

In the case $k_i \geq 1$, we take $i = k = 1$ and it is implied that $f(P_k^{(j)}) = 0$ for all j. So that X_j are disjoint up to a set of measure 0. Applying ${}^0\sigma$, we know the atomic surface has the partition (5) whose elements are disjoint each other up to a set of measure 0.

In case $k_i = 0$ for some i, for all j there exists N such that $\sigma^N(j) = 1v_j$ for some $v_j \in \mathcal{A}^*$. Then the disjointness is proved in the same way. $\qquad\square$

2.   NATURAL EXTENSIONS

Sections 2 and 3 base on the paper [10]. Therefore we would like to give an outline of the way to our main theorem. For the details, see [10].
Let

$$\beta = \beta^{(1)}, \beta^{(2)}, \ldots, \beta^{(r_1)}$$

be the real Galois conjugates and

$$\beta^{(r_1+1)}, \overline{\beta^{(r_1+1)}}, \beta^{(r_1+2)}, \overline{\beta^{(r_1+2)}}, \ldots, \beta^{(r_1+r_2)}, \overline{\beta^{(r_1+r_2)}}$$

be the complex Galois conjugates of β, where $r_1 + 2r_2 = n$ and $\bar{v}$ is the complex conjugate of a complex number v. The corresponding conjugates of $x \in \mathbb{Q}(\beta)$ are also denoted by

$$x = x^{(1)}, \ldots, x^{(r_1)}, x^{(r_1+1)}, \overline{x^{(r_1+1)}}, \ldots, x^{(r_1+r_2)}, \overline{x^{(r_1+r_2)}}.$$

Let $\widehat{X}_i \subset \mathbb{R}^d$ $(1 \leq i \leq d)$ be the domains

$$\widehat{X}_i = \{t \cdot \omega \alpha + \mathbf{x} \mid 0 \leq t < \gamma_i \text{ and } \mathbf{x} \in X_i\}$$

and let $\widehat{X} = \bigcup_{i=1}^d \widehat{X}_i$. Here $\omega = 1/(\alpha_1\gamma_1 + \cdots + \alpha_d\gamma_d)$ $(\in \mathbb{Q}(\beta))$ and $\alpha = {}^t(\alpha_1, \ldots, \alpha_d)$. Let $\widehat{T_\beta}$ be the transformation on $\widehat{X}$ given by

$$\widehat{T_\beta}(t \cdot \omega \alpha + \mathbf{x}) = (\beta t - \lfloor \beta t \rfloor) \cdot \omega \alpha + {}^0 \sigma \mathbf{x} - \lfloor \beta t \rfloor \pi \mathbf{e}_1.$$

Then we have the following result. Therefore $\widehat{T_\beta}$ is the natural extension of the transformation T_β.

Proposition 2.1. $\widehat{T_\beta}$ *is surjective and injective except the boundary on* $\widehat{X}$.

We put

$$Q := \begin{pmatrix} \alpha_1^{(1)} & \cdots & \alpha_1^{(r_1)} & \Re\alpha_1^{(r_1+1)} & -\Im\alpha_1^{(r_1+1)} & \cdots & \Re\alpha_1^{(r_1+r_2)} & -\Im\alpha_1^{(r_1+r_2)} \\ \alpha_2^{(1)} & \cdots & \alpha_2^{(r_1)} & \Re\alpha_2^{(r_1+1)} & -\Im\alpha_2^{(r_1+1)} & \cdots & \Re\alpha_2^{(r_1+r_2)} & -\Im\alpha_2^{(r_1+r_2)} \\ \vdots & & \vdots & \vdots & \vdots & & \vdots & \vdots \\ \alpha_d^{(1)} & \cdots & \alpha_d^{(r_1)} & \Re\alpha_d^{(r_1+1)} & -\Im\alpha_d^{(r_1+1)} & \cdots & \Re\alpha_d^{(r_1+r_2)} & -\Im\alpha_d^{(r_1+r_2)} \end{pmatrix}$$

where $\Re$ indicates the real part and $\Im$ indicates the imaginary part. Let us define the domains $\widehat{Y}$ and $\widehat{Y}_i$ $(1 \leq i \leq d)$ as follows:

$$\widehat{Y} := Q^{-1}(\widehat{X}) \quad \text{and} \quad \widehat{Y}_i := Q^{-1}(\widehat{X}_i).$$

Because of the change of bases, we can write $\widehat{Y_i}$ as domains in $\mathbb{R} \times \mathbb{R}^{d-1}$. And naturally, we can define the transformation $\widehat{S_\beta}$ on $\widehat{Y}$ as follows:

$$\widehat{S_\beta} := Q^{-1} \circ \widehat{T_\beta} \circ Q.$$

Then $\widehat{S_\beta}$ is also the natural extension of T_β. Here we put

$$R = \left(\beta^{(2)} \right) \oplus \cdots \oplus \left(\beta^{(r_1)} \right)$$

$$\oplus \left(\begin{array}{cc} \Re\beta^{(r_1+1)} & -\Im\beta^{(r_1+1)} \\ \Im\beta^{(r_1+1)} & \Re\beta^{(r_1+1)} \end{array} \right) \oplus \cdots \oplus \left(\begin{array}{cc} \Re\beta^{(r_1+r_2)} & -\Im\beta^{(r_1+r_2)} \\ \Im\beta^{(r_1+r_2)} & \Re\beta^{(r_1+r_2)} \end{array} \right)$$

and

$$\mathbf{v} = {}^t \left(\omega^{(2)}, \ldots, \omega^{(r_1)}, 2\Re\omega^{(r_1+1)}, 2\Im\omega^{(r_1+1)}, \ldots, 2\Re\omega^{(r_1+r_2)}, 2\Im\omega^{(r_1+r_2)} \right),$$

where $A \oplus B$ is a matrix of a form:

$$\left(\begin{array}{cc} A & 0 \\ 0 & B \end{array} \right).$$

Then we have the following result.

Proposition 2.2. *The transformation $\widehat{S_\beta}$ on $\widehat{Y}$ is given by*

$$\widehat{S_\beta}(t\omega, \mathbf{x}) = ((\beta t - \lfloor \beta t \rfloor)\omega, R\mathbf{x} - \lfloor \beta t \rfloor \mathbf{v})$$

and surjective.

3.   MAIN THEOREM

In this section, we would like to introduce the definition of "reduced" and give a survey of the proof of our main theorem. You can see the complete proof in [10].

Let $\widetilde{Y}$ ($\subset \mathbb{R} \times \mathbb{R}^{d-1}$) be the product space: $\widetilde{Y} := [0, \omega) \times \mathbb{R}^{d-1}$. Let $\widetilde{S_\beta}$ be the transformation on $\widetilde{Y}$ defined by

$$\widetilde{S_\beta}(\omega x, \mathbf{x}) := (\omega(\beta x - \lfloor \beta x \rfloor), R\mathbf{x} - \lfloor \beta x \rfloor \mathbf{v}), \quad x \in [0, 1).$$

Then the restriction of $\widetilde{S_\beta}$ on $\widehat{Y}$ ($\subset \widetilde{Y}$) is $\widehat{S_\beta}$.

Define a map $\rho : \mathbb{Q}(\beta) \to \mathbb{R} \times \mathbb{R}^{d-1}$ by

$$\rho(x) = \left(x, {}^t \left(x^{(2)}, \ldots, x^{(r_1)}, 2\Re x^{(r_1+1)}, 2\Im x^{(r_1+1)}, \ldots, 2\Re x^{(r_1+r_2)}, 2\Im x^{(r_1+r_2)} \right) \right).$$

Definition 3.1. A real number $x \in \mathbb{Q}(\beta) \cap [0, 1)$ is reduced if $\rho(\omega x) \in \widehat{Y}$.

We easily get the next result from the definitions of $\widetilde{S}_\beta$, ρ, and T_β.

Lemma 3.1. *Let* $x \in \mathbb{Q}(\beta) \cap [0, 1)$. *Then*

$$\widetilde{S}_\beta \left(\rho(\omega x) \right) = \rho \left(\omega \cdot T_\beta(x) \right).$$

Lemma 3.2. *Let* $x \in \mathbb{Q}(\beta) \cap [0, 1)$ *be reduced. Then*

1. $T_\beta(x)$ *is reduced,*

2. *there exists* x^* *such that* x^* *is reduced and* $T_\beta(x^*) = x$.

Proof. Since $x \in \mathbb{Q}(\beta) \cap [0, 1)$ is reduced, $\rho(\omega x) \in \widehat{Y}$.
1. From Lemma 3.1, $\widehat{S}_\beta \left(\rho(\omega x) \right) = \rho \left(\omega \cdot T_\beta(x) \right) \in \widehat{Y}$. Hence $T_\beta(x)$ is reduced.
2. From Proposition 2.2, $\widehat{S}_\beta$ is surjective on $\widehat{Y}$. Thus there exist $(\omega x^*, \mathbf{x}) \in \widehat{Y}$ such that $\widehat{S}_\beta (\omega x^*, \mathbf{x}) = \rho(\omega x)$. Hence $T_\beta(x^*) = x$. And we can show that $(\omega x^*, \mathbf{x}) = \rho(\omega x^*)$. Then $\rho(\omega x^*) \in \widehat{Y}$ implies x^* is reduced. $\qquad\square$

By the lemmas above, we can get sufficient condition of pure periodicity of β-expansions.

Proposition 3.3. *Let* $x \in \mathbb{Q}(\beta) \cap [0, 1)$ *be reduced. Then* x *has a purely periodic* β-*expansion.*

Proof. Lemma 3.2 2 shows that there exist x_i^* $(i \geq 0)$ such that x_i^* are reduced and $T_\beta(x_i^*) = x_{i-1}^*$, where we set $x_0^* = x$. As $\widehat{Y}$ is bounded, we can see the set $\{x_i^*\}_{i=0}^\infty$ is a finite set. Hence there exist j and k $(j > k)$ such that $x_j^* = x_{j-k}^*$. Hence $T_\beta^k(x) = x$. Therefore x has a purely periodic β-expansion. $\qquad\square$

Proposition 3.4. *Let* $x \in \mathbb{Q}(\beta) \cap [0, 1)$. *Then there exists* $N_1 > 0$ *such that* $T_\beta^N x$ *are reduced for any* $N \geq N_1$.

Proof. Simple computations show that $\widetilde{S}_\beta^k \left(\rho(\omega x) \right)$ exponentially comes near $\widehat{Y}$ as $k \to \infty$. Since there exist the finite number of $\rho \left(\omega \cdot T_\beta^k(x) \right)$ in a certain bounded domain, $\widetilde{S}_\beta^{N_1} \left(\rho(\omega x) \right) = \rho \left(\omega \cdot T_\beta^{N_1}(x) \right) \in \widehat{Y}$ for sufficiently large N_1. Then $T_\beta^{N_1}(x)$ is reduced. From Lemma 3.2 1, we see that $T_\beta^N(x)$ are reduced for any $N \geq N_1$. $\qquad\square$

Finally, we can get our Main Theorem.

Theorem 3.5. *Let $x \in [0,1)$. Then*

1. *$x \in \mathbb{Q}(\beta)$ if and only if x has an eventually periodic β-expansion,*

2. *$x \in \mathbb{Q}(\beta)$ is reduced if and only if x has a purely periodic β-expansion.*

Proof. 1. Assume that $x \in \mathbb{Q}(\beta)$. By Proposition 3.4, there exists $N > 0$ such that $T_\beta^N(x)$ is reduced. Proposition 3.3 says that $T_\beta^N(x)$ has a purely periodic β-expansion. Hence x has an eventually periodic β-expansion. The opposite side is trivial.

2. Necessity is obtained by Proposition 3.3. Conversely, assume that x has a purely periodic β-expansion. From *1*, we know $x \in \mathbb{Q}(\beta)$. According to Proposition 3.4, there exists $N > 0$ such that $T_\beta^N(x)$ is reduced. The pure periodicity of x implies that there exists $j > 0$ such that $T_\beta^{N+j}(x) = x$. Lemma 3.2 *1* says that x is reduced.   □

References

[1] S. Akiyama, Pisot numbers and greedy algorithm, Number Theory, Diophantine, Computational and Algebraic Aspects, Edited by K. Győry, A. Pethö and V. T. Sós, 9–21, de Gruyter 1998.

[2] P. Arnoux and Sh. Ito, Pisot substitutions and Rauzy fractals, Prétirage IML **98-18**, preprint submitted.

[3] A. Bertrand, Développements en base de Pisot et répartition modulo 1, C.R. Acad. Sci, Paris **285** (1977), 419–421.

[4] De Jun Feng, M. Furukado, Sh. Ito, and Jun Wu, Pisot substitutions and the Hausdorff dimension of boundaries of Atomic surfaces, preprint.

[5] Sh. Ito and H. Ei, Tilings from characteristic polynomials of β-expansion, preprint.

[6] S. Ito and Y. Sano, On periodic β-expansions of Pisot Numbers and Rauzy fractals, Osaka J. Math. **38** (2001), 1–20.

[7] W. Parry, On the β-expansions of real numbers, Acta Math. Acad. Sci. Hungar. **11** (1960), 401–416.

[8] M. Queffélec, Substitution Dynamical Systems - Spectral Analysis, Springer - Verlag Lecture Notes in Math. **1294**, New York, 1987.

[9] A. Rényi, Representations for real numbers and their ergodic properties, Acta Math. Acad. Sci. Hungar. **8** (1957), 477–493.

[10] Y. Sano, On purely periodic β-expansions of Pisot Numbers, preprint submitted.

[11] K. Schmidt, On periodic expansions of Pisot numbers and Salem numbers, Bull. London Math. Soc., **12** (1980), 269–278.

THE LARGEST PRIME FACTOR OF INTEGERS IN THE SHORT INTERVAL

Chaohua JIA

Institute of Mathematics, Academia Sinica, Beijing, China

Keywords: prime number, sieve method, Buchstab's identity

Abstract   In this report, the progress on the largest prime factor of integers in the short interval of two kinds is introduced.

2000 Mathematics Subject Classification 11N05.

I. THE DISTRIBUTION OF PRIME NUMBERS

In analytic number theory, one of important topics is the distribution of prime numbers. We often study the function

$$\pi(x) = \sum_{p \leq x} 1.$$

In 1859, Riemann connected $\pi(x)$ with the zeros of complex function $\zeta(s)$. He put forward his famous hypothesis that all non-trivial zeros of $\zeta(s)$ lie on the straight line $\mathrm{Re}(s) = \frac{1}{2}$. Along the direction pointed out by Riemann, Hadamard and Vallée Poussin proved the famous prime number theorem in 1896. It states that if $x \to \infty$, then

$$\pi(x) \sim \frac{x}{\log x}.$$

By this theorem, it is easy to prove the Bertrand Conjecture that in every interval $(x, 2x)$, there is a prime.

Assuming the Riemann Hypothesis (RH), we get

$$\pi(x) = \int_2^x \frac{dt}{\log t} + O(x^{\frac{1}{2}} \log x). \tag{1}$$

Then we can prove that in the short interval $(x, x + x^{\frac{1}{2}+\varepsilon})$, there is a prime. In fact, there are approximately $\frac{x^{\frac{1}{2}+\varepsilon}}{\log x}$ primes.

C. Jia and K. Matsumoto (eds.), Analytic Number Theory, 195–203.
© 2002 *Kluwer Academic Publishers. Printed in the Netherlands.*

If RH is not assumed, then we have to consider the weaker problem that there is a prime in the short interval $(x, x + y)$, where $y > x^{\frac{1}{2}+\varepsilon}$. Some famous mathematicians such as Montgomery, Huxley, Iwaniec, Heath-Brown made great contribution to this topic. The best result is due to Baker and Harman. They [3] proved that there is a prime in the short interval $(x, x + x^{0.535})$.

We also study the topic of probabilistic type. In 1943, assuming RH, Selberg [21] showed that, for almost all x, the short interval $(x, x + f(x) \log^2 x)$ contains a prime providing $f(x) \to \infty$ as $x \to \infty$. Here 'almost all' means that for $1 \leq x \leq X$, the measure of the exceptional set of x is $o(X)$. At the same time, Selberg [21] also proved an unconditional result. He showed that, for almost all x, $(x, x + x^{\frac{19}{77}+\varepsilon})$ contains a prime.

In 1990's, it was active on this topic. Jia, Li and Watt obtained some results. The last result is that for almost all x, $(x, x + x^{\frac{1}{20}+\varepsilon})$ contains a prime. One could see [16].

There is a difficult conjecture that there is a prime between two continuous squares. It can be expressed as that there is a prime in the short interval $(x, x + x^{\frac{1}{2}})$. But one can not prove it even on RH. Therefore we have to find the other way to study the existence of prime in the short intervals $(x, x + x^{\frac{1}{2}})$ and $(x, x + x^{\frac{1}{2}+\varepsilon})$.

II. THE LARGEST PRIME FACTOR OF INTEGERS IN THE SHORT INTERVAL $(x, x + x^{\frac{1}{2}})$

Firstly we decompose all integers in the interval $(x, x + x^{\frac{1}{2}})$. Among all prime factors, there is the largest one which is denoted as $P(x)$. We have an identity

$$\prod_{x < n \leq x + x^{\frac{1}{2}}} n = \prod_{p \leq P(x)} p^{h(p)}, \tag{2}$$

where

$$h(p) = \left(\left[\frac{x + x^{\frac{1}{2}}}{p} \right] - \left[\frac{x}{p} \right] \right) + \left(\left[\frac{x + x^{\frac{1}{2}}}{p^2} \right] - \left[\frac{x}{p^2} \right] \right) + \cdots .$$

Let

$$N(d) = \left[\frac{x + x^{\frac{1}{2}}}{d} \right] - \left[\frac{x}{d} \right].$$

Taking logarithm in both sides of the equation (2), we have

$$x^{\frac{1}{2}} \log x = \sum_d \Lambda(d) N(d) + O(x^{\frac{1}{2}})$$

$$= \sum_{d \le D} \Lambda(d) N(d) + \sum_{D < p \le x^\alpha} N(p) \log p \qquad (3)$$

$$+ \sum_{x^\alpha < p \le P(x)} N(p) \log p + O(x^{\frac{1}{2}}),$$

where $\Lambda(d)$ is the Mangoldt function.

If

$$\sum_{d \le D} \Lambda(d) N(d) + \sum_{D < p \le x^\alpha} N(p) \log p < (1 - \varepsilon) x^{\frac{1}{2}} \log x$$

is proved, then

$$P(x) > x^\alpha.$$

In the first sum

$$\sum_{d \le D} \Lambda(d) N(d),$$

we employ the Fourier expansion

$$((x)) = - \sum_{0 < |h| \le H} \frac{e(hx)}{2\pi i h} + O\left(\min\left(1, \frac{1}{H\|x\|}\right)\right)$$

to get the trigonometric sum

$$\sum_{0 < |h| \le H} \sum_d \Lambda(d) e\left(\frac{hx}{d}\right),$$

where $((x)) = \{x\} - \frac{1}{2}$ and $e(x) = e^{2\pi i x}$.

Using Vaughan's identity, we get two kinds of trigonometric sum. One is

$$\sum_{0 < |h| \le H} \sum_m a(m) \sum_n b(n) e\left(\frac{hx}{mn}\right),$$

where $a(m)$ is the complex number satisfying $a(m) = O(1)$, so does $b(n)$. The other is

$$\sum_{0 < |h| \le H} \sum_m a(m) \sum_n e\left(\frac{hx}{mn}\right).$$

The first sum is called the sum of type II and the second one is called the sum of type I. The original idea in Vaughan's identity came from Vinogradov. Vaughan made it clear and easy to use.

In the second sum
$$\sum_{D<p\leq x^{\alpha}} N(p)\log p,$$
we apply the sieve method. The error term in the sieve method can be transformed into the sum of type I or II.

One of the main tools which we use to estimate trigonometric sum is Weyl's inequality
$$\left|\sum_{n=1}^{N} a(n)\right|^2 \leq \frac{4N}{Q}\sum_{q=0}^{Q}(1-\frac{q}{Q})\operatorname{Re}\sum_{n=1}^{N-q} a(n)\overline{a(n+q)},$$

where $Q \leq N$. Weyl's inequality is a generalization of Cauchy's inequality.

The other is the reverse formula

$$\sum_{a<n\leq b} e(f(n)) = e\left(\frac{1}{8}\right)\sum_{\alpha<\nu\leq\beta}\frac{e(f(n_\nu)-\nu n_\nu)}{\sqrt{f''(n_\nu)}}$$
$$+O\left(\left(\frac{R}{U}+1\right)\left(\log(2+U)+\log(2+\frac{1}{R})\right)\right)$$
$$+O\left(\min(\sqrt{R},\frac{1}{\|\alpha\|})\right)+O\left(\min(\sqrt{R},\frac{1}{\|\beta\|})\right),$$

where $|f''(x)| \sim \frac{1}{R}$, $|f'''(x)| \sim \frac{1}{RU}$, $f'(n_\nu) = \nu$, $[\alpha, \beta]$ is the image of $[a, b]$ under the transformation $y = f'(x)$ and $\|x\|$ denotes the nearest distance from x to integers. One can refer to [15].

The reverse formula is based on Poisson's summation. A suitable combination of Weyl's inequality and reverse formula yields good estimate for trigonometric sum. On the new estimate for trigonometric sum, Theorem of Fouvry and Iwaniec [8] plays an important role, which transforms the trigonometric sum into suitable integral.

Ramachandra [20] proved that $P(x) > x^{\varphi}$, where $\varphi = \frac{5}{8}$. Graham [9] got $\varphi = 0.662$. He used the sieve method and the Fourier expansion. As the development of the estimate for trigonometric sum and the sieve method, some new exponents have appeared. There are exponents of Jia [14] ($\varphi = 0.69$), Baker [1] ($\varphi = 0.70$), Jia [15] ($\varphi = 0.728$), Baker and Harman [2] ($\varphi = 0.732$), Liu and Wu [19] ($\varphi = 0.738$).

If we can get $\varphi = 1$, then we can prove that there is a prime in the short interval $(x, x + x^{\frac{1}{2}})$. But the present results are far from the optimal one. The main reason is that on this topic, we can not use Perron's formula and the mean value estimate for Dirichlet polynomials.

III. THE LARGEST PRIME FACTOR OF INTEGERS IN THE SHORT INTERVAL $(x, x + x^{\frac{1}{2}+\varepsilon})$

Similarly as in the topic II, we define $Q(x)$ as the largest prime factor of integers in the short interval $(x, x + x^{\frac{1}{2}+\varepsilon})$. It is obvious that $Q(x) \geq P(x)$.

The first result on this topic is due to Jutila. He [18] proved that $Q(x) \geq x^{\vartheta - \varepsilon}$, where $\vartheta = \frac{2}{3}$. Bolog [4] improved it to $\vartheta = 0.772$. Balog, Harman and Pintz [5] obtained $\vartheta = 0.82$. They also used the basic frame in the topic II. But now they could use the Perron's formula

$$\sum_{n \leq x} a(n) = \frac{1}{2\pi i} \int_{b-iT}^{b+iT} \sum_{n=1}^{\infty} \frac{a(n)}{n^s} \cdot \frac{x^s}{s} \, ds + \left(\frac{x^{1+\varepsilon}}{T} \right)$$

to transform sums into integrals. Then they applied some mean value formulas to deal with these integrals.

Let

$$N_1(d) = \left[\frac{x + x^{\frac{1}{2}+\varepsilon}}{d} \right] - \left[\frac{x}{d} \right].$$

We have a formula similar to (3). In the first sum

$$\sum_{d \leq D} \Lambda(d) N_1(d),$$

Vaughan's identity, Perron's formula, mean value formulas

$$\int_1^T \left| \sum_{m \sim M} \frac{a(m)}{m^{\frac{1}{2}+it}} \right|^2 dt \ll (M + T) \log^c T \tag{4}$$

and

$$\int_1^T |\zeta(\tfrac{1}{2} + it)|^4 \, dt \ll T \log^c T \tag{5}$$

are used.

In the second sum

$$\sum_{D < p \leq x^\alpha} N_1(p) \log p,$$

the sieve method is applied. The error term in the sieve method can be dealt with by the estimate of Deshouillers and Iwaniec [6]

$$\int_1^T |\zeta(\tfrac{1}{2} + it)|^4 \left| \sum_{n \leq T^{\frac{1}{5}}} \frac{a(n)}{n^{\frac{1}{2}+it}} \right|^2 dt \ll T^{1+\varepsilon}. \tag{6}$$

This estimate which is based on the theory of modular forms is better than the classical one (5).

In 1996, Heath-Brown made a great progress on this topic. He [12] got $\vartheta = \frac{11}{12}$. Heath-Brown had an innovation on the basic frame. He considered the sum of the form

$$\sum = \sum_{x < pp_1 \cdots p_s \leq x + x^{\frac{1}{2}+\varepsilon}} 1,$$

where $p \sim P$, $p_1 \sim P^\varepsilon$, $\cdots$, $p_s \sim P^\varepsilon$.

If one can prove $\sum > 0$, then there is a prime factor $p \sim P$. Hence, the largest prime factor $Q(x) > P$. In the previous papers, the sum

$$\sum_{x < pl \leq x + x^{\frac{1}{2}+\varepsilon}} 1$$

is considered, where $p \leq x^\alpha$ and l is an integer. In Heath-Brown's device, the prime factors $p_1, \cdots, p_s$ are more flexible than l.

Heath-Brown introduced some new ideas on the sieve method. His ideas can be traced back to Linnik's identity. Let

$$\Pi(s) = \prod_{p < z} \left(1 - \frac{1}{p^s}\right).$$

We have

$$\log(\zeta(s)\Pi(s)) = \sum_{l=1}^{\infty} \sum_{p \geq z} \frac{1}{lp^{ls}} = \sum_{n=1}^{\infty} \frac{a(n)}{n^s}. \tag{7}$$

On the other hand,

$$\begin{aligned}
\log(\zeta(s)\Pi(s)) &= \log(1 + (\zeta(s)\Pi(s) - 1)) \\
&= \sum_{k=1}^{\infty} \frac{(-1)^{k-1}}{k} \cdot (\zeta(s)\Pi(s) - 1)^k \\
&= \sum_{k=1}^{\infty} \frac{(-1)^{k-1}}{k} \sum_{n=1}^{\infty} \frac{a_k(n)}{n^s}.
\end{aligned} \tag{8}$$

Comparing the equations (7) and (8), we can get some identities on the sieve method.

Now I give a quite rough explanation on the application of the above identity. For the sum

$$\sum_{p_1, p_2, p_3}{}^{*} 1, \tag{9}$$

where $\star$ means some conditions on p_1, p_2 and p_3, one considers the sum

$$\sum_{n_1, n_2, n_3}^{\star} \Lambda(n_1)\Lambda(n_2)\Lambda(n_3)$$

and uses the relationship

$$\Lambda(n) = \sum_{d|n} \mu(d) \log \frac{n}{d}$$

to get the expression

$$\sum_{d_1, d_2, d_3, l_1, l_2, l_3}^{\star} \mu(d_1)\mu(d_2)\mu(d_3) \log l_1 \log l_2 \log l_3.$$

Then in the different ranges of d and l, one can apply different mean value formulas.

In the joint work of Heath-Brown and Jia [13] in 1998, we got $\vartheta = \frac{17}{18}$. In this paper, we used Heath-Brown's innovation but applied the traditional sieve method. Harman's method was employed (see [10]). The major feature of Harman's method is that one can get an asymptotic formula for the sum

$$\sum_p S(\mathcal{A}_p, x^\tau),$$

where τ is a small positive constant, while usually one can only take $\tau = \varepsilon$. Harman's method works in some topics since there is the estimate for the sum of type II. By Harman's method, in the sum (9), we only decompose one of p_1, p_2, p_3 and keep the others unchanged. In this way, one can use the mean value estimate

$$\int_1^T |\zeta(\tfrac{1}{2} + it)|^2 \left| \sum_{n \leq T^{\frac{2}{7}}} \frac{a(n)}{n^{\frac{1}{2}+it}} \right|^4 dt \ll T^{1+\varepsilon}. \tag{10}$$

The estimate (10) is due to Deshouillers and Iwaniec [7], which depends on their work for the application of the theory of modular forms. It is difficult to apply formula (10) in Heath-Brown's original work [12].

Moreover, we used computer to deal with the complicated relationship among the sieve functions. By the help of computer, we can get good estimate for Buchstab's function $w(u)$ which is defined as

$$\begin{cases} w(u) = \frac{1}{u}, & 1 \leq u \leq 2, \\ (uw(u))' = w(u-1), & 2 < u. \end{cases}$$

We have
$$\begin{cases} 0.5607 \leq w(u) \leq 0.5644, & u \geq 3, \\ 0.5612 \leq w(u) \leq 0.5617, & u \geq 4. \end{cases}$$

One could refer to [16]. Before we only had Jingrun Chen's result that $w(u) \leq 0.5673$ for $u \geq 2$.

Recently Jia and M.-C. Liu [17] got a new exponent $\vartheta = \frac{25}{26}$. In this paper, we employed Harman's new idea on the sieve method (see [11]). In some sums of the form

$$\sum_p S(\mathcal{A}_p, p),$$

Harman applied the sieve method to the variable p, which is similar to Jingrun Chen's dual principle. In the sum $p_1 + p_2$, Chen applied the sieve method to p_1, then to p_2. We used the work of Deshouillers and Iwaniec [7] again in more delicate way and made complicated calculation in the sieve method. Then we got the new exponent $\vartheta = \frac{25}{26}$.

References

[1] R. C. Baker, The greatest prime factor of the integers in an interval, Acta Arith. **47** (1986), 193–231.

[2] R. C. Baker and G. Harman, Numbers with a large prime factor, Acta Arith. **73** (1995), 119–145.

[3] R. C. Baker and G. Harman, The difference between consecutive primes, Proc. London Math. Soc. (3) **72** (1996), 261–280.

[4] A. Balog, Numbers with a large prime factor II, Topics in classical number theory, Coll. Math. Soc. János Bolyai **34**, Elsevier North-Holland, Amsterdam (1984), 49–67.

[5] A. Balog, G. Harman and J. Pintz, Numbers with a large prime factor IV, J. London Math. Soc. (2) **28** (1983), 218–226.

[6] J.-M. Deshouillers and H. Iwaniec, Power mean values of the Riemann zeta-function, Mathematika **29** (1982), 202–212.

[7] J.-M. Deshouillers and H. Iwaniec, Power mean-values for Dirichlet's polynomials and the Riemann zeta-function, II, Acta Arith. **43** (1984), 305–312.

[8] E. Fouvry and H. Iwaniec, Exponential sums with monomials, J. Number Theory **33** (1989), 311–333.

[9] S. W. Graham, The greatest prime factor of the integers in an interval, J. London Math. Soc. (2) **24** (1981), 427–440.

[10] G. Harman, On the distribution of αp modulo one, J. London Math. Soc. (2) **27** (1983), 9–18.

[11] G. Harman, On the distribution of αp modulo one II, Proc. London Math. Soc. (3) **72** (1996), 241–260.

[12] D. R. Heath-Brown, The largest prime factor of the integers in an interval, Science in China, Series A, **39** (1996), 449–476.

[13] D. R. Heath-Brown and C. Jia, The largest prime factor of the integers in an interval, II, J. Reine Angew. Math. **498** (1998), 35–59.

[14] C. Jia, The greatest prime factor of the integers in a short interval (I), Acta Math. Sin. **29** (1986), 815–825, in Chinese.

[15] C. Jia, The greatest prime factor of the integers in a short interval (IV), Acta Math. Sin., New Series, **12** (1996), 433–445.

[16] C. Jia, Almost all short intervals containing prime numbers, Acta Arith. **76** (1996), 21–84.

[17] C. Jia and M.-C. Liu, On the largest prime factor of integers, Acta Arith. **95** (2000), 17–48.

[18] M. Jutila, On numbers with a large prime factor, J. Indian Math. Soc. (N. S.) **37** (1973), 43–53.

[19] H. Liu and J. Wu, Numbers with a large prime factor, Acta Arith. **89** (1999), 163–187.

[20] K. Ramachandra, A note on numbers with a large prime factor II, J. Indian Math. Soc. **34** (1970), 39–48.

[21] A. Selberg, On the normal density of primes in short intervals, and the difference between consecutive primes, Arch. Math. Naturvid. **47** (1943), 87–105.

A GENERAL DIVISOR PROBLEM IN LANDAU'S FRAMEWORK

S. KANEMITSU

Graduate School of Advanced Technology, University of Kinki, Iizuka, Fukuoka 820-8555, Japan

A. SANKARANARAYANAN

School of Mathematics, Tata Institute of Fundamental Research, Homi Bhabha Road, Mumbai–400 005, India

Dedicated to Professor Hari M. Srivastava on his sixtieth birthday

Keywords: divisor problem, mean value theorem, functional equation, zeta-function

Abstract   In this paper we shall consider the general divisor problem which arises by raising the generating zeta-fuction $Z(s)$ to the k-th power, where the zeta-functions in question are the most general E. Landau's type ones that satisfy the functional equations with multiple gamma factors.

Instead of simply applying Landau's colossal theorem to $Z^k(s)$, we start from the functional equation satisfied by $Z(s)$ and raise it to the k-th power. This, together with the strong mean value theorem of H. L. Montgomery and R. Vaughan, and K. Ramachandra's reasonings, enables us to improve earlier results of Landau and K. Chandrasekharan and R. Narasimhan in some range of intervening parameters.

2000 Mathematics Subject Classification: 11N37, 11M41.

1.   INTRODUCTION

In order to treat a general divisor problem for quadratic forms (first investigated by the second author [12]) in a more general setting, we shall work with well-known E. Landau's framework of Dirichlet series satisfying the functional equation with multiple gamma factors where the number of gamma factors may not necessarily be the same on both sides [6], [13].

The main feature is that we raise the generating Dirichlet series $Z(s)$ to the k-th power while both Landau [6] and Chandrasekharan and

205

Narasimhan [2] simply apply their theory of functional equation to $Z^k(s)$ and that we incorporate the mean value theorem in the estimation of the resulting integral, providing herewith some improvements over the results of Landau [6] and Chandrasekharan and Narasimhan [1] in certain ranges of intervening parameters. The main ingredient underlying the second feature is similar to K. Chandrasekharan and R. Narasimhan's approach; but in [2] they appeal to their famous approximate functional equation whereas in this paper we use a substitute for it (Lemma 3.3), following the idea of K. Ramachandra [8]–[11], we avoid the use of it thus giving a more direct approach to the general divisor problem.

To state main results, we shall fix the setting in which we work and some notation.

1. Let $\{a_n\}$ and $\{b_n\}$ be two sequences of complex numbers satisfying

$$a_n = O(n^{\alpha+\varepsilon}), \quad b_n = O(n^{\alpha+\varepsilon}) \qquad (1.1)$$

for every $\varepsilon > 0$, where $\alpha \geq 0$ is a fixed real number.

2. Form the Dirichlet series $(s = \sigma + it)$

$$Z(s) = \sum_{n=1}^{\infty} \frac{a_n}{n^s} \text{ and } \tilde{Z}(s) = \sum_{n=1}^{\infty} \frac{b_n}{n^s} \qquad (1.2)$$

which are absolutely convergent for $\sigma > \alpha + 1$ by Condition 1.

3. We suppose that $Z(s)$ can be continued to a meromorphic function in any finite strip $\sigma_1 \leq \sigma \leq \sigma_2$ such that $\sigma_2 \geq \alpha + 1$ with only real poles, and satisfies the convexity condition there:

$$Z(s) = O(e^{\gamma|t|}) \qquad (1.3)$$

for some constant $\gamma = \gamma(\sigma_1, \sigma_2) > 0$.

4. For $\sigma < 0$ suppose $Z(s)$ satisfies the functional equation

$$A_1^s Z(s)\Delta(s) = A_2 A_3^{-s} \tilde{Z}(\alpha + 1 - s)\tilde{\Delta}(-s), \qquad (1.4)$$

where A_1, A_2, A_3 are positive numbers and

$$\Delta(s) = \prod_{i=1}^{\mu} \Gamma(\alpha_i + \beta_i s) \text{ and}$$

$$\tilde{\Delta}(s) = \prod_{j=1}^{\nu} \Gamma(\gamma_j + \delta_j s) \qquad (1.5)$$

are gamma factors and where the real parameters α_i, β_i, $1 \leq i \leq \mu$, γ_j, δ_j, $1 \leq j \leq \nu$ are subject to the conditions

$$\beta_i > 0, \ 1 \leq i \leq \mu, \ \delta_j > 0, \ 1 \leq j \leq \nu,$$

$$\sum_{i=1}^{\mu} \beta_i = \sum_{j=1}^{\nu} \delta_j = \frac{H}{2} \text{ (say)}. \tag{1.6}$$

5. Let

$$\eta = \sum_{j=1}^{\nu} \gamma_j - \sum_{i=1}^{\mu} \alpha_i + \frac{\mu - \nu}{2} \tag{1.7}$$

and suppose that η satisfies

$$\eta \geq 1 \text{ and } \eta > \alpha + \frac{1}{2}. \tag{1.8}$$

Also put

$$\eta_1 = \max\{2\eta - 1, \ 2\eta + 1 - H, \ 2\alpha + 1\}. \tag{1.9}$$

6. For any fixed integer $k \geq 2$, we define $a_k(n)$ by

$$Z^k(s) = \sum_{n=1}^{\infty} \frac{a_k(n)}{n^s}, \tag{1.10}$$

so that by (1.2)

$$a_k(n) = \sum_{n_1 n_2 \cdots n_k = n} a_{n_1} a_{n_2} \cdots a_{n_k}. \tag{1.11}$$

Now we are in a position to state the main results of the paper.

Theorem 1. *We write*

$$\sum_{n \leq x} a_k(n) = M_k(x) + E_k(x), \tag{1.12}$$

where $M_k(x)$ is the sum of residues of the function $\frac{x^s}{s} Z^k(s)$ at all positive real poles of $\frac{Z^k(s)}{s}$ in the strip $0 < \sigma \leq \alpha + 1$.
Then for every $\varepsilon > 0$, we have

$$E_k(x) = O\left(x^{\alpha + 1 - \frac{2\alpha+1}{2\{(k-2)\lambda + \eta_1\}} + \varepsilon}\right),$$

where λ_1 is as in (3.3).

Theorem 2. *In addition to the conditions in Theorem 1 suppose also that for $s_0 = \sigma_0 + it$ (with fixed σ_0 satisfying $1/2 \leq \sigma_0 \leq \alpha + 1$),*

$$|Z(s_0)| \ll |t|^{\lambda_0},$$

that for some fixed integer $j \geq 2$ we have for every $\varepsilon > 0$

$$\int_T^{2T} |Z(s_0)|^j dt \ll T^{\eta_0(j,\alpha)+\varepsilon},$$

where $\eta_0 = \eta_0(j,\alpha)$ is a positive constant depending on j and α, and that $j\lambda_0 < \eta_0$. Then we have

$$E_k(x) = O\left(x^{\alpha+1-\frac{2\alpha+1}{2\{(k-j)\lambda_0+\eta_0\}}+\varepsilon}\right).$$

Acknowledgment. It gives us a great pleasure to thank Professor Yoshio Tanigawa for scrutinizing our paper thoroughly which resulted in this improved version. The authors would also like to thank the referee whose comments helped us to improve the presentation of the paper.

2.   NOTATION AND PRELIMINARIES

We use complex variables $s = \sigma + it$, $w = u + iv$. ε always denotes a small positive constant and $\varepsilon_1, \varepsilon_2, \ldots$ denote small positive constants which may depend on ε. $c_1, c_2, \ldots$ denote positive absolute constants.

The Stirling formula [15] states that in any fixed strip $\sigma_1 \leq \sigma \leq \sigma_2$ as $|t| \to \infty$ we have

$$|\Gamma(\sigma + it)| = \sqrt{2\pi}|t|^{\sigma-\frac{1}{2}}e^{-\frac{\pi}{2}|t|}\left(1 + O\left(\frac{1}{|t|}\right)\right).$$

If we write (1.4) as

$$Z(s) = \chi(s)\tilde{Z}(\alpha + 1 - s), \tag{2.1}$$

then by Stirling's formula above,

$$|\chi(s)| \asymp (A_1 A_3)^{-\sigma}|t|^{\eta-\sigma H}, \tag{2.2}$$

where the symbol $f \asymp g$ means that $f \gg g$ and $f \ll g$ ($\ll$ being Vinogradov's symbolism).

We make use of the standard truncated Perron formula, which states that for $0 < T \leq x$,

$$\sum_{n \leq x} a_k(n) = \frac{1}{2\pi i}\int_{c_0-iT}^{c_0+iT} Z^k(s)\frac{x^s}{s}ds + O\left(x^{c_0}T^{-1}\right) + O(x^{2\varepsilon}), \tag{2.3}$$

where we let

$$c_0 = c_0(\varepsilon) = \alpha + 1 + \varepsilon. \tag{2.4}$$

This follows from the approximate formula for the discontinuous integral (see Davenport [3]):

Let

$$J(x) = \begin{cases} 1 & x > 1 \\ 0 & 0 < x < 1 \\ \frac{1}{2} & x = 1. \end{cases}$$

Then for $x > 0$ and $c > 0$, we have

$$\left| \frac{1}{2\pi i} \int_{c-iT}^{c+iT} \frac{x^s}{s} ds - J(s) \right| \le \min \left\{ \frac{x^c}{T|\log x|}, \, 2x^c \right\}.$$

3. SOME LEMMAS AND A MEAN VALUE THEOREM

Lemma 3.1. *Let $\dot{c}_0 = c_0(\varepsilon) = \alpha + 1 + \varepsilon$ as in (2.4). Then we have*

$$Z(\sigma + it) \ll |t|^{\frac{(\eta+\varepsilon H)(c_0-\sigma)}{c_0+\varepsilon}} \tag{3.1}$$

uniformly in $-\varepsilon \le \sigma \le c_0$. In particular, we have

$$Z\left(\frac{1}{2} + it\right) \ll |t|^{\lambda_1}, \tag{3.2}$$

where

$$\lambda_1 = \frac{(2\alpha+1)\eta}{2(\alpha+1)} + \varepsilon_1. \tag{3.3}$$

We also have

$$Z(s) \ll |t|^{\frac{\lambda_1(c_0-\sigma)}{c_0-\frac{1}{2}}} \tag{3.4}$$

uniformly in $\frac{1}{2} \le \sigma \le c_0$.

Proof. Since

$$Z(c_0 - it) \ll 1,$$

it follows from (2.1) and (2.2) that

$$Z(-\varepsilon + it) \ll |t|^{\eta+\varepsilon H}. \tag{3.5}$$

In view of (1.3), the Phragmén-Lindelöf principle (see e.g. [15]) applies to infer that the exponent of t is a linear function $\mu(\sigma)$ connecting the points $(-\varepsilon, \eta + \varepsilon H)$ and $(c_0, 0)$, or

$$\mu(\sigma) = \frac{\eta + \varepsilon H}{c_0 + \varepsilon}(c_0 - \sigma) = \eta\left(1 - \frac{\sigma}{\alpha + 1}\right) + O(\varepsilon),$$

i.e. (3.1). □

Lemma 3.2 (Hilbert's inequality à la Montgomery et Vaughan [7]). *If $\{h_n\}$ is an infinite sequence of complex numbers such that $\sum_{n=1}^{\infty} n|h_n|^2$ is convergent, then*

$$\int_T^{T+H_1} \left|\sum_{n=1}^{\infty} h_n n^{-it}\right|^2 dt = \sum_{n=1}^{\infty} |h_n|^2 (H_1 + O(n)),$$

where $c_1 \le H_1 \le T$.

Lemma 3.3 (Substitute for an approximate functional equation). *For $T \le t \le 2T$ and a positive parameter $Y > 1$ we have the approximation*

$$Z\left(\frac{1}{2} + it\right)$$

$$= \sum_{n=1}^{\infty} \frac{a_n}{n^{\frac{1}{2}+it}} e^{-\frac{n}{Y}}$$

$$+ \frac{1}{2\pi i} \int_{\substack{u=-\varepsilon \\ |v| \le (\log T)^2}} \chi\left(\frac{1}{2} + it + w\right)\left(\sum_{n \le Y} b_n n^{\frac{1}{2}+it+w-\alpha-1}\right) Y^w \Gamma(w)dw$$

$$+ \frac{1}{2\pi i} \int_{\substack{u=-\frac{1}{2}-2\varepsilon \\ |v| \le (\log T)^2}} \chi\left(\frac{1}{2} + it + w\right)\left(\sum_{Y \le n} b_n n^{\frac{1}{2}+it+w-\alpha-1}\right) Y^w \Gamma(w)dw$$

$$+ O\left(Y^{\alpha+\frac{1}{2}+\varepsilon} T^{\eta+2\varepsilon H} e^{-c_2(\log T)^2}\right).$$

Proof. By the Mellin transform we have, after truncation using Stirling's formula,

$$S := \sum_{n=1}^{\infty} \frac{a_n}{n^{\frac{1}{2}+it}} e^{-\frac{n}{Y}} = I + O\left(Y^{c_0-\frac{1}{2}} e^{-c_3(\log T)^2}\right),$$

where

$$I = \frac{1}{2\pi i} \int_{\substack{u=c_0-\frac{1}{2} \\ |v| \le (\log T)^2}} Z\left(\frac{1}{2} + it + w\right) Y^w \Gamma(w)dw.$$

Now, we move the line of integration of I to $u = -\frac{1}{2} - 2\varepsilon$, encountering the pole of the integrand at $w = 0$ with residue $Z(\frac{1}{2} + it)$, and estimate the horizontal integrals by Stirling's formula and (3.1) to get

$$S = Z \left(\frac{1}{2} + it \right) + I' + O \left(Y^{c_0 - \frac{1}{2}} e^{-c_3 (\log T)^2} \right)$$
$$+ O \left(Y^{c_0 - \frac{1}{2}} T^{\eta + \varepsilon H} e^{-c_4 (\log T)^2} \right), \tag{3.6}$$

where

$$I' = \frac{1}{2\pi i} \int_{\substack{u = -\frac{1}{2} - 2\varepsilon \\ |v| \leq (\log T)^2}} Z \left(\frac{1}{2} + it + w \right) Y^w \Gamma(w) dw.$$

To transform I' we substitute the functional equation (2.1) for $Z(\frac{1}{2} + it + w)$ thereby dividing the sum

$$\tilde{Z} \left(\alpha + 1 - \frac{1}{2} - it - w \right) = \sum_{n=1}^{\infty} b_n n^{\frac{1}{2} + it + w - \alpha - 1}$$

into two parts I_1' ($n \leq Y$) and I_2' ($n > Y$): $I' = I_1' + I_2'$.

In I_1' we move the line of integration back to $u = -\varepsilon$ committing an error of order $O(T^{c_5} e^{-c_6 (\log T)^2})$.

Thus

$$I_1' = \frac{1}{2\pi i} \int_{\substack{u = -\varepsilon \\ |v| \leq (\log T)^2}} X \left(\frac{1}{2} + it + w \right) \left(\sum_{n \leq Y} b_n n^{\frac{1}{2} + it + w - \alpha - 1} \right) Y^w \Gamma(w) dw$$
$$+ O \left(T^{c_7} e^{-c_8 (\log T)^2} \right). \tag{3.7}$$

I' being the sum of I_1' and I_2', we substitute (3.7) into (3.6) to conclude the assertion. $\qquad \square$

Theorem 3. *For $T \geq T_0$, where T_0 denotes a large positive constant, we have for every $\varepsilon > 0$*

$$\int_T^{2T} \left| Z \left(\frac{1}{2} + it \right) \right|^2 dt \ll T^{\eta_1 + \varepsilon}.$$

In particular, we have

$$\int_1^T \left| Z \left(\frac{1}{2} + it \right) \right|^2 \frac{dt}{t} \ll T^{\eta_1 - 1 + \varepsilon}. \tag{3.8}$$

Proof. Choosing $Y = T$ in Lemma 3.3, we see that it suffices to compute the mean square of the first three terms S, I_1' and I_2', say (I_1' is not quite the same, but the namings are to be suggestive of their origins in

Lemma 3.3), the error term being negligible. In doing so, we shall make good use of Hilbert's inequality, Lemma 3.2.

By the Cauchy-Schwarz inequality, we infer that

$$|I_2'|^2 \ll \left(\int_{\substack{u=-\frac{1}{2}-2\varepsilon \\ |v| \le (\log T)^2}} \left| \chi\left(\frac{1}{2} + it + w \right) T^w \Gamma(w) \right|^2 |dw| \right) \times$$

$$\times \left(\int_{\substack{u=-\frac{1}{2}-2\varepsilon \\ |v| \le (\log T)^2}} \left| \sum_{T<n} b_n n^{\frac{1}{2}+it+w-\alpha-1} \right|^2 |dw| \right),$$

so that by (2.2)

$$\int_T^{2T} |I_2'|^2 dt$$

$$\ll T^{2\eta+4\varepsilon H + \varepsilon_2} T^{-1} \int_{-\log^2 T}^{\log^2 T} \left(\int_T^{2T} \left| \sum_{T<n} b_n n^{-\alpha-1-2\varepsilon+iv} n^{it} \right|^2 dt \right) dv.$$

By Lemma 3.2, the inner integral $\int_T^{2T}$ is

$$\ll \sum_{T<n} |b_n|^2 n^{-2\alpha-2-4\varepsilon} (T + O(n))$$

$$\ll \sum_{T<n} n^{-1-2\varepsilon} \ll T^{-2\varepsilon}$$

by (1.1).

Thus, substituting this, we conclude that

$$\int_T^{2T} |I_2'|^2 dt \ll T^{2\eta-1+\varepsilon_3}. \tag{3.9}$$

To estimate the mean square of I_1' and S, we apply Lemma 3.2. Direct use of the estimate (2.2) and then of that lemma yields

$$\int_T^{2T} |I_1'|^2 \ll T^{2\eta-H+\varepsilon_4} Y^{-2\varepsilon} \int_T^{2T} \left| \sum_{n \le T} b_n n^{\frac{1}{2}+it-\varepsilon+iv-\alpha-1} \right|^2 dt$$

$$\ll T^{2\eta-H+\varepsilon_5} T \sum_{n \le T} |b_n|^2 n^{-2\alpha-1-2\varepsilon} \tag{3.10}$$

$$\ll T^{2\eta+1-H+\varepsilon_6}$$

as above.

Similarly, we have

$$\int_T^{2T} |S|^2 dt = \sum_{n=1}^{\infty} \frac{|a_n|^2}{n} e^{-\frac{2n}{T}} (T + O(n)).$$

Dividing the sum on the right into two parts $n \le T$ and $n > T$ and substituting the approximation $e^{-\frac{2n}{T}} = 1 + O(\frac{2n}{T})$ in the former and $e^{-\frac{2n}{T}} = O((\frac{T}{n})^{2\alpha+2})$ in the latter, we conclude that

$$\int_T^{2T} |S|^2 dt = T \sum_{n \le T} \frac{|a_n|^2}{n} \left(1 + O\left(\frac{n}{T}\right)\right) + O(TT^{2\alpha+2}T^{-2})$$

$$= O\left(T \sum_{n \le T} \frac{|a_n|^2}{n}\right) + O(T^{2\alpha+1}) \qquad (3.11)$$

$$= O\left(T^{2\alpha+1+\varepsilon_7}\right).$$

Since

$$\int_T^{2T} |Z\left(\frac{1}{2} + it\right)|^2 dt = \int_T^{2T} \left(|S|^2 + |I_1'|^2 + |I_2'|^2\right) dt,$$

the assertion follows from (3.9)–(3.11). □

Lemma 3.4. *For the quantities η_1 and λ_1 defined by (1.9) and (3.3), respectively, we have*

$$2\lambda_1 \le \eta_1 + 2\varepsilon_1.$$

Proof. We distinguish two cases

(i) $1 \le \eta \le \alpha + 1$

(ii) $\alpha + 1 \le \eta$.

In Case (i), we get

$$2\lambda_1 = \eta \frac{2\alpha + 1}{\alpha + 1} + 2\varepsilon_1 \le 2\alpha + 1 + 2\varepsilon_1 \le \eta_1 + 2\varepsilon_1$$

by (1.9), and in Case (ii), we get

$$2\lambda_1 = \eta \left(2 - \frac{1}{\alpha + 1}\right) + 2\varepsilon_1 \le 2\eta - 1 + 2\varepsilon_1 \le \eta_1 + 2\varepsilon_1$$

again by (1.9). □

4.   PROOF OF THEOREM 1 AND 2

Proof of Theorem 1. We move the line of integration in the integral appearing in (2.3) to $\sigma = \frac{1}{2}$. By the theorem of residues we obtain

$$\sum_{n \le x} a_k(n) = M_k'(x) + I_{\pm h} + I_v$$
$$+ O\left(x^{c_0} T^{-1}\right) + O\left(x^{2\varepsilon}\right),$$
(4.1)

where $M_k'(x)$ is the main term which is the sum of residues of the function $\frac{x^s}{s} Z^k(s)$ at all possible poles of $\frac{Z^k(s)}{s}$ in the interval $\frac{1}{2} < \sigma \le \alpha + 1$, and $I_{\pm h}$ (resp. I_v) denotes the horizontal (resp. vertical) integrals:

$$I_{\pm h} = \frac{1}{2\pi i} \int_{\frac{1}{2} \pm iT}^{c_0 \pm iT} Z^k(s) \frac{x^s}{s} ds$$
$$= O\left(x^{c_0} T^{-1}\right) + O\left(x^{\frac{1}{2}} T^{k\lambda_1 - 1}\right)$$
(4.2)

by (3.4), while

$$I_v = \frac{1}{2\pi i} \int_{\frac{1}{2} - iT}^{\frac{1}{2} + iT} Z^k(s) \frac{x^s}{s} ds = O\left(x^{\frac{1}{2}}\right) + O\left(x^{\frac{1}{2}} \int_1^T \left| Z\left(\frac{1}{2} + it\right) \right|^k \frac{dt}{t}\right).$$

In the integral in the error term for I_v we factor the integrand $|Z|^k$ into $|Z|^{k-2}$ and $|Z|^2$ to which we apply (3.2) and (3.8) respectively. Then we get

$$I_v = O\left(x^{\frac{1}{2}} T^{(k-2)\lambda_1 + \eta_1 - 1 + \varepsilon}\right).$$
(4.3)

Since the contribution from possible poles in the interval $0 < \sigma \le \frac{1}{2}$ to the main term is $O(x^{\frac{1}{2}})$, we may duly write $M_k(x)$ instead of $M_k'(x)$. Hence from (4.1)–(4.3) and the above remark we conclude that

$$\sum_{n \le x} a_k(n) - M_k(x) = O\left(x^{c_0} T^{-1}\right) + O\left(x^{\frac{1}{2}} T^{k\lambda_1 - 1}\right)$$
$$+ O\left(x^{\frac{1}{2}} T^{(k-2)\lambda_1 + \eta_1 - 1 + \varepsilon}\right).$$
(4.4)

Now, by Lemma 3.4, the last error term dominates the middle one (apart from an ε-factor). Therefore we have

$$\sum_{n \le x} a_k(n) - M_k(x) = O\left(x^{\alpha + 1 + \varepsilon} T^{-1}\right)$$
$$+ O\left(x^{\frac{1}{2}} T^{(k-2)\lambda_1 + \eta_1 - 1 + \varepsilon}\right).$$
(4.5)

Choosing

$$T = x^{\frac{\alpha+\frac{1}{2}}{(k-2)\lambda_1+\eta_1}},\tag{4.6}$$

we get

$$\sum_{n\leq x} a_k(n) = M_k(x) + O\left(x^{\alpha+1-\frac{\alpha+\frac{1}{2}}{(k-2)\lambda_1+\eta_1}+\varepsilon}\right),\tag{4.7}$$

whence we conclude the assertion of Theorem 1. □

Proof of Theorem 2 is similar to that of Theorem 1. For simplicity, we assume $\sigma_0 = 1/2$. Indeed, instead of (4.3) we have

$$I_v = O\left(x^{\frac{1}{2}}T^{(k-j)\lambda_0+\eta_0-1+\varepsilon_0}\right).\tag{4.3$_0$}$$

Hence instead of (4.4) we have

$$\sum_{n\leq x} a_k(n) - M_k(x) = O\left(x^{c_0}T^{-1}\right) + O\left(x^{\frac{1}{2}}T^{k\lambda_0-1}\right)$$
$$+ O\left(x^{\frac{1}{2}}T^{(k-j)\lambda_0+\eta_0-1+\varepsilon_0}\right).\tag{4.4$_0$}$$

Now, instead of (4.4), the assumed inequality $j\lambda_0 < \eta_0$ implies that the last error term dominates the second one, whence it follows that

$$\sum_{n\leq x} a_k(n) - M_k(x) = O\left(x^{\alpha+1+\varepsilon}T^{-1}\right)$$
$$+ O\left(x^{\frac{1}{2}}T^{(k-j)\lambda_0+\eta_0-1+\varepsilon_0}\right).\tag{4.5$_0$}$$

Choosing

$$T = x^{\frac{\alpha+\frac{1}{2}}{(k-j)\lambda_0+\eta_0}},\tag{4.6$_0$}$$

we get

$$\sum_{n\leq x} a_k(n) = M_k(x) + O\left(x^{\alpha+1-\frac{\alpha+\frac{1}{2}}{(k-j)\lambda_0+\eta_0}+\varepsilon}\right),\tag{4.7$_0$}$$

thereby completing the proof. □

5.   SOME APPLICATIONS

5.1. Let $Q = Q(y_1, \ldots, y_l)$ be a positive definite quadratic form in l-variables ($l \geq 2$ an integer). Let $Z_Q(s)$ be the associated zeta-function,

$$Z_Q(s) = \sum_{(y_1,\ldots,y_l)\neq(0,\ldots,0)} (Q(y_1,\ldots,y_l))^{-s}, \quad \sigma > \frac{l}{2},$$

summation being extended over all integer l-tuples not all zero. $Z_Q(s)$ can be continued meromorphically over the whole plane with its unique simple pole at $s = \frac{l}{2}$ and satisfies the functional equation of type (1.4):

$$\left(\frac{d^{\frac{1}{l}}}{2\pi}\right)^s \Gamma(s) Z_Q(s) = \left(\frac{d^{1-\frac{1}{l}}}{2\pi}\right)^{\frac{l}{2}-s} \Gamma\left(\frac{l}{2}-s\right) Z_{\overline{Q}}\left(\frac{l}{2}-s\right), \qquad (5.1)$$

where d is the discriminant of Q and $\overline{Q}$ denotes the reciprocal of Q [14].
    Hence we may take in Theorem 1,

$$\eta = \frac{l}{2}, \quad H = 2.$$

Since $a_n = b_n = O(n^{\frac{l}{2}-1+\varepsilon})$, we may take $\alpha = \frac{l}{2} - 1$, so that $\eta_1 = l - 1$. Also we have $\lambda_1 = \frac{l}{2} - \frac{1}{2} + \varepsilon_1$.
    Hence Theorem 1 gives

$$\sum_{n \leq x} a_k(n) = \operatorname*{Res}_{s=\frac{l}{2}} Z_Q(s)\frac{x^s}{s} + O\left(x^{\frac{l}{2}-\frac{k}{2}+\varepsilon}\right),$$

which recovers the theorem of the second author [12].

Remark. For $l = 2$, Kober [5] has proved a mean value theorem slightly better than Theorem 3.

    5.2. In the special case where

$$Q(y_1, y_2) = y_1^2 + ay_2^2,$$

where a is a positive constant, it is known from [16] that

$$Z_Q\left(\frac{1}{2} + it\right) \ll t^{\frac{1}{3}}(\log t)^3$$

and that $a_n = b_n = O(n^\varepsilon)$. Hence we can take $\lambda_0 = \frac{1}{3} + \varepsilon$, $\alpha = 0$, $\eta = 1$, $H = 2$. Also we use Theorem 3, so that we take $j = 2$. Then $\eta_0 = 1$, and Theorem 2 implies that

$$\sum_{n \leq x} a_k(n) = M_k(x) + O\left(x^{1-\frac{3}{2(k+1)}+\varepsilon}\right).$$

    5.3. If $Z(s) = \zeta(s)$, then we have the most famous functional equation

$$\pi^{-\frac{s}{2}}\Gamma\left(\frac{s}{2}\right)\zeta(s) = \pi^{-\frac{1-s}{2}}\Gamma\left(\frac{1-s}{2}\right)\zeta(1-s),$$

so that we may take $\eta = 1/2$, $H = 1$, and also $\alpha = 0$. We remark here that this value η does not satisfy the assumption (1.8). So we cannot use Theorem 1. However we use Theorem 5.5 [15] to take $\lambda_0 = \frac{1}{6} + \varepsilon$. Instead of Theorem 3, we apply the 4-th power moment $(j = 4)$ with $\eta_0 = 1$.

In this way we can recover Theorem 12.3 of Titchmarsh [15], i.e.

$$\alpha_k \leq \frac{k-1}{k+2} \quad (k = 4, 5, \ldots),$$

where α_k is defined as the least exponent such that

$$\sum_{n \leq x} d_k(n) = M_k(x) + O\left(x^{\alpha_k + \varepsilon}\right),$$

with $d_k(n)$ denoting the k-fold divisor function due to Piltz $(a_k(n) = d_k(n))$. For $k \geq 12$, we can take $j = 12$, $\lambda_0 = 1/6$, $\eta_0 = 2$ (from a result of D.R. Heath-Brown). The condition $j\lambda_0 \leq \eta_0$ is satisfied and hence we get $\alpha_k \leq \frac{k-3}{k}$ for $k \geq 12$ which improves the earlier result slightly for $k \geq 12$.

6.   COMPARISON WITH THE RESULTS OF LANDAU AND OF CHANDRASEKHARAN AND NARASIMHAN

6.1. If we apply Landau's theorem [6], we get

$$\sum_{n \leq x} a_k(n) - M_k(x) = O\left(x^{\alpha + 1 - \frac{2(\alpha+1)}{2k\eta+1}}\right). \tag{6.1}$$

Hence, comparing this with Theorem 1, we are to show that

$$(4\eta + 1)(2\alpha + 1) > 4\eta_1(\alpha + 1) \tag{6.2}$$

in some cases. Recall Definition (1.9) of η_1 and consider three cases.
(i) $\eta_1 = 2\eta - 1$, i.e. $H \geq 2$ and $\eta \geq \alpha + 1$. In this case (6.2) is of the form

$$6\alpha + 5 > 4\eta,$$

which is true in view of (1.8).
(ii) $\eta_1 = 2\alpha + 1$, i.e. $\alpha \geq \eta - 1$ and $\alpha \geq \eta - \frac{H}{2}$. In this case we further suppose that

$$\alpha < \frac{4\eta - 3}{4}, \tag{6.3}$$

in which case necessarily $H \geq \frac{3}{2}$. Then (6.2) becomes

$$(4\eta + 1)\eta_1 > 4\eta_1(\alpha + 1), \text{ or } 4\eta + 1 > 4\alpha + 4,$$

i.e. (6.3).

(iii) $\eta_1 = 2\eta + 1 - H$, i.e. $H \leq \min\{2, 2(\eta - \alpha)\}$. In this case we must have $\eta \geq \alpha + \frac{3}{4}$. Then solving (6.2) in H gives $\frac{4\eta + 2\alpha + 3}{4(\alpha + 1)} < H$.

Thus we have proved

Proposition 6.1. *If one of the following conditions are satisfied, then our estimate supersedes Landau's bound* (6.1):

(i) $H \geq 2$ *and* $\eta \geq \alpha + 1$

(ii) $\max\left\{\eta - 1, \eta - \dfrac{H}{2}\right\} \leq \alpha < \dfrac{4\eta - 3}{4}$ *and* $H \geq \dfrac{3}{2}$

(iii) $\dfrac{4\eta + 2\alpha + 3}{4(\alpha + 1)} < H \leq \min\{2, 2(\eta - \alpha)\}$ *and* $\eta \geq \alpha + \frac{3}{4}$.

6.2. Chandrasekharan and Narasimhan [1], [2] considered the pair of Dirichlet series $\varphi(s) = \sum_{n=1}^{\infty} \frac{a_n}{\lambda_n^s}$ and $\psi(s) = \sum_{n=1}^{\infty} \frac{b_n}{\mu_n^s}$ that satisfy the functional equation with equal multiple gamma factor $\Delta(s)$:

$$\Delta(s)\varphi(s) = \Delta(\delta - s)\psi(\delta - s), \tag{1.4$'$}$$

with $\delta > 0$, where

$$\Delta(s) = \prod_{\nu=1}^{N} \Gamma(\alpha_\nu s + \beta_\nu). \tag{1.5$'$}$$

For comparison of the theory of Landau-Walfisz [17] and Bochner-Chandra-sekharan-Narasimhan, cf. [4].

Their theorem (Theorem 4.1 [1]) states that if the functional equation (1.4)$'$ is satisfied as in [1], in particular, $\alpha_\nu > 0$, $1 \leq \nu \leq N$, $A = \sum_{\nu=1}^{N} \alpha_\nu \geq 1$ (see (1.6)$'$ below), and the only singularities of the function φ are poles, then

$$A_\lambda^0(x) - Q_0(x) = O\left(x^{\frac{\delta}{2} - \frac{1}{4A} + 2A\eta_2 u}\right) + O\left(x^{q - \frac{1}{2A} - \eta_2}(\log x)^{r-1}\right)$$

$$+ O\left(\sum_{x < \lambda_n \leq x'} |a_n|\right) \tag{6.4}$$

for $\eta_2 \geq 0$ at our choice, where $x' = x + O(x^{1 - \eta_2 - \frac{1}{2A}})$, $q = $ maximum of the real parts of the singularities of φ, $r = $ maximum order of a pole with the real part ν, $u = \beta - \frac{\delta}{2} - \frac{1}{4A}$ with β satisfying

$$\sum_{n=1}^{\infty} \frac{|b_n|}{\mu_n^\beta} < \infty.$$

If in addition, $a_n \geq 0$, then we can dispense with the last error term:

$$A_\lambda^0(x) - Q_0(x) = O\left(x^{\frac{\delta}{2} - \frac{1}{4A} + 2A\eta_2 u}\right) + O\left(x^{q - \frac{1}{2A} - \eta_2}(\log x)^{r-1}\right). \quad (6.5)$$

There being essentially no difference between the sequences $\{n\}$ and $\{\lambda_n\}$ and $\{\mu_n\}$, we may compare Chandrasekharan and Narasimhan's result above with ours.

In our setting we must have

$$A_1 = A_2 = A_3 = 1, \quad \mu = \nu \ (= N), \quad \delta = A + 1,$$

$$1 \leq A = \sum_{i=1}^{\mu} \beta_i = \frac{H}{2}, \quad (1.6)'$$

and

$$\eta = \frac{\delta H}{2} = \delta A, \text{ etc.} \quad (1.7)'$$

Since to our situation, A is transformed into kA, our applications are mostly for $a_n \geq 0$, and $q \ (\geq \alpha + 1 = \delta)$ can be as big as δ, (6.5) reads

$$A_\lambda^0(x) - Q_0(x) = O\left(x^{\frac{\delta}{2} - \frac{1}{4kA} + 2kA\eta_2 u}\right) + O\left(x^{\delta - \frac{1}{2kA} - \eta_2}(\log x)^{r-1}\right).$$

We are to choose η_2 so that the two error terms be more or less equal:

$$\eta_2 \doteq \frac{\left(\frac{\delta}{2} - \frac{1}{4kA}\right)}{1 + 2kAu}.$$

By (1.2) and the condition on u, we must have $u > \frac{\delta}{2} - \frac{1}{4kA}$, so that

$$2kAu + 1 > \frac{1}{2}\left(1 + kH(\alpha + 1)\right).$$

It is enough to choose u so as to satisfy

$$2kAu + 1 \geq 1 + kH(\alpha + 1).$$

Thus we choose

$$\eta_2 = \frac{\left(\frac{\alpha+1}{2} - \frac{1}{2kH}\right)}{1 + kH(\alpha + 1)}. \quad (6.6)$$

We are to prove that

$$\alpha + 1 - \frac{1}{kH} - \frac{\frac{\alpha+1}{2} - \frac{1}{2kH}}{1 + kH(\alpha + 1)}$$

$$> \alpha + 1 - (\alpha + 1)\frac{2\alpha + 1}{(k - 2)\eta(2\alpha + 1) + 2\eta_1(\alpha + 1)},$$

or

$$\frac{\frac{3}{2} + \frac{\frac{1}{2}}{2kH(\alpha+1)}}{1 + kH(\alpha+1)} \frac{(k-2)\eta(2\alpha+1) + 2\eta_1(\alpha+1)}{\alpha+1} < 2\alpha + 1.$$

Hence it suffices to prove

$$\frac{2}{1 + kH(\alpha+1)} \frac{(k-2)\eta(2\alpha+1) + 2\eta(\alpha+1)}{\alpha+1} < 2\alpha + 1. \qquad (6.7)$$

As in 6.1, we distinguish two cases and solve (6.7) in H. Then we can immediately prove the following

Proposition 6.2. *If either of the following conditions are satisfied, then our estimate supersedes Chandrasekharan and Narasimhan's bound* (6.5):

(i) $H > \max\{2, \frac{1}{k(\alpha+1)}(\frac{2}{2\alpha+1}\{(k-2)\eta(2\alpha+1) + 2(2\eta-1)(\alpha+1)\} - 1)\}$ *and* $\eta \geq \alpha + 1$,

(ii) $\alpha \geq \eta - 1$ *and* $H > \frac{1}{k(\alpha+1)}(\frac{2}{2\alpha+1}\{(k-2)\eta(2\alpha+1) + 2(2\eta-1)(\alpha+1)\} - 1)$.

We note that Case (iii) *in Proposition 1 cannot occur in Proposition 2 in view of* (1.6)′.

References

[1] K. Chandrasekharan and R. Narasimhan, *Functional equations with multiple gamma factors and the average order of arithmetical functions*, Ann. of Math. (2) **76** (1962), 93–136.

[2] K. Chandrasekharan and R. Narasimhan, *The approximate functional equation for a class of zeta-functions*, Math. Ann. **152** (1963), 30–64.

[3] H. Davenport, *Multiplicative number theory*, Markham, Chicago 1967; second edition revised by H. L. Montgomery. Springer Verlag, New York-Berlin 1980.

[4] S. Kanemitsu and I. Kiuchi, *Functional equation and asymptotic formulas*, Mem. Fac. Gen. Edu. Yamaguchi Univ., Sci. Ser. **28** (1994), 9–54.

[5] H. Kober, *Ein Mittelwert Epsteinscher Zetafunktionen*, Proc. London Math. Soc. (2) **42** (1937), 128–141.

[6] E. Landau, *Über die Anzahl der Gitterpunkte in gewißen Bereichen*, Nachr. Akad. Wiss. Ges. Göttingen (1912), 687–770 = Collected Works, Vol. 5 (1985), 159–239, Thales Verl., Essen. (See also Parts II–IV).

[7] H. L. Montgomery and R. Vaughan, *Hilbert's inequality*, J. London Math. Soc. (2) **8** (1974), 73–82.

[8] K. Ramachandra, *Some problems of analytic number theory–I*, Acta Arith, **31** (1976), 313–324.

[9] K. Ramachandra, *Some remarks on a theorem of Montgomery and Vaughan*, J. Number Theory **11** (1979), 465–471.

[10] K. Ramachandra, *A simple proof of the mean fourth power estimate for $\zeta(\frac{1}{2} + it)$ and $L(\frac{1}{2} + it, \chi)$*, Ann. Scuola Norm. Sup. Pisa Cl. Sci. (4) **1** (1974), 81–97.

[11] K. Ramachandra, *Application of a theorem of Montgomery and Vaughan to the zeta-function*, J. London Math. Soc. (2) **10** (1975), 482–486.

[12] A. Sankaranarayanan, *On a divisor problem related to the Epstein zeta-function*, Arch. Math. (Basel) **65** (1995), 303–309.

[13] A. Sankaranarayanan, *On a theorem of Landau*, (unpublished).

[14] C. L. Siegel, *Lectures on advanced analytic number theory*, Tata Institute of Fundamental Research, Bombay 1961, 1981.

[15] E. C. Titchmarsh, *The theory of the Riemann zeta-function*, The Clarendon Press, Oxford 1951, second edition revised by Heath-Brown 1986.

[16] E. C. Titchmarsh, *On Epstein's zeta-function*, Proc. London Math. Soc. (2) **36** (1934), 485–500.

[17] A. Walfisz, *Über die summatorische Funktionen einiger Dirichletscher Reihen–II*, Acta Arith. **10** (1964), 71–118.

ON INHOMOGENEOUS DIOPHANTINE APPROXIMATION AND THE BORWEINS' ALGORITHM, II

Takao KOMATSU

Faculty of Education, Mie University, Mie, 514-8507 Japan

komatsu@edu.mie-u.ac.jp

Keywords: Inhomogeneous diophantine approximation, Borweins' algorithm, Continued fractions, quasi-periodic representation

Abstract   We obtain the values $\mathcal{M}(\theta, \phi) = \liminf_{|q| \to \infty} |q| \|q\theta - \phi\|$ by using the algorithm by Borwein and Borwein. Some new results for $\theta = e^{1/s}$ $(s \geq 1)$ are evaluated.

1991 Mathematics Subject Classification: 11J20, 11J80.

1.   INTRODUCTION

Let θ be irrational and ϕ real. Throughout this paper we shall assume that $q\theta - \phi$ is never integral for any integer q. Define the *inhomogeneous approximation constant* for the pair θ, ϕ

$$\mathcal{M}(\theta, \phi) = \liminf_{|q| \to \infty} |q| \|q\theta - \phi\|,$$

where $\| \cdot \|$ denotes the distance from the nearest integer. If we use the auxiliary functions

$$\mathcal{M}_+(\theta, \phi) = \liminf_{q \to +\infty} q\|q\theta - \phi\| \quad \text{and} \quad \mathcal{M}_-(\theta, \phi) = \liminf_{q \to +\infty} q\|q\theta + \phi\|,$$

then $\mathcal{M}(\theta, \phi) = \min(\mathcal{M}_+(\theta, \phi), \mathcal{M}_-(\theta, \phi))$. Several authors have treated $\mathcal{M}(\theta, \phi)$ or $\mathcal{M}_+(\theta, \phi)$ by using their own algorithms (See [2], [3], [4], [5] e.g.), but it has been difficult to find the exact values of $\mathcal{M}(\theta, \phi)$ for the concrete pair of θ and ϕ.

In [6] the author establishes the relationship between $\mathcal{M}(\theta, \phi)$ and the algorithm of Nishioka, Shiokawa and Tamura [8]. By using this result,

C. Jia and K. Matsumoto (eds.), Analytic Number Theory, 223–242.
© 2002 *Kluwer Academic Publishers. Printed in the Netherlands.*

we can evaluate the exact value of $\mathcal{M}(\theta, \phi)$ for any pair of θ and ϕ at least when θ is a quadratic irrational and $\phi \in \mathbb{Q}(\theta)$. Furthermore, in [7] the author demonstrates that the exact value of $\mathcal{M}(\theta, \phi)$ can be calculated even if θ is a Hurwitzian number, namely its continued fraction expansion has a quasi-periodic form.

In this paper we establish the relation between $\mathcal{M}(\theta, \phi)$ and the Borweins' algorithm, yielding some new results about the typical Hurwitzian numbers $\theta = e$ and $\theta = e^{1/s}$ ($s \geq 2$).

As usual, $\theta = [a_0; a_1, a_2, \ldots]$ denotes the simple continued fraction of θ, where

$$\theta = a_0 + \theta_0, \qquad a_0 = \lfloor \theta \rfloor,$$
$$1/\theta_{n-1} = a_n + \theta_n, \qquad a_n = \lfloor 1/\theta_{n-1} \rfloor \qquad (n = 1, 2, \ldots).$$

The n-th convergent $p_n/q_n = [a_0; a_1, \ldots, a_n]$ of θ is then given by the recurrence relations

$$p_n = a_n p_{n-1} + p_{n-2} \quad (n = 0, 1, \ldots), \quad p_{-2} = 0, \quad p_{-1} = 1,$$
$$q_n = a_n q_{n-1} + q_{n-2} \quad (n = 0, 1, \ldots), \quad q_{-2} = 1, \quad q_{-1} = 0.$$

Borwein and Borwein [1] use the algorithm as follows:

$$\phi = d_0 + \gamma_0, \qquad d_0 = \lfloor \phi \rfloor,$$
$$\gamma_{n-1}/\theta_{n-1} = d_n + \gamma_n, \qquad d_n = \lfloor \gamma_{n-1}/\theta_{n-1} \rfloor \quad (n = 1, 2, \ldots).$$

Then, ϕ is represented by

$$\phi = d_0 + d_1\theta_0 + d_2\theta_0\theta_1 + \cdots + d_i\theta_0\theta_1 \cdots \theta_{i-1} + \gamma_i\theta_0\theta_1 \cdots \theta_{i-1}$$
$$= d_0 + d_1 D_0 - d_2 D_1 + \cdots + (-1)^{i-1} d_i D_{i-1} + (-1)^{i-1}\gamma_i D_{i-1}$$
$$= d_0 + \sum_{i=1}^{\infty} (-1)^{i-1} d_i D_{i-1},$$

where $D_i = q_i\theta - p_i = (-1)^i \theta_0\theta_1 \ldots \theta_i$ ($i \geq 0$). Put $C_n = \sum_{i=1}^{n}(-1)^{i-1}d_i q_{i-1}$. Then $\|C_n\theta - \phi\| = 1 - \{C_n\theta - \phi\} = \|(-1)^n\gamma_n D_{n-1}\| = \gamma_n|D_{n-1}|$.

We can assume that $0 < \phi \leq 1/2$ without loss of generality. Then ϕ can be represented as $\phi = \sum_{i=1}^{\infty}(-1)^{i-1}d_i D_{i-1}$. $1 - \phi$ is also expanded by the Borweins' algorithm as

$$1 - \phi = d_0' + \gamma_0' = \gamma_0', \qquad d_0' = \lfloor 1 - \phi \rfloor = 0,$$
$$\gamma_{n-1}'/\theta_{n-1}' = d_n' + \gamma_n', \qquad d_n' = \lfloor \gamma_{n-1}'/\theta_{n-1}' \rfloor \quad (n = 1, 2, \ldots).$$

Then $1 - \phi = \sum_{i=1}^{\infty}(-1)^{i-1}d_i' D_{i-1}$. Hence, if we put $C_n' = \sum_{i=1}^{n}(-1)^{i-1}d_i' q_{i-1}$, then $\|C_n'\theta + \phi\| = \gamma_n'|D_{n-1}|$.

Theorem 1. *For any irrational θ and real ϕ so that $q\theta - \phi$ is never integral for any integer q, we have*

$$\mathcal{M}(\theta, \phi) = \liminf_{n \to +\infty} \min(|C_n|\gamma_n|D_{n-1}|, |C'_n|\gamma'_n|D_{n-1}|,$$

$$(|C_n| + q_{n-1})(1 - \gamma_n)|D_{n-1}|, (|C'_n| + q_{n-1})(1 - \gamma'_n)|D_{n-1}|).$$

Remark. As seen in the proof of Theorem 1, the last two values are considered only if $C_{2n-1} > 0$ (so, $C'_{2n-1} > 0$ by Lemmas below); $C_{2n} < 0$ (so, $C'_{2n} < 0$).

By applying Theorem 1 we can calculate $\mathcal{M}(\theta, \phi)$ for any concrete pair (θ, ϕ). In special we establish the following two theorems.

Theorem 2. *For any integer $l \geq 2$, we have*

$$\mathcal{M}\left(e, \frac{1}{l}\right) = \frac{1}{2l^2}.$$

Remark. It is known that this equation holds for $l = 2, 3$ and $l = 4$ ([6], [7]).

Theorem 3. *For any integers $l \geq 2$ and $s \geq 2$ with $s \equiv 0 \pmod{l}$, we have*

$$\mathcal{M}\left(e^{1/s}, \frac{1}{l}\right) = \frac{1}{2l^2}.$$

2.   LEMMAS

Lemma 1.
(1) If $a_n = d_n > 0$, then $d_{n+1} = 0$.
(2) If $d_n > 0$, then

$$\begin{cases} 1 \leq C_n \leq q_n, & \text{if } n \text{ is odd}; \\ -q_n + 1 \leq C_n \leq 0, & \text{if } n \text{ is even}. \end{cases}$$

Proof. It is easy to prove.

Lemma 2. *If $C_n + C'_n = (-1)^{n-1}q_n$, then $C_{n+1} + C'_{n+1} = (-1)^{n-1}q_n$.*
If $C_n + C'_n = (-1)^{n-1}(q_n - q_{n-1})$ or $C_n + C'_n = (-1)^n q_{n-1}$, then

$$C_{n+1} + C'_{n+1} = \begin{cases} (-1)^n q_{n+1}, & \text{if } \gamma_{n+1} + \gamma'_{n+1} = \theta_{n+1}; \\ (-1)^n (q_{n+1} - q_n), & \text{if } \gamma_{n+1} + \gamma'_{n+1} = \theta_{n+1} + 1. \end{cases}$$

Proof. When $n = 1$, we have $(d_1 + d'_1) + (\gamma_1 + \gamma'_1) = a_1 + \theta_1$ because

$$d_1 + \gamma_1 = \frac{\gamma_0}{\theta_0}, \quad d'_1 + \gamma'_1 = \frac{1 - \gamma_0}{\theta_0} \quad \text{and} \quad a_1 + \theta_1 = \frac{1}{\theta_0}.$$

Notice that d_1, d_1' and a_1 are integers, γ_1, γ_1' and θ_1 are non-integral real numbers. Thus, if $\theta_1 > \gamma_1$, then $d_1 + d_1' = a_1$ and $\gamma_1 + \gamma_1' = \theta_1$, yielding $C_1 + C_1' = (d_1 + d_1')q_0 = q_1$. If $\theta_1 < \gamma_1$, then $d_1 + d_1' = a_1 - 1$ and $\gamma_1 + \gamma_1' = \theta_1 + 1$, yielding $C_1 + C_1' = q_1 - q_0$.

Assume that $C_n + C_n' = (-1)^{n-1}q_n$ and $\gamma_n + \gamma_n' = \theta_n$. Since $0 < \gamma_n, \gamma_n' < \theta_n$, we have

$$d_{n+1} = \left\lfloor \frac{\gamma_n}{\theta_n} \right\rfloor = 0 \quad \text{and} \quad d_{n+1}' = \left\lfloor \frac{\gamma_n'}{\theta_n} \right\rfloor = 0 .$$

Hence, $C_{n+1} + C_{n+1}' = (-1)^{n-1}q_n + (-1)^n(d_{n+1} + d_{n+1}')q_n = (-1)^{n-1}q_n$ and $\gamma_{n+1} + \gamma_{n+1}' = \gamma_n/\theta_n + \gamma_n'/\theta_n = 1$.

Assume that $C_n + C_n' = (-1)^{n-1}(q_n - q_{n-1})$ and $\gamma_n + \gamma_n' = \theta_n + 1$. In this case $d_{n+1} = \lfloor \gamma_n/\theta_n \rfloor \geq 1$ and $d_{n+1}' = \lfloor \gamma_n'/\theta_n \rfloor \geq 1$ because $\gamma_n > \theta_n$ and $\gamma_n' > \theta_n$. Then

$$d_{n+1} + d_{n+1}' = \frac{\gamma_n}{\theta_n} - \gamma_{n+1} + \frac{\gamma_n'}{\theta_n} - \gamma_{n+1}' = (a_{n+1}+1) + (\theta_{n+1} - \gamma_{n+1} - \gamma_{n+1}').$$

Hence, if $\gamma_{n+1} + \gamma_{n+1}' = \theta_{n+1}$, then $d_{n+1} + d_{n+1}' = a_{n+1} + 1$ and

$$\begin{aligned} C_{n+1} + C_{n+1}' &= (-1)^{n-1}(q_n - q_{n-1}) + (-1)^n(d_{n+1} + d_{n+1}')q_n \\ &= (-1)^{n-1}(q_n - q_{n-1}) + (-1)^n(a_{n+1} + 1)q_n \\ &= (-1)^n q_{n+1} . \end{aligned}$$

If $\gamma_{n+1} + \gamma_{n+1}' = \theta_{n+1} + 1$, then $d_{n+1} + d_{n+1}' = a_{n+1}$ and

$$\begin{aligned} C_{n+1} + C_{n+1}' &= (-1)^{n-1}(q_n - q_{n-1}) + (-1)^n a_{n+1}q_n \\ &= (-1)^n(q_{n+1} - q_n) . \end{aligned}$$

Assume that $C_n + C_n' = (-1)^n q_{n-1}$ and $\gamma_n + \gamma_n' = 1$. Then

$$d_{n+1} + d_{n+1}' = \frac{\gamma_n}{\theta_n} - \gamma_{n+1} + \frac{\gamma_n'}{\theta_n} - \gamma_{n+1}' = a_{n+1} + (\theta_{n+1} - \gamma_{n+1} - \gamma_{n+1}').$$

Hence, if $\gamma_{n+1} + \gamma_{n+1}' = \theta_{n+1}$, then $d_{n+1} + d_{n+1}' = a_{n+1}$ and $C_{n+1} + C_{n+1}' = (-1)^n q_{n+1}$. If $\gamma_{n+1} + \gamma_{n+1}' = \theta_{n+1} + 1$, then $d_{n+1} + d_{n+1}' = a_{n+1} - 1$ and $C_{n+1} + C_{n+1}' = (-1)^n(q_{n+1} - q_n)$.

3.   PROOF OF THEOREM 1

First of all, any integer K can be uniquely expressed as

$$K = \sum_{i=1}^{n} (-1)^{i-1} z_i q_{i-1} ,$$

where z_i $(1 \leq i \leq n)$ is an integer with $0 \leq z_i \leq a_i$ and $z_n \neq 0$. If $0 \leq K \leq a_1$, put $n = 1$ and $z_1 = K$.

If $K > a_1$, then choose the odd number $n(\geq 3)$ satisfying $q_{n-2} + 1 \leq K \leq q_n$. Put $K_n = K$ and $z_n = \lceil (K_n - q_{n-2})/q_{n-1} \rceil$, so that $1 \leq z_n \leq a_n$. If $q_{n-2} - q_{n-3} + 1 \leq K_n - z_n q_{n-1} \leq q_{n-2}$, then put $z_{n-1} = 0$ and $K_{n-2} = K_n - z_n q_{n-1}$ (so, $z_{n-2} = a_{n-2}$). Otherwise, put $K_{n-1} = z_n q_{n-1} - K_n$.

If $K < 0$, then choose the even number $n(\geq 2)$ satisfying $-q_n + 1 \leq K \leq -q_{n-2}$. Put $K_n = -K$ and $z_n = \lceil (K_n - q_{n-2} + 1)/q_{n-1} \rceil$, so that $1 \leq z_n \leq a_n$. If $n \neq 2$ and $q_{n-2} - q_{n-3} \leq K_n - z_n q_{n-1} \leq q_{n-2} - 1$, then $z_{n-1} = 0$ and $K_{n-2} = K_n - z_n q_{n-1}$ (so, $z_{n-2} = a_{n-2}$). Otherwise, put $K_{n-1} = z_n q_{n-1} - K_n$.

By repeating these steps we can determine $z_n, z_{n-1}, \ldots, z_2$. Finally, put $z_1 = K_1$.

For general $i < n$ we have $0 \leq z_i \leq a_i$ and

$$\begin{aligned} -q_{i-1} + q_{i-2} + 1 \leq K_i \leq q_i & \qquad \text{if } i \text{ is odd}; \\ -q_{i-1} + q_{i-2} \leq K_i \leq q_i - 1 & \qquad \text{if } i \text{ is even}. \end{aligned}$$

Next, we can obtain

$$\{K\theta\} = \sum_{i=1}^{n} (-1)^{i-1} z_i D_{i-1} .$$

Notice that if $z_{i-1} = a_{i-1}$ then $z_i = 0$ $(2 \leq i \leq n)$. Put

$$T_i = (\cdots ((z_n \theta_{n-1} + z_{n-1})\theta_{n-2} + z_{n-2}) \cdots)\theta_i + z_i \quad (1 \leq i \leq n-1).$$

Since $z_n \neq 0$ is followed by $z_{n-1} \neq a_{n-1}$,

$$T_{n-1} = z_n \theta_{n-1} + z_{n-1} \leq \frac{a_n}{a_n + \theta_n} + (a_{n-1} - 1) < a_{n-1} .$$

If $z_{n-2} \neq a_{n-2}$, then

$$T_{n-2} = T_{n-1}\theta_{n-2} + z_{n-2} < \frac{a_{n-1}}{a_{n-1} + \theta_{n-1}} + z_{n-2} < z_{n-2} + 1 \leq a_{n-2} .$$

If $z_{n-2} = a_{n-2}$, then by $z_{n-1} = 0$ and $z_{n-3} \neq a_{n-3}$

$$T_{n-2} = (z_n \theta_{n-1} + z_{n-1})\theta_{n-2} + z_{n-2} \leq \frac{a_n}{a_n + \theta_n}\theta_{n-2} + a_{n-2} < a_{n-2} + \theta_{n-2}$$

and $T_{n-3} = T_{n-2}\theta_{n-3} + z_{n-3} < (a_{n-2} + \theta_{n-2})\theta_{n-3} + (a_{n-3} - 1) = a_{n-3}$. Hence, by induction, if $z_i \neq a_i$ then $T_i < a_i$. If $z_i = a_i$ then $T_i < a_i + \theta_i$

and $T_{i-1} < a_{i-1}$. Therefore, $T_i\theta_{i-1} < (a_i + \theta_i)\theta_{i-1} = 1$. Especially, we have

$$0 < \sum_{i=1}^{n}(-1)^{i-1}z_i D_{i-1} = T_1\theta_0 < 1.$$

We shall assume that $\|K\theta - \phi\| = \pm(\{K\theta\} - \phi)$. If $\|K\theta - \phi\| = 1 - (\{K\theta\} - \phi) > \phi$, then $|K|\|K\theta - \phi\| > |K|\phi \to \infty$ ($|K| \to \infty$). If $\|K\theta - \phi\| = 1 - (\phi - \{K\theta\}) > 1 - \phi$, then $|K|\|K\theta - \phi\| > |K|(1 - \phi) \to \infty$ ($|K| \to \infty$).

Suppose that $K \neq C_n$ or $d_i = z_i$ ($1 \leq i \leq s - 1$) and $d_s \neq z_s$ ($1 \leq s \leq n$). If $d_s > z_s$, then

$$\|K\theta - \phi\| = \left|\sum_{i=1}^{n}(-1)^{i-1}z_i D_{i-1} - \sum_{i=1}^{\infty}(-1)^{i-1}d_i D_{i-1}\right|$$

$$= \left|(-1)^{s-1}(z_s - d_s)D_{s-1} + \sum_{i=s+1}^{n}(-1)^{i-1}z_i D_{i-1} - \sum_{i=s+1}^{\infty}(-1)^{i-1}d_i D_{i-1}\right|$$

$$= |(-1)^{s-1}(z_s - d_s + \gamma_s^* - \gamma_s)D_{s-1}|$$

$$= (d_s - z_s + \gamma_s - \gamma_s^*)|D_{s-1}|$$

$$= (d_s - z_s - \gamma_s^*)|D_{s-1}| + \|C_s\theta - \phi\| > \|C_s\theta - \phi\|,$$

where $\gamma_s^* = T_{s+1}\theta_s = T_s - z_s(< 1)$. Since

$$\begin{cases} q_{n-2} + 1 \leq K \leq q_n & n : \text{odd}; \\ -q_n + 1 \leq K \leq -q_{n-2} & n : \text{even} \end{cases} \quad \text{and} \quad \begin{cases} 1 \leq C_s \leq q_s & s : \text{odd}; \\ -q_s + 1 \leq C_s \leq 0 & s : \text{even} \end{cases}$$

if $s \leq n - 2$, we have $|K| > |C_s|$, yielding $|K|\|K\theta - \phi\| > |C_s|\|C_s\theta - \phi\|$.

When $s = n-1$, $K = C_{n-1} + (-1)^{n-2}(z_{n-1} - d_{n-1})q_{n-2} + (-1)^{n-1}z_n q_{n-1}$. If $K > 0$ (so, n is odd), by $d_{n-1} > z_{n-1} \geq 0$ we have $-q_{n-1} + 1 \leq C_{n-1} \leq 0$, and $C_{n-1} + C'_{n-1} = -q_{n-1}$ or $-q_{n-1} + q_{n-2}$. If $K \geq q_{n-1} - 1$, then $K \geq |C_{n-1}|$, yielding $K\|K\theta - \phi\| > |C_{n-1}|\|C_{n-1}\theta - \phi\|$. Assume that $K < q_{n-1} - 1$. Then $z_n = 1$. Hence,

$$K = q_{n-1} - |C_{n-1}| + (d_{n-1} - z_{n-1})q_{n-2} \geq q_{n-1} - |C_{n-1}| + q_{n-2}$$

and

$$\|K\theta - \phi\| = (d_{n-1} - z_{n-1} + \gamma_{n-1} - \theta_{n-1})|D_{n-2}| \geq (1 + \gamma_{n-1} - \theta_{n-1})|D_{n-2}|.$$

If $\gamma_{n-1} + \gamma'_{n-1} = \theta_{n-1}$ and $|C_{n-1}| + |C'_{n-1}| = q_{n-1}$, then

$$K\|K\theta - \phi\| \geq (|C'_{n-1}| + q_{n-2})(1 - \gamma'_{n-1})|D_{n-2}|$$
$$= (|C'_{n-1}| + q_{n-2})\|(|C'_{n-1}| + q_{n-2})\theta - \phi\|.$$

If $\gamma_{n-1} + \gamma'_{n-1} = \theta_{n-1} + 1$ and $|C_{n-1}| + |C'_{n-1}| = q_{n-1} - q_{n-2}$, then

$$K\|K\theta - \phi\| \geq (|C'_{n-1}| + 2q_{n-2})(2 - \gamma'_{n-1})|D_{n-2}|$$
$$> (|C'_{n-1}| + q_{n-2})\|(1 - \gamma'_{n-1})|D_{n-2}|.$$

If $K < 0$ (so, n is even), $1 \leq C_{n-1} \leq q_{n-1}$ and $C_{n-1} + C'_{n-1} = q_{n-1}$ or $q_{n-1} - q_{n-2}$. If $|K| \geq q_{n-1}$, then $|K| \geq C_{n-1}$, so

$$|K|\|\|K|\theta + \phi\| = (-K)\|K\theta - \phi\| > C_{n-1}\|C_{n-1}\theta - \phi\|.$$

Assume that $|K| \leq q_{n-1} - 1$. Then $z_n = 1$. Hence,

$$|K| = q_{n-1} - C_{n-1} + (d_{n-1} - z_{n-1})q_{n-2} \geq q_{n-1} - C_{n-1} + q_{n-2}$$

and

$$\|K\theta - \phi\| = (d_{n-1} - z_{n-1} + \gamma_{n-1} - \theta_{n-1})|D_{n-2}| \geq (1 + \gamma_{n-1} - \theta_{n-1})|D_{n-2}|.$$

Therefore,

$$(-K)\|K\theta - \phi\| = |K|\|\|K|\theta + \phi\| \geq (C'_{n-1} + q_{n-2})(1 - \gamma'_{n-1})|D_{n-2}|.$$

When $s = n$, $K = C_{n-1} + (-1)^{n-1}z_n q_{n-1} = C_n + (-1)^{n-1}(z_n - d_n)q_{n-1}$ and $\|K\theta - \phi\| = (d_n - z_n + \gamma_n)|D_{n-1}|$. When $K > 0$ (so, n is odd), by $d_n > z_n > 0$ we have $1 \leq C_n \leq q_n$, and $C_n + C'_n = q_n$ or $q_n - q_{n-1}$. If $C_{n-1} > 0$ or $z_n \geq 2$, then

$$K\|K\theta - \phi\| - C_n\|C_n\theta - \phi\|$$
$$= (C_{n-1} + z_n q_{n-1})(d_n - z_n + \gamma_n)|D_{n-1}| - (C_{n-1} + d_n q_{n-1})\gamma_n|D_{n-1}|$$
$$= (C_{n-1} + (z_n - \gamma_n)q_{n-1})(d_n - z_n)|D_{n-1}| > 0.$$

If $C_{n-1} < 0$ and $z_n = 1$, then $|C_{n-1}| + |C'_{n-1}| = q_{n-1}$ or $q_{n-1} - q_{n-2}$. By $\gamma_{n-1}/\theta_{n-1} \geq d_n \geq z_n + 1 \geq 2$ we have $\gamma_{n-1} - \theta_{n-1} \geq \theta_{n-1} > 0$. Hence, $\gamma_{n-1} + \gamma'_{n-1} = \theta_{n-1}$ is impossible. Thus, $\gamma_{n-1} + \gamma'_{n-1} = \theta_{n-1} + 1$ and $|C_{n-1}| + |C'_{n-1}| = q_{n-1} - q_{n-2}$. Therefore,

$$K\|K\theta - \phi\| = (q_{n-1} - |C_{n-1}|)(d_n - 1 + \gamma_n)|D_{n-1}|$$
$$= (|C'_{n-1}| + q_{n-2})\left(\frac{\gamma_{n-1}}{\theta_{n-1}} - 1\right)|D_{n-1}|$$
$$= (|C'_{n-1}| + q_{n-2})(1 - \gamma'_{n-1})|D_{n-2}|.$$

When $K < 0$ (so, n is even), we have $-q_n + 1 \leq C_n \leq 0$, and $|C_n| + |C'_n| = q_n$ or $q_n - q_{n-1}$. If $C_{n-1} < 0$ or $z_n \geq 2$, then

$$|K|\|K\theta - \phi\| - |C_n|\|C_n\theta - \phi\|$$
$$= (z_n q_{n-1} - C_{n-1})(d_n - z_n + \gamma_n)|D_{n-1}| - (d_n q_{n-1} - C_{n-1})\gamma_n|D_{n-1}|$$
$$= ((z_n - \gamma_n)q_{n-1} - C_{n-1})(d_n - z_n)|D_{n-1}| > 0.$$

If $C_{n-1} > 0$ and $z_n = 1$, then $C_{n-1} + C'_{n-1} = q_{n-1}$ or $q_{n-1} - q_{n-2}$. By applying the same argument as the case where $K > 0$ (n is odd), $C_{n-1} < 0$ and $z_n = 1$ above, we have

$$|K|\|K\theta - \phi\| = (C'_{n-1} + q_{n-2})(1 - \gamma'_{n-1})|D_{n-2}|$$
$$= (C'_{n-1} + q_{n-2})\|(C'_{n-1} + q_{n-2})\theta - \phi\|.$$

If $z_s > d_s$, then

$$\|K\theta - \phi\| = (z_s - d_s + \gamma_s^* - \gamma_s)|D_{s-1}|$$
$$> (1 - \gamma_s)|D_{s-1}| = \|(C_s + q_{s-1})\theta - \phi\|.$$

If $s \le n - 3$, $|K| > |C_s| + q_{s-1}$ yields the result. Let n be odd. If n be even, the proof is similar. When $s = n-2$, we can assume that $d_{n-2} > 0$, so $C_{n-2} > 0$. Otherwise, there is a positive integer $s(< n - 2)$ such that $d_s > 0$ and $C_s = C_{s+1} = \cdots = C_{n-2}$. Then,

$$K = z_n q_{n-1} - z_{n-1}q_{n-2} + (z_{n-2} - d_{n-2})q_{n-3} + C_{n-2}$$
$$\ge q_{n-1} - (a_{n-1} - 1)q_{n-2} + q_{n-3} + C_{n-2}$$
$$= q_{n-2} + 2q_{n-3} + C_{n-2} > C_{n-2} + q_{n-3}.$$

When $s = n - 1$, we can assume that $d_{n-1} > 0$, so $C_{n-1} < 0$. Otherwise, this case is reduced to the case $s \le n - 2$. Then,

$$K = z_n q_{n-1} - (z_{n-1} - d_{n-1})q_{n-2} - C_{n-1}$$
$$\ge q_{n-1} - (a_{n-1} - 1)q_{n-2} + |C_{n-1}| > |C_{n-1}| + q_{n-2}.$$

When $s = n$, we can assume that $d_n > 0$, so $C_n > 0$. Otherwise, this case is reduced to the case $s \le n - 1$. Then, $K = (z_n - d_n)q_{n-1} + C_n \ge C_n + q_{n-1}$.

4.   SOME APPLICATIONS

We shall denote the representation of ϕ ($0 < \phi < 1$) through the expansion of θ by the Borweins' algorithm by $\phi =_\theta \langle d_1, d_2, \ldots, d_n, \ldots \rangle$ with omission of $d_0 = 0$. The overline means the periodic or quasi-periodic representation. For example,

$$\phi = _\theta\langle d_1, \ldots, d_n, \overline{d_{n+1}, \ldots, d_m} \rangle$$
$$= _\theta\langle d_1, \ldots, d_n, d_{n+1}, \ldots, d_m, d_{n+1}, \ldots, d_m, \ldots \rangle,$$
$$\phi = _\theta\left\langle d_1, \ldots, d_n, \overline{f_1(i), \ldots, f_m(i)} \right\rangle_{i=1}^\infty$$
$$= _\theta\langle d_1, \ldots, d_n, f_1(1), \ldots, f_m(1), \ldots, f_1(k), \ldots, f_m(k), \ldots \rangle.$$

The first example, Theorem 2, shows the case where θ is one of the typical Hurwitzian numbers, $\theta = e$.

Proof of Theorem 2. First of all, we shall look at the cases when $l = 5$ and $l = 6$.

When $\theta = e = [2; \overline{1, 2i, 1}]_{i=1}^{\infty}$, $\phi = 1/5$ is represented as

$$\frac{1}{5} = \theta\Big\langle 0, 4i - 4, 1, 0, 8i - 7, 1, 0, 4i - 3, 0, 0, 4i - 3, 1, 0, 16i - 8, 0,$$

$$0, 16i - 6, 1, 0, 12i - 4, 1, 0, 16i - 3, 0, 0, 16i - 1, 1, 0, 4i, 0\Big\rangle_{i=1}^{\infty}$$

$$= \theta\Big\langle 0, \frac{a_{30i-28} - 2}{5}, 1, 0, \frac{2a_{30i-25} - 3}{5}, 1, 0, \frac{a_{30i-22} - 1}{5}, 0, 0,$$

$$\frac{a_{30i-19} - 3}{5}, 1, 0, \frac{4}{5}a_{30i-16}, 0, 0, \frac{4a_{30i-13} + 2}{5}, 1, 0, \frac{3a_{30i-10} - 2}{5},$$

$$1, 0, \frac{4a_{30i-7} + 1}{5}, 0, 0, \frac{4a_{30i-4} + 3}{5}, 1, 0, \frac{1}{5}a_{30i-1}, 0\Big\rangle_{i=1}^{\infty}$$

and for $n = 1, 2, \ldots$

$$\gamma_{30n-29} = \frac{1+\theta_{30n-29}}{5}, \quad \gamma_{30n-28} = \frac{3+\theta_{30n-28}}{5}, \quad \gamma_{30n-27} = \frac{-1+3\theta_{30n-27}}{5},$$

$$\gamma_{30n-26} = \frac{2-\theta_{30n-26}}{5}, \quad \gamma_{30n-25} = \frac{2+2\theta_{30n-25}}{5}, \quad \gamma_{30n-24} = \frac{-1+2\theta_{30n-24}}{5},$$

$$\gamma_{30n-23} = \frac{1-\theta_{30n-23}}{5}, \quad \gamma_{30n-22} = \frac{\theta_{30n-22}}{5}, \quad \gamma_{30n-21} = \frac{1}{5},$$

$$\gamma_{30n-20} = \frac{1+\theta_{30n-20}}{5}, \quad \gamma_{30n-19} = \frac{4+\theta_{30n-19}}{5}, \quad \gamma_{30n-18} = \frac{4\theta_{30n-18}}{5},$$

$$\gamma_{30n-17} = \frac{4}{5}, \quad \gamma_{30n-16} = \frac{4\theta_{30n-16}}{5}, \quad \gamma_{30n-15} = \frac{4}{5},$$

$$\gamma_{30n-14} = \frac{4+4\theta_{30n-14}}{5}, \quad \gamma_{30n-13} = \frac{2+4\theta_{30n-13}}{5}, \quad \gamma_{30n-12} = \frac{1+2\theta_{30n-12}}{5},$$

$$\gamma_{30n-11} = \frac{3+\theta_{30n-11}}{5}, \quad \gamma_{30n-10} = \frac{3+3\theta_{30n-10}}{5}, \quad \gamma_{30n-9} = \frac{1+3\theta_{30n-9}}{5},$$

$$\gamma_{30n-8} = \frac{4+\theta_{30n-8}}{5}, \quad \gamma_{30n-7} = \frac{4\theta_{30n-7}}{5}, \quad \gamma_{30n-6} = \frac{4}{5},$$

$$\gamma_{30n-5} = \frac{4+4\theta_{30n-5}}{5}, \quad \gamma_{30n-4} = \frac{1+\theta_{30n-4}}{5}, \quad \gamma_{30n-3} = \frac{\theta_{30n-3}}{5},$$

$$\gamma_{30n-2} = \frac{1}{5}, \quad \gamma_{30n-1} = \frac{\theta_{30n-1}}{5}, \quad \gamma_{30n} = \frac{1}{5}.$$

Notice that

$$\theta_{3n-2} = [0; 2n, 1, 1, \ldots] \to 0,$$

$$\theta_{3n-1} = [0; 1, 1, 2n + 2, \ldots] \to \frac{1}{2},$$

$$\theta_{3n} = [0; 1, 2n + 2, 1, 1, \ldots] \to 1 \quad (n \to \infty).$$

Notice also that

$$q_{3n}|D_{3n}| = \frac{1}{a_{3n+1} + \theta_{3n+1} + q_{3n-1}/q_{3n}} \to \frac{1}{1 + 0 + 1} = \frac{1}{2} \quad (n \to \infty)$$

as well as

$$q_{3n+1}|D_{3n}| \to 1, \quad q_{3n-1}|D_{3n-1}| \to \tfrac{1}{2}, \quad q_{3n}|D_{3n-1}| \to \tfrac{1}{2},$$
$$q_{3n-2}|D_{3n-2}| \to 0, \quad q_{3n-1}|D_{3n-2}| \to 1 \quad (n \to \infty).$$

Since

$$
\begin{aligned}
C_{30n-1} = \sum_{i=1}^{n} \Bigg(&\frac{1}{5}a_{30i-1}q_{30i-2} + q_{30i-4} - \frac{4a_{30i-4}+3}{5}q_{30i-5} \\
&+ \frac{4a_{30i-7}+1}{5}q_{30i-8} + q_{30i-10} - \frac{3a_{30i-10}-2}{5}q_{30i-11} \\
&- q_{30i-13} + \frac{4a_{30i-13}+2}{5}q_{30i-14} - \frac{4}{5}a_{30i-16}q_{30i-17} \\
&- q_{30i-19} + \frac{a_{30i-19}-3}{5}q_{30i-20} - \frac{a_{30i-22}-1}{5}q_{30i-23} \\
&- q_{30i-25} + \frac{2a_{30i-25}-3}{5}q_{30i-26} + q_{30i-28} \\
&- \frac{a_{30i-28}-2}{5}q_{30i-29} \Bigg) \\
= &\frac{1}{5}\sum_{i=1}^{n}(q_{30i-1} - q_{30i-31}) = \frac{1}{5}q_{30n-1} ,
\end{aligned}
$$

one can have

$$
\begin{aligned}
C_{30n-1}|D_{30n-2}|\gamma_{30n-1} &= \frac{1}{5}q_{30n-1}|D_{30n-2}|\frac{1}{5}\theta_{30n-1} \\
&\rightarrow \frac{1}{5}\cdot 1 \cdot \frac{1}{5}\cdot\frac{1}{2} = \frac{1}{50} \quad (n\rightarrow\infty)
\end{aligned}
$$

and

$$
\begin{aligned}
(C_{30n-1} &+ q_{30n-2})|D_{30n-2}|(1-\gamma_{30n-1}) \\
&= \left(\frac{1}{5}q_{30n-1}|D_{30n-2}| + q_{30n-2}|D_{30n-2}|\right)\left(1 - \frac{1}{5}\theta_{30n-1}\right) \\
&\rightarrow \left(\frac{1}{5}\cdot 1 + 0\right)\left(1 - \frac{1}{10}\right) = \frac{9}{50} \quad (n\rightarrow\infty).
\end{aligned}
$$

For simplicity we put

$$
\Xi_n = 2l^2|C_n|\|C_n\theta - \phi\| = 2l^2|C_n D_{n-1}|\gamma_n
$$

and

$$
\Xi_n^* = 2l^2(|C_n| + q_{n-1})|D_{n-1}|(1-\gamma_n).
$$

In a similar manner one can find

$$\Xi_{30n} \to 1, \ \Xi_{30n-2} \to 1, \ \Xi_{30n-3} \to 1, \ \Xi^*_{30n-3} \to 24,$$
$$\Xi_{30n-4} \to 24, \ \Xi^*_{30n-4} \to 16, \ \Xi_{30n-5} \to 16, \ \Xi^*_{30n-5} \to 9,$$
$$\Xi_{30n-6} \to 16, \ \Xi_{30n-7} \to 16, \ \Xi^*_{30n-7} \to 24, \ \Xi_{30n-8} \to 8,$$
$$\Xi_{30n-9} \to 8, \ \Xi^*_{30n-9} \to 9, \ \Xi_{30n-10} \to 27, \ \Xi^*_{30n-10} \to 3,$$
$$\Xi_{30n-11} \to 3, \ \Xi_{30n-12} \to 3, \ \Xi^*_{30n-12} \to 12, \ \Xi_{30n-13} \to 32,$$
$$\Xi^*_{30n-13} \to 8, \ \Xi_{30n-14} \to 16, \ \Xi^*_{30n-14} \to 9, \ \Xi_{30n-15} \to 16,$$
$$\Xi_{30n-16} \to 16, \ \Xi^*_{30n-16} \to 24, \ \Xi_{30n-17} \to 16, \ \Xi_{30n-18} \to 16,$$
$$\Xi^*_{30n-18} \to 9, \ \Xi_{30n-19} \to 9, \ \Xi^*_{30n-19} \to 1, \ \Xi_{30n-20} \to 1,$$
$$\Xi_{30n-20} \to 24, \ \Xi_{30n-21} \to 1, \ \Xi_{30n-22} \to 1, \ \Xi^*_{30n-22} \to 9,$$
$$\Xi_{30n-23} \to 3, \ \Xi_{30n-24} \to 3, \ \Xi^*_{30n-24} \to 32, \ \Xi_{30n-25} \to 12,$$
$$\Xi^*_{30n-25} \to 8, \ \Xi_{30n-26} \to 8, \ \Xi_{30n-27} \to 8, \ \Xi^*_{30n-27} \to 27,$$
$$\Xi_{30n-28} \to 7, \ \Xi^*_{30n-28} \to 3, \ \Xi_{30n-29} \to 1, \ \Xi^*_{30n-29} \to 24 \ (n \to \infty).$$

$1 - \phi = 4/5$ is represented as

$$\frac{4}{5} = \langle 1,0,0,\overline{0, 12i-10, 1, 0, 16i-11, 0, 0, 16i-9, 1, 0, 4i-2, 0, 0, 4i-2,}$$
$$\overline{1, 0, 8i-3, 1, 0, 4i-1, 0, 0, 4i-1, 1, 0, 16i, 0, 0, 16i+2, 1} \rangle^{\infty}_{i=1}$$
$$= \langle 1,0,0,0, \frac{3a_{30i-25}-2}{5}, 1, 0, \frac{4a_{30i-22}+1}{5}, 0, 0, \frac{4a_{30i-19}+3}{5}, 1,$$
$$0, \frac{a_{30i-16}}{5}, 0, 0, \frac{a_{30i-13}-2}{5}, 1, 0, \frac{2a_{30i-10}-3}{5}, 1, 0, \frac{a_{30i-7}-1}{5},$$
$$0, 0, \frac{a_{30i-4}-3}{5}, 1, 0, \frac{4a_{30i-1}}{5}, 0, 0, \frac{4a_{30i+2}+2}{5}, 1 \rangle^{\infty}_{i=1}.$$

and $\gamma'_1 = (-1+4\theta_1)/5, \ \gamma'_2 = (2 - \theta_2)/5$ and for $n = 1, 2, \dots$

$$\gamma'_{30n-27} = \frac{1+2\theta_{30n-27}}{5}, \qquad \gamma'_{30n-26} = \frac{3+\theta_{30n-26}}{5}, \qquad \gamma'_{30n-25} = \frac{3+3\theta_{30n-25}}{5},$$
$$\gamma'_{30n-24} = \frac{1+3\theta_{30n-24}}{5}, \qquad \gamma'_{30n-23} = \frac{4+\theta_{30n-23}}{5}, \qquad \gamma'_{30n-22} = \frac{4\theta_{30n-22}}{5},$$
$$\gamma'_{30n-21} = \frac{4}{5}, \qquad \gamma'_{30n-20} = \frac{4+4\theta_{30n-20}}{5}, \qquad \gamma'_{30n-19} = \frac{1+4\theta_{30n-19}}{5},$$
$$\gamma'_{30n-18} = \frac{\theta_{30n-18}}{5}, \qquad \gamma'_{30n-17} = \frac{1}{5}, \qquad \gamma'_{30n-16} = \frac{\theta_{30n-16}}{5},$$
$$\gamma'_{30n-15} = \frac{1}{5}, \qquad \gamma'_{30n-14} = \frac{1+\theta_{30n-14}}{5}, \qquad \gamma'_{30n-13} = \frac{3+\theta_{30n-13}}{5},$$
$$\gamma'_{30n-12} = \frac{-1+3\theta_{30n-12}}{5}, \qquad \gamma'_{30n-11} = \frac{2+\theta_{30n-11}}{5}, \qquad \gamma'_{30n-10} = \frac{2+2\theta_{30n-10}}{5},$$
$$\gamma'_{30n-9} = \frac{-1+2\theta_{30n-9}}{5}, \qquad \gamma'_{30n-8} = \frac{1+\theta_{30n-8}}{5}, \qquad \gamma'_{30n-7} = \frac{\theta_{30n-7}}{5},$$
$$\gamma'_{30n-6} = \frac{1}{5}, \qquad \gamma'_{30n-5} = \frac{1+\theta_{30n-5}}{5}, \qquad \gamma'_{30n-4} = \frac{4+\theta_{30n-4}}{5},$$
$$\gamma'_{30n-3} = \frac{4\theta_{30n-3}}{5}, \qquad \gamma'_{30n-2} = \frac{4}{5}, \qquad \gamma'_{30n-1} = \frac{4\theta_{30n-1}}{5},$$
$$\gamma'_{30n} = \frac{4}{5}, \qquad \gamma'_{30n+1} = \frac{4+4\theta_{30n+1}}{5}, \qquad \gamma'_{30n+2} = \frac{2+4\theta_{30n+2}}{5}.$$

Since

$$
C'_{30n+3} = 1 + \sum_{i=1}^{n} \left(q_{30i+2} - \frac{4a_{30i+2}+2}{5} q_{30i+1} + \frac{4}{5} a_{30i-1} q_{30i-2} \right.
$$

$$
+ q_{30i-4} - \frac{a_{30i-4}-3}{5} q_{30i-5} - \frac{a_{30i-7}-1}{5} q_{30i-8} + q_{30i-10}
$$

$$
- \frac{2a_{30i-10}-3}{5} q_{30i-11} - q_{30i-13} + \frac{a_{30i-13}-2}{5} q_{30i-14}
$$

$$
- \frac{1}{5} a_{30i-16} q_{30i-17} - q_{30i-19} + \frac{4a_{30i-19}+3}{5} q_{30i-20}
$$

$$
\left. - \frac{4a_{30i-22}+1}{5} q_{30i-23} - q_{30i-25} - \frac{3a_{30i-25}-2}{5} q_{30i-26} \right)
$$

$$
= 1 + \frac{1}{5} \sum_{i=1}^{n} (q_{30i+2} + 2q_{30i+1} - q_{30i-28} - 2q_{30i-29})
$$

$$
= \frac{1}{5} (q_{30n+2} + 2q_{30n+1}) = \frac{1}{5} (2q_{30n+3} - q_{30n+2}),
$$

one can have

$$
C'_{30n+3} | D_{30n+2} | \gamma'_{30n+3}
$$

$$
= \frac{1}{5} (2q_{30n+3} | D_{30n+2} | - q_{30n+2} | D_{30n+2} |) \frac{1+2\theta_{30n+3}}{5}
$$

$$
\rightarrow \frac{1}{5} \left(2 \cdot \frac{1}{2} - \frac{1}{2} \right) \cdot \frac{3}{5} = \frac{3}{50} \quad (n \rightarrow \infty)
$$

and

$$
(C'_{30n+3} + q_{30n+2}) | D_{30n+2} | (1-\gamma'_{30n+3}) \rightarrow \left(\frac{1}{10} + \frac{1}{2} \right) \frac{2}{5} = \frac{12}{50} \quad (n \rightarrow \infty).
$$

For simplicity we put

$$
\Psi_n = 2l^2 |C'_n| \|C'_n\theta + \phi\| = 2l^2 |C'_n D_{n-1}| \gamma'_n
$$

and

$$
\Psi_n^* = 2l^2 (|C'_n| + q_{n-1}) | D_{n-1} | (1 - \gamma'_n).
$$

In a similar manner one can find

$$\Psi_{30n+2} \to 32, \ \Psi^*_{30n+2} \to 8, \ \Psi_{30n+1} \to 16, \ \Psi^*_{30n+1} \to 9,$$
$$\Psi_{30n} \to 16, \ \Psi_{30n-1} \to 16, \ \Psi^*_{30n-1} \to 24, \ \Psi_{30n-2} \to 16,$$
$$\Psi_{30n-3} \to 16, \ \Psi^*_{30n-3} \to 9, \ \Psi_{30n-4} \to 9, \ \Psi^*_{30n-4} \to 1,$$
$$\Psi_{30n-5} \to 1, \ \Psi^*_{30n-5} \to 24, \ \Psi_{30n-6} \to 1, \ \Psi_{30n-7} \to 1,$$
$$\Psi^*_{30n-7} \to 9, \ \Psi_{30n-8} \to 3, \ \Psi_{30n-9} \to 3, \ \Psi^*_{30n-9} \to 32,$$
$$\Psi_{30n-10} \to 12, \ \Psi^*_{30n-10} \to 8, \ \Psi_{30n-11} \to 8, \ \Psi_{30n-12} \to 8,$$
$$\Psi^*_{30n-12} \to 27, \ \Psi_{30n-13} \to 7, \ \Psi^*_{30n-13} \to 3, \ \Psi_{30n-14} \to 1,$$
$$\Psi^*_{30n-14} \to 24, \ \Psi_{30n-15} \to 1, \ \Psi_{30n-16} \to 1, \ \Psi^*_{30n-16} \to 9,$$
$$\Psi_{30n-17} \to 1, \ \Psi_{30n-18} \to 1, \ \Psi^*_{30n-18} \to 24, \ \Psi_{30n-19} \to 24,$$
$$\Psi^*_{30n-19} \to 16, \ \Psi_{30n-20} \to 16, \ \Psi^*_{30n-20} \to 9, \ \Psi_{30n-21} \to 16,$$
$$\Psi_{30n-22} \to 16, \ \Psi^*_{30n-22} \to 24, \ \Psi_{30n-23} \to 8, \ \Psi_{30n-24} \to 8,$$
$$\Psi^*_{30n-24} \to 7, \ \Psi_{30n-25} \to 27, \ \Psi^*_{30n-25} \to 3, \ \Psi_{30n-26} \to 3 \ (n \to \infty).$$

Therefore, we have $\mathcal{M}(e, 1/5) = 1/50$.

$\phi = 1/6$ is represented as

$$\frac{1}{6} = {}_\theta\langle \overline{0, 2i - 2, 1, 0, 2i - 2, 0, 0, 10i - 5, 1, 0, 10i - 3, 0,}$$
$$\overline{0, 10i - 1, 1, 0, 2i, 0}\rangle^\infty_{i=1}$$
$$= {}_\theta\Big\langle 0, \frac{a_{18i-16} - 2}{6}, 1, 0, \frac{a_{18i-13} - 4}{6}, 0, 0, \frac{5a_{18i-10}}{6}, 1,$$
$$0, \frac{5a_{18i-7} + 2}{6}, 0, 0, \frac{5a_{18i-4} + 4}{6}, 1, 0, \frac{a_{18i-1}}{6}, 0 \Big\rangle^\infty_{i=1}$$

and for $n = 1, 2, \ldots$

$$\gamma_{18n-17} = \frac{1+\theta_{18n-17}}{6}, \qquad \gamma_{18n-16} = \frac{3+\theta_{18n-16}}{6}, \qquad \gamma_{18n-15} = \frac{-2+3\theta_{18n-15}}{6},$$
$$\gamma_{18n-14} = \frac{1-2\theta_{18n-14}}{6}, \qquad \gamma_{18n-13} = \frac{2+\theta_{18n-13}}{6}, \qquad \gamma_{18n-12} = \frac{3+2\theta_{18n-12}}{6},$$
$$\gamma_{18n-11} = \frac{5+3\theta_{18n-11}}{6}, \qquad \gamma_{18n-10} = \frac{3+5\theta_{18n-10}}{6}, \qquad \gamma_{18n-9} = \frac{2+3\theta_{18n-9}}{6},$$
$$\gamma_{18n-8} = \frac{5+2\theta_{18n-8}}{6}, \qquad \gamma_{18n-7} = \frac{5\theta_{18n-7}}{6}, \qquad \gamma_{18n-6} = \frac{5}{6},$$
$$\gamma_{18n-5} = \frac{5+5\theta_{18n-5}}{6}, \qquad \gamma_{18n-4} = \frac{1+5\theta_{18n-4}}{6}, \qquad \gamma_{18n-3} = \frac{\theta_{18n-3}}{6},$$
$$\gamma_{18n-2} = \frac{1}{6}, \qquad \gamma_{18n-1} = \frac{\theta_{18n-1}}{6}, \qquad \gamma_{18n} = \frac{1}{6}.$$

Since

$$C_{18n} = \sum_{i=1}^{n} \left(\frac{a_{18i-1}}{6} q_{18i-2} + q_{18i-4} - \frac{5a_{18i-4}+4}{6} q_{18i-5} \right.$$

$$+ \frac{5a_{18i-7}+2}{6} q_{18i-8} + q_{18i-10} - \frac{5}{6} a_{18i-10} q_{18i-11}$$

$$\left. + \frac{a_{18i-13}-4}{6} q_{18i-14} + q_{18i-16} - \frac{a_{18i-16}-2}{6} q_{18i-17} \right)$$

$$= \frac{1}{6} \sum_{i=1}^{n} (q_{18i-1} - q_{18i-19}) = \frac{1}{6} q_{18n-1},$$

one can have

$$C_{18n-1}|D_{18n-2}|\gamma_{18n-1} = \frac{1}{6} q_{18n-1}|D_{18n-2}| \frac{\theta_{18n-1}}{6}$$

$$\rightarrow \frac{1}{6} \cdot 1 \cdot \frac{1}{12} = \frac{1}{72} \quad (n \rightarrow \infty)$$

and

$$(C_{18n-1} + q_{18n-2})|D_{18n-2}|(1 - \gamma_{18n-1}) \rightarrow \left(\frac{1}{6} + 0 \right) \frac{11}{12} = \frac{11}{72} \quad (n \rightarrow \infty).$$

In a similar manner one can find that

$$\Xi_{18n} \rightarrow 1, \ \Xi_{18n-2} \rightarrow 1, \ \Xi_{18n-3} \rightarrow 1, \ \Xi^*_{18n-3} \rightarrow 35,$$
$$\Xi_{18n-4} \rightarrow 35, \ \Xi^*_{18n-4} \rightarrow 25, \ \Xi_{18n-5} \rightarrow 25, \ \Xi^*_{18n-5} \rightarrow 11,$$
$$\Xi_{18n-6} \rightarrow 25, \ \Xi_{18n-7} \rightarrow 25, \ \Xi^*_{18n-7} \rightarrow 35, \ \Xi_{18n-8} \rightarrow 5,$$
$$\Xi_{18n-9} \rightarrow 5, \ \Xi^*_{18n-9} \rightarrow 7, \ \Xi_{18n-10} \rightarrow 55, \ \Xi^*_{18n-10} \rightarrow 5,$$
$$\Xi_{18n-11} \rightarrow 5, \ \Xi^*_{18n-11} \rightarrow 7, \ \Xi_{18n-12} \rightarrow 5, \ \Xi_{18n-13} \rightarrow 5,$$
$$\Xi^*_{18n-13} \rightarrow 7, \ \Xi_{18n-14} \rightarrow 5, \ \Xi_{18n-15} \rightarrow 5, \ \Xi^*_{18n-15} \rightarrow 55,$$
$$\Xi_{18n-16} \rightarrow 7, \ \Xi^*_{18n-16} \rightarrow 5, \ \Xi_{18n-17} \rightarrow 1, \ \Xi^*_{18n-17} \rightarrow 35 \quad (n \rightarrow \infty).$$

$1 - \phi = 5/6$ is represented as

$$\frac{5}{6} = \phi \langle 1,0,0,\overline{0,10i-6,0,0,2i-1,1,0,2i-1,0,}$$

$$\overline{0,2i-1,1,0,10i,0,0,10i+2,1} \rangle_{i=1}^{\infty}$$

$$= \phi \left\langle 1,0,0,0, \overline{\frac{5a_{18i-13}+4}{6}}, 0,0, \overline{\frac{a_{18i-10}}{6}}, 1,0, \overline{\frac{a_{18i-7}-2}{6}}, 0, \right.$$

$$\left. 0, \overline{\frac{a_{18i-4}-4}{6}}, 1,0, \overline{\frac{5a_{18i-1}}{6}}, 0,0, \overline{\frac{5a_{18i+2}+2}{6}}, 1 \right\rangle_{i=1}^{\infty}$$

and $\gamma_1' = (-1 + 5\theta_1)/6$, $\gamma_2' = (3 - \theta_2)/6$, for $n = 1, 2, \ldots$

$$\gamma_{18n-15}' = \tfrac{2+3\theta_{18n-15}}{6}, \quad \gamma_{18n-14}' = \tfrac{5+2\theta_{18n-14}}{6}, \quad \gamma_{18n-13}' = \tfrac{-2+5\theta_{18n-13}}{6},$$

$$\gamma_{18n-12}' = \tfrac{3-2\theta_{18n-12}}{6}, \quad \gamma_{18n-11}' = \tfrac{1+3\theta_{18n-11}}{6}, \quad \gamma_{18n-10}' = \tfrac{3+\theta_{18n-10}}{6},$$

$$\gamma_{18n-9}' = \tfrac{-2+3\theta_{18n-9}}{6}, \quad \gamma_{18n-8}' = \tfrac{1-2\theta_{18n-8}}{6}, \quad \gamma_{18n-7}' = \tfrac{\theta_{18n-7}}{6},$$

$$\gamma_{18n-6}' = \tfrac{1}{6}, \quad \gamma_{18n-5}' = \tfrac{1+\theta_{18n-5}}{6}, \quad \gamma_{18n-4}' = \tfrac{5+\theta_{18n-4}}{6},$$

$$\gamma_{18n-3}' = \tfrac{5\theta_{18n-3}}{6}, \quad \gamma_{18n-2}' = \tfrac{5}{6}, \quad \gamma_{18n-1}' = \tfrac{5\theta_{18n-1}}{6},$$

$$\gamma_{18n}' = \tfrac{5}{6}, \quad \gamma_{18n+1}' = \tfrac{5+5\theta_{18n+1}}{6}, \quad \gamma_{18n+2}' = \tfrac{3+5\theta_{18n+2}}{6}.$$

Since

$$C_{18n+3}' = 1 + \sum_{i=1}^{n} \left(q_{18i+2} - \frac{5a_{18i+2}+2}{6} q_{18i+1} + \frac{5}{6} a_{18i-1} q_{18i-2} \right.$$

$$+ q_{18i-4} - \frac{a_{18i-4}-4}{6} q_{18i-5} + \frac{a_{18i-7}-2}{6} q_{18i-8}$$

$$+ q_{18i-10} - \frac{5}{6} a_{18i-10} q_{18i-11} + \left. \frac{5a_{18i-13}+4}{6} q_{18i-14} \right)$$

$$= 1 + \frac{1}{6} \sum_{i=1}^{n} (3q_{18i+3} - 2q_{18i+2} - 3q_{18i-15} + 2q_{18i-16})$$

$$= \frac{1}{6}(3q_{18n+3} - 2q_{18n+2}),$$

one can have

$$C_{18n+3}' | D_{18n+2} | \gamma_{18n+3}'$$

$$= \frac{1}{6}(3q_{18n+3} | D_{18n+2} - 2q_{18n+2} | D_{18n+1}) | \frac{2+3\theta_{18n+3}}{6}$$

$$\to \frac{1}{6}\left(3 \cdot \frac{1}{2} - 2 \cdot \frac{1}{2}\right) \frac{2+3\cdot 1}{6} = \frac{5}{72} \quad (n \to \infty)$$

and

$$(C_{18n+3}' + q_{18n+2}) | D_{18n+2} | (1 - \gamma_{18n+3}') \to \left(\frac{1}{12} + \frac{1}{2}\right)\frac{1}{6} = \frac{7}{72} \quad (n \to \infty).$$

In a similar manner one can find that

$$\Psi_{18n+2} \to 55, \quad \Psi_{18n+2}^* \to 5, \quad \Psi_{18n+1} \to 25, \quad \Psi_{18n+1}^* \to 11,$$

$$\Psi_{18n} \to 25, \quad \Psi_{18n-1} \to 25, \quad \Psi_{18n-1}^* \to 35, \quad \Psi_{18n-2} \to 25,$$

$$\Psi_{18n-3} \to 25, \quad \Psi_{18n-3}^* \to 11, \quad \Psi_{18n-4} \to 11, \quad \Psi_{18n-4}^* \to 1,$$

$$\Psi_{18n-5} \to 1, \quad \Psi_{18n-5}^* \to 35, \quad \Psi_{18n-6} \to 1, \quad \Psi_{18n-7} \to 1,$$

$$\Psi_{18n-7}^* \to 11, \quad \Psi_{18n-8} \to 5, \quad \Psi_{18n-9} \to 5, \quad \Psi_{18n-9}^* \to 55,$$

$$\Psi_{18n-10} \to 7, \quad \Psi_{18n-10}^* \to 5, \quad \Psi_{18n-11} \to 5, \quad \Psi_{18n-11}^* \to 55,$$

$$\Psi_{18n-12} \to 5, \quad \Psi_{18n-13} \to 5, \quad \Psi_{18n-13}^* \to 55, \quad \Psi_{18n-14} \to 5, \quad (n \to \infty).$$

Therefore, we have $\mathcal{M}(e, 1/6) = 1/72$.

For general l, when l is even, we have $\gamma_{3ln-1} = \theta_{3ln-1}/l \to 1/(2l)$ $(n \to \infty)$, $C_{3ln-1} = q_{3ln-1}/l$ and $d_{3ln-1} = a_{3ln-1}/l$. Thus,

$$\mathcal{M}(e, 1/l) = \lim_{n\to\infty} C_{3ln-1}|D_{3ln-2}|\gamma_{3ln-1}$$

$$= \lim_{n\to\infty} \frac{1}{l} q_{3ln-1}|D_{3ln-2}|\gamma_{3ln-1} = \frac{1}{l} \cdot 1 \cdot \frac{1}{2l} = \frac{1}{2l^2} \, .$$

When l is odd, we have $\gamma_{6ln-1} = \theta_{6ln-1}/l \to 1/(2l)$ $(n \to \infty)$, $C_{6ln-1} = q_{6ln-1}/l$ and $d_{6ln-1} = a_{6ln-1}/l$. Thus,

$$\mathcal{M}(e, 1/l) = \lim_{n\to\infty} C_{6ln-1}|D_{6ln-2}|\gamma_{6ln-1}$$

$$= \lim_{n\to\infty} \frac{1}{l} q_{6ln-1}|D_{6ln-2}|\gamma_{6ln-1} = \frac{1}{l} \cdot 1 \cdot \frac{1}{2l} = \frac{1}{2l^2} \, . \qquad \square$$

Next example, Theorem 3, where $\theta = e^{1/s}$ with some integer $s \geq 2$, looks like similar but is much more complicated. Continued fraction expansion of $e^{1/s}$ $(s \geq 2)$ is given by

$$e^{1/s} = [\, 1; \, \overline{(2k-1)s-1, \, 1, \, 1} \,]_{k=1}^{\infty} \, .$$

In other words,

$$a_{3n+1} = (2n+1)s - 1, \qquad a_{3n} = a_{3n+2} = 1 \qquad (n = 0, 1, 2, \ldots)$$

and for $n = 1, 2, \ldots \to \infty$

$$\theta_{3n-2} = [0; 1, 1, (2n+1)s - 1, 1, \ldots] \to \frac{1}{2} \, ,$$
$$\theta_{3n-1} = [0; 1, (2n+1)s - 1, 1, 1, \ldots] \to 1 \, ,$$
$$\theta_{3n} = [0; (2n+1)s - 1, 1, 1, 1, (2n+3)s - 1, \ldots] \to 0 \, .$$

Notice that

$$\lim_{n\to\infty} q_{3n-1}|D_{3n-2}| = \tfrac{1}{2}, \qquad \lim_{n\to\infty} q_{3n-2}|D_{3n-2}| = \tfrac{1}{2},$$
$$\lim_{n\to\infty} q_{3n}|D_{3n-1}| = 1, \qquad \lim_{n\to\infty} q_{3n-1}|D_{3n-1}| = \tfrac{1}{2},$$
$$\lim_{n\to\infty} q_{3n+1}|D_{3n}| = 1, \qquad \lim_{n\to\infty} q_{3n}|D_{3n}| = 0.$$

It is, however, not difficult to see the specific case, where $l \geq 2$ and $s \geq 2$ with $s \equiv 0 \pmod{l}$.

Proof of Theorem 3. When $s \equiv 0 \pmod{l}$ $(l \geq 2)$, $\phi = 1/l$ is represented as

$$\phi = \frac{1}{l} = {}_\theta\left\langle \frac{4i-3}{l}s - 1, 1, 0, \frac{(l-1)(4i-1)}{l}s - 1, 1, 0 \right\rangle_{i=1}^{\infty}$$

$$= {}_\theta\left\langle \frac{a_{6i-5} - l + 1}{l}, 1, 0, \frac{(l-1)a_{6i-2} - 1}{l}, 1, 0 \right\rangle_{i=1}^{\infty}$$

and for $i = 1, 2, \cdots \to \infty$

$$\gamma_{6i-6} = \tfrac{1}{l}, \quad \gamma_{6i-5} = \tfrac{l-1}{l} + \tfrac{1}{l}\theta_{6i-5} \to \tfrac{2l-1}{2l}, \quad \gamma_{6i-4} = \tfrac{l-1}{l}\theta_{6i-4} \to \tfrac{l-1}{l},$$
$$\gamma_{6i-3} = \tfrac{l-1}{l}, \quad \gamma_{6i-2} = \tfrac{1}{l} + \tfrac{l-1}{l}\theta_{6i-2} \to \tfrac{l+1}{2l}, \quad \gamma_{6i-1} = \tfrac{1}{l}\theta_{6i-1} \to \tfrac{1}{l}.$$

Since

$$C_{6n} = C_{6n-1}$$

$$= \sum_{i=1}^{n}\left(\frac{a_{6i-5} - l + 1}{l}q_{6i-6} - q_{6i-5} - \frac{(l-1)a_{6i-2} - 1}{l}q_{6i-3} + q_{6i-2} \right)$$

$$= \frac{1}{l}\sum_{i=1}^{n}(q_{6i-1} - q_{6i-7}) = \frac{1}{l}q_{6n-1},$$

we have

$$C_{6n}\gamma_{6n}|D_{6n-1}| = \frac{1}{l}q_{6n-1}|D_{6n-1}|\gamma_{6n} \to \frac{1}{l}\cdot\frac{1}{2}\cdot\frac{1}{l} = \frac{1}{2l^2} \quad (n \to \infty).$$

In a similar manner one finds that

$$C_{6n-1}\gamma_{6n-1}|D_{6n-2}| \to \frac{1}{2l^2}, \quad |C_{6n-2}|\gamma_{6n-2}|D_{6n-3}| \to \frac{(l+1)(l-1)}{2l^2},$$

$$(C_{6n-1} + q_{6n-2})(1 - \gamma_{6n-1})|D_{6n-2}| \to \frac{(l+1)(l-1)}{2l^2},$$

$$(|C_{6n-2}| + q_{6n-3})(1 - \gamma_{6n-2})|D_{6n-3}| \to \frac{(l-1)^2}{2l^2},$$

$$|C_{6n-3}|\gamma_{6n-3}|D_{6n-4}| \to \frac{(l-1)^2}{2l^2}, \quad |C_{6n-4}|\gamma_{6n-4}|D_{6n-5}| \to \frac{(l-1)^2}{2l^2},$$

$$(|C_{6n-4}| + q_{6n-5})(1 - \gamma_{6n-4})|D_{6n-5}| \to \frac{2l-1}{2l^2},$$

$$C_{6n-5}\gamma_{6n-5}|D_{6n-6}| \to \frac{2l-1}{2l^2},$$

$$(C_{6n-5} + q_{6n-6})(1 - \gamma_{6n-5})|D_{6n-6}| \to \frac{1}{2l^2} \quad (n \to \infty).$$

Next, $1 - \phi = 1 - 1/l$ is represented as

$$1 - \phi = \frac{l-1}{l} = {}_\theta\left\langle \overline{\frac{(l-1)(4i-3)}{l}s - 1, 1, 0, \ \frac{4i-1}{l}s - 1, 1, 0} \right\rangle_{i=1}^{\infty}$$

$$= {}_\theta\left\langle \overline{\frac{(l-1)a_{6i-5} - 1}{l}, 1, 0, \ \frac{a_{6i-2} - l + 1}{l}, 1, 0} \right\rangle_{i=1}^{\infty}$$

and for $i = 1, 2, \cdots \to \infty$

$$\gamma'_{6i-6} = \tfrac{l-1}{l}, \quad \gamma'_{6i-5} = \tfrac{1}{l} + \tfrac{l-1}{l}\theta_{6i-5} \to \tfrac{l+1}{2l}, \quad \gamma'_{6i-4} = \tfrac{1}{l}\theta_{6i-4} \to \tfrac{1}{l},$$
$$\gamma'_{6i-3} = \tfrac{1}{l}, \quad \gamma'_{6i-2} = \tfrac{l-1}{l} + \tfrac{1}{l}\theta_{6i-2} \to \tfrac{2l-1}{2l}, \quad \gamma'_{6i-1} = \tfrac{l-1}{l}\theta_{6i-1} \to \tfrac{l-1}{l}.$$

Since

$$C'_{6n} = C'_{6n-1}$$

$$= \sum_{i=1}^{n}\left(\frac{(l-1)a_{6i-5} - 1}{l}q_{6i-6} - q_{6i-5} - \frac{a_{6i-2} - l + 1}{l}q_{6i-3} + q_{6i-2}\right)$$

$$= \frac{l-1}{l}\sum_{i=1}^{n}(q_{6i-1} - q_{6i-7}) = \frac{l-1}{l}q_{6n-1},$$

we have

$$C'_{6n}\gamma'_{6n}|D_{6n-1}| = \frac{l-1}{l}q_{6n-1}|D_{6n-1}|\gamma'_{6n}$$

$$\to \frac{l-1}{l}\cdot\frac{1}{2}\cdot\frac{l-1}{l} = \frac{(l-1)^2}{2l^2} \quad (n \to \infty).$$

In a similar manner one finds that

$$C'_{6n-1}\gamma'_{6n-1}|D_{6n-2}| \to \frac{(l-1)^2}{2l^2}, \quad |C'_{6n-2}|\gamma'_{6n-2}|D_{6n-3}| \to \frac{2l-1}{2l^2},$$

$$(C'_{6n-1} + q_{6n-2})(1 - \gamma'_{6n-1})|D_{6n-2}| \to \frac{2l-1}{2l^2},$$

$$(|C'_{6n-2}| + q_{6n-3})(1 - \gamma'_{6n-2})|D_{6n-3}| \to \frac{1}{2l^2},$$

$$|C'_{6n-3}|\gamma'_{6n-3}|D_{6n-4}| \to \frac{1}{2l^2}, \quad |C'_{6n-4}|\gamma'_{6n-4}|D_{6n-5}| \to \frac{1}{2l^2},$$

$$(|C'_{6n-4}| + q_{6n-5})(1 - \gamma'_{6n-4})|D_{6n-5}| \to \frac{(l+1)(l-1)}{2l^2},$$

$$C'_{6n-5}\gamma'_{6n-5}|D_{6n-6}| \to \frac{(l+1)(l-1)}{2l^2},$$

$$(C'_{6n-5} + q_{6n-6})(1 - \gamma'_{6n-5})|D_{6n-6}| \to \frac{(l-1)^2}{2l^2} \quad (n \to \infty).$$

Therefore, $\mathcal{M}(e^{1/s}, 1/l) = 1/(2l^2)$ if $s \equiv 0 \pmod{l}$ $(l \geq 2)$.   □

The other cases can be achieved in similar ways but the situations are more complicated.

Conjecture 1. *For any integer* $s(\geq 1)$ *and* $l(\geq 2)$

$$\mathcal{M}\left(e^{1/s}, \frac{1}{l}\right) = \frac{1}{2l^2} \quad or \quad 0.$$

When l is even with $l \leq 50$, it has been checked that $\mathcal{M}(e^{1/s}, 1/l) = 1/(2l^2)$. When l is odd with $l \leq 50$, the following table is obtained.

l	$\{s \pmod{l} : \mathcal{M}(e^{1/s}, 1/l) = 0\}$
3	2
5	3
7	1,3,4,5
9	5
11	1,3,5,6,8
13	1,3,4,5,6,7,9,11
15	8
17	4,5,6,8,9,11
19	5,6,8,9,10,11,12,18
21	5,8,11,17
23	2,5,6,10,11,15,16,19,21
25	3,8,18
27	5,14
29	5,16,19,25,27
31	1,2,3,6,9,16,19,20,27,28
33	5,8,14,17,23
35	3,8,18,33
37	2,6,8,9,10,12,13,14,16,17,18,19,21,22,25,27,34
39	5,11,14,17,20,29,32,35
41	1,2,3,5,8,21,22,23,25,28,30,31,32,38,40
43	4,7,8,11,15,17,18,20,22,23,26,30,31,35,38,41,42
45	23
47	3,10,12,15,19,24,38,39,41,42,45
49	3,4,8,10,11,15,17,18,24,25,29,31,32,36,38,39,43,45,47

References

[1] J. M. Borwein and P. B. Borwein, *On the generating function of the integer part:* $[n\alpha + \gamma]$, J. Number Theory **43** (1993), 293–318.

[2] J. W. S. Cassels, *Über* $\underline{\lim}_{x \to +\infty} x|\vartheta x + \alpha - y|$, Math. Ann. **127** (1954), 288–304.

[3] T. W. Cusick, A. M. Rockett and P. Szüsz, *On inhomogeneous Diophantine approximation*, J. Number Theory, **48** (1994), 259–283.

[4] R. Descombes, *Sur la répartition des sommets d'une ligne polygonale régulière non fermée*, Ann. Sci. École Norm Sup., **73** (1956), 283–355.

[5] T. Komatsu, *On inhomogeneous continued fraction expansion and inhomogeneous Diophantine approximation*, J. Number Theory, **62** (1997), 192–212.

[6] T. Komatsu, *On inhomogeneous Diophantine approximation and the Nishioka-Shiokawa-Tamura algorithm*, Acta Arith. **86** (1998), 305–324.

[7] T. Komatsu, *On inhomogeneous Diophantine approximation with some quasi-periodic expressions*, Acta Math. Hung. **85** (1999), 311–330.

[8] K. Nishioka, I. Shiokawa and J. Tamura, *Arithmetical properties of a certain power series*, J. Number Theory, **42** (1992), 61–87.

ASYMPTOTIC EXPANSIONS OF DOUBLE GAMMA-FUNCTIONS AND RELATED REMARKS

Kohji MATSUMOTO

Graduate School of Mathematics, Nagoya University, Chikusa-ku, Nagoya 464-8602, Japan

Keywords: double gamma-function, double zeta-function, asymptotic expansion, real quadratic field

Abstract Let $\Gamma_2(\beta, (w_1, w_2))$ be the double gamma-function. We prove asymptotic expansions of $\log \Gamma_2(\beta, (1, w))$ with respect to w, both when $|w| \to +\infty$ and when $|w| \to 0$. Our proof is based on the results on Barnes' double zeta-functions given in the author's former article [12]. We also prove asymptotic expansions of $\log \Gamma_2(2\varepsilon_n - 1, (\varepsilon_n - 1, \varepsilon_n))$, $\log \rho_2(\varepsilon_n - 1, \varepsilon_n)$ and $\log \rho_2(\varepsilon_n, \varepsilon_n^2 - \varepsilon_n)$, where ε_n is the fundamental unit of $K = \mathbf{Q}(\sqrt{4n^2 + 8n + 3})$. Combining those results with Fujii's formula [6] [7], we obtain an expansion formula for $\zeta'(1; v_1)$, where $\zeta(s; v_1)$ is Hecke's zeta-function associated with K.

1991 Mathematics Subject Classification: 11M41, 11M99, 11R42, 33B99.

1. INTRODUCTION

This is a continuation of the author's article [12]. We first recall Theorem 1 and its corollaries in [12].

Let $\beta > 0$, and w is a non-zero complex number with $|\arg w| < \pi$. The Barnes double zeta-function is defined by

$$\zeta_2(v; \beta, (1, w)) = \sum_{m=0}^{\infty} \sum_{n=0}^{\infty} (\beta + m + nw)^{-v}. \qquad (1.1)$$

This series is convergent absolutely for $\Re v > 2$, and can be continued meromorphically to the whole v-plane, holomorphic except for the poles at $v = 1$ and $v = 2$.

Let $\zeta(v)$, $\zeta(v, \beta)$ be the Riemann zeta and the Hurwitz zeta-function, respectively, $\mathbf{C}$ the complex number field, θ_0 a fixed number satisfying

C. Jia and K. Matsumoto (eds.), Analytic Number Theory, 243–268.
© 2002 Kluwer Academic Publishers. Printed in the Netherlands.

$0 < \theta_0 < \pi$, and put

$$\mathcal{W}_\infty = \{w \in \mathbf{C} \,|\, |w| \geq 1, \, |\arg w| \leq \theta_0\}$$

and

$$\mathcal{W}_0 = \{w \in \mathbf{C} \,|\, |w| \leq 1, \, w \neq 0, \, |\arg w| \leq \theta_0\}.$$

Define

$$\binom{v}{n} = \begin{cases} v(v-1)\cdots(v-n+1)/n! & \text{if } n \text{ is a positive integer,} \\ 1 & \text{if } n = 0. \end{cases}$$

Theorem 1 in [12] asserts that for any positive integer N we have

$$\zeta_2(v; \beta, (1, w)) = \zeta(v, \beta) + \frac{\zeta(v-1)}{v-1} w^{1-v}$$
$$+ \sum_{k=0}^{N-1} \binom{-v}{k} \zeta(-k, \beta)\zeta(v+k)w^{-v-k} + O(|w|^{-\Re v - N}) \quad (1.2)$$

in the region $\Re v > -N + 1$ and $w \in \mathcal{W}_\infty$, and also

$$\zeta_2(v; \beta, (1, w)) = \zeta(v, \beta) + \frac{\zeta(v-1, \beta)}{v-1} w^{-1}$$
$$+ \sum_{k=0}^{N-1} \binom{-v}{k} \zeta(v+k, \beta)\zeta(-k)w^k + O(|w|^N) \quad (1.3)$$

in the region $\Re v > -N + 1$ and $w \in \mathcal{W}_0$. The implied constants in (1.2) and (1.3) depend only on N, v, β and θ_0.

There are two corollaries of these results stated in [12]. Corollary 1 gives the asymptotic expansions of Eisenstein series, which we omit here. The detailed proof of Corollary 1 are described in [12]. On the other hand, Corollary 2 in [12] is only stated without proof. Here we state it as the following Theorem 1.

Let $\Gamma_2(\beta, (1, w))$ be the double gamma-function defined by

$$\log\left(\frac{\Gamma_2(\beta, (1, w))}{\rho_2(1, w)}\right) = \zeta_2'(0; \beta, (1, w)), \quad (1.4)$$

where 'prime' denotes the differentiation with respect to v and

$$-\log \rho_2(1, w) = \lim_{\beta \to 0} \{\zeta_2'(0; \beta, (1, w)) + \log \beta\}. \quad (1.5)$$

Let $\psi(v) = (\Gamma'/\Gamma)(v)$ and γ the Euler constant. Then we have

Theorem 1. *For any positive integer $N \geq 2$, we have*

$$\log \Gamma_2(\beta, (1, w)) = -\frac{1}{2}\beta \log w + \log \Gamma(\beta) + \frac{1}{2}\beta \log 2\pi$$
$$+(\zeta(-1, \beta) - \zeta(-1))w^{-1} \log w - (\zeta(-1, \beta) - \zeta(-1))\gamma w^{-1}$$
$$+ \sum_{k=2}^{N-1} \frac{(-1)^k}{k}(\zeta(-k, \beta) - \zeta(-k))\zeta(k)w^{-k}$$
$$+ O(|w|^{-N}) \tag{1.6}$$

for $w \in \mathcal{W}_\infty$, where the implied constant depends only on N, β and θ_0. Also we have

$$\log \Gamma_2(\beta, (1, w))$$
$$= \log \Gamma(\beta w^{-1} + 1) + \frac{1}{2}\log \Gamma(\beta + 1) - \log \beta + \beta w^{-1} \log w$$
$$+ \{\zeta(-1) + \zeta'(-1) - \zeta_1(-1, \beta) - \zeta_1'(-1, \beta)\}w^{-1}$$
$$- \frac{1}{12}(\gamma + \psi(\beta + 1))w$$
$$- \sum_{k=2}^{N-1} \frac{(-1)^k}{k}(\zeta(k) - \zeta_1(k, \beta))\zeta(-k)w^k + O(|w|^N) \tag{1.7}$$

for $w \in \mathcal{W}_0$, where $\zeta_1(v, \beta) = \zeta(v, \beta) - \beta^{-v}$ and the implied constant depends only on N and θ_0.

Note that when $w > 0$, the formula (1.6) was already obtained in [10] by a different method.

We will show the proof of Theorem 1 in Section 2, and will give additional remarks in Section 3. In Section 4 we will prove Theorem 2, which will give a uniform error estimate with respect to β. In Section 5 we will give our second main result in the present paper, that is an asymptotic expansion formula for $\zeta'(1; v_1)$, where $\zeta(s; v_1)$ is Hecke's zeta-function associated with the real quadratic field $\mathbf{Q}(\sqrt{4n^2 + 8n + 3})$. This can be proved by combining Fujii's result [6] [7] with certain expansions of double gamma-functions (Theorem 3), and the proof of the latter will be described in Sections 6 to 8. Throughout this paper, the empty sum is to be considered as zero.

2.   PROOF OF THEOREM 1

Double gamma-functions were first introduced and studied by Barnes [3] [4] and others about one hundred years ago. In 1970s, Shintani [13]

[14] discovered the importance of double gamma-functions in connection with Kronecker limit formulas for real quadratic fields. Now the usefulness of double gamma-functions in number theory is a well-known fact. For instance, see Vignéras [17], Arakawa [1] [2], Fujii [6] [7]. Therefore it is desirable to study the asymptotic behaviour of double gamma-functions. Various asymptotic formulas of double gamma-functions are obtained by Billingham and King [5]. Also, when $w > 0$, the formula (1.6) was already proved in the author's article [10], by using a certain contour integral. It should be noted that in [10], it is claimed that the error term on the right-hand side of (1.6) is uniform in β (or α in the notation of [10]), but this is not true. See [11], and also Section 4 of the present paper.

Here we prove Theorem 1 by using the results given in [12]. The special case $u = 0$, $\alpha = \beta$ of the formula (3.8) of [12] implies

$$\zeta_2(v; \beta, (1, w)) = \zeta(v, \beta) + \frac{\zeta(v-1)}{v-1} w^{1-v}$$
$$+ \sum_{k=0}^{N-1} \binom{-v}{k} \zeta(-k, \beta) \zeta(v+k) w^{-v-k}$$
$$+ R_N(v; \beta, w) \tag{2.1}$$

for any positive integer N, where

$$R_N(v; \beta, w) = \frac{1}{2\pi i} \int_{(c_N)} \frac{\Gamma(v+z)\Gamma(-z)}{\Gamma(v)} \zeta(v+z, \beta) \zeta(-z) w^z dz, \tag{2.2}$$

$c_N = -\Re v - N + \varepsilon$ with an arbitrarily small positive ε, and the path of the above integral is the vertical line $\Re z = c_N$. In Section 5 of [12] it is shown that (2.1) holds for $\Re v > -N + 1 + \varepsilon$, $w \in \mathcal{W}_\infty$. The result (1.2) is an immediate consequence of the above facts.

From (2.1) we have

$$\zeta_2'(v; \beta, (1, w)) = \zeta'(v, \beta) - \frac{\zeta(v-1)}{(v-1)^2} w^{1-v} + \frac{\zeta'(v-1)}{v-1} w^{1-v}$$
$$- \frac{\zeta(v-1)}{v-1} w^{1-v} \log w + \sum_{k=0}^{N-1} A_k(v; \beta, w)$$
$$+ R_N'(v; \beta, w), \tag{2.3}$$

where

$$A_k(v;\beta,w) = \left\{ \binom{-v}{k}' \zeta(v+k) + \binom{-v}{k} \zeta'(v+k) \right.$$
$$\left. - \binom{-v}{k} \zeta(v+k) \log w \right\} \zeta(-k,\beta) w^{-v-k}.$$

Noting the facts

$$\zeta(0) = -\frac{1}{2}, \quad \zeta(-1) = -\frac{1}{12}, \quad \zeta'(0) = -\frac{1}{2}\log 2\pi,$$
$$\zeta(0,\beta) = \frac{1}{2} - \beta, \quad \zeta'(0,\beta) = \log\Gamma(\beta) - \frac{1}{2}\log 2\pi, \tag{2.4}$$

and

$$\binom{-v}{k}'\bigg|_{v=0} = \begin{cases} 0 & \text{if } k = 0, \\ (-1)^k k^{-1} & \text{if } k \geq 1, \end{cases}$$

we find

$$A_0(0;\beta,w) = \frac{1}{2}\left(\beta - \frac{1}{2}\right)(\log 2\pi - \log w),$$
$$\lim_{v\to 0} A_1(v;\beta,w) = \zeta(-1,\beta)w^{-1}(\log w - \gamma),$$
$$A_k(0;\beta,w) = \frac{(-1)^k}{k}\zeta(-k,\beta)\zeta(k)w^{-k} \quad (k \geq 2),$$

and so

$$\zeta_2'(0;\beta,(1,w))$$
$$= -\frac{1}{12}w\log w + \left(\frac{1}{12} - \zeta'(-1)\right)w + \frac{1}{2}\left(\frac{1}{2} - \beta\right)\log w$$
$$+ \log\Gamma(\beta) + \left(\frac{1}{2}\beta - \frac{3}{4}\right)\log 2\pi$$
$$+ \zeta(-1,\beta)w^{-1}\log w - \zeta(-1,\beta)\gamma w^{-1}$$
$$+ \sum_{k=2}^{N-1} \frac{(-1)^k}{k}\zeta(-k,\beta)\zeta(k)w^{-k} + R_N'(0;\beta,w) \tag{2.5}$$

for $w \in W_\infty$ and $N \geq 2$. (The above calculations are actually the same as in pp.395–396 of [10].)

Put $z = -v - N + \varepsilon + iy$ in (2.2), and differentiate with respect to v.

We obtain

$$R'_N(v; \beta, w)$$
$$= \frac{1}{2\pi} \int_{-\infty}^{\infty} \Gamma(-N + \varepsilon + iy)\zeta(-N + \varepsilon + iy, \beta)w^{-v-N+\varepsilon+iy}$$
$$\times \left\{ \frac{\Gamma'(v + N - \varepsilon - iy)}{\Gamma(v)}\zeta(v + N - \varepsilon - iy) \right.$$
$$- \Gamma(v + N - \varepsilon - iy)\frac{\Gamma'(v)}{\Gamma(v)^2}\zeta(v + N - \varepsilon - iy)$$
$$+ \frac{\Gamma(v + N - \varepsilon - iy)}{\Gamma(v)}\zeta'(v + N - \varepsilon - iy)$$
$$\left. - \frac{\Gamma(v + N - \varepsilon - iy)}{\Gamma(v)}\zeta(v + N - \varepsilon - iy)\log w \right\} dy.$$

Noting

$$\frac{\Gamma'(v)}{\Gamma(v)^2} = \frac{1}{\Gamma(v)}\left(\psi(v + 1) - \frac{1}{v}\right) = \frac{1}{\Gamma(v + 1)}(v\psi(v + 1) - 1) \to -1$$

(as $v \to 0$), we have

$$R'_N(0; \beta, w) = \frac{1}{2\pi} \int_{-\infty}^{\infty} \Gamma(-N + \varepsilon + iy)\zeta(-N + \varepsilon + iy, \beta)w^{-N+\varepsilon+iy}$$
$$\times \Gamma(N - \varepsilon - iy)\zeta(N - \varepsilon - iy)dy. \tag{2.6}$$

Hence, using Stirling's formula and Lemma 2 of [12] we can show that

$$R'_N(0; \beta, w) = O(|w|^{-N+\varepsilon}), \tag{2.7}$$

and this estimate is uniform in β if $0 < \beta \leq 1$.

Consider (2.5) with $N + 1$ instead of N, and compare it with the original (2.5). Then we find

$$R'_N(0; \beta, w) = \frac{(-1)^N}{N}\zeta(-N, \beta)\zeta(N)w^{-N} + R'_{N+1}(0; \beta, w)$$
$$= \frac{(-1)^N}{N}\zeta(-N, \beta)\zeta(N)w^{-N} + O(|w|^{-N-1+\varepsilon}),$$

hence

$$R'_N(0; \beta, w) = O(|w|^{-N}), \tag{2.8}$$

which is again uniform in β if $0 < \beta \leq 1$. Noting this uniformity and the fact

$$\log \Gamma(\beta) = \log \Gamma(\beta + 1) - \log \beta,$$

from (2.5) we obtain

$$
\begin{aligned}
-\log \rho_2(1, w) = {}& -\frac{1}{12} w \log w + \left(\frac{1}{12} - \zeta'(-1)\right) w + \frac{1}{4} \log w \\
& -\frac{3}{4} \log 2\pi + \zeta(-1) w^{-1} \log w - \zeta(-1)\gamma w^{-1} \\
& + \sum_{k=2}^{N-1} \frac{(-1)^k}{k} \zeta(-k)\zeta(k) w^{-k} + O(|w|^{-N})
\end{aligned}
\tag{2.9}
$$

for $w \in \mathcal{W}_\infty$. The first assertion (1.6) of Theorem 1 follows from (2.5), (2.8) and (2.9).

Next we prove (1.7). Our starting point is the special case $u = 0$, $\alpha = \beta$ of (6.7) of [12], that is

$$
\begin{aligned}
& \zeta_2(v; \beta, (1, w)) \\
& = \zeta(v, \beta) + \zeta_1(v, \beta/w) w^{-v} + \frac{1}{v-1} \zeta_1(v-1, \beta) w^{-1} \\
& \quad + \sum_{k=0}^{N-1} \binom{-v}{k} \zeta_1(v+k, \beta)\zeta(-k) w^k + S_{1,N}(v; \beta, w),
\end{aligned}
\tag{2.10}
$$

where

$$
\begin{aligned}
& S_{1,N}(v; \beta, w) \\
& = \frac{1}{2\pi i} \int_{(N-\varepsilon)} \frac{\Gamma(v+z)\Gamma(-z)}{\Gamma(v)} \zeta_1(v+z, \beta)\zeta(-z) w^z dz.
\end{aligned}
\tag{2.11}
$$

In Section 6 of [12] it is shown that (2.10) holds for $\Re v > 1 - N + \varepsilon$, $w \in \mathcal{W}_0$. From (2.10) it follows that

$$
\begin{aligned}
& \zeta_2'(v; \beta, (1, w)) \\
& = \zeta'(v, \beta) + \zeta_1'(v, \beta/w) w^{-v} - \zeta_1(v, \beta/w) w^{-v} \log w \\
& \quad - \frac{1}{(v-1)^2} \zeta_1(v-1, \beta) w^{-1} + \frac{1}{v-1} \zeta_1'(v-1, \beta) w^{-1} \\
& \quad + \sum_{k=0}^{N-1} B_k(v; \beta)\zeta(-k) w^k + S_{1,N}'(v; \beta, w),
\end{aligned}
\tag{2.12}
$$

where

$$
B_k(v; \beta) = \binom{-v}{k}' \zeta_1(v+k, \beta) + \binom{-v}{k} \zeta_1'(v+k, \beta).
$$

It is clear that $B_0(0; \beta) = \zeta_1'(0, \beta)$ and

$$B_k(0; \beta) = \frac{(-1)^k}{k} \zeta_1(k, \beta) \qquad (k \geq 2).$$

Also, since

$$\lim_{v \to 0} \left(\zeta_1(v + 1, \beta) - \frac{1}{v} \right) = -\psi(\beta) - \beta^{-1},$$

we see that

$$\lim_{v \to 0} B_1(v; \beta) = \psi(\beta) + \beta^{-1}.$$

Hence from (2.12) (with (2.4)) we get

$$\zeta_2'(0; \beta, (1, w)) = \frac{1}{2} \log w + \frac{1}{2} \log \Gamma(\beta + 1) - \log \beta$$
$$- \frac{3}{4} \log 2\pi + \log \Gamma \left(\frac{\beta}{w} + 1 \right) + \beta w^{-1} \log w$$
$$- \{\zeta_1(-1, \beta) + \zeta_1'(-1, \beta)\} w^{-1} - \frac{1}{12} \psi(\beta + 1) w$$
$$+ \sum_{k=2}^{N-1} \frac{(-1)^k}{k} \zeta_1(k, \beta) \zeta(-k) w^k + S_{1,N}'(0; \beta, w) \qquad (2.13)$$

for $N \geq 2$.

The estimate

$$S_{1,N}'(0; \beta, w) = O(|w|^N) \qquad (2.14)$$

can be shown similarly to (2.8); this time, instead of Lemma 2 of [12], we use the fact that $\zeta_1(v, \beta)$ and $\zeta_1'(v, \beta)$ are uniformly bounded with respect to β in the domain of absolute convergence. Hence (2.14) is uniform for any $\beta > 0$. From (2.13), (2.14) and this uniformity, we obtain

$$- \log \rho_2(1, w)$$
$$= \frac{1}{2} \log w - \frac{3}{4} \log 2\pi - \{\zeta(-1) + \zeta'(-1)\} w^{-1} + \frac{1}{12} \gamma w$$
$$+ \sum_{k=2}^{N-1} \frac{(-1)^k}{k} \zeta(k) \zeta(-k) w^k + O(|w|^N) \qquad (2.15)$$

for $N \geq 2$, $w \in \mathcal{W}_0$. From (2.13), (2.14) and (2.15), the assertion (1.7) follows.

3. ADDITIONAL REMARKS ON THEOREM 1

In this supplementary section we give two additional remarks. First we mention an alternative proof of (1.6). Shintani [15] proved

$$
\Gamma_2(\beta, (1, w)) = (2\pi)^{\beta/2} \exp\left\{ \left(\frac{\beta - \beta^2}{2w} - \frac{\beta}{2} \right) \log w + \frac{(\beta^2 - \beta)\gamma}{2w} \right\}
$$
$$
\times \Gamma(\beta) \prod_{n=1}^{\infty} \frac{\Gamma(\beta + nw)}{\Gamma(1 + nw)} \exp\left\{ \frac{\beta - \beta^2}{2nw} + (1 - \beta) \log(nw) \right\} \quad (3.1)
$$

(see also Katayama-Ohtsuki [9], p.179). Shintani assumed that $w > 0$, but (3.1) holds for any complex w with $|\arg w| < \pi$ by analytic continuation. We recall Stirling's formula of the form

$$
\log \Gamma(w + a) = \left(w + a - \frac{1}{2} \right) \log w - w + \frac{1}{2} \log 2\pi
$$
$$
+ \sum_{m=1}^{M} \frac{(-1)^{m-1} B'_{m+2}(a)}{m(m+1)(m+2)w^m} + O(|w|^{-M-1/2}), \quad (3.2)
$$

given in p.278, Section 13.6 of Whittaker-Watson [18], where $B'_{m+2}(a)$ is the derivative of the $(m+2)$th Bernoulli polynomial and M is any positive integer. Noting

$$
\zeta(-m, a) = -\frac{B'_{m+2}(a)}{(m+1)(m+2)} \quad (3.3)
$$

(p.267, Section 13.14 of [18]), we obtain

$$
\log \left(\prod_{n=1}^{\infty} \frac{\Gamma(\beta + nw)}{\Gamma(1 + nw)} \exp\left\{ \frac{\beta - \beta^2}{2nw} + (1 - \beta) \log(nw) \right\} \right)
$$
$$
= \sum_{n=1}^{\infty} \left\{ \frac{\beta - \beta^2}{2nw} - \sum_{m=1}^{M} \frac{(-1)^{m-1}}{mn^m} (\zeta(-m, \beta) - \zeta(-m)) w^{-m} \right.
$$
$$
\left. + O\left((n|w|)^{-M-1/2} \right) \right\}. \quad (3.4)
$$

From (3.3) and the fact $B_3(a) = a^3 - (3/2)a^2 + (1/2)a$ it follows that

$$
\zeta(-1, a) = \frac{1}{2}(a - a^2) - \frac{1}{12}. \quad (3.5)
$$

Hence the coefficient of the term of order w^{-1} on the right-hand side of (3.4) vanishes, and so the right-hand side of (3.4) is equal to

$$-\sum_{m=2}^{M}\frac{(-1)^{m-1}}{m}(\zeta(-m,\beta)-\zeta(-m))\zeta(m)w^{-m}+O\left(|w|^{-M-1/2}\right).$$

Substituting this into the right-hand side of (3.1), and noting (3.5), we arrive at the formula (1.6).

Next we discuss a connection with the Dedekind eta-function

$$\eta(w)=e^{\pi iw/12}\prod_{n=1}^{\infty}(1-e^{2\pi inw}).\qquad(3.6)$$

In the rest of this section we assume $\pi/2\leq\theta_0\leq\pi$, and define

$$\mathcal{W}(\theta_0)=\{w\in\mathbf{C}\mid\pi-\theta_0\leq\arg w\leq\theta_0\}.$$

For $w\in\mathcal{W}(\theta_0)$ we have $\log(-w)=-\pi i+\log w$. Hence from (2.9) and the facts $\zeta(-1)=-1/12$ and $\zeta(-k)=0$ for even k it follows that

$$\begin{aligned}&\log\rho_2(1,w)+\log\rho_2(1,-w)\\&=\frac{1}{12}\pi iw-\frac{1}{2}\log w+\frac{1}{4}\pi i+\frac{3}{2}\log 2\pi+\frac{1}{12}\pi iw^{-1}\\&\quad+O(|w|^{-N})\end{aligned}\qquad(3.7)$$

for $w\in\mathcal{W}_\infty\cap\mathcal{W}(\theta_0)$ and $N\geq 2$. Similarly, from (2.15) we get

$$\begin{aligned}&\log\rho_2(1,w)+\log\rho_2(1,-w)\\&=-\log w+\frac{1}{2}\pi i+\frac{3}{2}\log 2\pi+O(|w|^N)\end{aligned}\qquad(3.8)$$

for $w\in\mathcal{W}_0\cap\mathcal{W}(\theta_0)$ and $N\geq 2$. The reason why the terms of order $w^{\pm k}$ ($2\leq k\leq N-1$) vanish in (3.7) and (3.8) can be explained by the modularity of $\eta(w)$, by using the formula

$$\rho_2(1,w)\rho_2(1,-w)=(2\pi)^{3/2}w^{-1/2}\eta(w)\exp\left(\pi i\left(\frac{1}{4}+\frac{1}{12w}\right)\right)\qquad(3.9)$$

due to Shintani [15]. In fact, in view of (3.9), we see that (3.7) is a direct consequence of the definition (3.6) of $\eta(w)$. Also (3.8) follows easily from (3.9) and the modular relation of $\eta(w)$.

4.   THE UNIFORMITY OF THE ERROR TERMS

A difference between (1.6) and (1.7) is that the error estimate in (1.7) is uniform in β, while that in (1.6) is not. From the proof it can be seen that the implied constant in (1.6) does not depend on β if $0 < \beta \leq 1$. For general β, it is possible to separate the parts depending on β from the error term on the right-hand side of (1.6). An application can be found in [11].

We write $\beta = A + \tilde{\beta}$, where A is a non-negative integer and $0 < \tilde{\beta} \leq 1$. Then we have

Theorem 2. *For any positive integer N and $\Re v > -N + 1$, we have*

$$R_N(v; \beta, w) = -\sum_{k=N}^{\infty} \binom{-v}{k} \zeta(v+k) \sum_{j=0}^{A-1} (\tilde{\beta}+j)^k w^{-v-k}$$
$$+ O(|w|^{-\Re v - N}) \tag{4.1}$$

if $w \in W_\infty$ and $|w| > \beta - 1$, where the implied constant depends only on v, N and θ_0.

In the case $w > 0$, this result has been proved in [11], but the proof presented below is simpler.

Corollary. *Let $N \geq 2$, $w \in W_\infty$ and $|w| > \beta - 1$. Then the error term on the right-hand side of (1.6) can be replaced by*

$$-\sum_{k=N}^{\infty} \frac{(-1)^k}{k} \zeta(k) \sum_{j=0}^{A-1} (\tilde{\beta}+j)^k w^{-k} + O(|w|^{-N}),$$

and the implied constant depends only on N and θ_0.

Now we prove the theorem. Since

$$\zeta(v+z, \beta) = \zeta(v+z, \tilde{\beta}) - \sum_{j=0}^{A-1} (\tilde{\beta}+j)^{-v-z},$$

from (2.2) we have

$$R_N(v; \beta, w) = -\sum_{j=0}^{A-1} \frac{1}{2\pi i} \int_{(c_N)} \frac{\Gamma(v+z)\Gamma(-z)}{\Gamma(v)} (\tilde{\beta}+j)^{-v-z} \zeta(-z) w^z dz$$
$$+ R_N(v; \tilde{\beta}, w)$$
$$= -\sum_{j=0}^{A-1} T_N(j) + R_N(v; \tilde{\beta}, w), \tag{4.2}$$

say. Let L be a large positive integer $(L > N)$, and shift the path of integration of $T_N(j)$ to $\Re z = c_L$. Counting the residues of the poles

$z = -v - k \ (N \leq k \leq L - 1)$, we obtain

$$T_N(j) = \sum_{k=N}^{L-1} \binom{-v}{k} \zeta(v+k)(\tilde{\beta}+j)^k w^{-v-k} + T_L(j).$$

Using Stirling's formula we can see that $T_L(j) \to 0$ as $L \to +\infty$ if $|w| > \tilde{\beta} + j$. The resulting infinite series expression of $-\sum_{j=0}^{A-1} T_N(j)$ coincides with the explicit term on the right-hand side of (4.1). The remainder term $R_N(v; \tilde{\beta}, w)$ can be estimated by (5.4) of [12]. Since $0 < \tilde{\beta} \leq 1$, the estimate is uniform in $\tilde{\beta}$. Hence the proof of Theorem 2 is complete.

5.   AN ASYMPTOTIC EXPANSION OF THE DERIVATIVE OF HECKE'S ZETA-FUNCTION AT $s = 1$

Let D be a square-free positive integer, $D \equiv 2$ or $3 \pmod 4$. Hecke [8] introduced and studied the zeta-function (following the notation of Hecke)

$$\zeta(s; v_1) = \sum_{(\mu)} \frac{\text{sgn}(\mu\mu')}{|N(\mu)|^s} \qquad (5.1)$$

associated with the real quadratic field $\mathbf{Q}(\sqrt{D})$, where (μ) runs over all non-zero principal integral ideals of $\mathbf{Q}(\sqrt{D})$, $N(\mu)$ is the norm of (μ), μ' is the conjugate of μ, and $\text{sgn}(\mu\mu')$ is the sign of $\mu\mu'$. Hecke's motivation is to study the Dirichlet series

$$Z_{\sqrt{D}}(s) = \sum_{n=1}^{\infty} \frac{\{n\sqrt{D}\} - 1/2}{n^s},$$

where $\{x\}$ is the fractional part of x. The coefficients $G_1(\sqrt{D})$ and $G_2(\sqrt{D})$ in the Laurent expansion

$$Z_{\sqrt{D}}(s) = G_1(\sqrt{D})s^{-1} + G_2(\sqrt{D}) + G_3(\sqrt{D})s + \cdots$$

are important in the study of the distribution of $\{n\sqrt{D}\} - 1/2$, a famous classical problem in number theory. Hecke's paper [8] implicitly includes the evaluation of $G_1(\sqrt{D})$ and $G_2(\sqrt{D})$ in terms of $\zeta(1; v_1)$ and $\zeta'(1; v_1)$. In particular,

$$G_2(\sqrt{D}) = \frac{\zeta(1; v_1)\sqrt{D}}{\pi^2 \log \varepsilon_D}(\gamma + \log 2\pi) - \frac{\zeta'(1; v_1)\sqrt{D}}{2\pi^2 \log \varepsilon_D} - \frac{1}{12}\sqrt{D} + \frac{1}{8} \quad (5.2)$$

if $N(\varepsilon_D) = 1$, where ε_D is the fundamental unit of $\mathbf{Q}(\sqrt{D})$.

Fujii [6] proved different expressions for $G_1(\sqrt{D})$ and $G_2(\sqrt{D})$. Combining them with Hecke's results, Fujii obtained new expressions for $\zeta(1;v_1)$ and $\zeta'(1;v_1)$. An interesting feature of Fujii's results is that double gamma-functions appear in his expressions. This is similar to Shintani's theorem [14]. Shintani [14] proved a formula which expresses the value $L_F(1,\chi)$ of a certain Hecke L-function (associated with a real quadratic field F) in terms of double gamma-functions. Combining Shintani's result with our expansion formula for double gamma-functions, we have shown an expansion for $L_F(1,\chi)$ in [11]. In a similar way, in this paper we prove an expansion formula for $\zeta'(1;v_1)$.

Fujii [6] proved his results for any $\mathbf{Q}(\sqrt{D})$, D is square-free, positive, $\equiv 2$ or $3 \pmod 4$. However, his general statement is very complicated. Therefore in this paper we content ourselves with considering a typical example, given as Example 2 in Fujii [7].

To state Fujii's results, we introduce more general form of double zeta and double gamma-functions. Let α, w_1, w_2 be positive numbers, and define

$$\zeta_2(v; \alpha, (w_1, w_2)) = \sum_{m=0}^{\infty} \sum_{n=0}^{\infty} (\alpha + mw_1 + nw_2)^{-v}, \tag{5.3}$$

$$-\log \rho_2(w_1, w_2) = \lim_{\alpha \to 0} \{\zeta_2'(0; \alpha, (w_1, w_2)) + \log \alpha\}, \tag{5.4}$$

and

$$\log \frac{\Gamma_2(\alpha, (w_1, w_2))}{\rho_2(w_1, w_2)} = \zeta_2'(0; \alpha, (w_1, w_2)). \tag{5.5}$$

Let n be a non-negative integer such that $4n^2 + 8n + 3$ is square-free. We consider the case $D = 4n^2 + 8n + 3$. Then $\varepsilon_n = \sqrt{D} + 2n + 2$ is the fundamental unit of $\mathbf{Q}(\sqrt{D})$. Example 2 of Fujii [7] asserts that

$$\zeta(1; v_1) = \frac{\pi^2}{12} \frac{4n + 1}{\sqrt{4n^2 + 8n + 3}} \tag{5.6}$$

and

$$
\begin{aligned}
&\zeta'(1; v_1) \\
&= \frac{2\pi^2}{\sqrt{4n^2 + 8n + 3}} \left\{ -\log \frac{\Gamma_2(\varepsilon_n^2, (\varepsilon_n, \varepsilon_n^2 - \varepsilon_n))\rho_2(\varepsilon_n - 1, \varepsilon_n)}{\rho_2(\varepsilon_n, \varepsilon_n^2 - \varepsilon_n)\Gamma_2(2\varepsilon_n - 1, (\varepsilon_n - 1, \varepsilon_n))} \right. \\
&\quad + \frac{4n + 1}{12} \left(\gamma + \log 2\pi - \log(\varepsilon_n - \varepsilon_n^{-1}) \right)
\end{aligned}
$$

$$-\frac{\log \varepsilon_n}{24}\left(\frac{\sqrt{4n^2 + 8n + 3}}{2n + 1} + 8n + 5\right)\right\}. \tag{5.7}$$

Let α, β be positive numbers with $\alpha < \beta$, and define

$$\zeta_2((u,v);(\alpha,\beta)) = \sum_{m=0}^{\infty}\sum_{n=0}^{\infty}(\alpha + m)^{-u}(\beta + m + n)^{-v}. \tag{5.8}$$

In [12] we have shown that $\zeta_2((u,v);(\alpha,\beta))$ can be continued meromorphically to the whole $\mathbf{C}^2$-space. In Section 7 we will show the facts that $\zeta_2((0,v);(\alpha,\beta))$ is holomorphic at $v = 0$, while $\zeta_2((-k,v+k);(\alpha,\beta))$ has a pole of order 1 at $v = 0$ for any positive integer k. Denote the Laurent expansion at $v = 0$ by

$$\zeta_2((-k,v+k);(\alpha,\beta))$$
$$= C_{-1}(k;(\alpha,\beta))\frac{1}{v} + C_0(k;(\alpha,\beta)) + C_1(k;(\alpha,\beta))v + \cdots \tag{5.9}$$

for $k \geq 1$. We shall prove

Theorem 3. *Let* $D = 4n^2 + 8n + 3$, $\varepsilon_n = \sqrt{D} + 2n + 2$, *and* $\xi = \xi_n = \varepsilon_n - 1$. *Then, for any positive integer* $N \geq 2$, *we have*

$$\log \frac{\Gamma_2(\varepsilon_n^2, (\varepsilon_n, \varepsilon_n^2 - \varepsilon_n))\rho_2(\varepsilon_n - 1, \varepsilon_n)}{\rho_2(\varepsilon_n, \varepsilon_n^2 - \varepsilon_n)\Gamma_2(2\varepsilon_n - 1, (\varepsilon_n - 1, \varepsilon_n))}$$
$$= -\frac{1}{12}\xi \log \xi - \frac{1}{12}\xi \log(1 + \xi) + \left(\frac{1}{12} - \zeta'(-1)\right)\xi + \frac{1}{6}\log \xi$$
$$\quad - \frac{1}{4}\log(1 + \xi) + \frac{1}{4}\log 2\pi - \zeta_2'((0,0);(1,2))$$
$$\quad - \frac{1}{12}\xi^{-1}\log \xi - \frac{1}{12}\xi^{-1}\log(1 + \xi) + \left(\frac{1}{12}\gamma + C_0(1;(1,2))\right)\xi^{-1}$$
$$\quad + \sum_{k=2}^{N-1}\frac{(-1)^k}{k}\left\{\zeta(-k)\zeta(k) - C_0(k;(1,2))\right.$$
$$\qquad \left. -\frac{k}{12}\left(1 + \frac{1}{2} + \cdots + \frac{1}{k-1} - \log \xi\right)\right\}\xi^{-k}$$
$$\quad + O\left(\xi^{-N}\log \xi\right). \tag{5.10}$$

From this theorem and (5.7), we obtain the asymptotic expansion of $\zeta'(1;v_1)$ with respect to $\xi = \xi_n$ (or with respect to ε_n) when $n \to +\infty$. Moreover, combining with (5.2) and (5.6), we can deduce the asmptotic expansion of $G_2(\sqrt{D}) = G_2(\sqrt{4n^2 + 8n + 3})$. It should be noted that, by expanding the factor $\log(1 + \xi)$ on the right-hand side of (5.10), we can write down the asymptotic expansion with respect to ξ in the most strict sense. This is an advantage of the above theorem; the formula for $L_F(1, \chi)$ proved in [11] is not the asymptotic expansion in the strict sense.

The rest of this paper is devoted to the proof of Theorem 3. It is desirable to extend our consideration to the case of Fujii's general formula (Fujii [6], Theorem 6 and Corollary 2). It is also an interesting problem to evaluate the quantities $\zeta_2'((0,0);(1,2))$ and $C_0(k;(1,2))$ appearing on the right-hand side of (5.10).

6. THE BEHAVIOUR OF $\Gamma_2(\varepsilon_n^2, (\varepsilon_n, \varepsilon_n^2 - \varepsilon_n))$ AND $\rho_2(\varepsilon_n, \varepsilon_n^2 - \varepsilon_n)$

Let $\beta = \alpha/w_1$ and $w = w_2/w_1$. From (5.3) we have

$$\zeta_2(v; \alpha, (w_1, w_2)) = w_1^{-v} \sum_{m=0}^{\infty} \sum_{n=0}^{\infty} (\beta + m + nw)^{-v}$$

$$= w_1^{-v} \zeta_2(v; \beta, (1, w)) \qquad (6.1)$$

for $\Re v > 2$. This formula gives the analytic continuation of the function $\zeta_2(v; \alpha, (w_1, w_2))$ to the whole complex v-plane, and yields

$$\zeta_2'(0; \alpha, (w_1, w_2)) = -\zeta_2(0; \beta, (1, w)) \log w_1 + \zeta_2'(0; \beta, (1, w)). \qquad (6.2)$$

From (5.4) and (6.2), we have

$$- \log \rho_2(w_1, w_2)$$
$$= \lim_{\alpha \to 0} \{1 - \zeta_2(0; \beta, (1, w))\} \log w_1 + \lim_{\alpha \to 0} \{\zeta_2'(0; \beta, (1, w)) + \log \beta\}$$
$$= \{1 - \zeta_2(0; 0, (1, w))\} \log w_1 - \log \rho_2(1, w), \qquad (6.3)$$

where the existence of the limit

$$\zeta_2(0; 0, (1, w)) = \lim_{\beta \to 0} \zeta_2(0; \beta, (1, w))$$

can be seen from the expression

$$\zeta_2(0; \beta, (1, w))$$

$$= -B_1(\beta) + \sum_{\ell=0}^{2} \frac{B_\ell(0) B_{2-\ell}(1 - \beta)}{\ell!(2 - \ell)!} w^{\ell-1}$$

$$= \frac{1}{12} w + \frac{1}{2} \left(\frac{1}{2} - \beta \right) + \frac{1}{2} \left(\beta^2 - \beta + \frac{1}{6} \right) w^{-1}, \qquad (6.4)$$

which is the special case $m = 0$ of Theorem 5 in [10]. From (6.4) it follows that

$$\zeta_2(0;0,(1,w)) = \frac{1}{12}w + \frac{1}{4} + \frac{1}{12}w^{-1}, \tag{6.5}$$

$$\zeta_2(0;1,(1,w)) = \frac{1}{12}w - \frac{1}{4} + \frac{1}{12}w^{-1}. \tag{6.6}$$

Now we consider the case $(w_1, w_2) = (\varepsilon_n, \varepsilon_n^2 - \varepsilon_n)$. Our aim in this section is to prove the following

Proposition 1. *We have*

$$\log \rho_2(\varepsilon_n, \varepsilon_n^2 - \varepsilon_n) = \left(\frac{1}{12}\xi - \frac{3}{4} + \frac{1}{12}\xi^{-1}\right)\log(1+\xi) + \frac{1}{12}\xi\log\xi$$

$$- \left(\frac{1}{12} - \zeta'(-1)\right)\xi - \frac{1}{4}\log\xi + \frac{3}{4}\log 2\pi + \frac{1}{12}\xi^{-1}\log\xi - \frac{1}{12}\gamma\xi^{-1}$$

$$- \sum_{k=2}^{N-1} \frac{(-1)^k}{k}\zeta(-k)\zeta(k)\xi^{-k} + O(\xi^{-N}) \tag{6.7}$$

for any $N \geq 2$, and

$$\log \Gamma_2(\varepsilon_n^2, (\varepsilon_n, \varepsilon_n^2 - \varepsilon_n)) = -\log(1+\xi) - \frac{1}{2}\log\xi + \log 2\pi. \tag{6.8}$$

Proof. Putting $(w_1, w_2) = (\varepsilon_n, \varepsilon_n^2 - \varepsilon_n)$, the formula (6.3) gives

$$\log \rho_2(\varepsilon_n, \varepsilon_n^2 - \varepsilon_n) = \{\zeta_2(0;0,(1,\xi)) - 1\}\log(1+\xi) + \log \rho_2(1,\xi) \tag{6.9}$$

because $w = (\varepsilon_n^2 - \varepsilon_n)/\varepsilon_n = \varepsilon_n - 1 = \xi$. Substituting (2.9) and (6.5) into the right-hand side of (6.9), we obtain (6.7).

Next, for $\Re v > 2$, we have

$$\zeta_2(v; \varepsilon_n^2, (\varepsilon_n, \varepsilon_n^2 - \varepsilon_n))$$

$$= \varepsilon_n^{-v} \sum_{m=0}^{\infty} \sum_{n=0}^{\infty} (\varepsilon_n + m + n(\varepsilon_n - 1))^{-v}$$

$$= (1+\xi)^{-v} \sum_{m=0}^{\infty} \sum_{n=0}^{\infty} (1 + m + (1+n)\xi)^{-v}$$

$$= (1+\xi)^{-v} \left\{ \sum_{m=0}^{\infty} \sum_{\ell=0}^{\infty} (1 + m + \ell\xi)^{-v} - \sum_{m=0}^{\infty} (1+m)^{-v} \right\}$$

$$= (1+\xi)^{-v} \{\zeta_2(v; 1, (1,\xi)) - \zeta(v)\}. \tag{6.10}$$

This formula is valid for any v by analytic continuation. Hence from this formula we obtain

$$\zeta_2'(0; \varepsilon_n^2, (\varepsilon_n, \varepsilon_n^2 - \varepsilon_n)) = -\left\{\zeta_2(0; 1, (1, \xi)) + \frac{1}{2}\right\} \log(1 + \xi)$$
$$+ \zeta_2'(0; 1, (1, \xi)) + \frac{1}{2} \log 2\pi, \qquad (6.11)$$

which with (6.9) and (1.4) yields

$$\log \Gamma_2(\varepsilon_n^2, (\varepsilon_n, \varepsilon_n^2 - \varepsilon_n))$$
$$= \left\{\zeta_2(0; 0, (1, \xi)) - \zeta_2(0; 1, (1, \xi)) - \frac{3}{2}\right\} \log(1 + \xi)$$
$$+ \log \Gamma_2(1, (1, \xi)) + \frac{1}{2} \log 2\pi. \qquad (6.12)$$

From (3.1) we find that

$$\Gamma_2(1, (1, \xi)) = \left(\frac{2\pi}{\xi}\right)^{1/2}. \qquad (6.13)$$

Substituting (6.5), (6.6) and (6.13) into the right-hand side of (6.12), we obtain (6.8). This completes the proof of Proposition 1.

7. AN AUXILIARY INTEGRAL

In this section we prove several properties of the integral

$$I(v; (\alpha, \beta)) = \frac{1}{2\pi i} \int_{(c)} \frac{\Gamma(v + z)\Gamma(-z)}{\Gamma(v)} \zeta_2((v + z, -z); (\alpha, \beta)) \xi^z dz \quad (7.1)$$

which are necessary in the next section. Here $\Re v > 2$, $1 - \Re v < c < -1$, and $0 < \alpha < \beta$.

Let ε be an arbitrarily small positive number. From (9.2) of [12] we have

$$\zeta_2((v + z, -z); (\alpha, \beta))$$
$$= -\frac{1}{1 + z}\zeta(v - 1, \alpha) + \sum_{j=0}^{J-1} \binom{z}{j} \zeta(v + j, \alpha)\zeta(-j, \beta - \alpha)$$
$$+ S_{0,J}((v + z, -z); (\alpha, \beta)), \qquad (7.2)$$

where J is any positive integer and $S_{0,J}((v+z, -z); (\alpha, \beta))$ is holomorphic in the region $\Re v > 1 - J + \varepsilon$ and $\Re z < J - \varepsilon$. Since J is arbitrary, (7.2) implies that $z = -1$ is the only pole of $\zeta_2((v+z, -z); (\alpha, \beta))$ as a function in z. This pole is irrelevant when we shift the path of integration on the right-hand side of (7.1) to $\Re z = c_N = -\Re v - N + \varepsilon$, where N is a positive integer ≥ 2. It is not difficult to see that $\zeta_2((v + z, -z); (\alpha, \beta))$ is of polynomial order with respect to $\Im z$ (for example, by using (7.2)), hence this shifting is possible. Counting the residues of the poles $z = -v - k$ $(0 \leq k \leq N - 1)$, we obtain

$$
I(v; (\alpha, \beta)) = \sum_{k=0}^{N-1} \binom{-v}{k} \zeta_2((-k, v + k); (\alpha, \beta)) \xi^{-v-k}
$$
$$
+ I_N(v; (\alpha, \beta)), \tag{7.3}
$$

where

$$
I_N(v; (\alpha, \beta)) = \frac{1}{2\pi i} \int_{(c_N)} \frac{\Gamma(v + z)\Gamma(-z)}{\Gamma(v)} \zeta_2((v + z, -z); (\alpha, \beta)) \xi^z \, dz
$$
$$
= \frac{1}{2\pi i} \xi^{-v} \int_{(-N+\varepsilon)} \frac{\Gamma(z)\Gamma(v - z)}{\Gamma(v)} \zeta_2((z, v - z); (\alpha, \beta)) \xi^z \, dz. \tag{7.4}
$$

Next, from (9.9) of [12] we have

$$
\zeta_2((u, v); (\alpha, \beta))
$$
$$
= \frac{\Gamma(1 - u)\Gamma(u + v - 1)}{\Gamma(v)} \zeta(u + v - 1, \beta - \alpha)
$$
$$
+ \sum_{j=0}^{J-1} \binom{-v}{j} \zeta(u - j, \alpha)\zeta(v + j, \beta - \alpha)
$$
$$
+ R_{0,J}((u, v); (\alpha, \beta)), \tag{7.5}
$$

where

$$
R_{0,J}((u, v); (\alpha, \beta)) = \frac{1}{2\pi i} \int_{(-J+\varepsilon)} \frac{\Gamma(z')\Gamma(v - z')}{\Gamma(v)}
$$
$$
\times \zeta(u + z', \alpha)\zeta(v - z', \beta - \alpha) \, dz'. \tag{7.6}
$$

The formula (7.5) is valid in the region

$$
\{(u, v) \in \mathbf{C}^2 \mid \Re u < J + 1 - \varepsilon, \Re v > -J + 1 + \varepsilon\},
$$

and in this region $R_{0,J}((u, v); (\alpha, \beta))$ are holomorphic. In particular, choosing $J = 2$ and $(u, v) = (z, v - z)$, we have

$$\zeta_2((z, v - z); (\alpha, \beta)) = \frac{\Gamma(1 - z)\Gamma(v - 1)}{\Gamma(v - z)} \zeta(v - 1, \beta - \alpha)$$
$$+ \zeta(z, \alpha)\zeta(v - z, \beta - \alpha) - (v - z)\zeta(z - 1, \alpha)\zeta(v - z + 1, \beta - \alpha)$$
$$+ R_{0,2}((z, v - z); (\alpha, \beta)) \tag{7.7}$$

for $\Re z < 3 - \varepsilon$ and $\Re(v - z) > -1 + \varepsilon$. If $\Re z = -N + \varepsilon$, then the right-hand side of (7.7) can be singular only if $v = 2, 1, 0, -1, -2, \ldots$ or $v = z + 1$. Hence the integrand of the right-hand side of (7.4) is not singular on the path $\Re z = -N + \varepsilon$ if $\Re v > 1 - N + \varepsilon$, which implies that the integral (7.4) can be continued meromorphically to $\Re v > 1 - N + \varepsilon$. Moreover, the (possible) pole of $\zeta_2((z, v - z); (\alpha, \beta))$ at $v = 0$ cancels with the zero coming from the factor $\Gamma(v)^{-1}$, hence (7.4) is holomorphic at $v = 0$.

Let k be a non-negative integer. Putting $z = -k$ in (7.7), we have

$$\zeta_2((-k, v + k); (\alpha, \beta)) = \frac{\Gamma(1 + k)\Gamma(v - 1)}{\Gamma(v + k)} \zeta(v - 1, \beta - \alpha)$$
$$+ \zeta(-k, \alpha)\zeta(v + k, \beta - \alpha) - (v + k)\zeta(-k - 1, \alpha)\zeta(v + k + 1, \beta - \alpha)$$
$$+ R_{0,2}((-k, v + k); (\alpha, \beta)) \tag{7.8}$$

for $\Re v > -k - 1 + \varepsilon$. From (7.8) it is easy to see that $\zeta_2((0, v); (\alpha, \beta))$ is holomorphic at $v = 0$, while $\zeta_2((-k, v + k); (\alpha, \beta))$ has a pole of order 1 at $v = 0$ for $k \geq 1$. We may write the Laurent expansion at $v = 0$ as (5.9) for $k \geq 1$. Then

$$\binom{-v}{k} \zeta_2((-k, v + k); (\alpha, \beta)) \xi^{-v-k}$$

is holomorphic at $v = 0$, and its Taylor expansion is

$$= \frac{(-1)^k}{k} C_{-1}(k; (\alpha, \beta)) \xi^{-k} + \frac{(-1)^k}{k} \left\{ C_0(k; (\alpha, \beta)) \right.$$
$$\left. + C_{-1}(k; (\alpha, \beta)) \left(1 + \frac{1}{2} + \cdots + \frac{1}{k - 1} - \log \xi \right) \right\} \xi^{-k} v + \cdots \tag{7.9}$$

for $k \geq 1$; recall that the empty sum is to be considered as zero. Now by (7.3), $I(v; (\alpha, \beta))$ can be continued to the region $\Re v > 1 - N + \varepsilon$, and

$$I'(0; (\alpha, \beta)) = \zeta_2'((0,0); (\alpha, \beta)) - \zeta_2((0,0); (\alpha, \beta)) \log \xi$$
$$+ \sum_{k=1}^{N-1} \frac{(-1)^k}{k} \Bigg\{ C_0(k; (\alpha, \beta))$$
$$+ C_{-1}(k; (\alpha, \beta)) \left(1 + \frac{1}{2} + \cdots + \frac{1}{k-1} - \log \xi \right) \Bigg\} \xi^{-k}$$
$$+ I_N'(0; (\alpha, \beta)). \tag{7.10}$$

We claim that the limit values of $\zeta_2'((0,0); (\alpha, \beta))$, $\zeta_2((0,0); (\alpha, \beta))$, $C_0(k; (\alpha, \beta))$, $C_{-1}(k; (\alpha, \beta))$ and $I_N'(0; (\alpha, \beta))$, when $\alpha \to +0$, all exist. We denote them by $\zeta_2'((0,0); (0, \beta))$, $\zeta_2((0,0); (0, \beta))$, $C_0(k; (0, \beta))$, $C_{-1}(k; (0, \beta))$ and $I_N'(0; (0, \beta))$, respectively.

To prove this claim, first recall that if $\Re v < 0$, then $\zeta(v, \alpha)$ is continuous with respect to α when $\alpha \to +0$. This fact can be seen from (2.17.3) of Titchmarsh [16]. Hence the existence of $\zeta_2'((0,0); (0, \beta))$ and $\zeta_2((0,0); (0, \beta))$ follows easily from the case $k = 0$ of (7.8). Similarly we can show the existence of $I_N'(0; 0, \beta)$ by using (7.4) and (7.7), and at the same time we find

$$I_N'(0; (\alpha, \beta)) = O(\xi^{-N+\varepsilon}) \tag{7.11}$$

and

$$I_N'(0; (0, \beta)) = O(\xi^{-N+\varepsilon}). \tag{7.12}$$

Next consider the case $k \geq 1$ of (7.8). The Laurent expansion at $v = 0$ of the first term on the right-hand side of (7.8) is

$$= -k\zeta(-1, \beta - \alpha)v^{-1} - P(k; (\alpha, \beta)) + O(|v|),$$

where

$$P(k; (\alpha, \beta)) = k \Bigg\{ \zeta'(-1, \beta - \alpha) + \zeta(-1, \beta - \alpha)$$
$$- \zeta(-1, \beta - \alpha) \left(1 + \frac{1}{2} + \frac{1}{3} + \cdots + \frac{1}{k-1} \right) \Bigg\}. \tag{7.13}$$

The other terms on the right-hand side of (7.8) are holomorphic at $v = 0$ if $k \geq 2$. If $k = 1$, one more pole is coming from the second term on the right-hand side of (7.8), whose Laurent expansion is

$$= \zeta(-1, \alpha)v^{-1} - \zeta(-1, \alpha)\psi(\beta - \alpha) + O(|v|).$$

Collecting the above facts, we obtain

$$C_{-1}(k;(\alpha,\beta)) = \begin{cases} -\zeta(-1,\beta-\alpha)+\zeta(-1,\alpha) & \text{if } k=1, \\ -k\zeta(-1,\beta-\alpha) & \text{if } k\geq 2, \end{cases} \quad (7.14)$$

and

$$C_0(k;(\alpha,\beta)) = -P(k;(\alpha,\beta))+Q(k;(\alpha,\beta))$$
$$-k\zeta(-k-1,\alpha)\zeta(k+1,\beta-\alpha)+R_{0,2}((-k,k);(\alpha,\beta)), \quad (7.15)$$

where $P(k;(\alpha,\beta))$ is defined by (7.13) and

$$Q(k;(\alpha,\beta)) = \begin{cases} -\zeta(-1,\alpha)\psi(\beta-\alpha) & \text{if } k=1, \\ \zeta(-k,\alpha)\zeta(k,\beta-\alpha) & \text{if } k\geq 2. \end{cases} \quad (7.16)$$

From the above expressions it is now clear that the values $C_0(k;(0,\beta))$ and $C_{-1}(k;(0,\beta))$ exist for any $k\geq 1$. We complete the proof of our claim, and therefore from (7.10) we obtain

$$I'(0;(0,\beta)) = \lim_{\alpha\to 0} I'(0;(\alpha,\beta))$$
$$= \zeta_2'((0,0);(0,\beta)) - \zeta_2((0,0);(0,\beta))\log\xi$$
$$+ \sum_{k=1}^{N-1}\frac{(-1)^k}{k}\left\{C_0(k;(0,\beta))\right.$$
$$\left.+C_{-1}(k;(0,\beta))\left(1+\frac{1}{2}+\cdots+\frac{1}{k-1}-\log\xi\right)\right\}\xi^{-k}$$
$$+ I_N'(0;(0,\beta)). \quad (7.17)$$

8.   THE BEHAVIOUR OF $\Gamma_2(2\varepsilon_n - 1,(\varepsilon_n - 1,\varepsilon_n))$ AND $\rho_2(\varepsilon_n - 1,\varepsilon_n)$; COMPLETION OF THE PROOF OF THEOREM 3

Let $\Re v > 2$ and $0 < \alpha < 1$. We have

$$\zeta_2(v;\alpha,(\varepsilon_n-1,\varepsilon_n)) = \sum_{m=0}^{\infty}\sum_{n=0}^{\infty}(\alpha+m\xi+n(1+\xi))^{-v}$$
$$= (1+\xi)^{-v}\sum_{n=0}^{\infty}\left(\frac{\alpha}{1+\xi}+n\right)^{-v}$$
$$+ (\alpha+n)^{-v}\sum_{m=1}^{\infty}\sum_{n=0}^{\infty}\left(1+\frac{(m+n)\xi}{\alpha+n}\right)^{-v}. \quad (8.1)$$

Using the Mellin-Barnes integral formula ((2.2) of [12]) we get

$$\Gamma(v) \left(1 + \frac{(m+n)\xi}{\alpha+n} \right)^{-v}$$
$$= \frac{1}{2\pi i} \int_{(c)} \Gamma(v+z)\Gamma(-z) \left(\frac{(m+n)\xi}{\alpha+n} \right)^z dz, \qquad (8.2)$$

where $-\Re v < c < 0$. We may assume $1 - \Re v < c < -1$. Then, summing up the both sides of (8.2) with respect to m and n, we obtain

$$(\alpha+n)^{-v} \sum_{m=1}^{\infty} \sum_{n=0}^{\infty} \left(1 + \frac{(m+n)\xi}{\alpha+n} \right)^{-v}$$
$$= \frac{1}{2\pi i} \int_{(c)} \frac{\Gamma(v+z)\Gamma(-z)}{\Gamma(v)} \sum_{m=1}^{\infty} \sum_{n=0}^{\infty} (\alpha+n)^{-v-z}(m+n)^z \xi^z dz$$
$$= \frac{1}{2\pi i} \int_{(c)} \frac{\Gamma(v+z)\Gamma(-z)}{\Gamma(v)} \zeta_2((v+z,-z);(\alpha,1))\xi^z dz.$$

Therefore, combining with (8.1), we have

$$\zeta_2(v;\alpha,(\varepsilon_n - 1,\varepsilon_n)) = (1+\xi)^{-v} \zeta \left(v, \frac{\alpha}{1+\xi} \right) + I(v;(\alpha,1)) \qquad (8.3)$$

for $\Re v > 2$, and by analytic continuation for $\Re v > 1 - N + \varepsilon$. Hence

$$\zeta_2'(0;\alpha,(\varepsilon_n - 1,\varepsilon_n))$$
$$= -\zeta \left(0, \frac{\alpha}{1+\xi} \right) \log(1+\xi) + \zeta' \left(0, \frac{\alpha}{1+\xi} \right) + I'(0;(\alpha,1)). \qquad (8.4)$$

Applying (2.4) to the right-hand side, we have

$$\zeta_2'(0;\alpha,(\varepsilon_n - 1,\varepsilon_n)) + \log \alpha$$
$$= - \left(\frac{1}{2} - \frac{\alpha}{1+\xi} \right) \log(1+\xi) + \log \Gamma \left(1 + \frac{\alpha}{1+\xi} \right) - \frac{1}{2} \log 2\pi$$
$$+ \log(1+\xi) + I'(0;(\alpha,1)).$$

Taking the limit $\alpha \to 0$, and using (7.12) and (7.17) with $\beta = 1$, we obtain

Proposition 2. *We have*

$$\log \rho_2(\varepsilon_n - 1, \varepsilon_n) = -\frac{1}{2}\log(1 + \xi) + \frac{1}{2}\log 2\pi$$
$$-\zeta_2'((0,0);(0,1)) + \zeta_2((0,0);(0,1))\log \xi$$
$$-\sum_{k=1}^{N-1}\frac{(-1)^k}{k}\Big\{C_0(k;(0,1))$$
$$+C_{-1}(k;(0,1))\left(1 + \frac{1}{2} + \cdots + \frac{1}{k-1} - \log \xi\right)\Big\}\xi^{-k}$$
$$+O(\xi^{-N+\varepsilon}). \tag{8.5}$$

Remark. The estimate of the error term in (8.5) can be strengthened to $O(\xi^{-N}\log \xi)$. (Consider (8.5) with $N + 1$ instead of N, and compare it with the original (8.5).)

Our next aim is to prove the following

Proposition 3. *We have*

$$\log \Gamma_2(2\varepsilon_n - 1, (\varepsilon_n - 1, \varepsilon_n)) = -\frac{1}{2}\log(1 + \xi) + \frac{1}{2}\log 2\pi$$
$$+\zeta_2'((0,0);(1,2)) - \zeta_2'((0,0);(0,1))$$
$$-\{\zeta_2((0,0);(1,2)) - \zeta_2((0,0);(0,1))\}\log \xi$$
$$+\sum_{k=1}^{N-1}\frac{(-1)^k}{k}\{C_0(k;(1,2)) - C_0(k;(0,1))\}\xi^{-k}$$
$$+\sum_{k=1}^{N-1}\frac{(-1)^k}{k}\{C_{-1}(k;(1,2)) - C_{-1}(k;(0,1))\}$$
$$\times \left(1 + \frac{1}{2} + \cdots + \frac{1}{k-1} - \log \xi\right)\xi^{-k}$$
$$+O(\xi^{-N}\log \xi). \tag{8.6}$$

Proof. For $\Re v > 2$, we have

$$\zeta_2(v; 2\varepsilon_n - 1, (\varepsilon_n - 1, \varepsilon_n))$$
$$= \sum_{m=0}^{\infty}\sum_{n=0}^{\infty}(2\xi + 1 + m\xi + n(1 + \xi))^{-v}$$
$$= \sum_{m=0}^{\infty}\sum_{n=0}^{\infty}(n + 1 + (m + n + 2)\xi)^{-v},$$

which is, again using the Mellin-Barnes integral formula,

$$= \frac{1}{2\pi i} \int_{(c)} \frac{\Gamma(v+z)\Gamma(-z)}{\Gamma(v)} \sum_{m=0}^{\infty}\sum_{n=0}^{\infty}(n+1)^{-v}\left(\frac{(m+n+2)\xi}{n+1}\right)^z dz$$

$$= \frac{1}{2\pi i} \int_{(c)} \frac{\Gamma(v+z)\Gamma(-z)}{\Gamma(v)} \zeta_2((v+z,-z);(1,2))\xi^z dz.$$

That is,

$$\zeta_2(v; 2\varepsilon_n - 1, (\varepsilon_n - 1, \varepsilon_n)) = I(v; (1,2)), \tag{8.7}$$

and this identity is valid for $\Re v > 1 - N + \varepsilon$ by analytic continuation. Hence

$$\zeta_2'(0; 2\varepsilon_n - 1, (\varepsilon_n - 1, \varepsilon_n)) = I'(0; (1,2)). \tag{8.8}$$

Therefore using (7.10), (7.11) (with $\alpha = 1$, $\beta = 2$) and (8.5) we obtain the assertion of Proposition 3. The error estimate $O(\xi^{-N}\log\xi)$ can be shown similarly to the remark just after the statement of Proposition 2.

Now we can easily complete the proof of Theorem 3, by combining Propositions 1, 2 and 3. Since

$$\zeta_2((0,v);(1,2)) = \sum_{m=0}^{\infty}\sum_{n=0}^{\infty}(2+m+n)^{-v} = \zeta_2(v; 2, (1,1))$$

(valid at first for $\Re v > 2$ but also valid for any v by analytic continuation), by using (6.4) we find

$$\zeta_2((0,0);(1,2)) = \frac{5}{12}.$$

Also, (7.14) implies that $C_{-1}(1;(1,2)) = 0$ and $C_{-1}(k;(1,2)) = k/12$ for $k \geq 2$. Noting these facts, we can deduce the assertion of Theorem 3 straightforwardly.

It should be remarked finally that if we only want to prove Theorem 3, we can shorten the way; in fact, since the left-hand side of (5.10) is equal to

$$\zeta_2'(0; \varepsilon_n^2, (\varepsilon_n, \varepsilon_n^2 - \varepsilon_n)) - \zeta_2'(0; 2\varepsilon_n - 1, (\varepsilon_n - 1, \varepsilon_n)),$$

the formulas (6.11), (6.6), (2.5), (2.8), (8.8), (7.10), (7.11) are sufficient to deduce the conclusion of Theorem 3. However the formulas of Propositions 1, 2 and 3 themselves are of interest, therefore we have chosen the above longer but more informative route.

References

[1] T. Arakawa, Generalized eta-functions and certain ray class invariants of real quadratic fields, Math. Ann. **260** (1982), 475–494.

[2] T. Arakawa, Dirichlet series $\sum_{n=1}^{\infty} \frac{\cot \pi n\alpha}{n^s}$, Dedekind sums, and Hecke L-functions for real quadratic fields, Comment. Math. Univ. St. Pauli **37** (1988), 209–235.

[3] E. W. Barnes, The genesis of the double gamma functions, Proc. London Math. Soc. **31** (1899), 358–381.

[4] E. W. Barnes, The theory of the double gamma function, Philos. Trans. Roy. Soc. (A) **196** (1901), 265–387.

[5] J. Billingham and A. C. King, Uniform asymptotic expansions for the Barnes double gamma function, Proc. Roy. Soc. London Ser. A **453** (1997), 1817–1829.

[6] A. Fujii, Some problems of Diophantine approximation and a Kronecker limit formula, in "Investigations in Number Theory", T. Kubota (ed.), Adv. Stud. Pure Math. **13**, Kinokuniya, 1988, pp.215–236.

[7] A. Fujii, Diophantine approximation, Kronecker's limit formula and the Riemann hypothesis, in "Théorie des Nombres/Number Theory", J.-M. de Koninck and C. Levesque (eds.), Walter de Gruyter, 1989, pp.240–250.

[8] E. Hecke, Über analytische Funktionen und die Verteilung von Zahlen mod. eins, Abh. Math. Sem. Hamburg. Univ. **1** (1921), 54–76 (= Werke, 313–335).

[9] K. Katayama and M. Ohtsuki, On the multiple gamma-functions, Tokyo J. Math. **21** (1998), 159–182.

[10] K. Matsumoto, Asymptotic series for double zeta, double gamma, and Hecke L-functions, Math. Proc. Cambridge Phil. Soc. **123** (1998), 385–405.

[11] K. Matsumoto, Corrigendum and addendum to "Asymptotic series for double zeta, double gamma, and Hecke L-functions", Math. Proc. Cambridge Phil. Soc., to appear.

[12] K. Matsumoto, Asymptotic expansions of double zeta-functions of Barnes, of Shintani, and Eisenstein series, preprint.

[13] T. Shintani, On evaluation of zeta functions of totally real algebraic number fields at non-positive integers, J. Fac. Sci. Univ. Tokyo Sect. IA Math. **23** (1976), 393–417.

[14] T. Shintani, On a Kronecker limit formula for real quadratic fields, ibid. **24** (1977), 167–199.

[15] T. Shintani, A proof of the classical Kronecker limit formula, Tokyo J. Math. **3** (1980), 191–199.

[16] E. C. Titchmarsh, "The Theory of the Riemann Zeta-Function", Oxford, 1951.

[17] M.-F. Vignéras, L'equation fonctionelle de la fonction zêta de Selberg du groupe modulaire $PSL(2, \mathbf{Z})$, in "Journées Arithmétiques de Luminy, 1978", Astérisque **61**, Soc. Math. France, 1979, pp.235–249.

[18] E. T. Whittaker and G. N. Watson, "A Course of Modern Analysis", 4th ed., Cambridge Univ. Press, 1927.

A NOTE ON A CERTAIN AVERAGE OF $L(\frac{1}{2} + it, \chi)$

Leo MURATA

Department of Mathematics, Meijigakuin University, Shirokanedai 1-2-37, Minato-ku, Tokyo, 108-8636, Japan

leo@eco.meijigakuin.ac.jp

Keywords: Character sum, q-estimate of $L(\frac{1}{2} + it, \chi)$

Abstract   We consider an average of the character sum $S(\chi; 0, N) = \sum_{n=1}^{N} \chi(n)$ (Proposition), and making use of this result, we obtain some average-type results on the q-estimate of $L(\frac{1}{2} + it, \chi)$.

1991 Mathematics Subject Classification: 11L40, 11M06.

1.   INTRODUCTION

Let p be an odd prime, q be a positive integer, χ be a Dirichlet character mod q. We define the sums

$$S(\chi; M, N) = \sum_{n=M+1}^{M+N} \chi(n),$$

$$L(s, \chi) = \sum_{n=1}^{\infty} \frac{\chi(n)}{n^s},$$

and we are interested in q-estimate of $L(\frac{1}{2} + it, \chi)$, i.e. we want to find a good upper bound for $|L(\frac{1}{2} + it, \chi)|$ as a function in q, and we are not concerned with its t-aspect.

In the long history of the study of $|S(\chi; M, N)|$ and $|L(\frac{1}{2}+it, \chi)|$, the first important estimate is Pólya-Vinogradov inequality ([13] [14], 1918), which states

$$S(\chi; M, N) \ll \sqrt{q} \log q$$

and, by making use of partial summation, this gives, for any $\varepsilon > 0$,

$$L(\frac{1}{2} + it, \chi) \ll q^{\frac{1}{4} + \varepsilon}.$$

C. Jia and K. Matsumoto (eds.), Analytic Number Theory, 269–276.
© 2002 *Kluwer Academic Publishers. Printed in the Netherlands.*

In 1962, Burgess ([1]) proved the famous estimate; for any natural number r and any positive ε,

$$S(\chi; M, N) \ll N^{1-\frac{1}{r}} q^{\frac{r+1}{4r^2}+\varepsilon},$$

and from this estimate, we have

$$L(\frac{1}{2} + it, \chi) \ll q^{\frac{3}{16}+\varepsilon}. \tag{1}$$

For a large N, it is well-known that Pólya-Vinogradov inequality is almost best possible (cf. [12]), but in many cases we expect a much better estimate on $|S(\chi; M, N)|$ as well as $|L(\frac{1}{2} + it, \chi)|$. Actually, it is conjectured in [2] that, for the case $q = p$,

$$S(\chi; M, N) \ll N^{\frac{1}{2}} p^{\varepsilon},$$

and from this we can derive

$$L(\frac{1}{2} + it, \chi) \ll p^{\varepsilon},$$

which can be compared with the *Lindelöf conjecture for L-functions*:

$$L(\frac{1}{2} + it, \chi) \ll t^{\varepsilon}.$$

In the context, Burgess [3] [4] recently considered some average values of $|S(\chi; M, N)|$. He introduced the set of characters

$$\mathcal{E}(p^3) = \{\chi; \text{Dirichlet character mod } p^3, \chi^{p^2} = \chi_0\},$$

where χ_0 denotes the principal character mod p^3, and proved

Theorem A. (Burgess [3]) *For almost all $\chi \in \mathcal{E}(p^3)$, we have, for any $\varepsilon > 0$,*

$$S(\chi; M, N) \ll \sqrt{N} p^{\frac{1}{2}+\varepsilon}. \tag{2}$$

Theorem B. (Burgess [4])

$$\frac{1}{p^2} \sum_{\chi \in \mathcal{E}(p^3), \chi \neq \chi_0} |S(\chi; M, N)| \ll N^{\frac{1}{4}} p^{\frac{3}{4}}, \qquad \text{if } N \leq p^{\frac{3}{2}}, \tag{3}$$

$$\ll N^{\frac{1}{2}} p^{\frac{3}{8}+\varepsilon}, \qquad \text{if } N \geq p^{\frac{3}{2}}, \tag{4}$$

$$\ll p^{1+\varepsilon}, \qquad \text{if } p < N \leq p^{\frac{6}{5}}, \tag{5}$$

$$\ll N^{\frac{2}{3}} p^{\frac{1}{5}+\varepsilon}, \qquad \text{if } N \geq p^{\frac{6}{5}}. \tag{6}$$

From (2) we have, for almost all character of $\mathcal{E}(p^3)$,

$$L(\frac{1}{2} + it, \chi) \ll q^{\frac{1}{6} + \varepsilon},$$

and Theorem B (3)+(4) imply that

$$\frac{1}{p^2} \sum_{\chi \in \mathcal{E}(p^3), \chi \neq \chi_0} |L(\frac{1}{2} + it, \chi)| \ll q^{\frac{1}{6} + \varepsilon}. \tag{7}$$

In the literature on the subject, many people succeeded in constructing an infinite sequence of $\chi \bmod q$, for which

$$L(\frac{1}{2} + it, \chi) \ll q^{\frac{1}{6} + \varepsilon},$$

holds ([6] [7] [8]). So, it seems that the exponent $\frac{1}{6}$ has a big interest in this topics. (Very recently, "the exponent $\frac{1}{6}$ " is proved for any real, nonprincipal character of modulus q, see [15].)

Independently from Burgess' works, the present author considered in [11] the set

$$\mathcal{E}(p^2) = \{\chi; \text{Dirichlet character mod } p^2, \chi^p = \chi_0\},$$

here χ_0 denotes the principal character mod p^2, and he proved

Theorem C.

$$\frac{1}{p-1} \sum_{\chi \in \mathcal{E}(p^2), \chi \neq \chi_0} S(\chi; M, N) \ll p^{\frac{7}{8}} \log p,$$

uniformly in M and N.

In the paper [11], we proved a weaker estimate

$$\frac{1}{p-1} \sum_{\chi \in \mathcal{E}(p^2), \chi \neq \chi_0} S(\chi; M, N) \ll p^{\frac{11}{12}} \log p.$$

This result is based on Heath-Brown's estimate on Heilbronn's exponential sum, and by virtue of his new estimate [9], we can improve $p^{\frac{11}{12}}$ into $p^{\frac{7}{8}}$.

In this paper, first we shall prove

Theorem 1.

$$\frac{1}{p-1} \sum_{\chi \in \mathcal{E}(p^2), \chi \neq \chi_0} L(\frac{1}{2} + it, \chi) \ll p^{\frac{5}{16}} (\log p)^{\frac{19}{8}}.$$

This is the average of $L(\frac{1}{2} + it, \chi)$, not the average of $|L(\frac{1}{2} + it, \chi)|$, but, since $p^{\frac{5}{16}} = q^{\frac{5}{32}} < q^{\frac{1}{6}}$, this indicates that, for many characters, the exponent of q in the right hand side of (1) might be much smaller than $\frac{1}{6}$.

Our second result is

Theorem 2. *Let r be a natural number, $q = p^r$ and $\mathcal{A}$ be an arbitrary subset of the set $\{\chi; \text{Dirichlet character mod } p^r, \chi \neq \chi_0\}$. We put $|\mathcal{A}| = p^s$, with $s \in \mathbf{R}$. Then we have*

$$\frac{1}{|\mathcal{A}|} \sum_{\chi \in \mathcal{A}} |L(\frac{1}{2} + it, \chi)| \ll p^{\frac{r}{4} - \frac{s}{8}} (\log p^r)^{\frac{19}{8}}.$$

First, when we apply this result to the case $r = 3$ and $s = 2$, then

$$\frac{1}{p^2} \sum_{\chi \in \mathcal{A}, \chi \neq \chi_0} |L(\frac{1}{2} + it, \chi)| \ll q^{\frac{1}{6}} (\log p)^{\frac{19}{8}},$$

and this shows that the estimate (7) holds not only for $\mathcal{E}(p^3)$ but also for any set of characters mod p^3 of p^2 elements.

Secondly, we take s as a very near number to r, for example $s = r - 2$, and let r tend to infinity, then Theorem 2 says, asymptotically, the mean value of $|L(\frac{1}{2} + it, \chi)|$ on an arbitrary set $\mathcal{A}$ with $|\mathcal{A}| = p^{r-2}$ is bounded by $q^{\frac{1}{8}} \times$ (log-part).

Recently, Katsurada-Matsumoto [10] discussed the mean square of Dirichlet L-function, and from their result we can derive

$$L(\frac{1}{2} + it, \chi) \ll (pt)^{\frac{1}{6}} (\log pt)^{C_1} + p^{\frac{1}{4}} (\log pt)^{C_2},$$

with some positive constants C_1, C_2. This inequality gives q-estimate and t-estimate at the same time, so-called *hybrid estimate*, and they said that, limiting to the q-estimate aspect, the theoretical limit of their method seems to be $p^{\frac{1}{6}}$. So, the exponent $\frac{1}{8}$ might be the next threshold.

2.   PROOFS

Lemma 1.

$$\sum_{\chi \bmod q} \left| \sum_{N < n \leq N+H} a_n \chi(n) \right|^2 \leq (H + q) \sum_{N < n \leq N+H} |a_n|^2$$

holds uniformly for natural numbers $N, H > 0$, $a_n \in \mathbf{C}$ and $q \geq 2$.

For the proof, see, for example [5], Lemma 2.

Lemma 2. *Let* $\tau(n)$ *be the divisor function. We have the following inequality, for any natural number* K:

$$\sum_{n \leq X} \tau(n)^K \leq X(\log X)^{2^K - 1}.$$

For natural number $r \geq 2$, we define the set

$$\mathcal{D}(p^r) = \{\chi; \text{Dirichlet character mod } p^r\}.$$

Proposition. *Let* $\mathcal{A}$ *be an arbitrary subset of* $\mathcal{D}(p^r)$, *and we put* $|\mathcal{A}| = p^s$, *with* $s \in \mathbf{R}$ *and* l *be a positive integer* ≥ 2. *When* $N^{l-1} \leq p^r < N^l$, *then we have*

$$\frac{1}{|\mathcal{A}|} \sum_{\chi \in \mathcal{A}} |S(\chi; 0, N)|$$

$$\ll \min\left\{ Np^{-\frac{s}{2l}} (\log N^l)^{\frac{2^{2l} - 1}{2l}}, \sqrt{N} p^{\frac{r-s}{2(l-1)}} (\log p^r)^{\frac{2^{2l-2} - 1}{2(l-1)}} \right\}.$$

Proof. We take a natural number m. Then Hölder's inequality gives

$$\left\{ \frac{1}{|\mathcal{A}|} \sum_{\chi \in \mathcal{A}} |S(\chi; 0, N)| \right\}^{2m} \leq \left\{ |\mathcal{A}|^{1 - \frac{2m}{2m-1}} \right\}^{2m-1} \left\{ \sum_{\chi \in \mathcal{A}} |S(\chi; 0, N)^m|^2 \right\}$$

$$= \frac{1}{|\mathcal{A}|} \sum_{\chi \in \mathcal{A}} \left| \sum_{n \leq N^m} a_n \chi(n) \right|^2,$$

where

$$a_n = \sum_{n_1 \cdots n_m = n, 1 \leq n_i \leq N} 1.$$

Lemma 1 implies

$$\left\{ \frac{1}{|\mathcal{A}|} \sum_{\chi \in \mathcal{A}} |S(\chi; 0, N)| \right\}^{2m} \leq \frac{1}{|\mathcal{A}|} (N^m + p^r) \sum_{n \leq N^m} |a_n|^2,$$

and Lemma 2 gives an estimate

$$\sum_{n \leq N^m} |a_n|^2 \ll N^m (\log N^m)^{2^{2m} - 1}.$$

Consequently, we have

$$\frac{1}{|\mathcal{A}|} \sum_{\chi \in \mathcal{A}} |S(\chi; 0, N)| \ll p^{-\frac{s}{2m}} N^{\frac{1}{2}} (N^m + p^r)^{\frac{1}{2m}} (\log N^m)^{\frac{2^{2m}-1}{2m}}.$$

We take $m = l$ and $m = l - 1$, then we get the desired inequality.

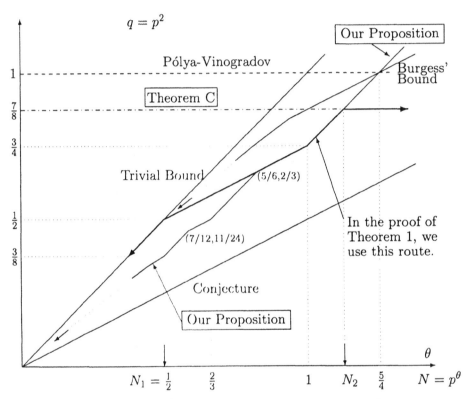

Figure 1 When, for $N = p^\theta$, an estimate "character sum" $\ll p^\alpha N^\beta (\log p)^\gamma$ holds, then we plot $(\theta, \alpha + \theta\beta)$.

As a direct consequence of Proposition, we have, for example,

Corollary 1. *For all except for at most $p^2 - 1$ of characters of $\mathcal{D}(p^3)$, we have*

$$S(\chi; 0, N) \ll N p^{-\frac{1}{3}} (\log p)^{\frac{63}{6}}, \quad if \quad p < N \le p^{\frac{7}{6}} (\log p)^{-\frac{81}{6}}, \qquad (1)$$
$$\ll N^{\frac{1}{2}} p^{\frac{1}{4}} (\log p)^{\frac{15}{4}}, \quad if \quad p^{\frac{7}{6}} (\log p)^{-\frac{81}{6}} < N < p^{\frac{3}{2}}. \qquad (2)$$

In fact, we arrange all $\chi \in \mathcal{D}(p^3)$ according to the size of $|S(\chi; 0, N)|$, and take the set $\mathcal{A} = \mathcal{A}(N)$ as the set of p^2 characters for these characters $|S(\chi; 0, N)|$ take the biggest p^2 values. Then Proposition gives our assertion.

Proof of Theorem 1. We take

$$N_1 = p^{\frac{1}{2}}, \quad N_2 = p^{\frac{9}{8}} (\log p)^{-\frac{11}{4}}.$$

We divide the sum into three parts,

$$\frac{1}{p-1} \sum_{\chi \in \mathcal{E}(p^2), \chi \neq \chi_0} L(\frac{1}{2} + it, \chi)$$

$$= \frac{1}{p-1} \sum_{\chi \in \mathcal{E}(p^2), \chi \neq \chi_0} \left(\sum_{n=1}^{N_1} + \sum_{n=N_1+1}^{N_2} + \sum_{n=N_2+1}^{\infty} \right) \frac{\chi(n)}{n^{\frac{1}{2}+it}},$$

and, according to the size of n, we estimate these three sums by

$$\text{trivial estimate} \longrightarrow \text{Proposition} \longrightarrow \text{Theorem C}$$

(see Fig). The trivial estimate $|\chi(n)| \leq 1$ gives

$$\frac{1}{p-1} \sum_{\chi} \sum_{n=1}^{N_1} \frac{\chi(n)}{n^{\frac{1}{2}+it}} \ll p^{\frac{1}{4}}.$$

As for the second sum $\sum_{n=N_1+1}^{N_2}$, applying partial summation and Proposition with $l = 2$, we have

$$\frac{1}{p-1} \sum_{\chi} \sum_{n=N_1+1}^{n=N_2} \frac{\chi(n)}{n^{\frac{1}{2}+it}} \ll \sqrt{N_2}\, p^{-\frac{1}{4}} (\log p)^{\frac{15}{4}} = p^{\frac{5}{16}} (\log p)^{\frac{19}{8}},$$

and finally applying partial summation and Theorem C, we can prove

$$\frac{1}{p-1} \sum_{\chi} \sum_{n=N_2+1}^{\infty} \frac{\chi(n)}{n^{\frac{1}{2}+it}} \ll p^{\frac{7}{8}} \log p \frac{1}{\sqrt{N_2}} = p^{\frac{5}{16}} (\log p)^{\frac{19}{8}},$$

this proves Theorem 1.

Proof of Theorem 2. We take

$$N_1 = p^{\frac{r}{4}}, \quad N_2 = p^{\frac{2r+s}{4}} (\log p^r)^{-\frac{11}{4}},$$

and divide the sum $\frac{1}{|A|} \sum_{\chi \in \mathcal{A}, \chi \neq \chi_0} |L(\frac{1}{2} + it, \chi)|$ into three parts as above. Now we estimate these three sums by

$$\text{trivial estimate} \longrightarrow \text{Proposition} \longrightarrow \text{Pólya-Vinogradov inequality},$$

then we get the desired result.

References

[1] Burgess D. A., On character sums and L-series II, *Proc. London Math. Soc.*, **13** (1963), pp. 524–536.

[2] Burgess D. A., Mean values of character sums, *Mathematika*, **33** (1986), pp. 1–5.

[3] Burgess D. A., Mean values of character sums III, *Mathematika*, **42** (1995), pp. 133–136.

[4] Burgess D. A., On the average of character sums for a group of characters, *Acta Arith.*, **79** (1997), pp. 313–322.

[5] Elliott P. D. T. A. and Murata Leo, The least primitive root mod $2p^2$, *Mathematika*, **45** (1998), pp. 371–379.

[6] Friedlander J. and Iwaniec H., A note on character sums, *Contemporary Math.*, **166** (1994), pp. 295–299.

[7] Fujii A., Gallagher P. X. and Montgomery H. L., Some hybrid bounds for character sums and Dirichlet L-series, *Colloquia Math Soc. Janos Bolyai*, Vol.13 (1974), pp. 41–57.

[8] Heath-Brown D. R., Hybrid bounds for Dirichlet L-functions II, *Quart. J. Math. Oxford* (2), **31** (1980), pp. 157–167.

[9] Heath-Brown D. R. and Konyagin S., New bounds for Gauss sums derived from k-th powers, and for Heilronn's exponential sum, *Quart. J. Math.*, **51** (2000), pp. 221–235.

[10] Katsurada M. and Matsumoto K., A weighted integral approach to the mean square of Dirichlet L-functions, in "Number Theory and its Applications", S.Kanemitsu and K.Györy (eds.), Developments in Mathematics Vol.2, Kluwer Academic Publishers, 1999, pp. 199–229.

[11] Murata Leo, On characters of order p (mod p^2), *Acta Arith.* **87** (1998), pp. 245–253.

[12] Paley R. E. A. C., A theorem on characters, *J. London Math. Soc.* **7** (1932), pp. 27–32.

[13] Pólya G., Über die Verteilung der quadratischen Reste und Nichtreste, *Göttingen Nachrichten* (1918), pp. 21–29.

[14] Vinogaradov I., On the distribution of power residues and non-residues, *J. Phys. Math. Soc. Perm Univ.* **1** (1918), pp. 94–98.

[15] Conrey J. B. and Iwaniec H., The cubic moment of central values of automorphic L-functions, *Ann. of Math.* **151** (2000), pp. 1175–1216.

ON COVERING EQUIVALENCE

Zhi-Wei SUN

Department of Mathematics, Nanjing University, Nanjing 210093,
the People's Republic of China
zwsun@nju.edu.cn

Keywords: Covering equivalence, uniform function, gamma function, Hurwitz zeta function, trigonometric function

Abstract   An arithmetic sequence $a(n) = \{a + nx : x \in \mathbb{Z}\}$ $(0 \le a < n)$ with weight $\lambda \in \mathbb{C}$ is denoted by $\langle \lambda, a, n \rangle$. For two finite systems $\mathcal{A} = \{\langle \lambda_s, a_s, n_s \rangle\}_{s=1}^k$ and $\mathcal{B} = \{\langle \mu_t, b_t, m_t \rangle\}_{t=1}^l$ of such triples, if $\sum_{n_s | x - a_s} \lambda_s = \sum_{m_t | x - b_t} \mu_t$ for all $x \in \mathbb{Z}$ then we say that $\mathcal{A}$ and $\mathcal{B}$ are covering equivalent. In this paper we characterize covering equivalence in various ways, our characterizations involve the Γ-function, the Hurwitz ζ-function, trigonometric functions, the greatest integer function and Egyptian fractions.

2000 Mathematics Subject Classification: Primary 11B25; Secondary 11B75, 11D68, 11M35, 33B10, 33B15, 33C05, 39B52.

1.   INTRODUCTION AND PRELIMINARIES

By $\mathbb{Z}^+$ we mean the set of positive integers. For $a \in \mathbb{Z}$ and $n \in \mathbb{Z}^+$ we let $a + n\mathbb{Z}$ represent the *arithmetic sequence*

$$\{\cdots,\ a - 2n,\ a - n,\ a,\ a + n,\ a + 2n,\ \cdots\}$$

and write $a(n)$ for $a + n\mathbb{Z}$ if $a \in R(n) = \{0, 1, \cdots, n - 1\}$. For a finite system

$$A = \{a_s(n_s)\}_{s=1}^k \qquad (n_s \in \mathbb{Z}^+ \text{ and } a_s \in R(n_s) \text{ for } s = 1, \cdots, k) \quad (1.1)$$

of such arithmetic sequences, we define the *covering function* $w_A : \mathbb{Z} \to \mathbb{N} = \{0, 1, 2, \cdots\}$ as follows:

$$w_A(x) = |\{1 \le s \le k : x \in a_s(n_s)\}|. \quad (1.2)$$

The research is supported by the Teaching and Research Award Fund for Outstanding Young Teachers in Higher Education Institutions of MOE, and the National Natural Science Foundation of P. R. China.

C. Jia and K. Matsumoto (eds.), Analytic Number Theory, 277–302.
© 2002 *Kluwer Academic Publishers. Printed in the Netherlands.*

Let m be a positive integer. If $w_A(x) \geq m$ for all $x \in \mathbb{Z}$, then we call (1.1) an *m-cover* (of $\mathbb{Z}$); when $w_A(x) = m$ for any $x \in \mathbb{Z}$, we say that A forms an *exact m-cover* (of $\mathbb{Z}$). The notion of cover (i.e. 1-cover) was introduced by P. Erdős ([Er]) in the early 1930's, a simple nontrivial cover of $\mathbb{Z}$ is $\{0(2), 0(3), 1(4), 5(6), 7(12)\}$. Covers of $\mathbb{Z}$ have many nice properties and interesting applications, for problems and results we recommend the reader to see R. K. Guy [G], and Š. Porubský and J. Schönheim [PS].

Let M be an additive abelian group. In 1989 the author [Sul] considered triples of the form $\langle \lambda, a, n \rangle$ where $\lambda \in M$, $n \in \mathbb{Z}^+$ and $a \in R(n)$, we can view $\langle \lambda, a, n \rangle$ as the arithmetic sequence $a(n)$ with *weight* (or *multiplier*) λ. Let $S(M)$ denote the set of all finite systems of such triples. For

$$\mathcal{A} = \{\langle \lambda_s, a_s, n_s \rangle\}_{s=1}^k \in S(M) \tag{1.3}$$

we associate it with the following periodic arithmetical map $w_A : \mathbb{Z} \to M$:

$$w_A(x) = \sum_{\substack{s=1 \\ x \in a_s(n_s)}}^k \lambda_s \quad \text{for } x \in \mathbb{Z}, \tag{1.4}$$

which is called the *covering map* of $\mathcal{A}$. ($w_\emptyset$ refers to the zero map into M.) Clearly w_A is periodic modulo the least common multiple $[n_1, \cdots, n_k]$ of all the moduli $n_1, \cdots, n_k$. For $\mathcal{A}, \mathcal{B} \in S(M)$, if $w_A = w_B$ then we say that $\mathcal{A}$ is *covering equivalent* to $\mathcal{B}$ and write $\mathcal{A} \sim \mathcal{B}$ for this. Observe that (1.3) and $\mathcal{B} = \{\langle \mu_t, b_t, m_t \rangle\}_{t=1}^l \in S(M)$ are equivalent if and only if

$$C = \{\langle \lambda_1, a_1, n_1 \rangle, \cdots, \langle \lambda_k, a_k, n_k \rangle, \langle -\mu_1, b_1, m_1 \rangle, \cdots, \langle -\mu_l, b_l, m_l \rangle\} \sim \emptyset. \tag{1.5}$$

We identify (1.1) with the system $\{\langle 1, a_s, n_s \rangle\}_{s=1}^k$. Thus, (1.1) forms an exact m-cover if and only if $A \sim \{\langle m, 0, 1 \rangle\}$.

Let M be an additive abelian group, and f a map of two complex variables into M such that $\{\langle \frac{x+r}{n}, ny \rangle : r \in R(n)\} \subseteq \text{Dom}(f)$ for all $\langle x, y \rangle \in \text{Dom}(f)$ and $n \in \mathbb{Z}^+$. If

$$\sum_{r=0}^{n-1} f\left(\frac{x+r}{n}, ny\right) = f(x, y) \quad \text{for any } \langle x, y \rangle \in \text{Dom}(f) \text{ and } n \in \mathbb{Z}^+, \tag{1.6}$$

then we call f a *uniform map* into M. Note that $f(x, y) = f(\frac{x+0}{1}, 1y)$ and $\{r(n)\}_{r=0}^{n-1} \sim \{0(1)\}$ for any $n \in \mathbb{Z}^+$.

In 1989 the author [Sul] showed the following

Theorem 1.1. *Let M be a left R-module where R is a ring with identity. Whenever two systems $\mathcal{A} = \{\langle \lambda_s, a_s, n_s \rangle\}_{s=1}^k$ and $\mathcal{B} = \{\langle \mu_t, b_t, m_t \rangle\}_{t=1}^l$ in $\mathrm{S}(R)$ are equivalent, for any uniform map f into M we have*

$$\sum_{s=1}^k \lambda_s f\left(\frac{x + a_s}{n_s}, n_s y\right) = \sum_{t=1}^l \mu_t f\left(\frac{x + b_t}{m_t}, m_t y\right) \quad \textit{for all } \langle x, y \rangle \in \mathrm{Dom}(f).$$

$$(1.7)$$

A simple proof of this remarkable theorem was given in Section 2 of [Su8].

Let f be a uniform map into the complex field $\mathbb{C}$ with $\mathrm{Dom}(f) = D \times D'$ and $f(x_0, y_0) \neq 0$ for some $\langle x_0, y_0 \rangle \in D \times D'$. If $f(x, y) = g(x)h(y)$ for all $x \in D$ and $y \in D'$, then for each $n \in \mathbb{Z}^+$ we have

$$\sum_{r=0}^{n-1} g\left(\frac{x_0 + r}{n}\right) h(ny_0) = g(x_0)h(y_0) \neq 0,$$

thus $\theta(n) = h(y_0)/h(ny_0) \neq 0$ and

$$\theta(n)h(ny) = \frac{h(ny)}{g(x_0)} \sum_{r=0}^{n-1} g\left(\frac{x_0 + r}{n}\right) = h(y) \quad \text{for every } y \in D',$$

therefore

$$\theta(mn) = \frac{h(y_0)}{h(my_0)} \cdot \frac{h(my_0)}{h(n(my_0))} = \theta(m)\theta(n) \quad \text{for any } m, n \in \mathbb{Z}^+$$

and

$$\sum_{r=0}^{n-1} g\left(\frac{x + r}{n}\right) = \theta(n)g(x) \quad \text{for all } n \in \mathbb{Z}^+ \text{ and } x \in D. \tag{1.8}$$

Such functions g with $D = \mathrm{Dom}(g) = [0, 1)$ or $\mathbb{Q} \cap [0, 1)$ were studied by H. Walum [W] in 1991 (where $\mathbb{Q}$ denotes the rational field), and investigated earlier by H. Bass [B], D. S. Kubert [K], S. Lang [La] and J. Milnor [M] in the case $\theta(n) = n^{1-s}$ where $s \in \mathbb{C}$.

In this paper by $\mathbb{R}$ and $\mathbb{R}^+$ we mean the field of real numbers and the set of positive reals. For a field F we use F^* to denote the multiplicative group of nonzero elements of F. When $x \in \mathbb{R}$, $[x]$ and $\{x\}$ denote the integral part and the fractional part of x respectively, if $n \in \mathbb{Z}^+$ then we write $\{x\}_n$ to mean $n\{x/n\}$. For convenience, we also set

$$q(0) = 0 \quad \text{and} \quad q(z) = \frac{1}{z} \text{ for } z \in \mathbb{C}^*. \tag{1.9}$$

2.   SOME EXAMPLES OF UNIFORM MAPS

For a uniform map f, the map $f^-(x,y) = f(1-x, y)$ is also a uniform map.

Example 2.1. Two simple uniform maps into the additive group $\mathbb{Z}$ are the functions $I, I_+ : \mathbb{C} \times \mathbb{C} \to \mathbb{Z}$ given below:

$$I(x,y) = \begin{cases} 1 & \text{if } x \in \mathbb{Z}, \\ 0 & \text{otherwise}; \end{cases} \qquad I_+(x,y) = \begin{cases} 1 & \text{if } x \in \mathbb{Z}^+, \\ 0 & \text{otherwise}. \end{cases} \qquad (2.1)$$

Another example is the function $[\ ] : \mathbb{R} \times \mathbb{R} \to \mathbb{Z}$ defined by

$$[\ ](x,y) = [x], \qquad (2.2)$$

the identity $\sum_{r=0}^{n-1} [\frac{x+r}{n}] = [x]$ (for $n \in \mathbb{Z}^+$) is well known and due to Hermite.

Now we turn to uniform maps into the multiplicative group $\mathbb{C}^*$.

Example 2.2. (i) Define $\gamma : \mathbb{C} \times \mathbb{R}^+ \to \mathbb{C}^*$ as follows:

$$\gamma(x,y) = \begin{cases} \Gamma(x) y^x / \sqrt{2\pi y} & \text{if } x \notin -\mathbb{N} = \{0, -1, -2, \cdots\}, \\ \frac{(-1)^x}{(-x)!} y^x \sqrt{2\pi y} & \text{otherwise}. \end{cases} \qquad (2.3)$$

When $n \in \mathbb{Z}^+$ and $y \in \mathbb{R}^+$, if $x \in \mathbb{C} \setminus -\mathbb{N}$, then by applying Gauss' multiplication formula

$$\prod_{r=0}^{n-1} \Gamma\left(z + \frac{r}{n}\right) = (2\pi)^{\frac{n-1}{2}} n^{\frac{1}{2} - nz} \Gamma(nz)$$

with $z = x/n$, we obtain that

$$\prod_{r=0}^{n-1} \left(\Gamma\left(\frac{x+r}{n}\right) \frac{(ny)^{(x+r)/n}}{\sqrt{2\pi ny}} \right) = \Gamma(x) \frac{y^x}{\sqrt{2\pi y}};$$

on the other hand, for $m = kn + l$ with $k \in \mathbb{N}$ and $l \in R(n)$, we have

$$\lim_{x \to -m} \prod_{\substack{r=0 \\ r \neq l}}^{n-1} \frac{\Gamma(\frac{x+r}{n})(ny)^{\frac{x+r}{n}}}{\sqrt{2\pi ny}} = \lim_{x \to -m} \frac{\Gamma(x) y^x / \sqrt{2\pi y}}{\Gamma(\frac{x+l}{n})(ny)^{\frac{x+l}{n}} / \sqrt{2\pi ny}}$$

$$= \lim_{x \to -m} \frac{\Gamma(x+m+1) \prod_{j=0}^{m} (x+j)^{-1} y^{x-1/2}}{\Gamma(\frac{x+m}{n}+1) \prod_{j=0}^{k} (\frac{x+l}{n}+j)^{-1} (ny)^{\frac{x+l}{n} - \frac{1}{2}}} = \frac{\frac{(-1)^m}{m!} y^{-m} \sqrt{2\pi y}}{\frac{(-1)^k}{k!} (ny)^{-k} \sqrt{2\pi ny}}.$$

Thus γ is a uniform map into $\mathbb{C}^*$.

(ii) As $\Gamma(z)\Gamma(1-z) = \pi/\sin \pi z$ for $z \notin \mathbb{Z}$, $\quad \gamma(x,y)\gamma^-(x,y) = (-1)^{I+(x,y)}/S(x,y)$ for $x \in \mathbb{C}$ and $y > 0$, where

$$S(x,y) = \begin{cases} 2\sin \pi x & \text{if } x \notin \mathbb{Z}, \\ (-1)^x y^{-1} & \text{if } x \in \mathbb{Z}. \end{cases} \tag{2.4}$$

It follows that the function $S : \mathbb{C} \times \mathbb{C}^* \to \mathbb{C}^*$ is a uniform map into $\mathbb{C}^*$.

(iii) For $\alpha, \beta, \gamma \in \mathbb{C}$ with $\text{Re}(\gamma) > \text{Re}(\alpha + \beta)$ and $\gamma \neq 0, -1, -2, \cdots$, we use $F(\alpha, \beta, \gamma, z)$ to denote the hypergeometric series given by

$$F(\alpha, \beta, \gamma, z) = 1 + \sum_{n=1}^{\infty} \frac{\alpha(\alpha+1)\cdots(\alpha+n-1)\beta(\beta+1)\cdots(\beta+n-1)}{n!\gamma(\gamma+1)\cdots(\gamma+n-1)} z^n$$

which converges absolutely for $|z| \leqslant 1$. Let $u, v, w \in \mathbb{C}$ and $D_{u,v,w}$ consist of all those $\langle x, y \rangle \in \mathbb{C} \times \mathbb{C}^*$ such that $x + \frac{w}{y}, x + \frac{w-u}{y}, x + \frac{w-v}{y} \notin -\mathbb{N}$ and $\text{Re}(x + \frac{w-u-v}{y}) > 0$. When $\langle x, y \rangle \in D_{u,v,w}$, by a formula in [Ba] we have

$$F\left(\frac{u}{y}, \frac{v}{y}, \frac{w}{y} + x, 1\right) = \frac{\Gamma(x_1)\Gamma(x_1 - u/y - v/y)}{\Gamma(x_1 - u/y)\Gamma(x_1 - v/y)} = \frac{\gamma(x_1, y)\gamma(x_2, y)}{\gamma(x_3, y)\gamma(x_4, y)} \neq 0$$

where $x_1 = x + \frac{w}{y}$, $x_2 = x + \frac{w-u-v}{y}$, $x_3 = x + \frac{w-u}{y}$ and $x_4 = x + \frac{w-v}{y}$; for any $n \in \mathbb{Z}^+$ clearly $\langle \frac{x+r}{n}, ny \rangle \in D_{u,v,w}$ for all $r \in R(n)$ (since $\frac{x+r}{n} + \frac{z}{ny} = \frac{x+z/y+r}{n}$) and

$$\prod_{r=0}^{n-1} F\left(\frac{u}{ny}, \frac{v}{ny}, \frac{w}{ny} + \frac{x+r}{n}, 1\right) = \prod_{r=0}^{n-1} F\left(\frac{u}{ny}, \frac{v}{ny}, \frac{x_1+r}{n}, 1\right)$$

$$= \prod_{r=0}^{n-1} \frac{\gamma(\frac{x_1+r}{n}, ny)\gamma(\frac{x_2+r}{n}, ny)}{\gamma(\frac{x_3+r}{n}, ny)\gamma(\frac{x_4+r}{n}, ny)} = \frac{\gamma(x_1, y)\gamma(x_2, y)}{\gamma(x_3, y)\gamma(x_4, y)} = F\left(\frac{u}{y}, \frac{v}{y}, \frac{w}{y} + x, 1\right).$$

So the function $F_{u,v,w} : D_{u,v,w} \to \mathbb{C}^*$ given by

$$F_{u,v,w}(x, y) = F\left(\frac{u}{y}, \frac{v}{y}, \frac{w}{y} + x, 1\right) \tag{2.5}$$

is a uniform map into $\mathbb{C}^*$.

Remark 2.1. Since the function S is a uniform map into $\mathbb{C}^*$, we have

$$\sum_{r=0}^{n-1} \log\left(2\sin \pi \frac{x+r}{n}\right) = \log(2\sin \pi x) \quad \text{for } n \in \mathbb{Z}^+ \text{ and } x \in (0,1).$$

In 1966 H. Bass [B] showed that every linear relation over $\mathbb{Q}$ among the numbers $\log(2\sin\pi x)$ with $x \in \mathbb{Q}\cap(0,1)$, is a consequence of the last identity, together with the fact that $\log(2\sin\pi(1-x)) = \log(2\sin\pi x)$. (See also V. Ennola [E].)

Example 2.3. (i) Define $G, H : \mathbb{C}\times\mathbb{R}^+ \to \mathbb{C}$ by

$$G(x,y) = \frac{1}{y}\left(\log y + \sum_{m=0}^{\infty}\left(\frac{1}{m+1} - q(m+x)\right)\right) \qquad (2.6)$$

and

$$H(x,y) = \frac{1}{y}\left(\log y + \sum_{m=1}^{\infty}\left(\frac{1}{m} + q(x-m)\right)\right). \qquad (2.7)$$

We assert that G is a uniform map into $\mathbb{C}$ and hence so is $H = G^-$.

Let $n \in \mathbb{Z}^+$ and $y > 0$. By Example 2.2(i),

$$\prod_{r=0}^{n-1}\left(\Gamma\left(\frac{x+r}{n}\right)(ny)^{\frac{x+r}{n}}(2\pi ny)^{-\frac{1}{2}}\right) = \Gamma(x)y^x(2\pi y)^{-\frac{1}{2}} \quad \text{for } x \in \mathbb{C}\setminus -\mathbb{N}.$$

Taking the logarithmic derivatives of both sides we get that

$$\frac{1}{n}\sum_{r=0}^{n-1}\left(\frac{\Gamma'((x+r)/n)}{\Gamma((x+r)/n)} + \log(ny)\right) = \frac{\Gamma'(x)}{\Gamma(x)} + \log y \quad \text{for } x \neq 0, -1, -2, \cdots.$$

Thus

$$\frac{1}{ny}\sum_{r=0}^{n-1}\left(\frac{\Gamma'((x+r)/n)}{\Gamma((x+r)/n)} + \log(ny) + \gamma\right)$$

$$= \frac{1}{y}\left(\frac{\Gamma'(x)}{\Gamma(x)} + \log y + \gamma\right) \quad \text{for } x \in \mathbb{C}\setminus -\mathbb{N}$$

where γ is the Euler constant. By a known formula (see, e.g., [Ba]), if $x \in \mathbb{C}\setminus -\mathbb{N}$ then

$$yG(x,y) - \log y = \sum_{m=0}^{\infty}\left(\frac{1}{m+1} - \frac{1}{m+x}\right)$$

$$= \sum_{m=0}^{\infty}\frac{x-1}{(m+1)(m+x)} = \frac{\Gamma'(x)}{\Gamma(x)} + \gamma.$$

So, for any $x \in \mathbb{C}\setminus -\mathbb{N}$ we have

$$\sum_{r=0}^{n-1}G\left(\frac{x+r}{n}, ny\right) = G(x,y).$$

When $x = -a - bn$ with $a \in R(n)$ and $b \in \mathbb{N}$,

$$\sum_{r=0}^{n-1} G\left(\frac{x+r}{n}, ny\right) - G(x, y)$$

$$= G\left(\frac{x+a}{n}, ny\right) - G(x, y) + \lim_{\substack{z \to x \\ z \notin \mathbb{Z}}} \sum_{\substack{r=0 \\ r \neq a}}^{n-1} G\left(\frac{z+r}{n}, ny\right)$$

$$= G(-b, ny) - G(x, y) + \lim_{\substack{z \to x \\ z \notin \mathbb{Z}}} \left(G(z, y) - G\left(\frac{z+a}{n}, ny\right)\right)$$

$$= \lim_{\substack{z \to x \\ z \notin \mathbb{Z}}} \left(\frac{1}{y} \sum_{\substack{m=0 \\ m \neq -x}}^{\infty} (q(m+x) - q(m+z)) + \frac{1}{y}(q(0) - q(-x+z))\right.$$

$$+ \frac{1}{ny} \sum_{\substack{m=0 \\ m \neq b}}^{\infty} \left(q\left(m + \frac{z+a}{n}\right) - q(m-b)\right)$$

$$\left. + \frac{1}{ny} \left(q\left(b + \frac{z+a}{n}\right) - q(0)\right)\right)$$

$$= \lim_{z \to x} \left(\frac{1}{y} \cdot \frac{1}{x-z} + \frac{1}{ny} \cdot \frac{1}{(z+a)/n+b}\right) = 0.$$

(Note that for any $v \in \mathbb{N}$ we have

$$\lim_{u \to v} \sum_{\substack{m=0 \\ m \neq v}}^{\infty} \left(\frac{1}{m-u} - \frac{1}{m-v}\right) = \sum_{m=v+1}^{\infty} \lim_{u \to v} \frac{u-v}{(m-u)(m-v)} = 0.$$

For, if $|u - v| < 1/2$ then

$$\left|\frac{1}{m-u} - \frac{1}{m-v}\right| = \left|\frac{u-v}{(m-u)(m-v)}\right| < \frac{1/2}{(m-v-1/2)^2}$$

for $m = v+1, v+2, \cdots$, thus the series $\sum_{m=v+1}^{\infty} \left(\frac{1}{m-u} - \frac{1}{m-v}\right)$ (with $|u - v| < 1/2$) converges uniformly.)

(ii) The Hurwitz zeta function $\zeta(s, v)$ is defined by the series

$$\zeta(s, v) = \sum_{m=0}^{\infty} \frac{1}{(m+v)^s}$$

for $s, v \in \mathbb{C}$ with $\mathrm{Re}(s) > 1$ and $\mathrm{Re}(v) > 0$ (or $v \notin -\mathbb{N}$), by Lemma 1 of [M] it extends to a function which is defined and holomorphic in both variables for all complex $s \neq 1$ and for all v in the simply connected region $\mathbb{C} \setminus (-\infty, 0]$. In addition, we set $\zeta(1, v) = -G(v, 1)$. For $v \in \mathbb{C} \setminus -\mathbb{N}$, if $\mathrm{Re}(s) > 1$ then

$$\zeta(s, v + 1) - \zeta(s, v) = \sum_{m=0}^{\infty} \frac{1}{(m + v + 1)^s} - \sum_{m=0}^{\infty} \frac{1}{(m + v)^s} = -\frac{1}{v^s};$$

we also have

$$\zeta(1, v + 1) - \zeta(1, v) = -\sum_{m=1}^{\infty} \left(\frac{1}{m} - \frac{1}{m + v} \right) + \sum_{m=0}^{\infty} \left(\frac{1}{m + 1} - \frac{1}{m + v} \right)$$

$$= -\frac{1}{v}.$$

If $\mathrm{Re}(s) > 1$ then

$$\frac{d}{dv} \zeta(s, v) = \sum_{m=0}^{\infty} \frac{d(m + v)^{-s}}{dv} = -s\zeta(s + 1, v) \quad \text{for all } v \in \mathbb{C} \setminus -\mathbb{N}.$$

Whenever $s \in \mathbb{C} \setminus \{0, 1\}$,

$$\frac{d}{dv} \zeta(s, v) = -s\zeta(s + 1, v) \quad \text{for all } v \in \mathbb{C} \setminus (-\infty, 0]$$

by analytic continuation. As $\zeta(0, v) = \frac{1}{2} - v$, $\frac{d}{dv} \zeta(0, v) = -1$. For those $v \in \mathbb{C} \setminus -\mathbb{N}$, we have

$$\frac{d}{dv} \zeta(1, v) = -\frac{d}{dv} \sum_{m=0}^{\infty} \left(\frac{1}{m + 1} - \frac{1}{m + v} \right) = -\sum_{m=0}^{\infty} \frac{1}{(m + v)^2} = -\zeta(2, v).$$

For $s \in \mathbb{C}$ we define $\zeta_s : (\mathbb{C} \setminus (-\infty, 0]) \times \mathbb{R}^+ \to \mathbb{C}$ by

$$\zeta_s(x, y) = \begin{cases} y^{-s} \zeta(s, x) & \text{if } s \neq 1, \\ y^{-1}(\zeta(1, x) - \log y) & \text{if } s = 1. \end{cases} \tag{2.8}$$

Let $x \in \mathbb{C} \setminus (-\infty, 0]$ and $y \in \mathbb{R}^+$. Then $\zeta_1(x, y) = -G(x, y)$. If $\mathrm{Re}(s) > 1$ and $n \in \mathbb{Z}^+$, then

$$\sum_{r=0}^{n-1} n^{-s} \zeta \left(s, \frac{x + r}{n} \right) = \sum_{r=0}^{n-1} \sum_{k=0}^{\infty} \frac{1}{(x + r + kn)^s} = \zeta(s, x).$$

By part (i) and analytic continuation, for any $s \in \mathbb{C}$ the function ζ_s is a uniform map into $\mathbb{C}$.

Remark 2.2. (a) In [M] Milnor observed that there is no constant c such that

$$\sum_{r=0}^{n-1} \frac{1}{n} \left(\frac{\Gamma'\left((x+r)/n\right)}{\Gamma\left((x+r)/n\right)} + c \right) = \frac{\Gamma'(x)}{\Gamma(x)} + c \text{ for } n \in \mathbb{Z}^+ \text{ and } x \in (0,1).$$

(b) By [Ba], if $x, y > 0$ then

$$\left. \frac{d\zeta_s(x,y)}{ds} \right|_{s=0} = y^{-s} \frac{d}{ds} \zeta(s,x) \bigg|_{s=0} - y^{-s} \log y \cdot \zeta(s,x) \bigg|_{s=0}$$

$$= \log \Gamma(x) - \frac{1}{2} \log(2\pi) + \left(x - \frac{1}{2} \right) \log y = \log \gamma(x,y).$$

(c) For $s \in \mathbb{C} \setminus -\mathbb{N}$ Milnor [M] proved in 1983 that the complex vector space of all those continuous functions $g : (0,1) \to \mathbb{C}$ which satisfy (1.8) with $D = (0,1)$ and $\theta(n) = n^s$, is spanned by linearly independent functions $\zeta(s,x)$ and $\zeta(s,1-x)$ where x ranges over $(0,1)$.

Example 2.4. (i) Let ψ be a function into $\mathbb{C}$ with $\text{Dom}(\psi) \subseteq \mathbb{C}$. Let D_ψ denote the set of those $\langle x, y \rangle \in \mathbb{C} \times \mathbb{C}$ such that $(x+k)y \in \text{Dom}(\psi)$ for all $k \in \mathbb{Z}$ and $\sum_{k \equiv a \pmod{n}} \psi((x+k)y)$ converges for any $a \in \mathbb{Z}$ and $n \in \mathbb{Z}^+$. If $\langle x, y \rangle \in D_\psi$ and $n \in \mathbb{Z}^+$, then $\langle \frac{x+r}{n}, ny \rangle \in D_\psi$ for all $r \in R(n)$ since $(\frac{x+r}{n} + k)ny = (x + (r+kn))y$, furthermore

$$\sum_{r=0}^{n-1} \sum_{k=-\infty}^{\infty} \psi\left(\left(\frac{x+r}{n} + k \right) ny \right) = \sum_{r=0}^{n-1} \sum_{l \equiv r \pmod{n}} \psi((x+l)y)$$

$$= \sum_{l=-\infty}^{+\infty} \psi((x+l)y).$$

Thus the function $\tilde{\psi} : D_\psi \to \mathbb{C}$ given by

$$\tilde{\psi}(x,y) = \sum_{k=-\infty}^{+\infty} \psi(xy + ky) \tag{2.9}$$

is a uniform map into $\mathbb{C}$. This fact was first mentioned by the author in [Sul].

(ii) Let $m \in \mathbb{N}$. We define $\cot_m : \mathbb{C} \times \mathbb{C}^* \to \mathbb{C}$ by

$$\cot_m(x,y) = \begin{cases} \frac{(-1)^m}{y^{m+1}} \cot^{(m)}(\pi x) & \text{if } x \notin \mathbb{Z}, \\ (-1)^{\frac{m-1}{2}} (\frac{2}{y})^{m+1} \frac{B_{m+1}}{m+1} & \text{if } x \in \mathbb{Z} \And 2 \nmid m, \\ 0 & \text{if } x \in \mathbb{Z} \And 2 \mid m, \end{cases} \tag{2.10}$$

where $\cot^{(m)}(z) = \frac{d^m \cot z}{dz^m}$ for $z \in \mathbb{C} \setminus \pi\mathbb{Z}$, and B_n is the nth Bernoulli number. Fix $x \in \mathbb{C}$ and $y \in \mathbb{C}^*$. As

$$\pi \cot \pi v = \sum_{k=-\infty}^{+\infty} \frac{1}{v+k} = \frac{1}{v} + 2v \sum_{k=1}^{\infty} \frac{1}{v^2 - k^2} \quad \text{for } v \in \mathbb{C} \setminus \mathbb{Z},$$

if $x \notin \mathbb{Z}$ then

$$\pi^{m+1} \cot^{(m)}(\pi x) = \frac{d^m}{dv^m}\left(\frac{1}{v}\right)\Big|_{v=x} + \sum_{k=1}^{\infty} \frac{d^m}{dv^m}\left(\frac{2v}{v^2 - k^2}\right)\Big|_{v=x}$$

$$= \sum_{k=-\infty}^{+\infty} \frac{d^m}{dv^m}\left(\frac{1}{v+k}\right)\Big|_{v=x} = (-1)^m m! \sum_{k=-\infty}^{+\infty} \frac{1}{(x+k)^{m+1}}$$

and so

$$\cot_m(x, y) = \frac{m!}{(\pi y)^{m+1}} \sum_{k=-\infty}^{+\infty} \frac{1}{(x+k)^{m+1}}.$$

If $x \in \mathbb{Z}$ and $2 \mid m$, then $\cot_m(x, y)$ vanishes and

$$\sum_{\substack{k=-\infty \\ k \neq -x}}^{+\infty} \frac{1}{(x+k)^{m+1}} = \sum_{\substack{l=-\infty \\ l \neq 0}}^{+\infty} \frac{1}{l^{m+1}} = 0.$$

If $x \in \mathbb{Z}$ and $2 \nmid m$, then

$$\cot_m(x, y) = (-1)^{\frac{m+1}{2}-1} \frac{(2\pi)^{m+1}}{(m+1)!} B_{m+1} \frac{m!}{(\pi y)^{m+1}} = \frac{m! 2\zeta(m+1)}{(\pi y)^{m+1}}$$

$$= \frac{m!}{(\pi y)^{m+1}} \sum_{\substack{l=-\infty \\ l \neq 0}}^{+\infty} \frac{1}{l^{m+1}} = \frac{m!}{(\pi y)^{m+1}} \sum_{\substack{k=-\infty \\ k \neq -x}}^{+\infty} \frac{1}{(x+k)^{m+1}}.$$

So we always have

$$\cot_m(x, y) = \frac{m!}{(\pi y)^{m+1}} \sum_{\substack{k=-\infty \\ k \neq -x}}^{+\infty} \frac{1}{(x+k)^{m+1}} = \frac{m!}{\pi^{m+1}} \sum_{k=-\infty}^{+\infty} q(xy + ky)^{m+1}.$$

By part (i) this implies that $\cot_m$ is a uniform map into $\mathbb{C}$.

Let $x \in \mathbb{C}$ and $y > 0$. Clearly

$$\pi \cot_0(x, y) = H(x, y) - H^-(x, y) = \zeta_1(x, y) - \zeta_1^-(x, y). \tag{2.11}$$

Also,

$$\cot_1(x,y) = \begin{cases} \frac{1}{y^2}\csc^2\pi x & \text{if } x \notin \mathbb{Z}, \\ \frac{1}{3y^2} & \text{if } x \in \mathbb{Z}; \end{cases} \tag{2.12}$$

$$\cot_2(x,y) = \begin{cases} \frac{2\cos\pi x}{y^3\sin^3\pi x} & \text{if } x \notin \mathbb{Z}, \\ 0 & \text{if } x \in \mathbb{Z}, \end{cases} = 2\cot_0(x,y)\cot_1(x,y); \tag{2.13}$$

and

$$\cot_3(x,y) = \begin{cases} \frac{2(1+2\cos^2\pi x)}{y^4\sin^4\pi x} & \text{if } x \notin \mathbb{Z}, \\ \frac{2}{15y^4} & \text{if } x \in \mathbb{Z}, \end{cases} = \frac{6}{(\pi y)^4}\sum_{\substack{k=-\infty\\k\neq -x}}^{+\infty}\frac{1}{(x+k)^4}. \tag{2.14}$$

Remark 2.3. In 1970 S. Chowla [C] proved that if p is an odd prime, then the $\frac{p-1}{2}$ real numbers $\cot 2\pi\frac{r}{p}$ $(r = 1, 2, \cdots, \frac{p-1}{2})$ are linearly independent over $\mathbb{Q}$, this was extended by T. Okada [O] in 1980. By Lemma 7 of Milnor [M], there is a unique function $g : \mathbb{R} \to \mathbb{R}$ periodic mod 1 for which $g(x) = \cot^{(m)}(\pi x) = (-1)^m\cot_m(x,1)$ for $x \in \mathbb{R} \setminus \mathbb{Z}$ and (1.8) holds with $\theta(n) = n^{m+1}$ and $D = \mathbb{R}$, we remark that $g(x) = (-1)^m\cot_m(x,1)$ for all $x \in \mathbb{R}$ because $\cot_m$ is a uniform map into $\mathbb{C}$. With the help of Dirichlet L-functions, Milnor [M] also showed that every $\mathbb{Q}$-linear relation among the values $g(x)$ with $x \in \mathbb{Q}$, follows from (1.8) with $\theta(n) = n^{m+1}$ and $D = \mathbb{Q}$, and the facts $g(x+1) = g(x)$ and $g(-x) = (-1)^{m+1}g(x)$ for $x \in \mathbb{Q}$. So, each $\mathbb{Q}$-linear relation among the values $\cot_m(x,n) = \frac{(-1)^m}{n^{m+1}}g(x)$ with $x \in \mathbb{Q}$ and $n \in \mathbb{Z}^+$, is a consequence of the fact that $\cot_m$ is a uniform map into $\mathbb{C}$, together with the trivial equalities $\cot_m(x+1,n) = \cot_m(x,n)$ and $\cot_m^- = (-1)^{m-1}\cot_m$.

3. CHARACTERIZATIONS OF COVERING EQUIVALENCE

In order to characterize covering equivalence, we need

Lemma 3.1. *Let f be a complex-valued function so that for any $n \in \mathbb{Z}^+$, $f(x,n)$ is defined and continuous at $x \in (-\infty, 1) \setminus \mathbb{Z}$, and*

$$\sum_{r=0}^{n-1} f\left(\frac{x+r}{n}, n\right) = f(x,1) \quad \text{for all } x < 1 \text{ with } x \notin \mathbb{Z}.$$

Suppose that

$$\lim_{x\to -m} f(x,1) = \infty \quad \text{for each } m \in \mathbb{N}.$$

Let $\mathcal{A} = \{\langle \lambda_s, a_s, n_s \rangle\}_{s=1}^k$ be such a system in $S(\mathbb{C})$ that

$$\sum_{s=1}^k \lambda_s f\left(\frac{x+a_s}{n_s}, n_s\right) = 0 \quad \text{for all } x < 1 \text{ with } x \notin \mathbb{Z}.$$

Then $\mathcal{A} \sim \emptyset$.

Proof. Since $w_\mathcal{A}$ is periodic mod $[n_1, \cdots, n_k]$, it suffices to show $w_\mathcal{A}(m) = 0$ for any $m \in \mathbb{N}$. As $\lim_{x \to -m} f(x, 1) = \infty$ there exists a $\delta \in (0, 1)$ such that $f(x, 1) \neq 0$ for all $x \in (-m - \delta, -m + \delta)$ with $x \neq -m$. If $n \in \mathbb{Z}^+$ then

$$\lim_{x \to -m} \left(f(x, 1) - f\left(\frac{x + \{m\}_n}{n}, n\right) \right)$$

$$= \lim_{x \to -m} \sum_{\substack{r=0 \\ r \neq \{m\}_n}}^{n-1} f\left(\frac{x+r}{n}, n\right) = \sum_{\substack{r=0 \\ r \neq \{m\}_n}}^{n-1} f\left(\frac{-m+r}{n}, n\right)$$

and hence

$$f\left(\frac{x + \{m\}_n}{n}, n\right) \sim f(x, 1) \qquad \text{as } x \to -m.$$

So

$$\lim_{x \to -m} \frac{f(\frac{x+a_s}{n_s}, n_s)}{f(x, 1)} = \begin{cases} 1 & \text{if } n_s \mid m - a_s, \\ 0 & \text{otherwise.} \end{cases}$$

Therefore

$$w_\mathcal{A}(m) = \sum_{s=1}^k \lambda_s \lim_{x \to -m} \frac{f(\frac{x+a_s}{n_s}, n_s)}{f(x, 1)}$$

$$= \lim_{x \to -m} \frac{1}{f(x, 1)} \sum_{s=1}^k \lambda_s f\left(\frac{x+a_s}{n_s}, n_s\right) = 0.$$

Let's now characterize the covering equivalence of two systems of arithmetic sequences.

Theorem 3.1. *Let* $n_s, m_t \in \mathbb{Z}^+$, $a_s \in R(n_s)$ *and* $b_t \in R(m_t)$ *for* $s = 1, \cdots, k$ *and* $t = 1, \cdots, l$. *Then the following statements are equivalent:*

$$A = \{a_s(n_s)\}_{s=1}^k \sim B = \{b_t(m_t)\}_{t=1}^l; \tag{3.1}$$

$$\sum_{\substack{I\subseteq\{1,\cdots,k\} \\ \sum_{s\in I}\frac{1}{n_s}=c}} (-1)^{|I|}e^{2\pi i\sum_{s\in I}\frac{a_s}{n_s}}$$

$$= \sum_{\substack{J\subseteq\{1,\cdots,l\} \\ \sum_{t\in J}\frac{1}{m_t}=c}} (-1)^{|J|}e^{2\pi i\sum_{t\in J}\frac{b_t}{m_t}} \quad \text{for all } c\geqslant 0; \tag{3.2}$$

$$2^k \prod_{\substack{s=1 \\ s\notin S_z}}^{k} \sin\pi\frac{a_s-z}{n_s} \cdot \prod_{s\in S_z} \frac{(-1)^{[\frac{z}{n_s}]}}{n_s}$$

$$= 2^l \prod_{\substack{t=1 \\ t\notin T_z}}^{l} \sin\pi\frac{b_t-z}{m_t} \cdot \prod_{t\in T_z} \frac{(-1)^{[\frac{z}{m_t}]}}{m_t} \quad \text{for } z\in\mathbb{C} \tag{3.3}$$

where $S_z = \{1\leqslant s\leqslant k:\ z\in a_s(n_s)\}$ and $T_z = \{1\leqslant t\leqslant l:\ z\in b_t(m_t)\}$;

$$\frac{\prod_{\substack{1\leq s\leq k \\ s\notin U_z}} \Gamma\left(\frac{a_s-z}{n_s}\right) n_s^{\frac{a_s-z}{n_s}-\frac{1}{2}}}{\prod_{\substack{1\leq t\leq l \\ t\notin V_z}} \Gamma\left(\frac{b_t-z}{m_t}\right) m_t^{\frac{b_t-z}{m_t}-\frac{1}{2}}}$$

$$= (2\pi)^{\frac{k-l}{2}} \frac{\prod_{s\in U_z} [\frac{z}{n_s}]!(-1)^{[\frac{z}{n_s}]} n_s^{[\frac{z}{n_s}]-\frac{1}{2}}}{\prod_{t\in V_z} [\frac{z}{m_t}]!(-1)^{[\frac{z}{m_t}]} m_t^{[\frac{z}{m_t}]-\frac{1}{2}}} \quad \text{for } z\in\mathbb{C} \tag{3.4}$$

where $U_z = \{1\leqslant s\leqslant k:\ z\in a_s+n_s\mathbb{N}\}$ and $V_z = \{1\leqslant t\leqslant l:\ z\in b_t+m_t\mathbb{N}\}$;

$$\prod_{s=1}^{k} F\left(\frac{u}{n_s},\frac{v}{n_s},\frac{w+a_s}{n_s},1\right) = \prod_{t=1}^{l} F\left(\frac{u}{m_t},\frac{v}{m_t},\frac{w+b_t}{m_t},1\right)$$

for $u,v,w\in\mathbb{C}$ with $\text{Re}(w)>\text{Re}(u+v)$ and $w, w-u, w-v\notin -\mathbb{N}$. $\tag{3.5}$

Proof. $(3.1)\Leftrightarrow(3.2)$. Let $N=[n_1,\cdots,n_k,m_1,\cdots,m_l]$. Set

$$f(z) = \prod_{s=1}^{k}\left(1-z^{N/n_s}e^{2\pi ia_s/n_s}\right) \quad \text{and} \quad g(z) = \prod_{t=1}^{l}\left(1-z^{N/m_t}e^{2\pi ib_t/m_t}\right).$$

Clearly any zero of $f(z)$ or $g(z)$ is an Nth root of unity. For each $a\in\mathbb{Z}$, $e^{2\pi ia/N}$ is a zero of $f(z)$ with multiplicity $w_A(-a)$, and a zero of $g(z)$ with multiplicity $w_B(-a)$. By Viéte's theorem, we have the identity

$f(z) = g(z)$ if and only if $w_A = w_B$. Note that $f(z) = g(z)$ if and only if

$$\sum_{I \subseteq \{1, \cdots, k\}} (-1)^{|I|} z^{\sum_{s \in I} N/n_s} e^{2\pi i \sum_{s \in I} a_s/n_s}$$
$$= \sum_{J \subseteq \{1, \cdots, l\}} (-1)^{|J|} z^{\sum_{t \in J} N/m_t} e^{2\pi i \sum_{t \in J} b_t/m_t}.$$

By comparing the coefficients of powers of z, we find that $w_A = w_B$ if and only if

$$\sum_{\substack{I \subseteq \{1, \cdots, k\} \\ \sum_{s \in I} N/n_s = a}} (-1)^{|I|} e^{2\pi i \sum_{s \in I} a_s/n_s} = \sum_{\substack{J \subseteq \{1, \cdots, k\} \\ \sum_{t \in J} N/m_t = a}} (-1)^{|J|} e^{2\pi i \sum_{t \in J} b_t/m_t}$$

for all $a = 0, 1, 2, \cdots$. This proves the equivalence of (3.1) and (3.2).

(3.1) $\Rightarrow$ (3.3),(3.4). We can view the multiplicative group $\mathbb{C}^*$ as a $\mathbb{Z}$-module with the scalar product $\langle m, z \rangle \mapsto z^m$. By Theorem 1.1, for any $z \in \mathbb{C}$ we have

$$\prod_{s=1}^{k} S\left(\frac{a_s - z}{n_s}, n_s\right) = \prod_{t=1}^{l} S\left(\frac{b_t - z}{m_t}, m_t\right)$$

and

$$\prod_{s=1}^{k} \gamma\left(\frac{a_s - z}{n_s}, n_s\right) = \prod_{t=1}^{l} \gamma\left(\frac{b_t - z}{m_t}, m_t\right).$$

Apparently $|S_z| = |T_z|$ and $|U_z| = |V_z|$. If $n \in \mathbb{Z}^+$, $a \in R(n)$ and $z \in a(n)$, then $-\frac{a-z}{n} = \frac{z-a}{n} = [\frac{z}{n}]$. Therefore (3.3) and (3.4) follow.

(3.3)$\Rightarrow$(3.1), and (3.4)$\Rightarrow$(3.1). For $n \in \mathbb{Z}^+$ and $x \in (-\infty, 1) \setminus \mathbb{Z}$, we put

$$f_1(x, n) = \log|2 \sin \pi x|, \quad f_2(x, n) = \log|\Gamma(x)| + \left(x - \frac{1}{2}\right) \log n - \frac{1}{2}\log(2\pi).$$

Let $j \in \{1, 2\}$. Then $\lim_{x \to -m} f_j(x, 1) = \infty$ for all $m \in \mathbb{N}$. Let $n \in \mathbb{Z}^+$. Then $f_j(x, n)$ is continuous for $x \in (-\infty, 1) \setminus \mathbb{Z}$. When $x < 1$ and $x \notin \mathbb{Z}$,

$$\sum_{r=0}^{n-1} f_1\left(\frac{x+r}{n}, n\right) = \log\left|\prod_{r=0}^{n-1}\left(2 \sin \pi \frac{x+r}{n}\right)\right| = \log|2 \sin \pi x| = f_1(x, 1)$$

and

$$\sum_{r=0}^{n-1} f_2\left(\frac{x+r}{n}, n\right) = \log\left|\prod_{r=0}^{n-1}\left(\Gamma\left(\frac{x+r}{n}\right)\frac{n^{\frac{x+r}{n}-\frac{1}{2}}}{\sqrt{2\pi}}\right)\right|$$

$$= \log\left|\frac{\Gamma(x)}{\sqrt{2\pi}}\right| = f_2(x,1).$$

If

$$\prod_{s=1}^{k}\left(2\sin\pi\frac{x+a_s}{n_s}\right)\cdot\prod_{t=1}^{l}\left(2\sin\pi\frac{x+b_t}{m_t}\right)^{-1} = 1$$

for $x \in (-\infty, 1) \setminus \mathbb{Z}$, or

$$\prod_{s=1}^{k}\left(\Gamma\left(\frac{x+a_s}{n_s}\right)\frac{n_s^{\frac{x+a_s}{n_s}-\frac{1}{2}}}{\sqrt{2\pi}}\right)\cdot\prod_{t=1}^{l}\left(\Gamma\left(\frac{x+b_t}{m_t}\right)\frac{m^{\frac{x+b_t}{m_t}-\frac{1}{2}}}{\sqrt{2\pi}}\right)^{-1} = 1$$

for $x \in (-\infty, 1) \setminus \mathbb{Z}$, then for $j = 1$ or 2 we have

$$\sum_{s=1}^{k} f_j\left(\frac{x+a_s}{n_s}, n_s\right) - \sum_{t=1}^{l} f_j\left(\frac{x+b_t}{m_t}, m_t\right) = 0 \quad \text{for all } x < 1 \text{ with } x \notin \mathbb{Z},$$

therefore $\{\langle 1, a_1, n_1\rangle, \cdots, \langle 1, a_k, n_k\rangle, \langle -1, b_1, m_1\rangle, \cdots, \langle -1, b_l, m_l\rangle\} \sim \emptyset$ by Lemma 3.1. So, each of (3.3) and (3.4) implies (3.1).

(3.1) $\Rightarrow$ (3.5). Let u, v, w be complex numbers with $\text{Re}(w) > \text{Re}(u+v)$ and $w, w - u, w - v \notin -\mathbb{N}$. By Example 2.2(iii), $F_{u,v,w}$ is a uniform map into the multiplicative group $\mathbb{C}^*$. Note that $\langle 0, 1\rangle \in D_{u,v,w}$. Since $A \sim B$, applying Theorem 1.1 we get that

$$\prod_{s=1}^{k} F_{u,v,w}\left(\frac{a_s}{n_s}, n_s\right) = \prod_{t=1}^{l} F_{u.v.w}\left(\frac{b_t}{m_t}, m_t\right),$$

i.e.,

$$\prod_{s=1}^{k} F\left(\frac{u}{n_s}, \frac{v}{n_s}, \frac{w+a_s}{n_s}, 1\right) = \prod_{t=1}^{l} F\left(\frac{u}{m_t}, \frac{v}{m_t}, \frac{w+b_t}{m_t}, 1\right).$$

(3.5) $\Rightarrow$ (3.1). Let $x \in \mathbb{R} \setminus \mathbb{Z}$. By a known formula in [Ba], if $n \in \mathbb{Z}^+$ and $a \in R(n)$ then

$$F\left(\frac{\frac{1}{2}-x}{n}, \frac{x-\frac{1}{2}}{n}, \frac{\frac{1}{2}+a}{n}, 1\right) = \frac{\Gamma\left(\frac{\frac{1}{2}+a}{n}\right)\Gamma\left(\frac{\frac{1}{2}+a}{n} - \frac{\frac{1}{2}-x}{n} - \frac{x-\frac{1}{2}}{n}\right)}{\Gamma\left(\frac{\frac{1}{2}+a}{n} - \frac{\frac{1}{2}-x}{n}\right)\Gamma\left(\frac{\frac{1}{2}+a}{n} - \frac{x-\frac{1}{2}}{n}\right)}$$

$$= \frac{\Gamma^2\left(\frac{1+2a}{2n}\right)}{\Gamma\left(\frac{x+a}{n}\right)\Gamma\left(\frac{1-x+a}{n}\right)}.$$

So, by (3.5) we have

$$
\prod_{s=1}^{k} \frac{\Gamma^2\left(\frac{1+2a_s}{2n_s}\right)}{\Gamma\left(\frac{x+a_s}{n_s}\right)\Gamma\left(\frac{1-x+a_s}{n_s}\right)} = \prod_{t=1}^{l} \frac{\Gamma^2\left(\frac{1+2b_t}{2m_t}\right)}{\Gamma\left(\frac{x+b_t}{m_t}\right)\Gamma\left(\frac{1-x+b_t}{m_t}\right)},
$$

i.e.,

$$
\frac{\prod_{s=1}^{k}\Gamma\left(\frac{x+a_s}{n_s}\right)}{\prod_{t=1}^{l}\Gamma\left(\frac{x+b_t}{n_s}\right)} = \frac{\prod_{s=1}^{k}\left(\Gamma^2\left(\frac{1+2a_s}{2n_s}\right)/\Gamma\left(\frac{1-x+a_s}{n_s}\right)\right)}{\prod_{t=1}^{l}\left(\Gamma^2\left(\frac{1+2b_t}{2m_t}\right)/\Gamma\left(\frac{1-x+b_t}{m_t}\right)\right)}. \qquad (*)
$$

Note that $\Gamma(1+z) = z\Gamma(z) = z(z-1)\cdots(z-n)\Gamma(z-n)$ for $n \in \mathbb{N}$ and $z \notin \mathbb{Z}$. Let $m \in \mathbb{N}$. Then

$$
\frac{\prod_{\substack{1\leqslant s\leqslant k \\ n_s|m-a_s}}\left(\Gamma\left(1+\frac{x+m}{n_s}\right)/\prod_{j=0}^{[m/n_s]}\left(\frac{x+m}{n_s}-j\right)\right)}{\prod_{\substack{1\leqslant t\leqslant l \\ m_t|m-b_t}}\left(\Gamma\left(1+\frac{x+m}{m_t}\right)/\prod_{j=0}^{[m/m_t]}\left(\frac{x+m}{m_t}-j\right)\right)}
$$

$$
= \frac{\prod_{\substack{1\leqslant s\leqslant k \\ n_s|m-a_s}}\Gamma\left(\frac{x+a_s}{n_s}\right)}{\prod_{\substack{1\leqslant t\leqslant l \\ m_t|m-b_t}}\Gamma\left(\frac{x+b_t}{m_t}\right)}. \qquad (*)
$$

Letting $x \to -m$ we obtain from $(\star)$ and $(*)$ that

$$
\prod_{\substack{1\leqslant s\leqslant k \\ n_s|m-a_s}}\left(\frac{x+m}{n_s}\right)^{-1} \Big/ \prod_{\substack{1\leqslant t\leqslant l \\ m_t|m-b_t}}\left(\frac{x+m}{m_t}\right)^{-1} \longrightarrow c \text{ for some } c \in \mathbb{C}^*.
$$

Thus $(x+m)^{w_B(m)-w_A(m)}$ tends to a nonzero number as $x \to -m$. This shows that $w_A(m) = w_B(m)$. We are done.

Remark 3.1. (a) When (3.1) holds with $n_1 \leqslant \cdots \leqslant n_{k-1} < n_k$ and $m_1 \leqslant \cdots \leqslant m_l$, by taking $c = 1/n_k$ in (3.2) we obtain that $1/n_k = \sum_{t\in J} 1/m_t$ for some $J \subseteq \{1, \cdots, l\}$ and hence $n_k \leqslant m_l$. From this we can deduce that if $A = \{a_s(n_s)\}_{s=1}^{k}$ and $B = \{b_t(m_t)\}_{t=1}^{l}$ are equivalent systems with $n_1 < \cdots < n_k$ and $m_1 < \cdots < m_l$ then $A = B$ (i.e. $k = l$, $n_s = m_s$ and $a_s = b_s$ for $s = 1, \cdots, k$), this uniqueness result was discovered by S. K. Stein [S] in a weaker form and presented by Znám [Z2] in the current form. When $B = \{b_t(m_t)\}_{t=1}^{l}$ is the system of m copies of $0(1)$, the right hand side of the formula in (3.2) turns out to be

$$
\sum_{\substack{J\subseteq\{1,\cdots,m\} \\ |J|=\sum_{t\in J}1=c}} (-1)^{|J|} = \begin{cases} (-1)^n\binom{m}{n} & \text{if } c = n \text{ for some } n = 0, 1, \cdots, m, \\ 0 & \text{otherwise.} \end{cases}
$$

Thus (1.1) forms an exact m-cover of $\mathbb{Z}$ if and only if

$$\sum_{\substack{I \subseteq \{1,\cdots,k\} \\ \sum_{s \in I} \frac{1}{n_s} = n}} (-1)^{|I|} e^{2\pi i \sum_{s \in I} \frac{a_s}{n_s}} = (-1)^n \binom{m}{n} \quad \text{for } n = 1, \cdots, m$$

and

$$\sum_{\substack{I \subseteq \{1,\cdots,k\} \\ \sum_{s \in I} \frac{1}{n_s} = \sum_{s \in J} \frac{1}{n_s}}} (-1)^{|I|} e^{2\pi i \sum_{s \in I} \frac{a_s}{n_s}} = 0$$

$$\text{for any } J \subseteq \{1, \cdots, k\} \text{ with } \sum_{s \in J} \frac{1}{n_s} \notin \mathbb{Z}.$$

For connections between covers of $\mathbb{Z}$ and Egyptian fractions, the reader may see [Su3], [Su4], [Su5], [Su6], [Su7].

(b) By the proof of '(3.3)⇒(3.1)' and '(3.4)⇒(3.1)', (3.1) is valid if the formula in (3.3) or (3.4) holds for any $z \in \mathbb{C} \setminus \mathbb{Z}$. Thus (3.1) has the following two equivalent forms by analytic continuation.

$$2^{k-l} \prod_{s=1}^{k} \sin \pi \frac{a_s - z}{n_s} = \prod_{t=1}^{l} \sin \pi \frac{b_t - z}{m_t} \quad \text{for all } z \in \mathbb{C}; \tag{†}$$

$$\prod_{s=1}^{k} \Gamma \left(\frac{z + a_s}{n_s} \right) n_s^{\frac{z+a_s}{n_s} - \frac{1}{2}}$$

$$= (2\pi)^{\frac{k-l}{2}} \prod_{t=1}^{l} \Gamma \left(\frac{z + b_t}{m_t} \right) m_t^{\frac{z+b_t}{m_t} - \frac{1}{2}} \quad \text{for } z \in \mathbb{C} \setminus -\mathbb{N}. \tag{‡}$$

That (†) implies (3.1) was first obtained by Stein [S] in the case $B = \{0(1)\}$; the converse in general case was noticed by the author in 1989 as a consequence of theorems in [Su1], when $B = \{b_t(m_t)\}_{t=1}^{l}$ is simply $\{0(1)\}$ it was found repeatedly by J. Beebee [Be1] in 1991. That (3.1) implies (‡) is essentially Corollary 3 of Sun [Su1] which extends the Gauss multiplication formula, the converse was mentioned in [Su1] as a conjecture. By the above, (1.1) is an exact m-cover of $\mathbb{Z}$ (i.e. $A \sim \{\langle m, 0, 1 \rangle\}$) if and only if

$$\prod_{s=1}^{k} \left(\Gamma \left(\frac{z + a_s}{n_s} \right) n_s^{\frac{z+a_s}{n_s} - \frac{1}{2}} \right) = (2\pi)^{\frac{k-m}{2}} \Gamma(z)^m \quad \text{for all } z \in \mathbb{C} \setminus -\mathbb{N}. \tag{3.6}$$

Consequently, if (1.1) is an exact 1-cover of $\mathbb{Z}$ with $a_1 = 0$, then

$$\frac{n_1^{\frac{z}{n_1}-\frac{1}{2}} \prod_{s=2}^{k} \left(\Gamma\left(\frac{z+a_s}{n_s}\right) n_s^{\frac{z+a_s}{n_s}-\frac{1}{2}} \right)}{n_1^{-\frac{1}{2}} \prod_{s=2}^{k} \left(\Gamma\left(\frac{a_s}{n_s}\right) n_s^{\frac{a_s}{n_s}-\frac{1}{2}} \right)} = \frac{(2\pi)^{\frac{k-1}{2}} \Gamma(z)/\Gamma\left(\frac{z}{n_1}\right)}{(2\pi)^{\frac{k-1}{2}} \lim_{z' \to 0} \frac{\Gamma(z')z'}{\Gamma\left(\frac{z'}{n_1}\right)\frac{z'}{n_1} \cdot n_1}}$$

for $z \neq 0, -1, -2, \cdots$, i.e.,

$$\Gamma(z) = \frac{\Gamma\left(\frac{z}{n_1}\right)}{n_1^{1-\frac{z}{n_1}}} \prod_{s=2}^{k} \frac{\Gamma\left(\frac{z+a_s}{n_s}\right)}{n_s^{-\frac{z}{n_s}} \Gamma\left(\frac{a_s}{n_s}\right)} \quad \text{for all } z \in \mathbb{C} \setminus -\mathbb{N} \qquad (3.7)$$

(by Theorem 3.1 we directly have $\prod_{s=2}^{k} \Gamma(\frac{a_s}{n_s}) n_s^{a_s/n_s - 1/2} = (2\pi)^{(k-1)/2} n_1^{-1/2}$). In 1994 Beebee [Be3] showed that the relative formula (3.7) holds for any exact 1-cover (1.1) of $\mathbb{Z}$, as we have seen this was actually rooted in [Sul] published in 1989. The new contribution of [Be3] is that if (3.7) holds then (1.1) must be an exact 1-cover of $\mathbb{Z}$, however we have a simpler equivalent form (‡) of (3.1).

Now we give

Theorem 3.2. *For every* $s = 1, \cdots, k$ *we let* $\lambda_s \in \mathbb{C}$, $n_s \in \mathbb{Z}^+$ *and* $a_s \in R(n_s)$. *Then the following statements are equivalent:*

$$\mathcal{A} = \{\langle \lambda_s, a_s, n_s \rangle\}_{s=1}^{k} \sim \emptyset; \qquad (3.8)$$

$$\sum_{s=1}^{k} \frac{(\lambda_s^{(t)})^2}{n_s} = -2 \sum_{\substack{1 \leqslant i < j \leqslant k \\ \gcd(n_i,n_j)|a_i-a_j}} \frac{\lambda_i^{(t)}\lambda_j^{(t)}}{[n_i, n_j]} \quad \text{for } t = 1, 2 \qquad (3.9)$$

where $\lambda_s^{(1)} = \text{Re}\lambda_s$ *and* $\lambda_s^{(2)} = \text{Im}\lambda_s$ *for any* $s = 1, \cdots, k$;

$$\sum_{s=1}^{k} \lambda_s \left(\left[\frac{x + ma_s}{n_s}\right] - \frac{m-1}{2} \right) = 0 \quad \text{for all } x \in \mathbb{Z} \qquad (3.10)$$

where m *is an integer prime to the moduli* $n_1, \cdots, n_k$;

$$\sum_{s=1}^{k} \lambda_s \cot_m \left(\frac{z + a_s}{n_s}, n_s \right) = 0 \quad \text{for all } z \in \mathbb{C} \qquad (3.11)$$

where m *is a nonnegative integer and* $\cot_z^{(m)}$ *is as in Example 2.4(ii);*

$$\sum_{t=1}^{k} \lambda_t \zeta_s \left(\frac{z + a_t}{n_t}, n_t \right) = 0 \quad \text{for all } z \in \mathbb{C} \setminus (-\infty, 0] \qquad (3.12)$$

where s *is a complex number not in* $-\mathbb{N}$ *and* ζ_s *is as in Example 2.3(ii).*

Proof. (3.8) ⇔ (3.9). For $t = 1, 2$ and $N = [n_1, \cdots, n_k]$ we can easily check that

$$\frac{1}{N} \sum_{r=0}^{N-1} \left(\sum_{\substack{1 \leqslant s \leqslant k \\ n_s | r - a_s}} \lambda_s^{(t)} \right)^2 = \sum_{s=1}^{k} \frac{(\lambda_s^{(t)})^2}{n_s} + 2 \sum_{\substack{1 \leqslant i < j \leqslant k \\ a_i(n_i) \cap a_j(n_j) \neq \emptyset}} \frac{\lambda_i^{(t)} \lambda_j^{(t)}}{[n_i, n_j]}.$$

So (3.8) and (3.9) are equivalent.

(3.8) ⇔ (3.10). Let $\mathcal{B} = \{\langle \lambda_s, b_s, n_s \rangle\}_{s=1}^{k}$ where b_s is the least nonnegative residue of $ma_s \bmod n_s$. As m is prime to $N = [n_1, \cdots, n_k]$, any integer z can be written in the form $mu + Nv$ (with $u, v \in \mathbb{Z}$) and hence $w_{\mathcal{B}}(z) = w_{\mathcal{B}}(mu) = w_{\mathcal{A}}(u)$. Thus $\mathcal{B} \sim \emptyset$ if $\mathcal{A} \sim \emptyset$. Clearly $f(x, y) = x - \frac{1}{2}$ and $g(x, y) = f(x, y) - [\ ](x, y) = \{x\} - \frac{1}{2}$ over $\mathbb{R} \times \mathbb{R}$ are uniform maps into $\mathbb{R}$. If $\mathcal{A} \sim \emptyset$ and $x \in \mathbb{R}$, then

$$\sum_{s=1}^{k} \lambda_s \left(\left[\frac{x + ma_s}{n_s} \right] - \frac{m-1}{2} \right)$$

$$= m \sum_{s=1}^{k} \lambda_s \left(\frac{x/m + a_s}{n_s} - \frac{1}{2} \right) - \sum_{s=1}^{k} \lambda_s \left(\left\{ \frac{x + b_s}{n_s} \right\} - \frac{1}{2} \right) = 0.$$

If (3.10) holds, then so does (3.8), because for any $x \in \mathbb{Z}$ we have

$$w_{\mathcal{A}}(x) = \sum_{\substack{1 \leqslant s \leqslant k \\ n_s | ma_s - mx}} \lambda_s = \sum_{s=1}^{k} \lambda_s \left(\left[\frac{ma_s - mx}{n_s} \right] - \left[\frac{ma_s - mx - 1}{n_s} \right] \right)$$

$$= \sum_{s=1}^{k} \lambda_s \left(\left[\frac{ma_s - mx}{n_s} \right] - \frac{m-1}{2} \right)$$

$$- \sum_{s=1}^{k} \lambda_s \left(\left[\frac{ma_s - mx - 1}{n_s} \right] - \frac{m-1}{2} \right)$$

$$= 0.$$

(3.8) ⇔ (3.11). Since $\cot_m$ is a uniform map into $\mathbb{C}$, (3.8) implies (3.11) by Theorem 1.1. By Example 2.4(ii),

$$\cot_m(z, 1) = \frac{m!}{\pi^{m+1}} \sum_{k=-\infty}^{+\infty} \frac{1}{(k + z)^{m+1}} \quad \text{for } z \in \mathbb{C} \setminus \mathbb{Z}.$$

Obviously $\cot_m(z, 1) \to \infty$ as z tends to an integer. If $n \in \mathbb{Z}^+$, then $\cot_m(z, n) = \cot_m(z, 1)/n^{m+1}$ is continuous for $z \in \mathbb{C} \setminus \mathbb{Z}$. In the light of Lemma 3.1, (3.11) implies (3.8).

(3.8) $\Leftrightarrow$ (3.12). By Example 2.3(ii), ζ_s is a uniform map into $\mathbb{C}$. So (3.12) is implied by (3.8).

As $s \neq 0, -1, -2, \cdots$, by Example 2.3(ii) we have

$$\frac{d}{dv}\zeta(s,v) = -s\zeta(s+1,v), \quad \frac{d}{dv}\zeta(s+1,v) = (-s-1)\zeta(s+2,v), \quad \cdots$$

in the region $\mathbb{C} \setminus (-\infty, 0]$. Let $m = [2 - \operatorname{Re}(s)]$ if $\operatorname{Re}(s) \leqslant 1$, and $m = 0$ if $\operatorname{Re}(s) > 1$. Put $\bar{s} = s + m$. Then $\operatorname{Re}(\bar{s}) > 1$ and

$$\frac{d^m}{dv^m}\zeta(s,v) = \prod_{0 \leqslant j < m} (-s-j) \cdot \zeta(\bar{s},v) \quad \text{for } v \in \mathbb{C} \setminus (-\infty, 0].$$

If $n \in \mathbb{Z}^+$ and $a \in R(n)$, then

$$\frac{d^m}{dz^m}\zeta_s\left(\frac{z+a}{n}, n\right) = (-1)^m s(s+1)\cdots(\bar{s}-1)\zeta_{\bar{s}}\left(\frac{z+a}{n}, n\right)$$

$$\text{for } z \in \mathbb{C} \setminus (-\infty, 0].$$

Apparently $\zeta_{\bar{s}}(z,1) = \zeta(\bar{s},z) \to \infty$ as z tends to an integer in $-\mathbb{N}$.

Let's assume (3.12). Then

$$\sum_{t=1}^{k} \lambda_t \zeta_{\bar{s}}\left(\frac{z+a_t}{n_t}, n_t\right) = \frac{(-1)^m}{s(s+1)\cdots(\bar{s}-1)} \cdot \frac{d^m}{dz^m} \sum_{t=1}^{k} \lambda_t \zeta_s\left(\frac{z+a_t}{n_t}, n_t\right)$$

$$= 0$$

for $z \in \mathbb{C} \setminus (-\infty, 0]$, and hence by analytic continuation

$$\sum_{t=1}^{k} \lambda_t \zeta_{\bar{s}}\left(\frac{z+a_t}{n_t}, n_t\right) = 0 \quad \text{for all } z \neq 0, -1, -2, \cdots.$$

Applying Lemma 3.1 we then get (3.8).

So far we have completed the proof of Theorem 3.2.

Remark 3.2. In the case $m = 1$, that (3.8) implies (3.10) was first realized by the author [Sul] in 1989 and later refound by Porubský [P4] in 1994. If (3.8) holds, then the formula in (3.10) in the case $m = 1$ and $x = [n_1, \cdots, n_k]$ yields the equality $\sum_{s=1}^{k} \frac{\lambda_s}{n_s} = 0$. In 1989 the author [Sul] obtained Theorem 1.1 and noted that $f(x,y) = \frac{1}{y}\cot \pi x$ over $(\mathbb{C} \setminus \mathbb{Z}) \times \mathbb{C}^*$ is a uniform map into $\mathbb{C}$, thus for any exact 1-cover (1.1) we have

$$\sum_{s=1}^{k} \frac{1}{n_s} \cot\left(\pi\frac{z+a_s}{n_s}\right) = \cot \pi z \quad \text{and} \quad \sum_{s=1}^{k} \frac{1}{n_s^2} \csc^2\left(\pi\frac{z+a_s}{n_s}\right) = \csc^2(\pi z)$$

for all $z \in \mathbb{C} \setminus \mathbb{Z}$, this was also given by Beebee [Be1] in 1991.

Corollary 3.1. *Let* (1.1) *be a finite system of arithmetic sequences, and m a positive integer. Then*
(i) (1.1) *forms an exact m-cover of* $\mathbb{Z}$ *if and only if*

$$\sum_{s=1}^{k} \frac{1}{n_s} = m \quad and \quad \sum_{\substack{1 \leqslant i < j \leqslant k \\ (n_i, n_j) | a_i - a_j}} \frac{1}{[n_i, n_j]} = \frac{m(m-1)}{2}, \qquad (3.13)$$

also (1.1) *forms an exact m-cover of* $\mathbb{Z}$ *if and only if*

$$\sum_{s=1}^{k} \frac{1}{n_s} = m \quad and \quad \sum_{s=1}^{k} \left[\frac{a + na_s}{n_s} \right] = am + \frac{(k-m)(n-1)}{2} \quad for\ a \in R(N)$$

$$(3.14)$$

where n is any fixed integer prime to $N = [n_1, \cdots, n_k]$.
(ii) *Suppose that* (1.1) *is an m-cover of* $\mathbb{Z}$. *Then*

$$\sum_{t=1}^{k} n_t^{-s} \zeta\left(s, \frac{x + a_t}{n_t}\right) \geqslant m\zeta(s, x) \quad for\ s > 1 \quad and\ x > 0,$$

and for any $n \in \mathbb{Z}^+$ *we have*

$$\sum_{\substack{1 \leqslant s \leqslant k \\ x + a_s \notin n_s \mathbb{Z}}} \frac{1}{n_s^{2n}} \cot^{(2n-1)}\left(\pi \frac{x + a_s}{n_s}\right)$$

$$\leqslant \begin{cases} m \cot^{(2n-1)}(\pi x) & if\ x \in \mathbb{R} \setminus \mathbb{Z}, \\ (-4)^n \frac{B_{2n}}{2n}\left(m - \sum_{\substack{1 \leqslant s \leqslant k \\ n_s | x + a_s}} \frac{1}{n_s^{2n}}\right) & if\ x \in \mathbb{Z}. \end{cases}$$

$$(3.16)$$

Proof. Let $\mathcal{A} = \{\langle 1, a_1, n_1\rangle, \cdots, \langle 1, a_k, n_k\rangle, \langle -m, 0, 1\rangle\}$.
i) Clearly $\mathcal{A} = \{a_s(n_s)\}_{s=1}^{k}$ forms an exact m-cover if and only if $\mathcal{A} \sim \emptyset$. If $\mathcal{A} \sim \emptyset$, then $\sum_{s=1}^{k} 1/n_s - m = 0$ by Remark 3.2. (That $\sum_{s=1}^{k} 1/n_s = m$ for any exact m-cover (1.1) is actually a well-known result, it can be found in [P2].)
By the equivalence of (3.8) and (3.9), $\mathcal{A} \sim \emptyset$ if and only if

$$\sum_{s=1}^{k} \frac{1}{n_s} + \frac{(-m)^2}{1} = -2\left(\sum_{\substack{1 \leqslant i < j \leqslant k \\ (n_i, n_j) | a_i - a_j}} \frac{1}{[n_i, n_j]} + \sum_{\substack{1 \leqslant s \leqslant k \\ (n_s, 1) | a_s - 0}} \frac{-m}{n_s} \right). \quad (\triangle)$$

Under the condition $\sum_{s=1}^{k} 1/n_s = m$, $(\triangle)$ reduces to the latter equality in (3.13). So, $\mathcal{A} \sim \emptyset$ if and only if (3.13) holds.

By Theorem 3.2, $\mathcal{A} \sim \emptyset$ if and only if for all $x \in \mathbb{Z}$ we have

$$\sum_{s=1}^{k} \left(\left[\frac{x + n a_s}{n_s} \right] - \frac{n-1}{2} \right) - m \left(\left[\frac{x + n \cdot 0}{1} \right] - \frac{n-1}{2} \right) = 0,$$

i.e.

$$\sum_{s=1}^{k} \left[\frac{x + n a_s}{n_s} \right] = mx + (k - m)\frac{n-1}{2}.$$

Any integer x can be written in the form $a + qN$ where $a \in R(n)$ and $q \in \mathbb{Z}$, thus the last equality holds for all $x \in \mathbb{Z}$ if and only if (3.14) is valid. This ends the proof of part (i).

ii) As $w_A(x) \geqslant m$ for all $x \in \mathbb{Z}$, $w_A(x) \geqslant 0$ for any $x \in \mathbb{Z}$. Obviously $\mathcal{A} \sim \{\langle w_A(r), r, N \rangle\}_{r=0}^{N-1}$. When $s > 1$ and $x > 0$, by Theorem 3.2

$$\sum_{t=1}^{k} \zeta_s \left(\frac{x + a_t}{n_t}, n_t \right) - m \zeta_s \left(\frac{x + 0}{1}, 1 \right) = \sum_{r=0}^{N-1} w_A(r) \zeta_s \left(\frac{x + r}{N}, N \right),$$

therefore

$$\sum_{t=1}^{k} \frac{1}{n_t^s} \zeta \left(s, \frac{x + a_t}{n_t} \right) - m \zeta(s, x) = \sum_{r=0}^{N-1} \frac{w_A(r)}{N^s} \zeta \left(s, \frac{x + r}{N} \right) \geqslant 0$$

since $\zeta(s, \frac{x+r}{N}) = \sum_{j=0}^{\infty} (j + \frac{x+r}{N})^{-s} > 0$.

By Example 2.4(ii),

$$\cot_{2n-1}(x, N) = \frac{(2n - 1)!}{(\pi N)^{2n}} \sum_{\substack{j=-\infty \\ j \neq -x}}^{+\infty} \frac{1}{(j + x)^{2n}} > 0 \quad \text{for all } x \in \mathbb{R}.$$

As in the last paragraph, now we have

$$\sum_{s=1}^{k} \cot_{2n-1} \left(\frac{x + a_s}{n_s}, n_s \right) - m \cot_{2n-1}(x, 1) \geqslant 0 \quad \text{for any } x \in \mathbb{R}.$$

Clearly this is equivalent to (3.16). We are done.

Remark 3.3. (3.16) in the case $n = 1$ gives the following inequality:

$$\sum_{\substack{1 \leqslant s \leqslant k \\ x + a_s \notin n_s \mathbb{Z}}} \frac{1}{n_s^2} \csc^2 \left(\pi \frac{x + a_s}{n_s} \right) \geqslant \begin{cases} m \csc^2(\pi x) & \text{if } x \in \mathbb{R} \setminus \mathbb{Z}, \\ \frac{1}{3} (m - \sum_{\substack{1 \leqslant s \leqslant k \\ n_s | x + a_s}} \frac{1}{n_s^2}) & \text{if } x \in \mathbb{Z}. \end{cases}$$

$$(3.17)$$

Let $\mathcal{A} = \{\langle \lambda_s, a_s, n_s \rangle\}_{s=1}^k \in S(\mathbb{C})$ and $z \in \mathbb{C}$. For

$$y \in (-2\pi/\max\{n_1, \cdots, n_k\}, 0)$$

we have

$$- e^{yz} \sum_{n=0}^{\infty} w_{\mathcal{A}}(n)(e^y)^n = -\sum_{s=1}^k \lambda_s \frac{(e^y)^{z+a_s}}{1 - (e^y)^{n_s}}$$

$$= \sum_{s=1}^k \frac{\lambda_s}{n_s y} \cdot \frac{n_s y e^{n_s y \cdot \frac{z+a_s}{n_s}}}{e^{n_s y} - 1} = \sum_{s=1}^k \frac{\lambda_s}{n_s y} \sum_{n=0}^{\infty} \frac{B_n(\frac{z+a_s}{n_s})}{n!}(n_s y)^n$$

where $B_n(x)$ is the Bernoulli polynomial of degree n. So, $\mathcal{A} \sim \emptyset$ if and only if

$$\sum_{s=1}^k \lambda_s n_s^{n-1} B_n \left(\frac{z + a_s}{n_s} \right) = 0 \quad \text{for all } n = 0, 1, 2, \cdots. \qquad (3.18)$$

For the system $\mathcal{B} = \{\langle 1, a_1, n_1 \rangle, \cdots, \langle 1, a_k, n_k \rangle, \langle -m, 0, 1 \rangle\}$, this was proved by A. S. Fraenkel [F1], [F2] in the case $m = 1$ and $z = 0$, by Beebee [Be2] in the case $m = 1$, and by Porubský [P2] in the case $m \in \mathbb{Z}^+$ and $z = 0$. See also Porubský [P1], [P3] and Znám [Z1] for the case $z = 0$ with the weights 1 in $\mathcal{B}$ replaced by real weights. In 1994 Porubský [P4] essentially established the above general result. However, before the works of Beebee [Be2] and Porubský [P4], in 1989 the author [Su1] proved Theorem 1.1 and observed that the function $b_n(x, y) = y^{n-1} B_n(x)$ is a uniform map into $\mathbb{C}$ for each $n \in \mathbb{N}$. In 1988 D. H. Lehmer [Le] showed that $B_n(x)$ is the only monic polynomial of degree n such that

$$\sum_{r=0}^{d-1} d^{n-1} B_n \left(\frac{x + r}{d} \right) = B_n(x) \quad \text{for all } d \in \mathbb{Z}^+.$$

For any $n \in \mathbb{N}$ clearly

$$\sum_{s=1}^k \lambda_s n_s^{n-1} B_n \left(\frac{z + a_s}{n_s} \right) = \sum_{l=0}^n \binom{n}{l} B_l \sum_{s=1}^k \lambda_s n_s^{l-1} (z + a_s)^{n-l}.$$

Thus $\mathcal{A} \sim \emptyset$ if and only if

$$\sum_{l=0}^n \binom{n}{l} B_l \sum_{s=1}^k \lambda_s n_s^{l-1} a_s^{n-l} = 0 \quad \text{for } n = 0, 1, 2, \cdots. \qquad (3.19)$$

In 1991 E. Y. Deeba and D. M. Rodriguez [DR] found this for the trivial system $\{\langle 1, 0, d \rangle, \langle 1, 1, d \rangle, \cdots, \langle 1, d-1, d \rangle, \langle -1, 0, 1 \rangle\}$ where $d \in \mathbb{Z}^+$, later

Beebee [Be2] obtained the result for system $\mathcal{B}$ with $m = 1$, and Porubský [P5] observed the generalization to $\mathcal{A} \in S(\mathbb{R})$.

For the covering equivalence between systems in $S(\mathbb{R})$, Porubský [P5] provided some characterizations involving Euler polynomials and recursions for Euler numbers.

In [Su2] the author announced several results closely related to this paper, proofs of them are presented in a recent paper [Su9].

References

[B]   H. Bass, *Generators and relations for cyclotomic units*, Nagoya Math. J. **27** (1966), 401–407. MR 34#1298.

[Ba]   H. Bateman, Higher Transcendental Functions (edited by Erdélyi et al), Vol I, McGraw-Hill, 1953, Chapters 1 and 2.

[Be1]   J. Beebee, *Some trigonometric identities related to exact covers*, Proc. Amer. Math. Soc. **112** (1991), 329–338. MR 91i:11013.

[Be2]   J. Beebee, *Bernoulli numbers and exact covering systems*, Amer. Math. Monthly **99** (1992), 946–948. MR 93i:11025.

[Be3]   J. Beebee, *Exact covering systems and the Gauss-Legendre multiplication formula for the gamma function*, Proc. Amer. Math. Soc. **120** (1994), 1061–1065. MR 94f:33001.

[C]   S. D. Chowla, *The nonexistence of nontrivial linear relations between the roots of a certain irreducible equation*, J. Number Theory **2** (1970), 120–123. MR 40#2638.

[DR]   E. Y. Deeba and D. M. Rodriguez, *Stirling's series and Bernoulli numbers*, Amer. Math. Monthly **98** (1991), 423–426. MR 92g:11025.

[E]   V. Ennola, *On relations between cyclotomic units*, J. Number Theory **4** (1972), 236–247. MR 45#8633 (E 46#1754).

[Er]   P. Erdös, *On integers of the form $2^k + p$ and some related problems*, Summa Brasil. Math. **2** (1950), 113–123. MR 13, 437.

[F1]   A. S. Fraenkel, *A characterization of exactly covering congruences*, Discrete Math. **4** (1973), 359–366. MR 47#4906.

[F2]   A. S. Fraenkel, *Further characterizations and properties of exactly covering congruences*, Discrete Math. **12** (1975), 93–100, 397. MR 51#10276.

[G]   R. K. Guy, Unsolved Problems in Number Theory (2nd edition), Springer-Verlag, New York, 1994, Sections A19, B21, E23, F13, F14.

[K]   D. S. Kubert, *The universal ordinary distribution*, Bull. Soc. Math. France **107** (1979), 179–202. MR 81b:12004.

[La]   S. Lang, Cyclotomic Fields II (Graduate Texts in Math.; 69), Springer -Verlag, New York, 1980, Ch. 17.

[Le]   D. H. Lehmer, *A new approach to Bernoulli polynomials*, Amer. Math. Monthly **95** (1988), 905–911. MR 90c:11014.

[M]   J. Milnor, *On polylogarithms, Hurwitz zeta functions, and the Kubert identities*, Enseign. Math. **29** (1983), 281–322. MR 86d:11007.

[O]   T. Okada, *On an extension of a theorem of S. Chowla*, Acta Arith. **38** (1981), 341–345. MR 83b:10014.

[P1]   Š. Porubský, *Covering systems and generating functions*, Acta Arith. **26** (1975), 223–231. MR 52#328.

[P2]   Š. Porubský, *On m times covering systems of congruences*, Acta Arith. **29** (1976), 159–169. MR 53#2884.

[P3]   Š. Porubský, *A characterization of finite unions of arithmetic sequences*, Discrete Math. **38** (1982), 73–77. MR 84a:10057.

[P4]   Š. Porubský, *Identities involving covering systems I*, Math. Slovaca **44** (1994), 153–162. MR 95f:11002.

[P5]   Š. Porubský, *Identities involving covering systems II*, Math. Slovaca **44** (1994), 555–568.

[PS]   Š. Porubský and J. Schönheim, *Covering systems of Paul Erdös: past, present and future*, preprint, 2000.

[S]   S. K. Stein, *Unions of arithmetic sequences*, Math. Ann. **134** (1958), 289–294. MR 20#17.

[Su1] Z. W. Sun, *Systems of congruences with multipliers*, Nanjing Univ. J. Math. Biquarterly **6** (1989), no. 1, 124–133. Zbl. M. 703.11002, MR 90m:11006.

[Su2] Z. W. Sun, *Several results on systems of residue classes*, Adv. in Math. (China) **18** (1989), no. 2, 251–252.

[Su3] Z. W. Sun, *On exactly m times covers*, Israel J. Math. **77** (1992), 345–348. MR 93k:11007.

[Su4] Z. W. Sun, *Covering the integers by arithmetic sequences*, Acta Arith. **72** (1995), 109–129. MR 96k:11013.

[Su5] Z. W. Sun, *Covering the integers by arithmetic sequences II*, Trans. Amer. Math. Soc. **348** (1996), 4279–4320. MR 97c:11011.

[Su6] Z. W. Sun, *Exact m-covers and the linear form $\sum_{s=1}^{k} x_s/n_s$*, Acta Arith. **81** (1997), 175–198. MR 98h:11019.

[Su7] Z. W. Sun, *On covering multiplicity*, Proc. Amer. Math. Soc. **127** (1999), 1293–1300. MR 99h:11012.

[Su8] Z. W. Sun, *Products of binomial coefficients modulo p^2*, Acta Arith. **97** (2001), 87–98.

[Su9] Z. W. Sun, *Algebraic approaches to periodic arithmetical maps*, J. Algebra **240** (2001), no. 2, 723–743.

[W]   H. Walum, *Multiplication formulae for periodic functions*, Pacific J. Math. **149** (1991), 383–396. MR 92c:11019.

[Z1]   Š. Znám, *Vector-covering systems of arithmetic sequences*, Czech. Math. J. **24** (1974), 455–461. MR 50#4520.

[Z2]   Š. Znám, *On properties of systems of arithmetic sequences*, Acta Arith. **26** (1975), 279–283. MR 51#329.

CERTAIN WORDS, TILINGS, THEIR NON-PERIODICITY, AND SUBSTITUTIONS OF HIGH DIMENSION

Jun-ichi TAMURA

3-3-7-307 Azamino Aoba-ku Yokohama 225-0011 Japan

Keywords: automaton; higher dimensional substitution, word, and tiling; non-periodicity; p-adic number

Abstract We consider certain partition of a lattice into c ($1 < c \leqq \infty$) parts, and its characteristic word $W(A; c)$ on the lattice, which are determined uniquely by c and a given square matrix A of size $s \times s$ with integer entries belonging to a class (Bdd). We give a definition of substitutions for s-dimensional words in a general situation, and then, define special ones, and automata of dimension s together with their conjugates. We show that the word $W(A; c)$ can be described by iterations of a substitution for A belonging to a subclass of (Bdd). The hermitian canonical forms of integer matrices play an important role in some cases for finding substitutions. We give a theorem which discloses a p-adic link with hermitian canonical forms. We give two definitions for periodicity: Ψ-periodicity, and Σ-periodicity, both for words and for tilings of dimension s. The non-Σ-periodicity strongly requires non-periodicity, so that it excludes some non-periodic words in the usual sense. We also consider certain Voronoi tessellations coming from a word $W(A; c)$. We show that some of the words $W(A; c)$, and the tessellations are non-Σ-periodic.

1991 Mathematics Subject Classification: 11B85, 11F85.

0.   INTRODUCTION

In this paper, we intend to give proofs for Theorems 1–5 reported without giving proofs in [3]. For completeness, we repeat all the definitions and remarks given there, and correct some errors*.

---

*The author would like to express his deep gratitude to World Scientific Co. Pte. Ltd., who returned the copyright of [3] to him.

C. Jia and K. Matsumoto (eds.), Analytic Number Theory, 303–348.
© 2002 *Kluwer Academic Publishers. Printed in the Netherlands.*

For any integers $c > 1$, $d > 1$, there exists a unique partition

$$\overset{\circ}{\underset{0 \leqq j < c}{\bigcup}} d^j \Gamma_+ = \mathbb{N} \setminus \{0\} \tag{1}$$

of the set of positive integers into c parts, where $\overset{\circ}{\cup}$ indicates a disjoint union, $m\Gamma := \{mn; \, n \in \Gamma\}$, $\mathbb{N}$ is the set of non-negative integers. Then the word $W = w_1 w_2 w_3 \cdots$ over

$$K_c := \{0, 1, 2, \ldots, c - 1\}$$

defined by $w_n := j$ $(n \in d^j \Gamma)$ becomes a non-periodic word, which is the fixed point of a substitution σ over K_c given by $\sigma(i) := 0^{d-1}(i+1)$ $(0 \leqq i < c - 1)$, $\sigma(c - 1) := 0^d$, cf. [2]. We can consider an s-dimensional version of (1), that is

$$\overset{\circ}{\underset{0 \leqq j < c}{\bigcup}} A^j \Gamma = \mathbb{Z}^s \setminus \{o\}, \quad s \geqq 1, \; c > 1, \tag{2}$$

where $\mathbb{Z}$ is the set of integers, Γ is a subset of $\mathbb{Z}^s$, A is an $s \times s$ matrix with integer entries, $A^j \Gamma$ is a set $\{A^j x; \, x \in \Gamma\}$, $o := {}^T(0, \ldots, 0) \in \mathbb{Z}^s$, and the "$T$" indicates the transpose. We also consider (2) with $c = \infty$, which becomes a partition of the lattice into infinite parts. We define a word W for a partition (2):

$$W = W(A; c) = (w_x)_{x \in \mathbb{Z}^s} \in \tilde{K}_c^{\mathbb{Z}^s}, \; w_x := j \text{ if } x \in A^j \Gamma \; (w_o := \infty),$$

where
$$\tilde{K}_c := K_c \cup \{\infty\} \text{ for } 1 < c \leqq \infty \; (K_\infty := \mathbb{N}).$$

The word $W(A; c)$ will be referred to as the characteristic word of a partition (2).

We remark that a set Γ satisfying (2) can be considered as one of the discrete versions of a self-similar set for regular matrix A. For any given $1 < c \leqq \infty$, the partition (2) with $s = 1$ exists and uniquely determined by numbers c, a if and only if $A = (a)$, $a \in \mathbb{Z} \setminus \{0, \pm 1\}$. Note that, in the case of $s = 1$, we get (1) (resp., (2)) from (2) (resp., (1)) by setting $\Gamma_+ = \Gamma \cap \mathbb{N}$ (resp., $\Gamma = \Gamma_+ \cup (-\Gamma_+)$). Hence, we get nothing new for $s = 1$. But, the situation for $s > 1$ turns out to be quite changed from that for $s = 1$. For instance, if $s > 1$, for some matrices A having an n-th root $\zeta \neq 1$ as their eigenvalues, partitions (2) do exist, but are not uniquely determined; while a partition (2) is uniquely determined even for some singular matrices A. In such a singular case, we can consider also partitions (2) with $\mathbb{Z}^s$ in place of $\mathbb{Z}^s \setminus \{o\}$. So, there appear some new phenomena that do never occur in the case of $s = 1$. Among them, Theorem 6 related to p-adic numbers may be of interest.

The main objective of this paper is to investigate words and tilings of higher dimension, and their non-periodicity together with its new definitions. We give a definition of substitutions for words of dimension s in a general situation, and then, define special ones together with their conjugates, and automata of dimension s. We describe the word $W(A; c)$ $(1 < c < \infty)$ by a substitution of dimension s for A belonging to a class of matrices. We give new definitions for non-periodicity for words, and tilings of dimension s. One of them requires non-periodicity in a very strong sense. We shall see that some of the words $W(A; c)$ are non-periodic in the strong sense.

We deote by $M(s; \mathbb{Z})$ the set of matrices of size $s \times s$ with integer entries; by $M_r(s; \mathbb{Z})$ (resp., $GL(s; \mathbb{Z})$) the set of matrices $A \in M(s; \mathbb{Z})$ with $\det A \neq 0$ (resp., $\det A = \pm 1$). We say that a matrix $A \in M(s; \mathbb{Z})$ satisfies a condition $(P(c))$ (resp., $(P_u(c))$) if a partition of the form (2) exists (resp., if a partition exists and it is uniquely determined); we mean by $A \in (P)$ that A satisfies a condition (P).

We consider (2) for regular (resp., singular) matrices A in Sections 1, 3 (resp., Section 4). We shall give some necessary and sufficient conditions for $(P_u(c))$ in Theorem 1, and in Theorem 2, we give a necessary, and a sufficient condition for $(P_u(c))$ in terms of characteristic polynomials. In Section 2, we define s-dimensional substitutions, and automata together with their conjugates, and give a class of matrices A such that the word $W(A; c)$ can be described by a limit of iterations of a substitution of dimension s, cf. Theorem 3. We give some examples of Theorem 3 in Section 3, where we consider only regular matrices. In Section 3, the so called hermitian canonical forms of integer matrices play an important role. Examples of partitions (2) with singular matrices A will be given in Section 4. In Section 5, we give two definitions for non-periodicity. One of them strongly requires non-periodicity. We also consider certain Voronoi tessellations coming from a word $W(A; c)$. We show that some of the words $W(A; c)$, and the tessellations are non-periodic in our strong sense, cf. Theorems 4–5. In Section 6, we give some problems and conjectures together with Theorem 6, which shows a beautiful connection between hermitian canonical forms and p-adic numbers. We shall give, in a forthcoming paper, the proof of Theorem 6 together with a theorem on a higher dimensional continued fraction in the p-adic sense.

Remark 1. Let us suppose $A \in (P(c))$. Then, setting $\Delta := U^{-1}\Gamma$ for a unimodular matrix $U \in GL(s; \mathbb{Z})$, we have by (2)

$$\overset{\circ}{\underset{0 \leq j < c}{\bigcup}} U^{-1}A^j U\Delta = \mathbb{Z}^s \setminus \{o\}, \quad 1 < c \leq \infty.$$

Hence, $A \in (P(c))$ (resp., $A \in (P_u(c))$) implies $U^{-1}AU \in (P(c))$ (resp., $U^{-1}AU \in (P_u(c))$), and vice versa. This fact can be applied also for a singular matrix A.

1.   THE CONDITION $P_u(c)$

In this section, we suppose $A \in M_r(s; \mathbb{Z})$, $s \geq 1$; we denote by L_s the set $\mathbb{Z}^s \setminus \{o\}$.

Lemma 1. *Let there exist a partition* (2) *for a matrix* $A \in M_r(s; \mathbb{Z})$ *with* $1 < c < \infty$, $\Gamma \supset S$. *Then*

$$\Gamma \supset \bigcup_{0 \leq m} A^{cm} S.$$

Proof. The assumption of the lemma implies $A^j \Gamma \supset A^j S$ for all $0 \leq j < c$. Suppose that an element $x \in A^c S$ belongs to a set $A^j \Gamma$ with $0 < j < c$. Then $0 < c - j < c$, so that $A^{-j} x \in A^{-j}(A^c S) = A^{c-j} S \subset A^{c-j}\Gamma$, which contradicts $A^{-j} x \in A^{-j}(A^j \Gamma) = \Gamma$. Hence, we get

$$\Gamma \supset A^c S,$$

so that $A^{j-c}\Gamma \supset A^{j-c}(A^c S) = A^j S$ for all $c \leq j < 2c$. Suppose that an element $x \in A^{2c} S$ belongs to a set $A^j \Gamma$ with $0 < j < c$. Then $c < 2c - j < 2c$, so that $A^{-j} x \in A^{-j}(A^{2c} S) = A^{2c-j} S \subset A^{c-j}\Gamma$ ($0 < c - j < c$), which contradicts $A^{-j} x \in A^{-j}(A^j \Gamma) = \Gamma$. Hence we get

$$\Gamma \supset A^{2c} S.$$

Repeating the argument, we get the lemma.   ∎

Lemma 2. *Let there exist a partition* (2) *for a matrix* $A \in M_r(s; \mathbb{Z})$ *with* $1 < c < \infty$, $x \in \Gamma$ *such that* $A^{-n}x \in L_s$ *for all* $n \in \mathbb{N}$. *Then* $\Gamma \supset \{A^{cm}x; \ m \in \mathbb{Z}\}$.

Proof. We put $x_n := A^n x$ ($n \in \mathbb{Z}$). Lemma 1 implies $\Gamma \supset \{x_{cm}; \ m \in \mathbb{N}\}$, so that

$$A^j \Gamma \supset \{x_{cm+j}; \ m \in \mathbb{N}\} \text{ for all } 0 \leq j < c. \tag{3}$$

Suppose $x_{-cm} \in A^j \Gamma$ $(m \geqq 0)$ for an integer $0 < j < c$. Then

$$x_{-cm-j} = A^{-j} x_{-cm} \in A^{-j}(A^j \Gamma) = \Gamma.$$

Hence, by Lemma 1, we get

$$x_{c-j} = x_{c(m+1)-cm-j} \in \Gamma \text{ with } 0 < c - j < c,$$

which contradicts (3). Therefore, we obtain $x_{-cm} \in \Gamma$ for all $m \geqq 0$, which together with Lemma 1 implies Lemma 2.   ∎

We consider two conditions (Bdd), (Emp) on A:
   (Bdd)   The set $\{j \in \mathbb{N}; \ A^{-j} x \in \mathbb{Z}^s\}$ is bounded for any $x \in L_s$,
   (Emp) $\bigcap_{0 \leqq n < \infty} A^n L_s = \phi$.

Theorem 1. *The condition* (Bdd) *holds if and only if one of the conditions* (Emp), $(P_u(2))$, $(P_u(3))$, ..., $(P_u(\infty))$ *holds.*

Corollary 1. *The conditions* $(P_u(2))$, $(P_u(3))$, ..., $(P_u(\infty))$ *are equivalent.*

Proof of Theorem 1. Following a diagram

$$(\text{Emp}) \overset{(\text{i})}{\Longleftrightarrow} (\text{Bdd}) \overset{(\text{ii})}{\Longrightarrow} (P_u(c)) \ (1 < c \leqq \infty),$$
$$\text{not (Bdd)} \overset{(\text{iii})}{\Longrightarrow} \text{not } (P_u(\infty)), \quad \text{not (Bdd)} \overset{(\text{iv})}{\Longrightarrow} \text{not } (P_u(c)) \ (1 < c < \infty),$$

we prove the theorem according to the number (i–iv):

(i) We denote by X^C for $X \subset L_s$ the complement of X in L_s. Since $A^{-m} x \in \mathbb{Z}^s$ implies $A^{-n} x \in \mathbb{Z}^s$ $(\forall n < m)$, we have

$$A^{-n} x \notin \mathbb{Z}^s \ (\exists n \in \mathbb{N}) \Longrightarrow A^{-m} x \notin \mathbb{Z}^s \ (\forall m > n). \tag{4}$$

Hence, noting

$$(\text{Emp}) \Longleftrightarrow \bigcup_{0 \leqq n < \infty} (A^n L_s)^C = L_s \Longleftrightarrow L_s \subset \bigcup_{0 \leqq n < \infty} (A^n L_s)^C$$
$$\Longleftrightarrow \forall x \in L_s \ \exists n \in \mathbb{N} \text{ such that } x \in (A^n L_s)^C$$
$$\Longleftrightarrow \forall x \in L_s \ \exists n \in \mathbb{N} \text{ such that } x \notin A^n \mathbb{Z}^s$$
$$\Longleftrightarrow \forall x \in L_s \ \exists n \in \mathbb{N} \text{ such that } A^{-n} x \notin \mathbb{Z}^s,$$

we get the equivalence (Emp)$\Longleftrightarrow$(Bdd).

(ii) We suppose (Bdd). Then, we can define a map ind $_A$ by

$$\text{ind }_A : \mathbb{Z}^s \rightarrow \mathbb{N} \cup \{\infty\},$$
$$\text{ind }_A(\boldsymbol{x}) := \min\{n \in \mathbb{N}; \ A^{-n-1}\boldsymbol{x} \notin \mathbb{Z}^s\} \ (\boldsymbol{x} \in L_s),$$
$$\text{ind }_A(\boldsymbol{o}) := \infty.$$

Note that (4) implies

$$A^{-n}\boldsymbol{x} \in \mathbb{Z}^s \ (n \leq \text{ind }_A(\boldsymbol{x})), \ \ A^{-n}\boldsymbol{x} \notin \mathbb{Z}^s \ (n > \text{ind }_A(\boldsymbol{x})).$$

We put

$$S_n = S_n(A) := \{\boldsymbol{x} \in L_s; \ \text{ind }_A(\boldsymbol{x}) \geq n\},$$
$$T_n = T_n(A) := S_n \setminus S_{n+1} = \{\boldsymbol{x} \in L_s; \ \text{ind }_A(\boldsymbol{x}) = n\}, \ n \geq 0.$$

The condition (Bdd) implies

$$\overset{\circ}{\underset{0 \leq n < \infty}{\bigcup}} T_n = L_s. \tag{5}$$

Since

$$\boldsymbol{x} \in A^j T_0 \Longleftrightarrow \boldsymbol{x} = A^j \boldsymbol{y}, \ \boldsymbol{y} \in T_0$$
$$\Longleftrightarrow A^{-j}\boldsymbol{x} = \boldsymbol{y}(\in \mathbb{Z}^s) \ \& \ A^{-j-1}\boldsymbol{x} = A^{-1}\boldsymbol{y} \notin \mathbb{Z}^s$$
$$\Longleftrightarrow \boldsymbol{x} \in T_j,$$

we get

$$A^j T_0 = T_j \ (0 \leq j < \infty). \tag{6}$$

Setting

$$\Gamma = \underset{0 \leq j < \infty}{\bigcup} T_{cj} \ (c < \infty), \ \Gamma = T_0 \ (c = \infty),$$

we have a partition (2). We show the uniqueness of the partition. It is clear that (2) implies $\Gamma \supset T_0$. First, we assume (2) with $c = \infty$. Then $\Gamma \supset T_0$ implies $A^n \Gamma \supset A^n T_0 = T_n$ for all $n \geq 0$. In view of (5), we must have

$$A^n \Gamma = T_n$$

for all $n \geq 0$, since (2) is also a disjoint union. Secondly, we assume (2) with $c < \infty$. From Lemma 1 together with (6), it follows

$$\Gamma \supset \underset{0 \leq m < \infty}{\bigcup} A^{cm} T_0 = \underset{0 \leq m < \infty}{\bigcup} T_{cm},$$

so that

$$A^j \Gamma \supset \bigcup_{0 \leq m < \infty} A^{cm+j} T_0 = \bigcup_{0 \leq m < \infty} T_{cm+j}.$$

Therefore, we get by (2), (5)

$$A^j \Gamma = \bigcup_{0 \leq m < \infty} T_{cm+j} \quad (0 \leq j < c),$$

which says the uniqueness.

(iii) We assume that (Bdd) does not hold, *i.e.*, there exists an element $x_0 \in L_s$ such that $A^{-n} x_0 \in \mathbb{Z}^s$ for all $n \geq 1$. We put

$$x_n := A^n x_0 (\in L_s), \ n \in \mathbb{Z},$$

and consider two cases:

(a) The sequence $\{x_n\}_{n \in \mathbb{Z}}$ is periodic, *i.e.*, $x_0 = x_k$ for some $k \in \mathbb{Z} \setminus \{0\}$.

(b) The sequence $\{x_n\}_{n \in \mathbb{Z}}$ is not periodic, *i.e.*, $x_0 \neq x_k$ for all $k \in \mathbb{Z} \setminus \{0\}$.

We suppose (a). We may assume that the number k is chosen to be the smallest positive number satisfying $x_k = x_0$. Then $x_0 \in A^j \Gamma$ $(j \geq 0)$ implies $x_0 \in A^{k+j} \Gamma$, so we can not have a partition (2) with $c = \infty$. We suppose (b) and that there exists a partition (2) for $c = \infty$ together with $x_0 \in A^j \Gamma$ for an integer $j \geq 0$. Then $x_{-j} \in \Gamma$, so that

$$x_{n-j} = A^{n-j} x_0 = A^n (A^{-j} x_0) = A^n x_{-j} \in A^n \Gamma$$

for all $n \geq 0$. Now, consider an element of x_{-j-1}, which can be supposed to be an element of $A^m \Gamma$ for an integer $m \geq 0$. Then

$$x_{-j} = A^{-j} x_0 = A(A^{-j-1} x_0) = A x_{-j-1} \in A(A^m \Gamma) = A^{m+1} \Gamma, \ m+1 > 0,$$

which contradicts $x_{-j} \in \Gamma = A^0 \Gamma$.

(iv) We assume that (Bdd) does not hold, and that there exists a partition (2) for $1 < c < \infty$. We consider two cases (a), (b) as in the proof (iii). First, we suppose (b). Then, setting

$$\Gamma^{(i)} := (\Gamma \setminus X) \cup \{x_{i+cm}; \ m \in \mathbb{Z}\}, \ X := \{x_m; \ m \in \mathbb{Z}\},$$

we obtain by (2)

$$\overset{\circ}{\bigcup_{0 \leq j < c}} A^j \Gamma^{(i)} = L_s \tag{7}$$

for all $i \in \mathbb{Z}$, which says that we can not have the uniqueness of the partition (2). Secondly, we suppose (a). If $x_i \in A^j\Gamma$ $(i \geq 0)$ for an integer $0 \leq j = j(i) < c$, then $x_{i-j} \in \Gamma$. Hence, by Lemma 2 we get $x_{i+cm} = A^j(x_{cm+i-j}) \in A^j\Gamma$, $m \in \mathbb{Z}$. Hence, x_i, and x_j belong to the same set $A^h\Gamma$ $(0 \leq h < c)$ if and only if $i \equiv j \pmod{c}$. In view of (a), we have $x_k = x_0$, so that k must be a multiple of c. Hence, we obtain partitions (7) with $\Gamma^{(i)}$ given above for all $0 \leq i < c$. Therefore, noting $\Gamma^{(i)} \neq \Gamma^{(j)}$ for integers i, j satisfying $0 \leq i < j < c$, $c > 1$, we can not have the uniqueness of the partition (2).   ∎

By the proof of Theorem 1, we get the following

Corollary 2. *Let (2) have a unique solution $\Gamma(\subset L_s)$. Then the matrix A satisfies* (Bdd), *and*

$$A^j\Gamma = \bigcup_{0 \leq m < \infty} T_{cm+j}(A) \ (0 \leq j < c) \ \text{for } c < \infty,$$

$$A^j\Gamma = T_j(A) \ (0 \leq j) \ \text{for } c = \infty.$$

holds.

In some cases, for a given matrix A, it is not so easy to conclude whether A satisfies the condition (Bdd). We aim at giving a class (K) of matrices such that we can make sure of the implication $A \in (K) \Rightarrow A \in$ (Bdd).

Lemma 3. *Let $A_i \in M_r(s_i; \mathbb{Z})$ $(1 \leq i \leq t)$ be matrices satisfying the condition* (Bdd), *and T a matrix*

$$T = \begin{pmatrix} A_1 & & * \\ & \ddots & \\ O & & A_t \end{pmatrix}, \ \text{or} \ \begin{pmatrix} A_1 & & O \\ & \ddots & \\ * & & A_t \end{pmatrix}.$$

Let $A \in M_r(s; \mathbb{Z})$ $(s = s_1 + \cdots + s_t)$ be a matrix given by $A = UTU^{-1}$ with a matrix $U \in GL(s; \mathbb{Z})$. Then A also satisfies (Bdd).

Proof. We may assume $s > 1$. It suffices to show the lemma for $t = 2$. Changing the basis of $\mathbb{Z}^s$ as a $\mathbb{Z}$-module if necessary, we may assume that T is an "upper triangular" matrix. Then, we can write

$$U^{-1}AU = \begin{pmatrix} P & Q \\ O & R \end{pmatrix}, \ U^{-1}A^{-1}U = \begin{pmatrix} P^{-1} & -P^{-1}QR^{-1} \\ O & R^{-1} \end{pmatrix}.$$

By induction, one can show

$$U^{-1}A^{-n}U = \begin{pmatrix} P^{-n} & -(P^{-n}QR^{-1} + P^{-n+1}QR^{-2} + \cdots + P^{-1}QR^{-n}) \\ O & R^{-n} \end{pmatrix}.$$

We get the equivalence:

$$A \notin (\text{Bdd}) \Leftrightarrow U^{-1}AU \notin (\text{Bdd})$$
$$\Leftrightarrow \exists x \in \mathbb{Z}^{s_1}, \exists y \in \mathbb{Z}^{s_2} \text{ with } {}^T\left({}^T x, {}^T y\right) \neq o \ \&$$
$$U^{-1}A^{-n}U \ {}^T\left({}^T x, {}^T y\right) \in \mathbb{Z}^s \text{ for } \forall n \in \mathbb{N}.$$
$$\Leftrightarrow P^{-n}x - \left(P^{-n}QR^{-1} + P^{-n+1}QR^{-2} + \cdots \right.$$
$$\left. + P^{-1}QR^{-n}\right)y \in \mathbb{Z}^{s_1} \ \&$$
$$R^{-n}y \in \mathbb{Z}^{s_2} \text{ for } \forall n \in \mathbb{N}.$$

Hence, noting that

$$y \neq o \Longrightarrow R \notin (\text{Bdd}),$$
$$y = o \Longrightarrow x \neq o \Longrightarrow P \notin (\text{Bdd}),$$

we obtain

$$A \notin (\text{Bdd}) \Longrightarrow P \notin (\text{Bdd}), \text{ or } R \notin (\text{Bdd}),$$

which implies the lemma.   ∎

We denote by $\Phi_A(x)$ the characteristic polynomial of a square matrix A, by $\mathbb{Q}$ the set of rational numbers.

Lemma 4. *Let* $A \in M_r(s; \mathbb{Z}^t)$ *be a matrix such that* $|\det A| > 1$, *and* $\Phi_A(x)$ *is irreducible over* $\mathbb{Z}[x]$. *Then* A *satisfies* (Bdd).

Proof. Since $\Phi_A(x)$ is irreducible, so is $F(x) := \Phi_{A^{-1}}(x)$. We may suppose that $\alpha = \alpha_1, \ldots, \alpha_t$ $(t > 1)$ are simple roots of $F(x)$, which are the conjugates of a root $\alpha \neq 0$. We put $A^{-n} = (a_{ij}^{(n)})_{1 \leq i, j \leq t}$. By Cayley-Hamilton's theorem, $\{a_{ij}^{(n)}\}_{n=1,2,\ldots}$ becomes a linear recurrence sequence with $F(x)$ as its characteristic polynomial for each $1 \leq i \leq t$, $1 \leq j \leq t$, so that

$$a_{ij}^{(n)} = \sum_{1 \leq k \leq t} \lambda_{ij}^{(k)} \alpha_k^n,$$

holds for all $n \in \mathbb{Z}$, where $\lambda_{ij}^{(k)}$ are numbers independent of n, belonging to the splitting field $\mathbb{K} := \mathbb{Q}(\alpha_1, \ldots, \alpha_t)$ of $F(x)$. Then, for $x = {}^T(x_1, \ldots, x_t) \in \mathbb{Z}^t \setminus \{o\}$,

$$x_n := A^{-n}x = \left(\sum_{1 \leq k \leq t} \lambda_{ij}^{(k)} \alpha_k^n\right)_{1 \leq i, j \leq t} x$$

$$= \left(\sum_{1 \leq j \leq t} \left(\sum_{1 \leq k \leq t} \lambda_{ij}^{(k)} \alpha_k^n\right) x_j\right)_{1 \leq i \leq t}$$

$$= \left(\sum_{1 \leq k \leq t} \left(\sum_{1 \leq j \leq t} \lambda_{ij}^{(k)} x_j \right) \alpha_k^n \right)_{1 \leq i \leq t}$$

$$= \Lambda^{T} (\alpha_1^n, \ldots, \alpha_t^n),$$

holds where $\Lambda := (\sum_{1 \leq j \leq t} \lambda_{ij}^{(k)} x_j)_{1 \leq i, k \leq t}$. Hence, we get

$$\det (x_n, x_{n+1}, \ldots, x_{n+t-1}) = (\det \Lambda) V \left(N_{\mathbb{K}/\mathbb{Q}}(\alpha) \right)^n, \qquad (8)$$

where $V := \prod_{1 \leq j < i \leq t} (\alpha_i - \alpha_j) \neq 0$, and $N_{\mathbb{K}/\mathbb{Q}}(\alpha)$ is the norm of α over $\mathbb{Q}$.

Now, we suppose that $x_n \in \mathbb{Z}^t$ for all $n \geq 0$. Then

$$\det (x_n, x_{n+1}, \ldots, x_{n+t-1}) \in \mathbb{Z} \qquad (9)$$

follows. By the assumption of the lemma, we have

$$N_{\mathbb{K}/\mathbb{Q}}(\alpha) = \det A^{-1} = d^{-1}, \; d := \det A, \; |d| > 1.$$

Hence, from (8), (9), it follows $\det \Lambda = 0$, so that

$$\det(x_n, x_{n+1}, \ldots, x_{n+t-1}) = 0$$

holds for all n. The matrix $A^{-1} (\in GL(t; \mathbb{Q}))$ gives rize to a linear transformation over $\mathbb{Q}$ on the linear space $\mathbb{Q}^t$. Since $x_0 = x \neq o$,

$$\dim_{\mathbb{Q}} \langle x_0, x_1, \ldots \rangle \geq 1$$

follows, where $\langle x_0, x_1, \ldots \rangle$ denotes the linear subspace of $\mathbb{Q}^t$ generated by elements $x_0, x_1, \ldots \in \mathbb{Q}^t$. Since $\det(x_0, x_1, \ldots, x_{t-1}) = 0$, there exists an integer $1 \leq i \leq t - 1$ such that

$$x_i = c_0 x_0 + c_1 x_1 + \cdots + c_{i-1} x_{i-1} \; (c_0, c_1, \ldots, c_{i-1} \in \mathbb{Q}).$$

Multiplying the both sides of the equality obtained above by A^{-1}, we get

$$\begin{aligned} x_{i+1} &= c_0 x_1 + c_1 x_2 + \cdots + c_{i-1} x_i \\ &= c_0 x_1 + c_1 x_2 + \cdots + c_{i-2} x_{i-1} + c_{i-1} (c_0 x_0 + c_1 x_1 + \cdots + c_{i-1} x_{i-1}) \\ &= c_0' x_0 + c_1' x_1 + \cdots + c_{i-1}' x_{i-1} \end{aligned}$$

where $c_0', c_1', \cdots, c_{i-1}' \in \mathbb{Q}$. Repeating the argument, we see that

$$x_n \in \langle x_0, x_1, \cdots, x_{i-1} \rangle$$

for all $n \geq 0$. Therefore, we obtain

$$1 \leq u := \dim_{\mathbb{Q}} \langle x_0, x_1, \ldots \rangle \leq i < t.$$

We may suppose $y_1 = x_{i_1}, \ldots, y_u = x_{i_u}$ form a basis of the space $\langle x_0, x_1, \ldots \rangle$. Extending the basis to a basis of $\mathbb{Q}^t$, we may assume that

$$y_1, \ldots, y_u, z_1, \ldots, z_{t-u} \quad (1 \leq u < t)$$

is a basis of $\mathbb{Q}^t$. Since $\langle x_0, x_1, \ldots \rangle$ is an invariant space, considering the linear transformation A^{-1} with respect to the basis $y_1, \ldots, y_u, z_1, \ldots, z_{t-u}$, we can find $S \in GL(t; \mathbb{Q})$ such that

$$S^{-1} A^{-1} S = \begin{pmatrix} P & Q \\ O & R \end{pmatrix}, \quad P \in GL(u; \mathbb{Q}), \; R \in GL(t - u; \mathbb{Q}),$$

which contradicts the irreducibility of $F(x)$. ∎

Lemmas 3–4 imply the assertion (ii) in the following theorem.

Theorem 2. (i) *Let $A \in M_r(s; \mathbb{Z}^s)$ be a matrix satisfying the condition* (Bdd). *Then A has no algebraic units as its eigenvalues.*

(ii) *Let $A = UTU^{-1} \in M_r(s; \mathbb{Z})$ $(U \in GL(s; \mathbb{Z}))$ be a matrix having no units as its eigenvalues with T given by*

$$T = \begin{pmatrix} A_1 & & * \\ & \ddots & \\ O & & A_t \end{pmatrix}, \quad or \quad \begin{pmatrix} A_1 & & O \\ & \ddots & \\ * & & A_t \end{pmatrix},$$

$$A_i \in M_r(s_i; \mathbb{Z}^s) \; (s_1 + \cdots + s_t = s)$$

such that all the characteristic polynomials $\Phi_{A_i}(x)$ are irreducible over $\mathbb{Z}[x]$. Then A satisfies the condition (Bdd).

Proof of (i). Let $A \in$ (Bdd). Suppose that A has an algebraic unit ε as an eigenvalue. Then we can choose a factor $g(x) = x^u + g_{u-1} x^{u-1} + \cdots + g_1 x + g_0 \in \mathbb{Z}[x]$ of $\Phi_A(x)$ such that

$$\Phi_A(x) = g(x) h(x), \quad g_0 = \pm 1, \; h(x) \in \mathbb{Z}[x], \; GCD(g(x), h(x)) = 1.$$

Notice that $\deg h(x) \geq 0$. Let $\varphi_A(x)$ be the minimal polynomial of A, and $e_i(x)$ $(1 \leq i \leq s, e_{i+1}(x)$ is divisible by $e_i(x)$ $(1 \leq i \leq n-1))$ be the elementary divisors of the matrix $xE - A$. Then $\Phi_A(x) = e_1(x) \cdots e_s(x)$, and $\varphi_A(x) = e_s(x)$, so that any irreducible factor of $\Phi_A(x)$ is a factor of $\varphi_A(x)$. Hence, $h(x)$ is not divisible by $\varphi_A(x)$. Therefore $h(A) \neq O$, so that there exists a lattice point $\boldsymbol{x} \in h(A)\mathbb{Z}^s$ such that $\boldsymbol{x} \neq \boldsymbol{o}$. Since, $g(A)\boldsymbol{x} = \boldsymbol{o}$, we get

$$A^n\boldsymbol{x} = -g_0^{-1}(g_1 A^{n+1}\boldsymbol{x} + \cdots + g_{u-1}A^{n+u-1} + A^{n+u}\boldsymbol{x}) \in \mathbb{Z}^s$$
$$(n = -1, -2, \ldots, \ g_0 = \pm 1),$$

which contradicts $A \in$ (Bdd). ∎

In what follows, $\Lambda(A)$ denotes the set of eigenvalues of a square matrix A.

Remark 2. $A \in M_r(s; \mathbb{Z})$ with $1 \leq s \leq 3$ satisfies (Bdd) if $\Lambda(A)$ contains no algebraic units.

Sketch of the proof. If $s = 1$, it is trivial. We may assume that Φ_A is reducible. If $s = 2$, or $= 3$, then by the reducibility of Φ_A, we may assume $a \in \Lambda(A) \cap \mathbb{Z}$, $|a| > 1$. For $s = 2$, we can take an eigenvector $\boldsymbol{x} = (x_1, x_2) \in \mathbb{Z}^2$ satisfying $GCD(x_1, x_2) = 1$. Then we can choose $\boldsymbol{y} \in \mathbb{Z}^2$ such that $U := (\boldsymbol{x} \ \boldsymbol{y}) \in GL(2; \mathbb{Z})$. Then $U^{-1}AU = T$ turns out to be of the form as in Theorem 2 with $s_1 = s_2 = 1$, so that $A \in$ (Bdd). For $s = 3$, we can choose an eigenvector $\boldsymbol{x} = (x_1, x_2, x_3) \in \mathbb{Z}^3$ with $GCD(x_1, x_2, x_3) = 1$. Applying a modified Jacobi-Perron algorithm, we can find $\boldsymbol{y}, \boldsymbol{z} \in \mathbb{Z}^3$ such that $U := (\boldsymbol{x} \ \boldsymbol{y} \ \boldsymbol{z}) \in GL(3; \mathbb{Z})$. The first column of $U^{-1}AU$ is $^T(a, 0, 0)$. Using Lemma 3 together with the result for $s = 2$ that we have obtained, we get $A \in$ (Bdd). ∎

2.   SUBSTITUTIONS OF HIGH DIMENSION, AND AUTOMATA

We shall give a definition of substitutions of high dimension in a general situation, and then define substitutions of a special form.

An element W of a set K^X ($K \neq \phi$) for a subset X of a lattice $L := P(\mathbb{Z}^s)$ ($P \in GL(s; \mathbb{R})$) will be referred to as a **word over** K **on the set** X, and we write $X := \mathrm{Dom}\,(W)$. The matrix P will be fixed, and we shall mainly consider the case where $P = E$. It is convenient, and natural to think that the set K^ϕ consists of one element for the empty set ϕ. We denote by λ the element of K^ϕ, which is referred to as the empty word. If $K = \{a\}$ and $X = \{\boldsymbol{x}\}$ ($a \in K$, $\boldsymbol{x} \in L$), the set K^X also consists of one word, which will be referred to as an atomic word, and denoted by $\langle a, \boldsymbol{x} \rangle$. We say that W is a **finite** (resp., **infinite**) word if $\mathrm{Dom}\,(W)$ is a finite (resp., infinite) set. We say two words W_1, W_2 over K are **collative** if they agree on $\mathrm{Dom}\,(W_1) \cap \mathrm{Dom}\,(W_2)$ as maps. If W_1 and W_2 are collative words over K, then we can define a word

$W_1 \vee W_2$ over K on Dom $(W_1) \cup$ Dom (W_2) by

$$(W_1 \vee W_2)(\boldsymbol{x}) := W_1(\boldsymbol{x}) \ (\boldsymbol{x} \in \text{Dom } (W_1)), \ := W_2(\boldsymbol{x}) \ (\boldsymbol{x} \in \text{Dom } (W_2)).$$

We call $W_1 \vee W_2$ the join of W_1 and W_2. It is clear that $W \vee \lambda = \lambda \vee W = W$ holds for any W, and the join operation $\vee$ satisfies the associative, the commutative, and the idempotent laws. Note that $W_1 \vee W_2$ is not always defined, so that $\cup_{X \subset L} K^X$ does not form a monoid with respect to the join operation if the cardinality of K is bigger than 1. If any two of finite, or infinitely many words W_ι ($\iota \in I$) are collative, then we can define the join $\vee_{\iota \in I} W_\iota$. Notice that the cardinality of the index set I can be continuous. Notice also that W_1 and W_3 can be non-collative even if W_1, W_2 are collative and so are W_2, W_3. For any $W = (w_{\boldsymbol{x}})_{\boldsymbol{x} \in X} \in K^X$, $W = \vee_{\boldsymbol{x} \in X} \langle w_{\boldsymbol{x}}, \boldsymbol{x} \rangle$ holds. We mean by a subword of $W \in K^X$ a restriction V of W as a map, and we write $V \neg W$. We denote by Sub (W) the set of all subwords of W. Sub (W) becomes a monoid having λ as its unit with respect to the operation $\vee$. We say $\Xi (\subset \cup_{X \subset L} K^X)$ is **generative** set (of words) if $W_\iota \in \Xi$ ($\iota \in I$) are mutually collative then $\vee_{\iota \in I} W_\iota \in \Xi$, and if $\vee_{\iota \in I} W_\iota \in \Xi$ then $W_\iota \in \Xi$ ($\iota \in I$). For instance, Sub (W) ($W \in K^X$), and

$$\text{Gen } (K, X) := \bigcup_{W \in K^X} \text{Sub } (W) = \bigcup_{Y \subset X} K^Y \ (X \subset L)$$

are generative set. For a word $W = (w_{\boldsymbol{x}})_{\boldsymbol{x} \in X} \in K^X$ and $t \in L$, $W \uparrow t$ denotes the word $(w_{\boldsymbol{x}+t})_{\boldsymbol{x} \in X} \in K^{t+X}$, the translation of W by t. If there exists $t \in L$ such that $W_1 \uparrow t = W_2$ for two words $W_1, W_2 \in$ Gen (K, L), we write $W_1 \equiv W_2$, which gives an equivalence relation on Gen (K, L). Let Ξ be a generative set consisting of some words in Gen (K, L). We say that a map $\sigma : \Xi \to \Xi$ is a substitution (over K on $X := \cup_{W \in \Xi} \text{Dom } (W)$) if it satisfies the following two conditions:

(i) **additivity**: If $W_\iota \in \Xi$ ($\iota \in I$) are mutually collative, then so are $\sigma(W_\iota)$ ($\iota \in I$), and

$$\sigma \left(\bigvee_{\iota \in I} W_\iota \right) = \bigvee_{\iota \in I} \sigma(W_\iota)$$

holds.

(ii) **context uniformity**: For any $a \in K$, we can find a finite word $F = F(a) \in$ Gen (K, L) independent of $\boldsymbol{x}$ satisfying that there exists an element $t = t(a, \boldsymbol{x}) \in \text{Dom } (\sigma(\langle a, \boldsymbol{x} \rangle))$ such that

$$\sigma(\langle a, \boldsymbol{x} \rangle) \uparrow -t \ \neg \ F$$

holds for all x provided that $\langle a, x \rangle \in \Xi$.

We remark that K can be an infinite set. Notice that a substitution $\sigma : \Xi \to \Xi$ is determined only by the values $\sigma(\langle a, x \rangle)$ for all $\langle a, x \rangle \in \Xi$. Notice also that for any generative Ξ, the identity map on Ξ is a substitution; so that any word W can be a fixed point of some trivial substitutions on Sub (W).

Our main objective in this section is to describe the characteristic words $W(A; c)$ $(1 < c \leqq \infty)$ of some of the partitions given by (2) as a limit starting from a word $\langle a, x \rangle$, i.e.,

$$\langle a, x \rangle \neg \sigma(\langle a, x \rangle) \neg \sigma^2(\langle a, x \rangle) \neg \ldots \neg \sigma^n(\langle a, x \rangle) \neg \ldots,$$
$$\lim \sigma^n(\langle a, x \rangle) = W(A; c),$$

where σ^n is the n-fold iteration of σ. Note that if $W(A; c)$ has such a description as a limit starting from $\langle a, x \rangle$, then it becomes the fixed point of σ with $\langle a, x \rangle$ as its subword, since $W(A; c) = \vee_{n \geq 0} \sigma^n(\langle a, x \rangle)$ holds. We shall construct a substitution σ describing $W(A; c)$ for some of $A \in$ (Bdd) in this sense, cf. Theorem 3. Here, in general, we mean by

$$\lim \sigma^n(W) = Z \ (\in \text{Gen} \ (K, L))$$

that $\sigma^n(W) \in$ Sub (Z) for all $n \geqq 0$, and

$$\liminf_{n \to \infty} \text{Dom} \ (\sigma^n(W)) := \lim_{n \to \infty} \left(\bigcup_{n \geq 0} \left(\bigcap_{n \leq m} \text{Dom} \ (\sigma^m(W)) \right) \right) = \text{Dom} \ (Z).$$

Our definition of a word "substitution" differs from the usual one in the case of dimension $s = 1$, but the new definition can describe any fixed point of a usual substitution. We give two typical examples: Let $W = abaababaab \ldots$ be the Fibonacci word, and consider it as an element of $\{a, b\}^{\mathbb{N}}$. If we define a substitution σ by

$$\sigma : \Xi \to \Xi, \ \Xi := \text{Sub} \ (W)$$
$$\sigma(\langle a, x \rangle) := \langle a, y \rangle \vee \langle b, y + 1 \rangle, \ \sigma(\langle b, x \rangle) := \langle a, y \rangle \ (x \geqq 0),$$
$$y := \rho^{-1}(\rho(x)0),$$

then $\lim \sigma^n(\langle a, 0 \rangle) = W$ holds, where $\rho(x)$ is the Fibonacci representation of an integer $x \geqq 0$. ($\rho(0) := \lambda$, $\rho(x)$ is obtained by the greedy algorithm using the Fibonacci numbers $1, 2, 3, 5, \ldots$ instead of using 2^n $(n \geqq 0)$ in the base-2 expansion, so that $\rho(x)$ becomes a word over $\{0, 1\}$ in the usual sense of words such that $w := \rho(x)$ has not 11 (resp., 0) as its subword (resp., prefix). If w is such a word, then $\rho^{-1}(0^i w)$ is defined

to be a number $\rho^{-1}(w)$ $(i \geq 0)$). In this case, all the finite subwords
Sub$_*(W)$ of W form a monoid, and σ becomes a monoid morphism on
Sub$_*(W)$. In general, σ may not be so explicit as the Fibonacci case, but
it is clear that, for arbitrary one of the fixed points of a substitution σ_0
(in the usual sense), we can define, according to σ_0, a substitution (in our
sense) that describes the fixed point. But, in general, we have difficulty
in describing plural fixed points by a substitution in our sense according to a given substitution in the usual sense. (Consider, for instance,
a substitution σ_0 in the usual sense over $\{a, b\}$ defined by $\sigma_0(a) = ab$,
$\sigma_0(b) = baa$ together with its two fixed points $W = abbaabaaabab\ldots$,
$V = baaababababbaa\ldots$, which are considered to be words on $\mathbb{N}$. Then,
the substitution $\sigma : \Xi \to \Xi$ $(W, V \in \Xi)$ in our sense according to σ_0
should satisfy $\sigma(\langle a, 3 \rangle) = \langle a, 8 \rangle \vee \langle b, 9 \rangle$ and $\sigma(\langle a, 3 \rangle) = \langle a, 7 \rangle \vee \langle b, 8 \rangle$ at
a time by $W(3) = V(3) = a$, which is impossible.) For substitutions of
constant length (in the usual sense), the situation turns out to be very
simple. We can take $\Xi = \text{Gen}(\{a, b\}, \mathbb{N})$, and describe plural fixed points
by one substitution in our sense. For instance, let $W_a = abbabaab\ldots$,
$W_b = baababba\ldots$ be the fixed points of the usual Thue-Morse substitution $a \to ab$, $b \to ba$, and consider them as elements of $\{a, b\}^{\mathbb{N}}$. If we
define a substitution σ over $\{a, b\}$ on $\mathbb{N}$ by

$$\sigma : \Xi \to \Xi, \ \Xi = \text{Gen}(\{a, b\}, \mathbb{N}),$$
$$\sigma(\langle a, x \rangle) := \langle a, 2x \rangle \vee \langle b, 2x + 1 \rangle,$$
$$\sigma(\langle b, x \rangle) := \langle b, 2x \rangle \vee \langle a, 2x + 1 \rangle \ (x \geq 0),$$

then $W_a, W_b \in \Xi$, and they become the fixed points of σ, which are
obtained by taking limit of the iteration $\sigma^n : W_a = \lim \sigma^n(\langle a, 0 \rangle)$, $W_b = \lim \sigma^n(\langle b, 0 \rangle)$.

For a substitution σ describing $W(A; \infty)$ that we shall give later, the
word $W(A; c)$ $(1 < c < \infty)$ can be described by a substitution over $\tilde{K}_c$
obtained by taking the reduction by modulo c as we shall see. Notice that
σ is not uniquely determined by $W(A; c)$ $(A \in \text{(Bdd)}, 1 < c \leq \infty)$, since
σ^m $(m > 1)$ works as well as σ. We remark that, in some cases, we can
give substitutions σ, τ such that $\sigma^p \neq \tau^q$ for any integers $p, q > 0$, and
$\lim \sigma^n(\langle \infty, o \rangle) = \lim \tau^n(\langle \infty, o \rangle) = W(A; c)$. For instance, consider the
word $W(A; \infty) \in \tilde{K}_{\infty}^{\mathbb{Z}^s}$ with $A = [1, -1//1, 1]$, see the notation in Section
3. Then, we can define a substitution $\sigma : \Xi \to \Xi$, $\Xi := \text{Gen}(\mathbb{N} \cup \{\infty\}, \mathbb{Z}^s)$
by

$$\text{Dom}(\sigma(\langle a, o \rangle)) := \{-1, 0, 1\}^2 \ni y = {}^T(y_1, y_2),$$
$$\sigma(\langle a, o \rangle)(o) := a, \ \sigma(\langle a, o \rangle)(y) := 1, \ y_1 y_2 \neq 0,$$
$$\sigma(\langle a, o \rangle)(y) := 0, \ \text{otherwise};$$

$$\text{Dom}\,(\sigma(\langle a, \boldsymbol{x}\rangle)) := 2\boldsymbol{x} + \{0, \varepsilon_1\} \times \{0, \varepsilon_2\}, \quad x_1 x_2 \neq 0,$$
$$\text{Dom}\,(\sigma(\langle a, \boldsymbol{x}\rangle)) := 2\boldsymbol{x} + \{0, \varepsilon_1\} \times \{-1, 0, 1\}, \quad x_1 \neq 0, x_2 = 0,$$
$$\text{Dom}\,(\sigma(\langle a, \boldsymbol{x}\rangle)) := 2\boldsymbol{x} + \{-1, 0, 1\} \times \{0, \varepsilon_2\}, \quad x_1 = 0, \ x_2 \neq 0$$

for $\boldsymbol{x} = {}^T(x_1, x_2) \in \mathbb{Z}^2 \setminus \{\boldsymbol{o}\}$, and for $\text{Dom}\,(\sigma(\langle a, \boldsymbol{x}\rangle)) \ni \boldsymbol{y} = 2\boldsymbol{x} + \boldsymbol{z}$, $\boldsymbol{z} = {}^T(z_1, z_2)$

$$\sigma(\langle a, \boldsymbol{x}\rangle)(\boldsymbol{y}) := a + 2; \ \boldsymbol{z} = \boldsymbol{o}, \ \sigma(\langle a, \boldsymbol{x}\rangle)(\boldsymbol{y}) := 1, \ z_1 z_2 \neq 0,$$
$$\sigma(\langle a, \boldsymbol{x}\rangle)(\boldsymbol{y}) := 0, \ \text{otherwise},$$

where $a \in \mathbb{N} \cup \{\infty\}$ ($\infty + a := \infty$), and $\varepsilon_i := \text{sgn}\,(x_i)$ ($i = 1, 2$) ($\text{sgn}\,(x) := 1, 0, -1$ accoring to $x > 0$, $x = 0$, $x < 0$). Then, as we shall see, $W := \lim \sigma^n(\langle \infty, \boldsymbol{o}\rangle)$, cf. (ii), Section 3. In this case, $\text{Dom}\,(\sigma(\langle a, \boldsymbol{x}\rangle)) \cap \text{Dom}\,(\sigma(\langle b, \boldsymbol{y}\rangle)) = \phi$ holds for all $\boldsymbol{x} \neq \boldsymbol{y}(\in \mathbb{Z}^2)$, $a, b \in \mathbb{N} \cup \{\infty\}$, so that we can extend σ to $\text{Gen}\,(\mathbb{N} \cup \{\infty\}, \mathbb{Z}^s)$ from the values $\sigma(\langle a, \boldsymbol{x}\rangle)(\boldsymbol{y})$ given above. We can define another substitution $\tau : \Xi \to \Xi$, $\Xi := \text{Sub}\,(W(A; \infty))$ for the same A as above by setting

$$\text{Dom}\,(\tau(\langle a, \boldsymbol{x}\rangle)) := A\boldsymbol{x} + (\{\boldsymbol{o}\} \cup D), \ D := \{{}^T(0, 1), {}^T(0, -1)\},$$
$$\tau(\langle a, \boldsymbol{x}\rangle)(A\boldsymbol{x} + \boldsymbol{y}) := a + 1 \ \text{if } \boldsymbol{y} = \boldsymbol{o}, \ := 0 \ \text{if } \boldsymbol{y} \in D \ (\langle a, \boldsymbol{x}\rangle \in \Xi),$$

we can extend τ to $\text{Sub}\,(W(A; \infty))$, and check that $\tau^n(\langle \infty, \boldsymbol{o}\rangle)$ tends to $W(A; \infty)$. Notice that $\text{Dom}\,(\tau(\langle a, \boldsymbol{x}\rangle)) \cap \text{Dom}\,(\tau(\langle b, \boldsymbol{y}\rangle)) \neq \phi$ for some $\boldsymbol{x} \neq \boldsymbol{y} \in \mathbb{Z}^2$, but in such a case, $\tau(\langle a, \boldsymbol{x}\rangle)(\boldsymbol{z}) = \tau(\langle b, \boldsymbol{y}\rangle)(\boldsymbol{z}) = 0$ holds for $\boldsymbol{z} \in \text{Dom}\,(\tau(\langle a, \boldsymbol{x}\rangle)) \cap \text{Dom}\,(\tau(\langle b, \boldsymbol{y}\rangle))$, so that $\tau(\langle a, \boldsymbol{x}\rangle)$ and $\tau(\langle b, \boldsymbol{y}\rangle)$ $(a, b \in \mathbb{N} \cup \{\infty\})$ are collative. Such a construction of τ is valid for $A \in (\text{Bdd})$ if and only if we can find a finite subset D of the set $T_0(A)(= \{\boldsymbol{x} \in \mathbb{Z}^s; \ \text{ind}_A(\boldsymbol{x}) = 0\})$ such that

$$\lim_{m \to \infty} \sum_{0 \leq m \leq n} A^m(\{\boldsymbol{o}\} \cup D) = \mathbb{Z}^s$$

holds. But, in general, we can not find such a set D for some $A \in (\text{Bdd})$ having a number ε satisfying $|\varepsilon| < 1$ as its eigenvalue, so that $\lim \tau^n(\langle \infty, \boldsymbol{o}\rangle)$ can not be a word on $\mathbb{Z}^s$. For instance, such a phenomenon takes place for $A = [2, 2//2, 3] \in (\text{Bdd})$. In fact, for $A = [2, 2//2, 3]$, we have difficulty in finding any substitution like subsitutions σ, τ given above, cf. Remark 8, Section 6. Note that if we take a trivial substitution υ, for instance, defined by

$$\upsilon : \text{Sub}\,(W(A; c)) \to \text{Sub}\,(W(A; c)), \ A = [2, 2//2, 3],$$
$$\text{Dom}\,(\upsilon(\langle a, \boldsymbol{x}\rangle)) := \boldsymbol{x} + \{\boldsymbol{o}\} \cup D, \ \langle a, \boldsymbol{x}\rangle \in \text{Sub}\,(W(A; c)),$$
$$\upsilon(\langle a, \boldsymbol{x}\rangle)(\boldsymbol{x}) := a, \ \upsilon(\langle a, \boldsymbol{x}\rangle)(\boldsymbol{x} + \boldsymbol{y}) := 0, \ \boldsymbol{y} \in D,$$

where D is any finite subset of $T_0(A)$, then $W(A; c)$ is a fixed point of v, but $v^n(\langle \infty, o \rangle)$ does not tend to a word on $\mathbb{Z}^2$.

In what follows, we mainly consider words on a set $X \subset \mathbb{N}^s$, and we shall define substitutions σ_* on $\mathbb{N}^s$ of special type together with its conjugates ${}^\tau \sigma_*$ ($\tau \in \{1, -1\}^s$) on $\mathbb{N}^s$. We shall see later that for a given matrix $U \in GL(s; \mathbb{Z})$ and a substitution σ_* on $\mathbb{N}^s$, we extend σ_*, by using the conjugates of σ_*, to a substitution $\tilde{\sigma}_* = \tilde{\sigma}_*(U)$ on $\mathbb{Z}^s$, which takes each $\langle a, x \rangle$ to a word on lattice points in "s-dimensional parallelepiped".

Recall the definition of the sets K_c, and $\tilde{K}_c$ ($c \in \mathbb{N} \cup \{\infty\}$). We denote by G the subset of $\mathbb{N}^s$ given by

$$G = G(b) = G(b_1, \ldots, b_s) := K_{b_1} \times K_{b_2} \times \cdots \times K_{b_s}$$

for $b = {}^T(b_1, \ldots, b_s) \in \mathbb{N}^s$. Note that $G(o) = \phi$. For a non-empty set K, we define

$$K^{*(s)} := \bigcup_{b \in \mathbb{N}^s} K^{G(b)}.$$

Note that $K^{*(s)}$ has the empty word λ as its element, and it is not a monoid with respect to the join operation except for the cases where $s = 1$, or the cardinality of K is one. The set $K^{*(1)}$ differs from K^*, which denotes the set of all finite words over K in the usual sense. In other words, K^* is a free monoid generated by K with λ as its unit with respect to the concatenation as its binary operation. For $u \in K^*$, u^{-1} denotes the inverse element of u in the free group generated by K, so that $uu^{-1} = u^{-1}u = \lambda$. For $S \subset K^*$, we denote by aS (resp., Sa) the set $\{as; s \in S\}$ (resp., $\{sa; s \in S\}$), and by $S_1 \cdots S_n$ the set consisting of the words $u_1 \cdots u_n$ ($u_i \in S_i \subset K^*$, $1 \leq i \leq n$). We put $S^n := S_1 \cdots S_n$ when $S_i = S$ ($1 \leq i \leq n$). Note that in some context, S^n denotes the usual cartesian product as it did before. In the case where we have to distinguish them, we denote by $S^{(n)}$ the cartesian product. Note also that $K^{*(s)}$ should not be read as $(K^*)^{(s)}$. We put

$$K^{\#(s)} := K^{*(s)} \cup K^{\mathbb{N}^s}.$$

For an integer $b > 1$, we denote the base-b expansion of $x \in \mathbb{N}$ as a word over K_b by $\rho_b(x) = \rho(x; b) \in K_b^* \setminus (0K_b^*)$. Note that $\rho_b(0) = \lambda$. The base-b expansion $\rho_b(x)$ of $x = {}^T(x_1, \ldots, x_s) \in \mathbb{N}^s$ with $b = {}^T(b_1, \ldots, b_s) \in \mathbb{N}^s$ $b_i > 1$ ($1 \leq i \leq s$) is a finite word over $G(b)$ defined by

$$\rho_b(x) := \begin{pmatrix} 0^{r-|\rho(x_1;b_1)|}\rho(x_1; b_1) \\ \cdots \\ 0^{r-|\rho(x_s;b_s)|}\rho(x_s; b_s) \end{pmatrix} \in G(b)^* \setminus (oG(b)^*)$$

where $r = r(x) := \max\{|\rho(x_i; b_i)|;\ 1 \leq i \leq s\}$, $|u|$ denotes the length of a finite word u, and $o = {}^T(0,\ldots,0) \in G(b)$. Note that $\rho_b(o) = \lambda = {}^T(\lambda,\ldots,\lambda)$. The map

$$\rho_b : \mathbb{N}^s \longrightarrow G(b)^* \setminus (oG(b)^*)$$

is a bijection. The inverse ρ_b^{-1} of the map ρ_b can be extended to $G(b)^*$ by

$$\rho_b^{-1}(o^n u) := \rho_b^{-1}(u),\ n \geqq 0,\ u \in G(b)^* \setminus (oG(b)^*).$$

For $x \in \mathbb{N}^s$, we define $\iota_b(x) \in G(b) \cup \{\lambda\}$ by

$$\iota_b(x) := u_0 \in G(b)\ (\iota_b(o) := \lambda)$$

with u_0 determined by $\rho_b(x) = u_j u_{j-1} \cdots u_0$ ($x \neq o$, $u_i \in G(b)$, $0 \leqq i \leqq j$).

Suppose a map

$$\sigma : K \longrightarrow K^{G(b)},\ \sigma(a) = (\sigma_x(a))_{x \in G(b)},\ \sigma_x(a) \in K\ (a \in K)$$

is given. Then we can define a substitution

$$\sigma_* : \mathrm{Gen}\,(K, \mathbb{N}^s) \to \mathrm{Gen}\,(K, \mathbb{N}^s)$$

by

$$\sigma_*(\langle a, x \rangle) := \sigma(a) \uparrow \rho_b^{-1}(\rho_b(x)o)\ (a \in K,\ x \in \mathbb{N}^s).$$

In fact, for any $W = (w_x)_{x \in H} \in \mathrm{Gen}\,(K, \mathbb{N}^s)$, W can be written as a join $\vee_{x \in H}\langle w_x, x \rangle$, so that $\sigma_*(W) = \vee_{x \in H}\sigma_*(\langle w_x, x \rangle) = \vee_{x \in H}(\sigma(w_x) \uparrow \rho_b^{-1}(\rho_b(x)o))$, since $\sigma(w_x) \uparrow \rho_b^{-1}(\rho_b(x)o)$ ($x \in H$) are mutually collative. In particular, for $W \in K^{\sharp(s)}$, the resulting word $\sigma_*(W)$ can be written by the following:

$$\sigma_*(W) \in K^{G(\rho_b^{-1}(\rho_b(c)o))}\ (\text{resp.},\ \sigma_*(W) \in K^{\mathbb{N}^s}),$$
$$\text{if } W \in K^{G(c)},\ c \in \mathbb{N}^s\ (\text{resp.},\ W \in K^{\mathbb{N}^s})$$

with $\sigma_*(W) = (z_x)$ defined by

$$z_x := \sigma_{\rho_b^{-1}(\iota_b(x))}\left(y_{\rho_b^{-1}(\rho_b(x)(\iota_b(x))^{-1})}\right)$$
$$\text{for } x \in G(\rho_b^{-1}(\rho_b(c)o))\ (\text{resp. } x \in \mathbb{N}^s).$$

Note that $G(\rho_b^{-1}(\rho_b(c)o)) = G(b_1 c_1,\ldots, b_s c_s)$ for $c = {}^T(c_1, \cdots, c_s) \in \mathbb{N}^s$, $\sigma_*(\lambda) = \lambda$. The substitution σ_* will be referred to as a **substitution** over K of dimension s of **size** $G(b)$.

In what follows, we mean by a word "substitution" such a substitution σ_* determined by a map $\sigma : K \to K^{G(b)}$ unless otherwise mentioned. Let σ_* be a substitution over K of size $G(b)$ with $b = T(b_1, \ldots, b_s)$, $b_i > 1$ $(1 \leqq i \leqq s)$ satisfying $(\sigma(a))(o) = \sigma_o(a) = a$ for an element $a \in K$. Then,

$$\sigma_*^n(\langle a, o \rangle) \neg \sigma_*^{n+1}(\langle a, o \rangle),$$

and $\sigma_*^n(\langle a, o \rangle)$ is a word on $G(b_1^n, \cdots, b_s^n)$, so that the limit of $\sigma_*^n(\langle a, o \rangle)$ always exists, which is an infinite word on the set $\mathbb{N}^s$, and it becomes the fixed point of σ_* with $\langle a, o \rangle$ as its subword.

We define an **automaton** $\mathcal{M}$ **of dimension** s:

$$\mathcal{M} = (@, K, G(b), \zeta),$$

which is a finite automaton with its initial state $@$, the set of states $K \ni @$, the set of input symbols $G(b) \neq \phi$, and a transition function $\zeta : K \times G(b) \to K$. In some cases, we consider a map π from the set K to a non-empty finite set F, which will be referred to as a projection. If we distinguish an element $h \in F$, we can specify a set $H := \pi^{-1}(h) \subset K$ as a final states of $\mathcal{M}$. The map ζ can be extended to the set $K \times G(b)^*$ as well as in the usual case $s = 1$ by

$$\zeta(a, i) := \zeta(\cdots \zeta(\zeta(a, i_1), i_2) \cdots, i_n),$$
$$i = i_1 i_2 \cdots i_n \quad (i_j \in G(b), n > 0);$$
$$\zeta(a, \lambda) := a, \quad a \in K.$$

If K is a finite set, then $\mathcal{M}$ becomes a finite automaton. An infinite word $W \in K^{\mathbb{N}^s}$ can be generated by $\mathcal{M}$:

$$W = (w_x), \quad w_x := \zeta(@, \rho_b(x)).$$

For a given substitution σ_* over $K \ni @$ of size $G(b)$, $b = {}^T(b_1, \ldots, b_s)$, $b_i > 1$ $(1 \leqq i \leqq s)$ determined by

$$\sigma(a) = (\sigma_i(a))_{i \in G(b)}, \quad \sigma_i(a) \in K \quad (a \in K)$$

satisfying $\sigma_o(@) = @$, we can define an automaton $\mathcal{M} = (@, K, G(b), \zeta)$ **corresponding to** σ by setting

$$\zeta(a, i) = \sigma_i(a). \tag{10}$$

Then the word V generated by $\mathcal{M}$ coincides with the fixed point $W = \lim \sigma_*^n(\langle @, o \rangle)$ of σ_*. Conversely, for a given automaton $\mathcal{M} = (@, K, G(b), \zeta)$ with $\zeta(@, o) = @$, $G(b) = {}^T(b_1, \ldots, b_s)$, $b_i > 1$ $(1 \leqq i \leqq s)$, if we define a map σ by (10), then the substitution σ_* determined by σ

has a fixed point $W = \lim \sigma_*^n(\langle @, o \rangle)$, and W coincides with the word V generated by $\mathcal{M}$.

We use the following notation: Let x, y, $b \in \mathbb{Z}^s$, $x = {}^T(x_1, \ldots, x_s)$, $y = {}^T(y_1, \ldots, y_s)$, $b = {}^T(b_1, \ldots, b_s)$, $b_i > 1$ $(1 \leq i \leq s)$, and $\tau = {}^T(\tau_1, \ldots, \tau_s) \in \{1, -1\}^s$. An equivalence relation $\equiv$ on $\mathbb{Z}^s$ can be given by

$$x \equiv y \pmod{G(b)} \overset{\text{def.}}{\Longleftrightarrow} x_i \equiv y_i \pmod{b_i} \text{ holds for all } 1 \leq i \leq s;$$

$\tau \otimes x := {}^T(\tau_1 x_1, \ldots, \tau_s x_s) \in \mathbb{Z}^s$; $\tau \otimes_{G(b)} x$ is defined to be an element $y \in G(b)$ satisfying $\tau \otimes x \equiv y \pmod{G(b)}$, which is uniquely determined by x, b, τ; $\tau \otimes S := \{\tau \otimes x; \ x \in S\}$, $\tau \otimes_{G(b)} S := \{\tau \otimes_{G(b)} x; \ x \in S\}$ for $S \subset \mathbb{Z}^s$. For a given substitution σ_* over K of size $G(b)$ determined by a map

$$\sigma : K \to K^{G(b)}, \ \sigma(y) := (\sigma_x(y))_{x \in G(b)}, \ \sigma_x(y) \in K \ (y \in K),$$

we mean by the τ-conjugate of σ_* the substitution ${}^T\sigma_*$ defined by

$$ {}^T\sigma : K \to K^{G(b)}, \ {}^T\sigma(y) := \left(\sigma_{\tau \otimes_{G(b)} x}(y) \right)_{x \in G(b)}.$$

Note that the map $\tau \otimes_{G(b)} x$ on $x \in G(b)$ is a bijection. For an automaton $\mathcal{M} = (@, K, G(b), \zeta)$, we denote by ${}^T\mathcal{M} = (@, K, G(b), {}^T\zeta)$ its τ-conjugate, which is an automaton with a transition function ${}^T\zeta$ defined by

$$ {}^T\zeta(a, i) := \zeta(a, \tau \otimes_{G(b)} i) \ (a \in K, \ i \in G(b)).$$

For a word $W = (w_x)_{x \in \mathbb{Z}^s} \in K^{\mathbb{Z}^s}$, $\tau \in \{1, -1\}^s$, and a unimodular matrix $U \in GL(s; \mathbb{Z})$, a word

$$(w_x)_{x \in U(\tau \otimes \mathbb{N}^s)}$$

over K on the set $U(\tau \otimes \mathbb{N}^s)$ can be identified with a word

$$(v_x)_{x \in \mathbb{N}^s}, \ v_x := w_{U(\tau \otimes x)}$$

over K on the set $\mathbb{N}^s$, which will be denoted by ${}^{\tau,U}W$, and referred to as the (τ, U)-sector of W. Noting that the column vectors of U form a basis of $\mathbb{Z}^s$ as a $\mathbb{Z}$-module, we say that a word $W \in K^{\mathbb{Z}^s}$ is a fixed point of a substitution σ_* (on $\mathbb{N}$) with respect to a basis U, if the (τ, U)-sector ${}^{\tau,U}W$ is a fixed point of ${}^T\sigma_*$ for all $\tau \in \{1, -1\}^s$.

We remark that if a word W over K on $\mathbb{Z}^s$ is a fixed point of a substitution σ_* (on $\mathbb{N}^s$) with respect to a basis U, then using the conjugates of σ_*, we can construct a substitution $\tilde{\sigma}_*$ (on $\mathbb{Z}^s$) such that

$\lim \tilde{\sigma}_*^n(\langle a, o \rangle) = W$ for an element $a \in K$. Such a substitution $\tilde{\sigma}_*$ can be given by the following way: Let σ_* be a substitution determined by a map $\sigma : K \to K^{G(b)}$. We put $\text{Sgn}(x) := \{\tau \in \{-1, 1\}^s; \ \tau \otimes x \in \mathbb{N}^s\}$ for $x \in \mathbb{Z}^s$, $\Xi := \text{Gen}(K, \mathbb{Z}^s)$, $U(W) := (w_x)_{x \in U(X)}$ for $W = (w_x)_{x \in X} \in \Xi$. Then we can define an additive map $\tilde{\sigma}_*$, which takes each $\langle a, x \rangle$ to a word on the lattice points in certain parallelepiped of dimension s, defined by

$$\text{Dom}\,(\tilde{\sigma}_*(\langle a, x \rangle)) := \bigcup_{\tau \in \text{Sgn}(x)} U(\tau \otimes (G(b) + \rho_b^{-1}(\rho_b(x)o)))$$

$$\tilde{\sigma}_*(\langle a, x \rangle)(U(\tau \otimes y + \rho_b^{-1}(\rho_b(x)o)))$$
$$:= {}^{\tau}\sigma(y) \ (\tau \in \text{Sgn}(x), \ y \in G(b)),$$

which can be extended to a substitution $\tilde{\sigma} : \text{Gen}\,(K, \mathbb{Z}^s) \to \text{Gen}\,(K, \mathbb{Z}^s)$. Note that $z \in \cap_{\tau \in S} U(\tau \otimes G(b))\,(S \subset \text{Sgn}(x))$ implies ${}^{\tau_1}\sigma(z) = {}^{\tau_2}\sigma(z)$ for all $\tau_1, \tau_2 \in S$, so that $\tilde{\sigma}_*$ is well-defined.

We say a word $W \in K^{\mathbb{Z}^s}$ is automatic with respect a basis U if there exists an automaton $\mathcal{M} = (@\,, K, G(b), \zeta)$ and a projection $\pi : K \to F$ such that ${}^{\tau, U}W$ coincides with $\pi({}^{\tau}V) := (\pi({}^{\tau}v_x))_{x \in \mathbb{N}^s}$ for the word ${}^{\tau}V = ({}^{\tau}v_x)_{x \in \mathbb{N}^s}$ generated by ${}^{\tau}\mathcal{M}$ for all $\tau \in \{1, -1\}^s$. We say a word W is finitely automatic if it is automatic with a finite set K.

We denote by $[x]$ the greatest integer not exceeding a real number x. We define $[i]_c \in \tilde{K}_c$ for $i \in \mathbb{Z} \cup \{\infty\}$, $c \in \tilde{K}_\infty$, $c \neq 0$ by

$$[i]_c := i - c[i/c] \ (i \neq \infty, \ c \neq \infty),$$
$$[i]_\infty := i, \quad [\infty]_c := \infty.$$

Note that if $i \neq \infty$, $c \neq \infty$, then $0 \leq [i]_c < c$ and $[i]_c \equiv i \pmod{c}$.

Now, we suppose $A \in (P_u(c))$, and consider $W(A; c) = (w(A; c)_x)_{x \in \mathbb{Z}^s} \in \tilde{K}_c^{\mathbb{Z}^s}$, the characteristic word of the partition (2) for $2 \leq c \leq \infty$. By Corollary 2, we have

Lemma 5.
$$w(A; c)_x = [\text{ind}\,_A(x)]_c \quad (x \in \mathbb{Z}^s)$$

holds for all $A \in (P_u(c)) \cap M_r(s; \mathbb{Z})$, $1 < c \leq \infty$.

We denote by $D = D(d)$ with $d = {}^T(d_1, \cdots, d_s) \in \mathbb{Q}^s$ the diagonal matrix with d_i as its (i, i)-component for $1 \leq i \leq s$. We define an equivalence relation $\sim$ on $M(s; \mathbb{Q})$ by

$$A \sim B \ (A, B \in M(s; \mathbb{Q})) \iff \exists V \in GL(s; \mathbb{Z}^s) \text{ such that } VA = B.$$

Theorem 3. *Let* $A \in M_r(s)$ *be a matrix such that there exists an integer* $k > 0$, *and a unimodular matrix* $U \in GL(s; \mathbb{Z}^s)$ *satisfying* $A^{-kn-j}U \sim$

$A^{-j}UD^n$ for all $n \geq 0$, $0 \leq j < k$ with $D := D(\boldsymbol{d})$, $\boldsymbol{d} := {}^T(d_1^{-1}, \cdots, d_s^{-1})$, $d_i \in \mathbb{Z}$, $|d_i| > 1$ for all $1 \leq i \leq s$. Then $A \in (P_u(c))$, and the word $W(A; c)$ becomes a fixed point, with respect to the basis U, of a substitution $\tilde{\sigma}_*$ determined by a map $\sigma = \sigma^{(c)} = \sigma^{(U^{-1}AU,c)}$ over $\tilde{K}_c$ of size $G := G(|d_1|, \cdots, |d_s|)$, so that $W(A; c) = \lim \tilde{\sigma}_*^n(\langle \infty, \boldsymbol{o}\rangle)$ holds. The map σ is defined by the following:

$$\sigma^{(c)} : \tilde{K}_c \to \tilde{K}_c^G, \ \sigma^{(c)}(h) = \left(\sigma_i^{(c)}(h)\right)_{i \in G} \in \tilde{K}_c^G,$$

(*case* $c = \infty$)

$$\sigma_o^{(\infty)}(\infty) := \infty, \ \sigma_o^{(\infty)}(h) := h + k, \ (0 \leq h < \infty),$$
$$\sigma_i^{(\infty)}(h) := j \ (i \in I_j, \ 0 \leq j < k, \ 0 \leq h < \infty),$$

(*case* $1 < c < \infty$)

$$\sigma_i^{(c)}([h]_c) := [\sigma_i^{(\infty)}(h)]_c \ (i \in G, \ 0 \leq h \leq \infty),$$

where

$$I_j := \{\boldsymbol{x} \in G \setminus \{\boldsymbol{o}\}; \ \text{ind}_{U^{-1}AU}(\boldsymbol{x}) = j\} \ (0 \leq j < k).$$

N.B.: The map σ given above is determined by $U^{-1}AU$, k and c. We remark that $\bigcup_{0 \leq j < k} I_j = G \setminus \{\boldsymbol{o}\}$ forms a disjoint union, and $I_j \neq \phi$ for all $0 \leq j < k$, as we shall see in the proof. Notice that $\sigma_i^{(\infty)}(h)$ does not depend on h if $i \in G \setminus \{\boldsymbol{o}\}$, and that $[h_1]_c = [h_2]_c \ (1 < c < \infty)$ implies $[h_1 + k]_c = [h_2 + k]_c$, so that $\sigma^{(c)} \ (1 < c \leq \infty)$ given in Theorem 3 is well-defined, which is extended to a substitution strictly over $\tilde{K}_c$ of size G.

Proof. The assumption of the lemma implies $U^{-1}A^{-k}U \sim D$, so that $A^k \in (\text{Bdd})$. Hence, we get $A \in (\text{Bdd})$, which together with Theorem 1 implies $A \in (P_u(c))$ for all $1 < c \leq \infty$. We put $B := U^{-1}AU$. It suffices to show that $W(B; c)$ becomes a fixed point of a substitution $\sigma_*(B; c)$ over $\tilde{K}_c$ of size G with respect to the basis given by the unit matrix $E = E_s$. From $A^{-kn-j}U \sim A^{-j}UD^n$ it follows

$$B^{-kn-j} = V_n B^{-j} D^n \ (V_n \in GL(s; \mathbb{Z})), \tag{11}$$

in particular, $B^{-k} = V_1 D$, so that for $\boldsymbol{x} \in \mathbb{Z}^s$

$$\boldsymbol{x} \equiv \boldsymbol{o} \pmod{G} \Longleftrightarrow D\boldsymbol{x} \in \mathbb{Z}^s \Longleftrightarrow B^{-k}\boldsymbol{x} \in \mathbb{Z}^s \Longleftrightarrow \text{ind}_B(\boldsymbol{x}) \geq k.$$

Since

$$x \in G \setminus \{o\} \Longrightarrow Dx \notin \mathbb{Z}^s \Longleftrightarrow \text{ind}_B(x) < k,$$

we get

$$G \setminus \{o\} = \bigcup_{0 \leq j < k}^{\circ} I_j, \tag{12}$$

where I_j is the subset of $G \setminus \{o\}$ given in Theorem 3. We shall soon see that $I_j \neq \phi$ for all $0 \leq j < k$. We put $b := {}^T(|d_1|, \ldots, |d_s|)$. Noting that $D(m \otimes b) \in \mathbb{Z}^s$, i.e., $B^{-k}(m \otimes b) \in \mathbb{Z}^s$ for any $m \in \mathbb{Z}^s$, we get

$$\begin{aligned}
& x \in I_j, \text{ and } x \equiv y \pmod{G}, \ y \in \mathbb{Z}^s \\
\Longleftrightarrow & B^{-j}x \in \mathbb{Z}^s, \ B^{-j-1}x \notin \mathbb{Z}^s, \text{ and} \\
& x = y + m \otimes b \text{ for an element } m \in \mathbb{Z}^s \\
\Longleftrightarrow & B^{-j}(y + m \otimes b) \in \mathbb{Z}^s, \text{ and } B^{-j-1}(y + m \otimes b) \notin \mathbb{Z}^s \\
\Longleftrightarrow & B^{-j}y \in \mathbb{Z}^s, \text{ and } B^{-j-1}y \notin \mathbb{Z}^s
\end{aligned}$$

for any $0 \leq j < k$. Hence we get for $y \in \mathbb{Z}^s \setminus \{o\}$

$$\text{ind}_B(y) = j \Longleftrightarrow x \in I_j = I_j(B), \text{ and } x \equiv y \pmod{G}$$

for any $0 \leq j < k$. Therefore, we obtain

$$T_j = T_j(B) = I_j(B) + (\mathbb{Z}d_1 e_1 + \cdots + \mathbb{Z}d_s e_s), \ 0 \leq j < k, \tag{13}$$

where e_i denotes the i-th fundamental vector, and $\mathbb{Z}d_1 e_1 + \cdots + \mathbb{Z}d_s e_s$ is the $\mathbb{Z}$-submodule of $\mathbb{Z}^s$ generated by $\{d_i e_i; \ 1 \leq i \leq s\}$. In view of Corollary 2, we have $T_j \neq \phi$, which implies $I_j \neq \phi$. We write $u \succ v$ for $u, v \in G^*$ if v is a suffix of u. By (13), we get for $0 \leq j < k$

$$x \in T_j \cap \mathbb{N}^s \Longleftrightarrow \rho_b(x) \succ i \in I_j \subset G,$$

and by (11)

$$x \in T_{kn+j} \cap \mathbb{N}^s \Longleftrightarrow \rho_b(x) \succ io^n \in I_j o^n \subset G^*$$

for all $0 \leq j < k, \ n \geq 0$. Therefore, if we consider an automaton

$$\mathcal{M} = (\infty, \tilde{K}_\infty, G, \zeta)$$

with $\zeta : \tilde{K}_\infty \times G \to \tilde{K}_\infty$ defined by

$$\begin{aligned}
& \zeta(\infty, o) = \infty, \ \zeta(\infty, i) = j \text{ if } i \in I_j \ (0 \leq j < k), \\
& \zeta(h, o) = h + k, \ \zeta(h, i) = j \text{ if } i \in I_j \ (0 \leq j < k, \ 0 \leq h < \infty),
\end{aligned} \tag{14}$$

then for $x \in \mathbb{N}^s$,

$$\text{ind}_B(x) = j \iff \zeta(\infty, \rho_b(x)) = j \ (0 \leqq j \leqq \infty)$$

holds. We get by (12)

$$\tau \otimes (G \setminus \{o\}) = \bigcup_{0 \leqq j < k}^{\circ} \tau \otimes I_j,$$

and by (13)

$$\begin{aligned}
x \in T_j \cap \tau \otimes \mathbb{N}^s &\iff \tau \otimes x \in (\tau \otimes T_j) \cap \mathbb{N}^s \\
&\iff \tau \otimes x \in (\tau \otimes I_j + \mathbb{Z}d_1 e_1 + \cdots + \mathbb{Z}d_s e_s) \cap \mathbb{N}^s \\
&\iff \tau \otimes_G x \in \tau \otimes_G I_j,
\end{aligned}$$

since $S \subset \tau \otimes \mathbb{N}^s$ implies $\tau \otimes S \subset \mathbb{N}^s$. Therefore, we obtain for $x \in \tau \otimes \mathbb{N}^s$

$$\text{ind}_B(x) = j \iff {}^\tau \zeta(\infty, \rho_b(x)) = j.$$

Hence, by Lemma 5, we see that ${}^\tau \mathcal{M}$ generates the (τ, E)-sector ${}^{\tau, E}W(B; \infty)$ of $W(B; \infty)$ for all $\tau \in \{1, -1\}$. Therefore, if we define a map $\sigma^{(\infty)} = \sigma^{(B;\infty)}$ by σ in (10) with ζ given by (14), then we can construct a substitution $\sigma_*^{(\infty)}$ on $\mathbb{N}^s$ strictly over $\tilde{K}_\infty$ of size G determined by $\sigma^{(\infty)}$. The word $W(B; \infty)$ becomes the fixed point of $\sigma_*^{(\infty)}$ with respect to the basis E. We can define a map $\sigma^{(c)} = \sigma^{(B;c)}$ by

$$\sigma_x^{(c)}([a]_c) =: [\sigma_x^{(B;\infty)}(a)]_c, \ x \in G.$$

Then, we can define a substitution $\sigma_*^{(c)}$ $(1 < c < \infty)$ on $\mathbb{N}^s$ determined by $\sigma^{(c)}$, which becomes a substitution strictly over $\tilde{K}_c$. Then, by Lemma 5, we see that $W(B; c)$ $(1 < c < \infty)$ becomes the fixed point of the substitution $\sigma_*^{(c)}$ of size G with respect the basis E. As we have mentioned, we can construct a subtitution $\tilde{\sigma}_*^{(c)}$ on $\mathbb{Z}^s$, and $W = \lim \tilde{\sigma}_*^{(c)n}(\langle \infty, o \rangle)$ $(1 < c \leqq \infty)$.   ∎

Remark 3. $\text{ind}_A(x) = \text{ind}_A(-x)$, ${}^{-\tau, U}W(A; c) = {}^{\tau, U}W(A; c)$, and ${}^{-\tau}\sigma = {}^\tau \sigma$ always hold for all $A \in (P_u(c))$, $\tau \in \{1, -1\}^s$, $U \in GL(s; \mathbb{Z})$. If $d = {}^T(2^{-1}, \cdots, 2^{-1})$, and $k = 2$, then ${}^\tau \sigma = \sigma$ holds for all $\tau \in \{1, -1\}^s$ by the proof of Theorem 3, so that all the (τ, U)-sectors of $W(A; c)$ are identical; while in general, such a symmetry can not be found, cf. (vii), Section 3.

In what follows, we write by σ, instead of writing σ_* (or $\tilde{\sigma}_*$), the substitution on $\mathbb{N}^s$ (or on $\mathbb{Z}^s$) over K of size G determined by a map

$\sigma : K \to K^G$ as in the usual case of dimension one. So, for a given map $\sigma : K \to K^G$, we say the substitution σ (on $\mathbb{N}^s$), or the substitution σ (on $\mathbb{Z}^s$) of size G with respect to a basis U, etc.

We denote by $\mathcal{D} = \mathcal{D}(k) = \mathcal{D}(k; d_1, \cdots, d_s) = \mathcal{D}(k; d_1, \cdots, d_s; s)$ the class of matrices A satisfying the assumption stated in Theorem 3; and we say A is of the class $\mathcal{D}$ with respect to a basis U. By $\mathcal{C} = \mathcal{C}(k) = \mathcal{C}(k; d) = \mathcal{C}(k; d; s)$, we denote the class of matrices defined by

$$\mathcal{C} := \{ A \in M_r(s; \mathbb{Z}); \; \exists k \in \mathbb{N} \setminus \{0\} \text{ such that } A^k \sim dE_s \}, \quad d > 1.$$

We can easily see the following

Remark 4. If $A \in \mathcal{C}$, then $A \in \mathcal{D}$ with respect to any basis. If $A \in \mathcal{C}(k; d; s)$, then $(\det A)A^{-1} \in \mathcal{C}(k; d^{s-1}; s)$. Theorem 3 gives an algorithm of a construction of σ as far as $A \in \mathcal{D}$, since I_j can be determined effectively, and so can be the σ.

We denote by $A_1 \oplus \cdots \oplus A_t$ the direct sum of $A_1, \ldots, A_t$ for square matrices, and by $\sigma_1 \oplus \cdots \oplus \sigma_s$ the substitution of size $G(d_1) \times \cdots \times G(d_t)$ over $K \subset \mathbb{N} \bigcup \{\infty\}$ defined by

$$(\sigma_1 \oplus \cdots \oplus \sigma_s)_{T(T_{x_1}, \ldots, T_{x_t})}(a) := \min\{\sigma_{i_{x_i}}(a); \; 1 \leqq i \leqq t\}$$

for given substitutions σ_i of size $G(d_i)$ over K. Then we get the following

Corollary 3. *Let* $A_i \in \mathcal{D}(k_i; d_1^{(i)}, \ldots, d_{s_i}^{(i)})$ *with respect to a basis* U_i *for all* $1 \leqq i \leqq t$, *and let* $A = A_1 \oplus \cdots \oplus A_t$, $s = s_1 + \cdots + s_t$. *Then*

$$A \in \mathcal{D}(m; d_1^{(1)^{h_1}}, \ldots, d_{s_1}^{(1)^{h_1}}, \ldots, d_1^{(t)^{h_t}}, \ldots, d_{s_t}^{(t)^{h_t}}; s)$$

with respect to a basis $U_1 \oplus \cdots \oplus U_t$, *and* $\sigma^{(U^{-1}AU; c)}$ *in Theorem 3 can be given by*

$$\sigma_x^{(U^{-1}AU; c)}(a) = [(\sigma^{(U_1^{-1}A_1U_1; \infty)^{h_1}} \oplus \cdots \oplus \sigma^{(U_t^{-1}A_tU_t; \infty)^{h_t}})_x(a)]_c,$$

which is of size $G = G^{(h_1)} \times \cdots \times G^{(h_t)}$, *where* $h_i := m/k_i$, $m :=$ LCM$\{k_i; \; 1 \leqq i \leqq t\}$, $x \in G$, *and* $G^{(h_i)} := G(d_1^{(i)^{h_i}}, \ldots, d_{s_i}^{(i)^{h_i}})$ $(1 \leqq i \leqq t)$.

3. EXAMPLES (NON-SINGULAR CASE)

We use the notation

$$[a_{11}, \ldots, a_{1s} // \cdots // a_{s1}, \ldots, a_{ss}] \quad (s \geqq 2)$$

for a matrix $A = (a_{ij})_{1 \leq i,j \leq s} \in M(s; \mathbb{Q})$ in some cases. If $\pm 1 \notin \Lambda(A) \subset \mathbb{Q}$, and some of the eigenvectors of A form a basis U of $\mathbb{Z}^s$ as a $\mathbb{Z}$-module, then it is clear that $A \in \mathcal{D}$; and by Corollary 3, it is easy to find a substitution having $W(A; c)$ as its fixed point with respect to the basis U. While, in some cases, we can construct a substitution for A with $\pm 1 \notin \Lambda(A) \subset \mathbb{Q}$ such that any eigenvectors can not form a $\mathbb{Z}$-basis. For instance,

(i) $A = [2, -4// - 2, 0]$ $(\Lambda(A) = \{-2, 4\})$. $x_1 = {}^T(1, 1)$, $x_2 = {}^T(2, -1)$ are eigenvectors, so that any pair of eigenvectors in $\mathbb{Z}^2$ can not be a $\mathbb{Z}$-basis, while, for $U = [1, 1//0, 1]$, $A^{-n}U \sim D(4^{-n}, 2^{-n})$ $(n \geq 0)$ can be shown by induction. Hence, $W(A; c)$ becomes the fixed point, with respect to the basis U, of a substitution σ of size 4×2 defined by

$$\sigma(\infty) = \begin{matrix} 0 & 0 & 0 & 0 \\ \infty & 0 & 0 & 0 \end{matrix}, \ \sigma(a) = \begin{matrix} 0 & 0 & 0 & 0 \\ b & 0 & 0 & 0 \end{matrix} \ (a \in K_c, \ b = [a + 1]_c).$$

The right-hand sides given above denote 2-dimensional words of size 4×2; for instance, $\sigma_o(a) = b$, and $\sigma_x(a) = 0$ for all ${}^T x \neq o$, $a \in K_c$.

For some A, $A \in \mathcal{C}$ (or $A \in \mathcal{D}$) can be seen by its shape. See the examples (ii–v) below. Note that, in general, $\Lambda(A) \subset \mathbb{Q}$ does not hold, cf. (ii, v, vi) given below.

(ii) $A = [1, -1//1, 1]$. $\Lambda(A) = \{1 \pm \sqrt{-1}\}$. $A \in \mathcal{C}(2; 2; 2)$. (It is clear that A^4 is a scalar matrix, so that $A \in \mathcal{C}(4)$.) For the set Γ satisfying (2), the sets Γ, $A\Gamma$, $A^2\Gamma, \ldots$ are of high symmetry similar to each other. $W(A; c)$ becomes the fixed point, with respect to the basis E, of a substitution σ given by

$$\sigma(\infty) = \begin{matrix} 0 & 1 \\ \infty & 0 \end{matrix}, \ \sigma(a) = \begin{matrix} 0 & 1 \\ b & 0 \end{matrix} \ (a \in K_c, \ b = [a + 2]_c).$$

(iii) $aE_s + T \in \mathcal{C}(|a|; |a|^{|a|}; s)$ for $a \in \mathbb{Z}$, $|a| > 1$, $T = (t_{ij})_{1 \leq i,j \leq s}$, $t_{ij} = 0$ $(i \geq j)$ satisfying $(|a| - 1)(|a| - 2) \cdots (|a| - m + 1)T_m \equiv O$ $(\bmod \ a^{m-1} \cdot m!)$ for all $1 < m < s$. In particular, $[a, b//0, a] \in \mathcal{C}$.

(iv) $(a_i \delta_{[i+k]_s, j})_{1 \leq i,j \leq s} \in \mathcal{C}(s; |a_1 \cdots a_s|; s)$ $(a_i \in \mathbb{Z}, |a_1 \cdots a_s| > 1, k \in \mathbb{Z}, GCD(k, s) = 1)$, where $\delta_{i,j}$ is Kronecker's delta.

(v) We put

$$A(a, b) = \begin{pmatrix} {}^T o & a \\ E_{s-1} & b \end{pmatrix}, \ U = \begin{pmatrix} 1 & & 1 \\ & \ddots & \vdots \\ & & 1 \end{pmatrix} \in M_r(s; \mathbb{Z}), \ |a| > 1.$$

Then, $A(a, ac) \in \mathcal{C}(s; |a|; s)$ for $c \in \mathbb{Z}^{s-1}$. In fact

$$A^{-sn-j} \sim |a|^{-n} \begin{pmatrix} |a|^{-1} E_j & O \\ O & E_{s-j} \end{pmatrix}, \quad 0 \leq j < s, \ n \geq 0,$$

which can be shown by induction. The word $W(A; c)$ becomes the fixed point, with respect the basis E, of the substitution σ defined by the following:

$\sigma_x(t) := 0$ for $x_1, \cdots, x_s \neq 0$,
$\sigma_x(t) := [j]_c$ for $x_1 = x_2 = \cdots = x_j = 0$, $x_{j+1} \neq 0$ with $1 \leq j < s$,
$\sigma_o(t) := [t + s]_c$ $(t \neq \infty)$, $\sigma_o(\infty) := \infty$

where $G := G(|a|, \ldots, |a|) \subset \mathbb{N}^s$, $x = {}^T(x_1, \ldots, x_s) \in G$, $t \in \tilde{K}_c$. Let B be the adjugate matrix of $A(a, b)$ with $b := {}^T(a - 1, \ldots, a - 1) \in \mathbb{Z}^{s-1}$. Then $cB \in \mathcal{D}(1; ac, \ldots, ac, c; s)$ with respect to the basis U for $c \in \mathbb{Z}$, $|c| > 1$.

As we have already seen in Lemma 5 that the word $W(A; c)$ with $c < \infty$ can be obtained from $W(A; \infty) = (\text{ind}_A(x))_{x \in \mathbb{Z}^s}$ by taking residues with mod c. For the computation of the values $\text{ind}_A(x)$, it is convenient, in some cases, to consider the Hermitian canonial form $H_n = H(d^n A^{-n})$ for $d^n A^{-n}$ with $d = \det A$. Here, the Hermitian canonical form $H = H(A)$ for a matrix $A \in M_r(s; \mathbb{Z})$ is defined to be a matrix H, that is uniquely determined by

$$A \sim H = (h_{ij})_{1 \leq i, j \leq s}, \ h_{11} > 0, \ 0 \leq h_{ij} < h_{jj} \ (1 \leq i < j \leq s),$$
$$h_{ij} = 0 \ (1 \leq j < i \leq s),$$

which can be obtained by elementary transformations by multiplying A by unimodular matrices from the left. We put

$$C_n(A, U) := d^{-n} H(d^n U^{-1} A^{-n} U)$$
$$(A \in M_r(s; \mathbb{Z}), \ d := \det A, \ U \in GL(s; \mathbb{Z})).$$

Then $C_n(A, U) \sim A^{-n} U$ $(n \in \mathbb{Z})$, so that

$$A^{-n} U x \in \mathbb{Z}^s \iff C_n(A, U) x \in \mathbb{Z}^s,$$

by which we can get the value $\text{ind}_{U^{-1} A U}(x)$. In some cases, for $A \notin \mathcal{C}$, we can conclude that $A \in \mathcal{D}$ by considering $C_n(A, E)$. For instance,

(vi) $A = [0, 4, -2//2, 2, -2// - 1, -2, 2]$. $\Lambda(A) = \{-2, 3 \pm \sqrt{7}\}$. By induction, we can show

$$C_{2n}(A, E) = 2^{-2n}[1, 2^n - 1, 0//0, 2^n, 0//0, 0, 2^n],$$

$$C_{2n+1}(A, E) = 2^{-2n}[1, 2^n - 1, 0//0, 2^n, 0//0, 0, 2^{n+1}].$$

Hence, setting $U = [1, 1, 0//0, 1, 0//0, 0, 1]$, we obtain

$$C_{2n}(A, U) = 2^{-2n}D(1, 2^n, 2^n), \quad C_{2n+1}(A, U) = 2^{-2n}D(1, 2^n, 2^{n+1}),$$

which implies $A \in \mathcal{D}(2; 4, 2, 2; 3)$

In the following examples, C_n indicates the matrix $C_n(A; E)$. We put

$$H_n(\{a_m\}_{m \in \mathbb{Z}}) := \begin{pmatrix} a_n & a_{n+1} & a_{n+2} \\ a_{n+1} & a_{n+2} & a_{n+3} \\ a_{n+2} & a_{n+3} & a_{n+4} \end{pmatrix} \in M(3; \mathbb{Z}).$$

(vii) $A_n = H_{4n}(\{a_m\}_{m \in \mathbb{Z}}) \in \mathcal{C}(3; 3; 3)$ holds for all $n \in \mathbb{Z}$, where $\{a_m\}_{m \in \mathbb{Z}}$ is a linear recurrence sequence having $x^3 - x^2 - 1$ as its characteristic polynomial determined by $a_0 = a_1 = 1$, $a_2 = 0$. In fact, using $A_{n+1} = U_L A_n = A_n U_R = A_0 U_R^{n+1}$ ($U_R = U^4$, $U_L = {}^T U_R$), we get

$$A_{n+1}^{-1} = U_R^{-1} A_n^{-1} \sim A_n^{-1}, \ A_n^{-1} \sim A_0^{-1} \sim C_1, \ A_0^{-2} \sim C_2;$$
$$A_{n+1}^{-2} = U_R^{-1} A_n^{-2} U_L^{-1} \sim A_n^{-2} U_L^{-1}, \ C_2 U_L^{-1} \sim C_2;$$
$$A_{n-1}^{-2} = U_R A_n^{-2} U_L \sim A_n^{-2} U_L, \ C_2 U_L \sim C_2,$$

where

$$U = [0, 0, 1//1, 0, 0//0, 1, 1] \in GL(3; \mathbb{Z}^s),$$
$$C_1 = 3^{-1}[1, 2, 2//0, 3, 0//0, 0, 3],$$
$$C_2 = 3^{-1}[1, 0, 1//0, 1, 2//0, 0, 3]$$

which implies $A_n^{-1} \sim C_1$, $A_n^{-2} \sim C_2$ for all $n \in \mathbb{Z}$. By the similar manner, we get $A_n^2 \sim 3C_1$ for all $n \in \mathbb{Z}$, so that $A_n^3 \sim 3C_1 A_n = 3C_1 A_0 U_R^n$, which together with $3C_1 A_0 = [3, 3, 6//3, 0, 3//0, 3, 6] \sim 3E$, so that $A_n \in \mathcal{C}(3; 3; 3)$. Hence, using $A_n^{-1} \sim C_1$ and $A_n^{-2} \sim C_2$, we get $\sigma = \sigma^{(U^{-1}AU;c)}$ over $\tilde{K}_c$:

	$z=0$				$z=1$			$z=2$		
	2	0	0	1	1 0 0			0 2 0		
$\sigma(a) =$	1	0	1	0	0 0 2			1 0 0		$(b = [a+3]_c)$.
	0	b	0	0	0 1 0			0 0 1		
y/x	0	1	2							

The right-hand side given above denotes a 3-dimensional word of size $3 \times 3 \times 3$; for instance, $\sigma_o(a) = b$, and $\sigma_x(a) = 2$ for all $a \in \tilde{K}_c$

if $^T x = (x, y, z) = (2, 1, 1)$, $(1, 2, 2)$. We remark that $^T \sigma \neq \sigma$ holds for $\tau = (-1, 1, 1)$, $(1, -1, 1)$, $(1, 1, -1)$ for $c \geqq 3$.

It is remarkable that Example (vii) gives infinitely many matrices $A(n)$ with nontrivial one parameter n such that $W(A(n); c)$ is independent of n. We can make some variants of such an example. For instance, the adjugate matrix B_n of $H_{2n}(\{b_m\}_{m \in \mathbb{Z}})$ belongs to $\mathcal{C}(3; 4; 3)$ if $\{b_m\}_{m \in \mathbb{Z}}$ is a linear recurrence sequence having $x^3 - x^2 - x - 1$ as its characteristic polynomial with $b_0 = b_2 = 0$, $b_1 = 1$.

4. EXAMPLES (SINGULAR CASE)

In this section, we suppose $A \in M(s; \mathbb{Z})$, $\det A = 0$. We consider partitions (2) with $s > 1$ together with their variants

$$\overset{\circ}{\underset{0 \leqq j < c}{\bigcup}} A^j \Gamma = \mathbb{Z}^s, \ s > 1, \ c > 1. \tag{15}$$

If $c = \infty$, or $s = 1$ then there does not exist a partition (2), nor (15) for any singular matrix A. Thus, we assume $c < \infty$ in this section. We denote by T_A the endomorphism on $\mathbb{Z}^s$ induced by $A \in M(s; \mathbb{Z})$ as a $\mathbb{Z}$-module. Then, we easily get the following

Remark 5. If there exists a partition (2), then

$$\ker T_A \subset A^{c-1} \Gamma \cup \{o\}, \ (\mathbb{Z}^s \setminus \operatorname{Im} T_A) \subset \Gamma$$

holds; if there exists a partition (15), then

$$o \in A^{c-1} \Gamma, \ \ker T_A \subset A^{c-2} \Gamma \cup A^{c-1} \Gamma, \ (\mathbb{Z}^s \setminus \operatorname{Im} T_A) \subset \Gamma$$

holds. In particular, $\ker T_A \subset \operatorname{Im} T_A$ follows.

(i) Let $A = (\delta_{i,j-1})_{1 \leqq i,j \leqq s}$ $(s \geqq 2)$. Then there exists a partition (2) if and only if c is a divisor of s. If $s = ct$ for an integer t, such a partition is uniquely determined:

$$A^j \Gamma = \underset{0 \leqq i < t}{\bigcup} \Xi_{ci+j}, \ 0 \leqq j < c \tag{16}$$

holds, where $\Xi_i := \mathbb{Z}^{s-i-1} \times (\mathbb{Z} \setminus \{o\}) \times \{o\}^i$, $0 \leqq i < s$.

Proof. One can make sure that if we define Γ by (16) with $j = 0$, then (16) holds for all $1 < j < c$, and the sets $A^j \Gamma$ form a partition (2). Suppose that there exists a partition (2) for a set $\Gamma \subset \mathbb{Z}^s \setminus \{o\}$.

$A^s = O$ implies $c \leq s$, since, otherwise, $A^{c-1}\Gamma \ni o$ follows. Since $\Xi_{c-1} \supset \mathbb{Z}^s \setminus \text{Im } T_A$, we get $\Gamma \supset \Xi_0$ by Remark 5, so that

$$A^j\Gamma \supset \Xi_j \text{ for all } 0 \leq i < c \tag{17}$$

holds. If $c = s$, then (16) follows. So, we may suppose $c < s$. Let us assume that an element $x \in \Xi_c$ does not belong to Γ. Then $x \in A^i\Gamma$ for an integer $0 < i < c$, so that there exists a $y \in \Gamma$ such that $A^iy = x \in \Xi_c$. Hence, setting $y = {}^T(y_1, \cdots, y_s)$, we get $y_{s-c+i} \neq 0$, $y_{s-c+i+1} = \cdots = y_s = 0$, so that $y \in \Xi_{c-i}$. Hence, $y \in \Xi_{c-i} \subset A^{c-i}\Gamma$ with some $0 < i < c$ follows from (17), which contradicts $y \in \Gamma$. Therefore, in the case $c < s$, we must have $\Gamma \supset \Xi_c$. Now, suppose $s - c = i$ with $0 < i < c$. Then $A^{i-1}\Gamma \supset \Xi_{c+i-1} = \Xi_{s-1} = (\mathbb{Z}\setminus\{0\}) \times \{0\}^{s-1}$ follows from $\Gamma \supset \Xi_c$, so that $A_i\Gamma \ni o$, consequently we can not have (2). Thus, we get $s - c = i \geq c$, i.e., $s \geq 2c$. If $s = 2c$ holds, then we get (16). Suppose $s > 2c$. Then, we can show $\Gamma \supset \Xi_{2c}$, and then a contradiction by setting $i = s - 2c$ with $0 < i < c$ in the same way as above. Thus, we get $s = 3c$, or $s > 3c$. If $s = 3c$, then (16) follows. Repeating the argument, we can arrive at (i). ∎

(ii) Let $A = (a_i\delta_{i,j-1})_{1 \leq i,j \leq s}$ with integers a_i, $|a_i| > 1$ for all $1 \leq i \leq s - 1$ ($s \geq 2$). Then there exists a partition (15) if and only if $c = 2$, and such a partition is uniquely determined:

$$\Gamma = \left(\mathbb{Z}^{s-1} \times (\mathbb{Z} \setminus \{0\})\right) \cup \left(\bigcup_{1 \leq i < s} (\mathbb{Z}^{i-1} \times (\mathbb{Z} \setminus a_i\mathbb{Z}) \times \mathbb{Z}^{s-i})\right),$$
$$A\Gamma = a_1\mathbb{Z} \times \cdots \times a_{s-1}\mathbb{Z} \times \{0\}.$$

Proof. $x := {}^T(1, 0, \ldots, 0) \notin \text{Im } T_A$, so that $x \in \Gamma$ follows from Remark 5, and so, $Ax = A^2x = A^3x = \cdots = o$, which implies $c \leq 2$. Suppose that (15) holds with $c = 2$. Since any element of $X := (\mathbb{Z}^{s-1} \times (\mathbb{Z}\setminus\{0\})) \cup (\bigcup_{1 \leq i < s}(\mathbb{Z}^{i-1} \times (\mathbb{Z} \setminus a_i\mathbb{Z}) \times \mathbb{Z}^{s-i}))$ can not be an element of Im T_A, we get $\Gamma \supset X$ by Remark 5. Hence, we obtain

$$A\Gamma \supset AX = Y := a_1\mathbb{Z} \times \cdots \times a_{s-1}\mathbb{Z} \times \{o\}.$$

Note that $s \geq 2$ implies $AX \ni o$. Noting $X \cup Y = \mathbb{Z}^s$ is a disjoint union, we get $\Gamma = X$ and $A\Gamma = Y$. ∎

We can prove the following (iii) by almost the same fashion as (ii).

(iii) Let $A = (a_i\delta_{i,j-1})_{1 \leq i,j \leq s}$ with integers a_i satisfying $a_i \neq 0$ ($1 \leq i \leq t$), $|a_1| > 1$, $a_j = 0$ ($t + 1 \leq j \leq s - 1$), $s \geq 3$, $1 \leq t \leq s - 2$. Then there exists a partition (15) if and only if $c = 2$, and such a partition is uniquely determined:

$$\Gamma = \left(\mathbb{Z}^t \times (\mathbb{Z}^{s-t} \setminus \{o\})\right) \cup \left(\bigcup_{\substack{1 \leq i \leq t, \\ |a_i| \neq 1}} (\mathbb{Z}^{i-1} \times (\mathbb{Z} \setminus a_i \mathbb{Z}) \times \mathbb{Z}^{t-i} \times \{0\}^{s-t}) \right),$$

$A\Gamma = a_1 \mathbb{Z} \times \cdots \times a_{s-1} \mathbb{Z} \times \{0\}^{s-t}$.

5.   NON-Σ-PERIODICITY OF WORDS AND TESSELLATIONS

We give some new definitions of non-periodicity for words and sets of dimension s as follows in a general situation.

In this section, $s \geq 1$ denotes a fixed integer. We denote by $\mathbb{R}^s$, the Euclidean space with the norm $\| * \|$ induced by the usual inner product $(*, *)$. For a set $S \subset \mathbb{R}^s$, we denote by $\overline{S}$ (resp., S°) the closure (resp., the interior) of S with respect to the usual topology. We mean by a **cone** J a closed convex set satisfying the following conditions:

  (i) $J^\circ \neq \phi$,

  (ii) If $x \in J$, then $rx \in J$ for all $r \in \mathbb{R}_+$,

where $\mathbb{R}_+$ is the set of non-negative numbers. We denote by Ψ_s the set of all cones in $\mathbb{R}^s$. Note that $\mathbb{R}^s$, $\mathbb{R}^s_+$, a half-space in $\mathbb{R}^s$ are cones $\in \Psi_s$, and that any cone $J \in \Psi_t$, $J \subset \mathbb{R}^t \times \{0\}^{s-t} \subset \mathbb{R}^s$ $(t < s)$ can not be an element of Ψ_s. Any cone J becomes a monoid with o as its unit with respect to the addition, so that $J \supset x + S$ holds for any $x \in J$, $S \subset J$. We say a set $S \subset \mathbb{R}^s$ is **spreading** if for any bounded set $B \subset \mathbb{R}^s$, there exists an element $x \in \mathbb{R}^s$ such that $x + B \subset S$. For instance, $\bigcup_{n \geq 1} B(\log(1+n);\ ^T(n^2, n^3, \ldots, n^{s+1}))$ is a spreading set, where $B(r, a)$ denotes the open ball $\{x \in \mathbb{R}^s;\ \|a - x\| < r\}$; and so is a set $x + J$ for all $x \in \mathbb{R}^s$, $J \in \Psi_s$. In what follows, Σ_s denotes the set of all spreading sets in $\mathbb{R}^s$. We denote by L the fixed lattice $L = L(P) := P(\mathbb{Z}^s)$ $(P \in GL(s; \mathbb{R}))$ as in Section 2. We say a subset S of $\mathbb{R}^s$ is **spreading with respect to** L if for any bounded set $B \subset L$, there exists an element $x \in \mathbb{R}^s$ such that $x + B \subset S$. $\Sigma(L)$ denotes the set of all spreading set with respect to L. For instance, $X \cap L$ is spreading with respect to L for any $X \in \Sigma_s$.

First, we give definitions for non-periodicity for words $W = (w_x) \in K^X$ $(K \neq \phi)$ on a set $X \subset L$. Note that K can be an infinite set. We denote by $W|_S$ its restriction to S:

$$W|_S := (w_x)_{x \in S \cap X},$$

which is a word on $S \cap X$. For words $W = (w_x)_{x \in X}$, $V = (v_x)_{x \in Y}$, we write $W = V$ if they coincide as maps; and write $W \equiv V$ if there exists $t \in L$ such that $W = V \uparrow t$ as in Section 2. We say a word $W \in K^X$ ($X \subset L$) is **non-Ψ-periodic** if

$$W|_{p+J} \equiv W|_{q+J} \ (p, q \in L, \ p + J \cap L \subset X, \ J \in \Psi_s) \Longrightarrow p = q.$$

We say that a word $W \in K^X$ ($X \in L$) is **non-Σ-periodic** if

$$W|_Y \equiv W|_{t+Y} \ (t \in L, \ Y \subset X, \ Y \in \Sigma(L)) \Longrightarrow t = o.$$

Note that, in general, $W|_S \equiv W|_T$ with $S \subset X$ implies $T \subset X$ for $W \in K^X$ ($X \subset L$). We say a word is **Ψ-periodic** (resp., **Σ-periodic**) if it is not non-Ψ-periodic (resp., not non-Σ-periodic). Note that, by the definition, any word on $X \subset L$ is Σ-periodic (resp., Ψ-periodic) if $X \notin \Sigma(L)$ (resp., if $X \supset p + J \cap L$ does not hold for all $p \in L$, $J \in \Psi_s$). We remark that both definitions of non-periodicity given above are considerably strong. In fact, the non-Ψ-periodicity excludes some trivially "non-periodic" words on $J \in \Psi_s$ for all $s \geq 2$; for $s = 1$, the definition requires the non-periodicity for both directions for words on $\mathbb{Z}$, while the definition is equivalent to the usual one for words on $\mathbb{N}$. In general, the non-Σ-periodicity implies the non-Ψ- periodicity. Note that some non-Ψ-periodic words are Σ-periodic. For instance, any non-Ψ -periodic word over $K \neq \phi$ on $\mathbb{N}$ of the form

$$u_1 v^{a_1} u_2 v^{a_2} \cdots u_n v^{a_n} \cdots \ \left(\lim_{n \to \infty} a_n = \infty; \ u_n, v \in K^* \setminus \{\lambda\} \right)$$

is Σ-periodic. In particular, any word coming from a normal number is non-Ψ-periodic, but it is Σ-periodic. On the other hand, for instance, the Thue-Morse word and the Fibonacci word are non-Σ-periodic, cf. Remarks 6–7.

Remark 6. A word W over $K \neq \phi$ on $\mathbb{N}$ is non-Σ-periodic if there exists an integer $m > 1$ such that W is m-th power free (*i.e.*, W has no subwords of the form u^m, $u \in K^* \setminus \{\lambda\}$). In general, any non-periodic (in the usual sense) fixed point of a primitive substitution is m-th power free for all sufficiently large m (cf. [1]), so that it is non-Σ-periodic.

Remark 7. All the Sturmian words (*i.e.*, the words having the complexity $p(n) = n+1$) on $\mathbb{N}$ are non-Σ-periodic. In particular, the Fibonacci word is non-Σ-periodic.

We can show that the words $W(A; c)$ are non-Σ-periodic for some $A \in (\text{Bdd})$, $c > 1$. For instance, the non-Σ-periodicity of the word

$W([1, -1//1, 1]; c)$ for even c follows from Theorem 5 below, cf. the example (ii), Section 3.

Secondly, we give definitions for non-periodicity for tilings. A set $\theta \subset \mathbb{R}^s$ is called a **tile** if θ is an arcwise connected closed set such that $\overline{\theta^\circ} = \theta$. We say that a set $\theta \subset \mathbb{R}^s$ is a **tessera** if θ is a compact tile. We say that $\Theta = \{\theta_\mu; \ \mu \in M\}$ is a **tiling** (resp., a **tessellation**) of J ($J \in \Psi_s$) if all the sets θ_μ ($\mu \in M$) are tiles (resp., tesserae) such that $\cup_{\mu \in M} \theta_\mu = J$ and $\theta_\mu^\circ \cap \theta_\nu^\circ = \phi$ ($\mu \neq \nu$) hold. We say that a set $\Theta = \{\theta_\mu; \ \mu \in M\}$ ($\theta_\mu \subset \mathbb{R}^s$) is Σ-distributed on a set $X \subset \mathbb{R}^s$ if $X \setminus (\cup_{\mu \in M} \theta_\mu) \notin \Sigma_s$. A set $\Theta = \{\theta_\mu; \ \mu \in M\}$ will be referred to as a **mosaic** on $X \in \Sigma_s$ if $\theta_\mu \subset X$ are relatively closed sets with respect to X such that $\theta_\mu^\circ \cap \theta_\nu^\circ = \phi$ ($\mu \neq \nu$), and Θ is Σ-distributed on $X \in \Sigma_s$. Any tiling of J is a mosaic on J. Let $\Theta = \{\theta_\mu; \ \mu \in M\}$ be a mosaic on X. We denote by $\Theta|_S$ the S-restriction:

$$\Theta|_S := \{\theta_\mu \cap S; \ \mu \in M\} \quad (S \subset \mathbb{R}^s).$$

Note that for any mosaic Θ, its restriction $\Theta|_Y$ is always a mosaic for a spreading set $Y \subset X$. For two sets $\Theta = \{\theta_\mu; \ \mu \in M\}$, $\Phi = \{\varphi_\nu; \ \nu \in N\}$ ($\theta_\mu, \ \varphi_\nu \subset \mathbb{R}^s$), we write

$$\Theta|_S \equiv \Phi|_{t+S} \ (t \in \mathbb{R}^s, \ S \subset \mathbb{R}^s)$$

if there exists a $z \in \mathbb{R}^s$ such that

$$\{z + (\theta_\mu \cap S); \ \mu \in M\} = \{\varphi_\nu \cap (t + S); \ \nu \in N\}.$$

We say that a mosaic $\{\theta_\mu; \ \mu \in M\}$ on X is **non-Ψ-periodic** if

$$\{\theta_\mu; \ \mu \in M\}|_{p+J} \equiv \{\theta_\mu; \ \mu \in M\}|_{q+J}$$
$$(p, q \in \mathbb{R}^s, \ p + J \subset X, \ J \in \Psi_s)$$
$$\Longrightarrow \{\theta_\mu; \ \mu \in M\}|_{p+J} = \{\theta_\mu; \ \mu \in M\}|_{q+J}.$$

We say that a mosaic $\{\theta_\mu; \ \mu \in M\}$ on X is **non-Σ-periodic** if

$$\{\theta_\mu; \ \mu \in M\}|_Y \equiv \{\theta_\mu; \ \mu \in M\}|_{t+Y} \ (t \in \mathbb{R}^s, \ Y \subset X, \ Y \in \Sigma_s)$$
$$\Longrightarrow \{\theta_\mu; \ \mu \in M\}|_Y = \{\theta_\mu; \ \mu \in M\}|_{t+Y}.$$

Note that any mosaic $\{\theta_\mu; \ \mu \in M\}$ such that one of the sets θ_μ is a spreading set is Σ-periodic, which can be easily seen.

Finally, we give a definition of non-periodicity for a discrete set $S = \{x_\mu; \ \mu \in M\} \subset \mathbb{R}^s$. We say that S is a mosaic on X if so is the set $\{\{x_\mu\}; \ \mu \in M\}$; and that S is non-Ψ-periodic (resp., non-Σ-periodic) if so is the mosaic $\{\{x_\mu\}; \ \mu \in M\}$. For instance, $A(\mathbb{Z}^s)$ is a Ψ-periodic mosaic for any $A \in GL(s; \mathbb{R})$; $\mathbb{Z} \cup \alpha\mathbb{Z}$ is a non-Ψ-periodic mosaic for any $\alpha \in \mathbb{R} \setminus \mathbb{Q}$; while $(\mathbb{Z} \cup \alpha\mathbb{Z}) \times \mathbb{Z}$ is a Ψ-periodic mosaic.

For a discrete set $X \subset \mathbb{R}^s$, the Voronoi cell $\Upsilon(X, x)$ with respect to $x \in X$ is defined by

$$\Upsilon(X, x) := \{y \in \mathbb{R}^s; \ \|x - y\| \leq \|y - z\| \text{ for all } z \in X \setminus \{x\}\}.$$

For $A \in (P_u(c))$, we denote by $\Gamma(A, c)$ the set determined by (2). Since any Voronoi cell Υ is a finite intersection of closed half-spaces, Υ is a closed convex set, so that it is arcwise connected. Hence, $\Upsilon(A^j\Gamma(A, c), x)$ is a tile for any $A \in (P_u(c))$, $0 \leq j < c$, and $x \in A^j\Gamma(A, c)$, so that the sets

$$\Omega(A, c; j) := \{\Upsilon(A^j\Gamma(A, c), x); \ x \in A^j\Gamma(A, c)\} \ (0 \leq j < c)$$

become tilings of $\mathbb{R}^s$. Note that $\Omega(A, c; j)$ is not always a tessellation. In fact, some of the cells $\Upsilon(A^j\Gamma(A, c), x)$ are unbounded, cf. the examples (i–iii) for the singular case in Section 4. We can show the following

Theorem 4. *Let* $A \in (\mathrm{Bdd})$, $0 \leq j < c$, $1 < c \leq \infty$. *Then the set* $\Omega(A, c; j)$ *is always a tessellation consisting of tesserae which are polyhedra of dimension* s. *For* $1 < c < \infty$, *all the tessellations* $\Omega(A, c; j)$ $(0 \leq j < c)$ *are non-Σ-periodic if and only if so is the word* $W(A; c)$. *The word* $W(A; \infty)$ *is non-Σ-periodic; while,* $\Omega(A, \infty; j)$ $(j \geq 0)$ *are* Ψ-*periodic.*

Proof. We show that $\Omega(A, c; j)$ is always a tessellation for $A \in (\mathrm{Bdd})$. We suppose $A \in (\mathrm{Bdd})$, so that $\det A \neq 0$. Then, $A^j(\mathbb{Z}^s)$ becomes a free $\mathbb{Z}$-module of rank s. In fact, the elements

$$b_j^{(i)} := A^j e_i \in A^j(\mathbb{Z}^s) \ (1 \leq i \leq s)$$

form a basis of $A^j(\mathbb{Z}^s)$ as a $\mathbb{Z}$-module, *i.e.*,

$$A^j(\mathbb{Z}^s) = \mathbb{Z}b_j^{(1)} + \cdots + \mathbb{Z}b_j^{(s)}, \tag{18}$$

where e_i is the i-th fundamental vector. The vectors $b_j^{(i)}$ are linearly independent over $\mathbb{R}$, so that $\mathbb{R}^s \setminus A^j(\mathbb{Z}^s)$ is not a spreading set. Hence, $A^j(\mathbb{Z}^s)$ is a mosaic. In view of Corollary 2, we have $T_j(A) = A^j\Gamma(A, \infty)$, which is not empty by Theorem 1. Since

$$x \in T_j(A) \iff A^{-j}x \in \mathbb{Z}^s, \ A^{-j-1}x \notin \mathbb{Z}^s \iff x \in A^j(\mathbb{Z}^s) \setminus A^{j+1}(\mathbb{Z}^s),$$

we get

$$T_j(A) = A^j\Gamma(A, \infty) = A^j(\mathbb{Z}^s) \setminus A^{j+1}(\mathbb{Z}^s) \neq \phi, \ 0 \leq j < \infty. \tag{19}$$

Suppose $1 < c < \infty$. Choose any $x_0 \in A^j \Gamma(A, c)$ $(0 \leq j < c)$. We shall show that $\Upsilon(A^j \Gamma(A, c), x_0)$ is a compact set. By Corollary 2, we have $x_0 \in A^j \Gamma(A, c) = \bigcup_{0 \leq m < \infty} T_{cm+j}(A)$, so that $x_0 \in T_{cm+j}(A)$, i.e., $\mathrm{ind}\,_A(x_0) = cm + j$ for some $m = m(x_0) \geq 0$, and $0 \leq j < c$. On the other hand, from the definition of $\mathrm{ind}\,_A$ it follows

$$\mathrm{ind}\,_A(x + y) = \mathrm{ind}\,_A(x) \text{ for all } x, y \in \mathbb{Z}^s \text{ with } \mathrm{ind}\,_A(x) < \mathrm{ind}\,_A(y).$$

Hence, noting that $\mathrm{ind}\,_A(x) > cm + j$ for any $x \in A^{cm+j+1}(\mathbb{Z}^s)$, we get

$$x_0 + \mathbb{Z}b^{(1)}_{cm+j+1} + \cdots + \mathbb{Z}b^{(s)}_{cm+j+1} \subset T_{cm+j}(A) \subset A^j \Gamma(A, c). \tag{20}$$

Thus, setting

$$F := \{x_0 + \{\pm b_1, \ldots, \pm b_s\}\} \cup \{x_0\} \text{ with } b_i := b^{(i)}_{cm+j+1} \ (1 \leq i \leq s),$$

we obtain

$$F \subset A^j \Gamma(A, c),$$

so that

$$\Upsilon(F, x_0) \supset \Upsilon(A^j \Gamma(A, c), x_0).$$

In addition, $\Upsilon(A^j \Gamma(A, c), x_0)$ is a closed set by the definition of Voronoi cells. Hence, it suffices to show that the set $\Upsilon(F, x_0)$ is bounded for the proof of the compactness of $\Upsilon(A^j \Gamma(A, c), x_0)$. For simplicity, we consider $\Upsilon(-x_0 + F, o)$ that is congruent to $\Upsilon(F, x_0)$. By the definition of a Voronoi cell, we have

$$\Upsilon(-x_0 + F, o) = \bigcap_{f \in F_*} (f + H_-(f)),$$

where $F_* := \{\pm 2^{-1} b_1, \ldots, \pm 2^{-1} b_s\}$, and $H_-(b)$ denotes a closed half space

$$H_-(b) := \{x \in \mathbb{R}^s; \ (b, x) \leq 0\}, \ b \in \mathbb{R}^s \setminus \{o\}.$$

Since b_i $(1 \leq i \leq s)$ are linearly independent over $\mathbb{R}$, by the orthonormalization of Schmidt, we can take an orthonormal basis of $\mathbb{R}^s$ as a vector space over $\mathbb{R}$ such that we can write

$$b_i = {}^T(b_{i1}, \ldots, b_{ii}, 0, \ldots, 0) \ (b_{ii} \neq 0, \ 1 \leq i \leq s)$$

with respect to the basis. Then, for any $\varepsilon = {}^T(\varepsilon_1, \cdots, \varepsilon_s) \in \{1, -1\}^s$, there exists a unique element $v(\varepsilon) = {}^T(v_1(\varepsilon), \cdots, v_s(\varepsilon)) \in \mathbb{R}^s$ satisfying the equations:

$$2\varepsilon_1 b_{11} v_1(\varepsilon) = b_{11}^2,$$

$$2\varepsilon_1 b_{21} v_1(\varepsilon) + 2\varepsilon_2 b_{22} v_2(\varepsilon) = b_{21}^2 + b_{22}^2,$$

$$\cdots$$

$$2\varepsilon_1 b_{s1} v_1(\varepsilon) + 2\varepsilon_2 b_{s2} v_2(\varepsilon) + \cdots + 2\varepsilon_s b_{ss} v_s(\varepsilon) = b_{s1}^2 + b_{s2}^2 + \cdots + b_{ss}^2,$$

namely

$$\|v(\varepsilon)\| = \|v(\varepsilon) - \varepsilon_1 b_1\| = \|v(\varepsilon) - \varepsilon_2 b_2\| = \cdots = \|v(\varepsilon) - \varepsilon_s b_s\|.$$

Thus, $\Upsilon(-x + F, o)$ becomes a parallelepiped of dimension s with 2^s vertices $v(\varepsilon)$ ($\varepsilon \in \{1, -1\}^s$), so that

$$\Upsilon(-x + F, o) \subset \overline{B(\max\{\|v(\varepsilon)\|; \ \varepsilon \in \{1, -1\}^s\}, o)}.$$

Thus, we have shown that the set $\Omega(A, c; j)$ is always a tessellation for $1 < c < \infty$. Note that, since $A^j(\Gamma(A, c)(\subset \mathbb{Z}^s)$ is a discrete set, each set $\Upsilon(A^j(\Gamma(A, c), x_0)$ has non-empty interior, so that it becomes an s dimensional body. Hence, the tessellation $\Omega(A, c; j)$ consists of polyhedra of dimension s. Now, we suppose $c = \infty$. Choose any $x_0 \in A^j \Gamma(A, \infty)$ for a given number $0 \le j < \infty$. Then, by (19), we get (20) with $cm = 0$, $c = \infty$. Hence, we can show that $\Upsilon(A^j \Gamma(A, \infty), x_0)$ is compact as we have shown in the case $1 < c < \infty$, which completes the proof of the first statement of the theorem.

We prove the second statement. Suppose $c < \infty$. We put

$$W_j(A; c) := \left(\chi_{A^j \Gamma(A, c)}(x)\right)_{x \in \mathbb{Z}^s}, \quad (0 \le j < c), \tag{21}$$

which is a word over $\{0, 1\}$ on the set $\mathbb{Z}^s$, where χ_S is the characteristic map with respect to a set S. If $\Omega(A, c; j)$ is Σ-periodic, then so is the set $A^j \Gamma$. Applying A^{-j} to the set $A^j \Gamma$, we see that the set Γ is Σ-periodic. Hence, we get the Σ-periodicity of $A^j \Gamma$ for all $0 \le j < c$, which implies the Σ-periodicity of $W_j(A; c)$ for all $0 \le j < c$. Then, we can conclude that $W(A; c)$ is Σ-periodic. Conversely, the Σ-periodicity of $W(A; c)$ implies that of $W_j(A; c)$ for each $0 \le j < c$, so that all the sets $A^j \Gamma(A; c)$ are Σ-periodic, and so are the tessellations $\Omega(A, c; j)$, $0 \le j < c$.

We prove the third assertion. First, we prove the latter half. In view of (18), we see that $A^j(\mathbb{Z}^s)$ is a Ψ-periodic set for all $0 \le j < \infty$. Hence, (19) together with $A^j(\mathbb{Z}^s) \supset A^{j+1}(\mathbb{Z}^s)$ implies that for any $0 \le j < \infty$, the set $A^j \Gamma(A, \infty)$ is a Ψ-periodic set as a mosaic with periods coming from (18) with $j + 1$ in place of j, so that the tesselation $\Omega(A, \infty; j)$ is Ψ-periodic for any $j \ge 0$. Secondly, we prove the first half. We choose an element $g_n \in A^n(\mathbb{Z}^s) \setminus \{o\}$ among elements having the smallest norm in the set $A^n(\mathbb{Z}^s) \setminus \{o\}$ for each n. Then we can show that

$$\lim_{n \to \infty} \|g_n\| = \infty. \tag{22}$$

Suppose the contrary. Then we can choose a bounded subsequence $\{g_{p(n)}\}_{n=0,1,2,...}$ $(\lim p(n) = \infty)$ of the sequence $\{g_n\}_{n=0,1,2,...}$. Then we can choose a subsequence $\{g_{q(n)}\}_{n=0,1,2,...}$ $(\lim q(n) = \infty)$ of $\{g_{p(n)}\}_{n=0,1,2,...}$ such that it converges to an element $g \in A^n(\mathbb{Z}^s) \setminus \{o\}$. Then $A^{q(n)}(\mathbb{Z}^s) \ni g_{q(n)} = g \neq o$ holds for all sufficiently large n. Hence, we get $A^{-q(n)}g \in \mathbb{Z}^s$, $g \neq o$ for infinitely many n, which contradicts $A \in$ (Bdd). Thus, we get (22). Hence, we see that any choice $x_n, y_n \in A^n\Gamma(A, \infty) = T_n(A) = A^n(\mathbb{Z}^s) \setminus A^{n+1}(\mathbb{Z}^s) \subset A^n(\mathbb{Z}^s)$,

$$x_n \neq y_n, \ n = 0, 1, 2, \cdots$$

implies

$$\|x_n - y_n\| \geq \|g_n\|. \tag{23}$$

Now, suppose that $W = W(A; \infty)$ is Σ-periodic, *i.e*, $W|_X \equiv W|_{t+X}$, $X \in \Sigma(\mathbb{Z}^s)$, $t \neq o$. Then we have

$$W_n(A; \infty)|_X \equiv W_n(A; \infty)|_{t+X}, \ X \in \Sigma(\mathbb{Z}^s), \ t \neq o, \tag{24}$$

where $W_n(A; \infty)$ is the word defined by (21). Recalling that $A^n\Gamma(A, \infty)$ is a Ψ-periodic set as a mosaic with periods coming from (18) with $j = n+1$, we can find $x_n, y_n \in A^n\Gamma(A, \infty) \cap X$ ($x_n \neq y_n$), $n = 0, 1, 2, \ldots$ for any spreading set $X \in \Sigma(\mathbb{Z}^s)$. Therefore, we get $\|t\| \geq \|g_n\|$ by (23), (24). Taking $n \to \infty$, we have a contradiction by (22). ∎

We remark that some tilings $\Omega(A, c; j)$ for singular matrices A are Ψ-periodic, cf. the examples (i–iii), Section 4.

Now, we are intending to show that some of the words $W(A; c)$ and tessellations $\Omega(A, c; j)$ are non-Σ-periodic for $1 < c < \infty$, $A \in$ (Bdd).

Lemma 6. *Let W be a word on $\mathbb{Z}^s$, and $U \in GL(s; \mathbb{Z})$. Then W is non-Ψ-periodic (resp., non-Σ-periodic) if and only if so are the (τ, U)-sectors for all $\tau \in \{1, -1\}^s$.*

Proof. By our definitions, any (τ, U)-sector of W is non-Ψ-periodic (resp., non-Σ-periodic) if so is W.

We assume that all the (τ, U)-sector of W are non-Ψ-periodic (resp., non-Σ-periodic). It suffices to show that $W|_X$ is non-Ψ-periodic (resp., non-Σ-periodic) for any $X = p + J$ with $p \in \mathbb{Z}^s$ and $J \in \Psi_s$ (resp., $X \in \Sigma_s$). Suppose the contrary. Then $W|_X$ is Ψ-periodic (resp., Σ-periodic) for some $X = p + J$ with $p \in \mathbb{Z}^s$ and $J \in \Psi_s$ (resp., $X \in \Sigma_s$). Then $J \cap U(\tau \otimes \mathbb{N}^s) \in \Psi_s \cap \mathbb{Z}^s$ (resp., $X \cap U(\tau \otimes \mathbb{N}^s) \in \Sigma(\mathbb{Z}^s)$) holds at least one element $\tau \in \{1, -1\}^s$. Hence, (τ, U)-sector of W is Ψ-periodic (resp., Σ-periodic), which contradicts our assumption. ∎

Apart from the partitions (2), we may consider substitutions of dimension s having fixed points which are non-Σ-periodic.

Theorem 5. *Let σ be a substitution over $K = \{@\} \cup F_0 \cup F_\times$ of size $G(b)$ with $F_0 \cap F_\times = \phi$, $b = {}^T(b_1, \cdots, b_s)$, $b_i > 1$ $(1 \le i \le s)$, $s > 1$, $\sigma(a) = (v_x(a))_{x \in G}$ satisfying*

$$v_o(@) = @\,;$$

$v_o(a) \in F_0$ *for all* $a \in F_0$, $v_o(b) \in F_\times$ *for all* $b \in F_\times$;

$v_x(a) \in F_0$ *for all* $a \in K$,

$$\text{and } x \in G_0 := \{{}^T(x_1, \cdots, x_s) \in G(b) \setminus \{o\};\ x_1 \cdots x_s = 0\},$$

$v_x(a) \in F_\times$ *for all* $a \in K$,

$$\text{and } x \in G_\times := \{{}^T(x_1, \cdots, x_s) \in G(b);\ x_1 \cdots x_s \ne 0\}.$$

Then the fixed point $W \in K^{\mathbb{Z}^s}$ $(W(o) = @)$ of σ is non-Σ-periodic.

N.B.: We mean by $x_1 \cdots x_s = 0$ not a word but the product of numbers $x_1, \ldots, x_s$ equals zero.

Proof. Let σ have the property stated in Theorem 5. Since, for any $\tau \in \{1, -1\}^s$, and $x \in G(b) \setminus \{o\}$,

$$\tau \otimes_{G(b)} x \in G_0 \text{ (resp. } \tau \otimes_{G(b)} x \in G_\times) \iff x \in G_0 \text{ (resp. } x \in G_\times),$$

all the conjugates of ${}^\tau \sigma$ have the same property. We consider the automata ${}^\tau \mathcal{M} = (@, K, G(b), {}^\tau \zeta)$ with ${}^\tau \zeta$ corresponding to ${}^\tau \sigma$. Then, all the conjugates ${}^\tau \mathcal{M}$ can be described as in Fig.1 given below. We mean the transition function ${}^\tau \zeta$ by the arrows labeled by G_0, $G_\times$, or o there. For instance, for $a \in F_\times$, ${}^\tau \zeta(a, i) \in F_0$ if $i \in G_0$, and ${}^\tau \zeta(a, i) \in F_\times$ if $i \in G_\times$, or $i = o$.

We consider a word $W^\% \in K^{\mathbb{N}^s}$ generated by the automaton $\mathcal{M}$ with a projection $\pi : K \to \{\alpha, \beta\}$ defined by

$$\pi(a) := \alpha \text{ if } a = @\,, \text{ or } a \in F_0;\ \pi(a) := \beta \text{ if } a \in F_\times,$$

namely,

$$W^\% = (w_x^\%)_{x \in \mathbb{N}^s} \in \{\alpha, \beta\}^{\mathbb{N}^s},\ w_x^\% := \pi({}^\tau \zeta(@\,, \rho_b(x))),\ x \in \mathbb{N}^s.$$

Note that $W^\%$ does not depend on τ, so that all the (τ, E)-sectors ${}^{(\tau,E)}W = W|_{\tau \otimes \mathbb{N}^s}$ of the fixed point $W \in K^{\mathbb{Z}^s}$ of σ are identical to $W^\%$ if we identify the symbols in K according to the projection π. It is clear that if $W^\%$ is non-Σ-periodic, then $W|_{\mathbb{N}^s}$ is non-Σ-periodic (in general, the converse is not valid). By Lemma 6, it suffices to show that $W^\%$ is

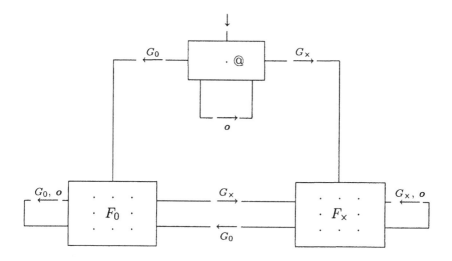

Figure 1:

non-Σ-periodic. In view of Fig. 1, recalling $s > 1$, we get the following facts (i), (ii):

Let $x \in \mathbb{N}^s$, $\rho_b(x) = u_m u_{m-1} \cdots u_0 \in G(b)^*$ with $u_j = {}^T(u_j^{(1)}, \ldots, u_j^{(s)}) \in G(b)$. Then

(i) $w_x^{\%} = \alpha$ if there exists a k $(1 \leq k \leq s)$ such that $u_h^{(k)} \neq 0$, $u_h^{(1)} \cdots u_h^{(s)} = 0$ for an integer $h = h(k)$ with $0 \leq h \leq m$ satisfying $u_j^{(k)} = 0$ for all j $(0 \leq j < h)$.

(ii) $w_x^{\%} = \beta$ if and only if there exists an h $(0 \leq h \leq m)$ such that $u_h^{(i)} \neq 0$ for all i $(1 \leq i \leq s)$ satisfing $u_j = o$ for all j $(0 \leq j < h)$.

Now suppose that $W^{\%}$ is Σ-periodic, i.e.,

$$W|_X \equiv W|_{p+X}, \tag{25}$$

holds for a set $X \in \Sigma(\mathbb{Z}^s)$, $X \subset \mathbb{N}^s$, and a fixed vector $p \in \mathbb{N}^s \setminus \{o\}$. For any integer $h \geq 2$, we can take $q = q(h) \in \mathbb{N}^s$ such that

$$Z := q + G(2b_1^{h+1}, \ldots, 2b_s^{h+1}) \subset X.$$

Hence, we can choose an $x = x(h) \in Z$ such that

$$\rho_b(x) = t_n t_{n-1} \cdots t_0 \in G(b)^*, \quad t_j = {}^T(t_j^{(1)}, \ldots, t_j^{(s)}) \in G(b),$$

$$t_h^{(1)} \neq 0, \ldots, t_h^{(s)} \neq 0, \quad t_i = o \text{ for all } 0 \leqq i < h, \qquad (26)$$

$$x + y \in X \text{ for all } y \in H_h^{(k)} \text{ for each } 1 \leqq k \leqq s,$$

where

$$H_h^{(k)} := \{{}^T(y_1, \ldots, y_s) \in \mathbb{Z}^s; \ y_k = 0, \ -b_i^h + 1 \leqq y_i \leqq b_i^h - 1$$
$$\text{for all } 1 \leqq i \leqq s \text{ with } i \neq k\}.$$

Since

$$\rho_b(x + y) \succ v_h v_{h-1} \cdots v_0 \succ v_j o^j \ (o^0 := \lambda)$$
$$0 \leqq j < h, \ v_j \in G_0$$

holds for an integer $j = j(y)$ for any given $y \in H_h^{(k)} \setminus \{o\}$, we get by (i,ii)

$$w_{x+y}^{\%} = \alpha \text{ for all } y \in \bigcup_{1 \leqq k \leqq s} H_h^{(k)} \setminus \{o\},$$

$$w_x^{\%} = \beta. \qquad (27)$$

We put

$$\text{wall}\,(x) := (w_{x+y}^{\%})_{y \in \bigcup_{1 \leqq k \leqq s} H_h^{(k)}}.$$

Noting $x + \bigcup_{1 \leqq k \leqq s} H_h^{(k)} \subset \mathbb{N}^s$, we see by (27) that wall (x) is a subword of $W^{\%}$ consisting of symbols identical to α except for an occurrence of a symbol β centered at x as far as x satisfies (26). Consider any $z \in \mathbb{N}^s$ satisfying

$$z + \bigcup_{1 \leqq k \leqq s} H_h^{(k)} \subset \mathbb{N}^s \text{ with } \rho_b(z) = u_n u_{n-1} \cdots u_0 \in G(b)^* \ (h \geqq 2).$$

$$u_j = {}^T(u_j^{(1)}, \ldots, u_j^{(s)}) \in G(b) \ (0 \leqq j \leqq n = n(z)).$$

We put

$$j_0 = j_0(z) := \min\{0 \leqq j \leqq n(z); \ u_j \neq o\}.$$

We have two cases:

(a) $u_{j_0}^{(i)} \neq 0$ for all $1 \leqq i \leqq s$.

(b) $u_{j_0}^{(i_1)} \neq 0$, $u_{j_0}^{(i_2)} = 0$ for some $1 \leqq i_1 \leqq s$, $1 \leqq i_2 \leqq s$.

Now, we assume $0 \leq j_0 \leq h - 2$ $(h \geq 2)$. Suppose the case (a). Then, taking $\boldsymbol{y} = {}^T(y_1, \ldots, y_s)$ with $y_1 = 0$, $y_i = b_i^{h-1}$ for all $2 \leq i \leq s$, we have by the definition of j_0

$$\rho_b(\boldsymbol{y} + \boldsymbol{z}) \succ \boldsymbol{u}_{j_0} o^{j_0}, \ \boldsymbol{u}_{j_0} \neq o,$$

and by (a), $u_{j_0}^{(i)} \neq 0$ for all $1 \leq i \leq s$. Hence, we get by (ii)

$$w_{z+y}^{\%} = \beta, \ \boldsymbol{y} \in H_h^{(1)},$$

so that

$$\text{wall}\,(\boldsymbol{z}) \neq \text{wall}\,(\boldsymbol{x})$$

for $\boldsymbol{x}$ satisfying (26). Suppose the case (b). Then taking $y_i = 0$ for all i with $u_{j_0}^{(i)} \neq 0$, $y_i = b_i^{j_0}$ for all i with $u_{j_0}^{(i)} = 0$, we get by (ii)

$$\rho_b(\boldsymbol{y} + \boldsymbol{z}) \succ \boldsymbol{v}_{j_0} o^{j_0}, \ \boldsymbol{v}_{j_0} \in G_\times.$$

Hence, we obtain

$$w_{z+y}^{\%} = \beta, \ \boldsymbol{y} \in H_h^{(i_2)},$$

so that

$$\text{wall}\,(\boldsymbol{x}) \neq \text{wall}\,(\boldsymbol{z})$$

for $\boldsymbol{x}$ satisfying (26). Therefore, $0 \leq j_0(\boldsymbol{z}) \leq h - 2$ implies wall $(\boldsymbol{x}) \neq$ wall $(\boldsymbol{z})$ for $\boldsymbol{x}$ satisying (26). Hence, wall $(\boldsymbol{x}) =$ wall $(\boldsymbol{z})$ with $\boldsymbol{x} \neq \boldsymbol{z}$ for $\boldsymbol{x}$ satisying (26) implies

$$\|\boldsymbol{x} - \boldsymbol{z}\| \geq \min\{b_i^{h-1}; \ 1 \leq i \leq s\}.$$

On the other hand, (25) implies

$$\text{wall}\,(\boldsymbol{x}) = \text{wall}\,(\boldsymbol{p} + \boldsymbol{x}), \ \boldsymbol{p} \neq o,$$

so that we get

$$\|\boldsymbol{p}\| \geq \min\{b_i^{h-1}; \ 1 \leq i \leq s\}.$$

Since we can take h arbitrarily large, we have a contradiction, and therefore, we have completed the proof. ∎

Applying Theorems 4–5 for the word $W(A; c)$ with $A = [1, -1//1, 1]$, $c = 2m$, $1 < m < \infty$, we get non-Σ-periodic tessellations $\Omega(A, c; j)$ $(0 \leq j < c)$. We can show that the tessellations $\Omega(A, 2; j)$ consist of three kinds of incongruent tesserae: a quadrangular, a pentagon, and a hexagon. In this case, each tessera of $\Omega(A, 2; 0)$ is similar to a tessera of $\Omega(A, 2; 1)$, cf. Fig. 2 below. Note that, in general, such similarity does not hold. Using Corollary 3, and Theorems 5–6, we can give a non-Σ-periodic word $W(A \oplus \cdots \oplus A; 2m)$ of dimension $2s$, and non-Σ-periodic tessellations $\Omega(A \oplus \cdots \oplus A, 2m; j)$ $(0 \leqq j < 2m < \infty)$ of Euclidean space $\mathbb{R}^{2s}$ for $A = [1, -1//1, 1]$.

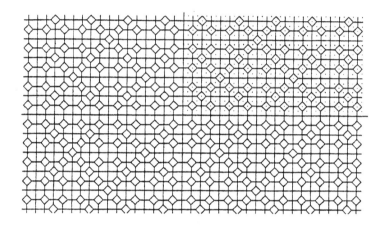

The dots are elements of $A\Gamma$ in the first section; the midpoints of the edges of length 2 and all the vertices are elements of $\Gamma \cup \{o\}$.

Figure 2: (Tessellation $\Omega(A, 2; 1)$ for $A = [1, -1//1, 1]$)

6.   PROBLEMS AND A THEOREM

We say a substitution of dimension s of size G over $\tilde{K}_c$ is **admissible** if there exists a matrix $A \in M(s; \mathbb{Z})$ such that the substitution comes from a partition (2). Can we give a criterion by which we can determine whether a given substitution is admissible, or not ?

We have already seen that a partition (2) exists and uniquely determined for A belonging to the class (Bdd), and have given its subclass $\mathcal{D}$ such that for any $A \in \mathcal{D}$, the characteristic word $W(A; c)$ corresponding to (2) can be described in terms of nontrivial substitutions. By the proof of Theorem 3, we can effectively determine the substitution for any given A belonging to $\mathcal{D}(k)$. Probably, we can not extend Theorem 3 to a class strictly including the class $\mathcal{D}$. While, we have not a good algorithm to determine whether a given matrix A belongs to $\mathcal{D}$. Can we find an effective criterion for that? For instance, it is easy to see $[2, 3//2, 2]$, and

$[2, 2//3, 2] \in \mathcal{C}(2)$, so that they belong to $\mathcal{D}(2)$; but we have difficulty for $[2, 2//2, 3]$, cf. Remark 6 below. It seems very likely that there exist some matrices $A \in$ (Bdd) for which $W(A; c)$ $(1 < c < \infty)$ can not be finitely automatic. We conjecture that $W(A; c)$ with $A = [4, -1//0, 2]$, $1 < c < \infty$ can not be a fixed point with respect to any base b, and basis $U \in GL(2; \mathbb{Z})$. We can show, for $A = [4, -1//0, 2]$

$$H((\det A)^n A^{-n}) = [2^n, 2^{n-1}(2^n - 1)//0, 4^n] \ (n \geqq 0).$$

Related to such problems, we can prove the following:

Remark 8. Let $A = [2, 2//2, 3]$. Then $A \in$ (Bdd). Let C_n be the hermitian canonical form $H((\det A)^n A^{-n})$. Then

$$A^{-n} \sim 2^{-n} C_n = 2^{-n}[1, t_n//0, 2^n] \ (n \geqq 1),$$
$$t_n \leqq t_{n+1} \ (n \geqq 1), \quad \lim_{n \to \infty} t_n = \infty.$$

If we put $u_0 := 0$, and for $n \geqq 1$,

$$u_n := 1 \text{ if } t_{n+1} > t_n; \quad u_n := 0 \text{ if } t_{n+1} = t_n,$$

then the word $u_n u_{n-1} \cdots u_0$ becomes the base-2 expansion of t_{n+1}; to be accurate,

$$t_{n+1} = \rho_2^{-1}(u_n u_{n-1} \cdots u_0) \ (n \geqq 1).$$

The sequence $\{t_n\}_{n=0,1,2,\dots}$ can be given by the following algorithm:

(i) $t_1 = 0$.

(ii) Find the continued fraction of $2^{-n-1}(2t_n - 3)$ $(n \geqq 1)$:

$$2^{-n-1}(2t_n - 3) = [a_0; a_1, \cdots, a_{2m(n)}] \ (a_0 \in \mathbb{Z}, \ a_i \in \mathbb{N} \setminus \{0\}).$$

(iii) Find the value $[a_0; a_1, \cdots, a_{2m(n)-1}] = p/q$, where $q = q(n) > 0$, $p = p(n)$ are coprime integers.

(iv) Find the residue t_{n+1} by $t_{n+1} \equiv 2(t_n - 1)q(n) \pmod{2^{n+1}}$, $0 \leqq t_{n+1} < 2^{n+1}$.

A few symbols of the begining of the word $u := u_0 u_1 \cdots$ are

$$01^2 0^3 10^2 1^2 01^2 0^2 1^2 010^4 1^2 0^2 1^2 0^2 1^2 0^5 1^5 01^2 01^3 010^2 101^4 01^2 0^3 1^5 0^2 10 \cdots,$$

which is almost the same as

$$v :=$$
$$01^3 0^3 10^2 1^2 01^2 0^2 1^2 010^4 1^2 0^2 1^2 0^2 1^2 0^5 1^5 01^2 01^3 010^2 101^4 01^2 0^3 1^5 0^2 10 \cdots,$$

where the word v comes from one of the eigenvalues of A in the p-adic sense. To be precise, if $\Phi_A(\alpha) = 0$, $|\alpha|_2 < 1$, $\alpha \in \mathbb{Z}_2$ (Hensel's lemma says that such an α uniquely exists), then the 2-adic expansion of the α is given by $v = v_0 v_1 v_2 \cdots$ ($v_n \in \{0,1\}$), i.e., $\alpha = \sum_{n \geq 0} v_n 2^n$, where $|*|_p$ denotes the p-adic valuation. (In what follows, we mean by $|*| = |*|_\infty$ the ordinal absolute value as before.) Comparing u with v, we might say that they should be the same except for one missing symbol 1 in u (in fact, this is valid as we shall see).

Related to such a phenomenon, we can show the following

Theorem 6. *Let* $f(x) := x^s - c_{s-1}x^{s-1} - \cdots - c_1 x - c_0 \in \mathbb{Z}[x]$ *be a polynomial satisfying* $s > 1$, $|c_0| > 1$, $GCD(c_0, c_1) = 1$ *with the decomposition* $|c_0| = \prod_{1 \leq k \leq t} p_k^{e_k}$, *where* p_k *are distinct primes. Let* C *be the companion matrix of* $f(x)$:

$$C := \begin{pmatrix} {}^T\boldsymbol{o} & c_0 \\ E_{s-1} & \boldsymbol{c} \end{pmatrix}, \quad \boldsymbol{c} = {}^T(c_1, \ldots, c_{s-1}),$$

where E_{s-1} *is the* $(s-1) \times (s-1)$ *unit matrix. Then there exists a unique root* $\alpha_{p_k} \in \mathbb{Z}_{p_k}$ *of* $f(x)$ *satisfying* $|\alpha_{p_k}|_{p_k} < 1$ *for each* p_k, *and the following statements are valid:*

(i) *The hermitian canonical form of* $(\det C \cdot C^{-1})^n$ *is of the shape*

$$\begin{pmatrix} 1 & {}^T\boldsymbol{x}_n \\ \boldsymbol{o} & d^n E_{s-1} \end{pmatrix} \in M(s; \mathbb{Z}), \quad \boldsymbol{x}_n = {}^T(x_n^{(1)}, \cdots, x_n^{(s-1)}), \quad 0 \leq x_n^{(j)} < d^n$$

for all $n \geq 1$, $1 \leq j \leq s-1$, *where* $d := |c_0|$.

(ii) $|\alpha_{p_k}^j - x_n^{(j)}|_{p_k} \leq p_k^{-e_k n}$ *holds for all* $n \geq 1$, $1 \leq j \leq s-1$, $1 \leq k \leq t$.

Using Theorem 6, we can give an explicit formula of a continued fraction of dimension $s-1$ whose p_k-adic value converges to ${}^T(\beta, \beta^2, \cdots, \beta^{s-1})$ with $\beta = \beta(k) = c_0^{-1}\alpha_{p_k}$ for all $1 \leq k \leq t$. The proof of this fact and Theorem 6 will appear in a forthcoming paper.

If we use Theorem 6, we can show that the sequence t_n in Remark 8 converges, in $\mathbb{Z}_2$, to the root γ of $g := 2x^2 - x - 2$ satisfying $|\gamma|_2 < 1$, so that $2\gamma + 2 = \alpha$. Hence, the last phrase in Remark 8 is valid. Since g is irreducible over $\mathbb{Q}$, so that $\beta \notin \mathbb{Q}$. Hence we see that the word u is a non-periodic word in the usual sense. We give the following

Conjecture 1: Let $f(x) \in \mathbb{Z}[x]$, and $C \in M(s; \mathbb{Z})$ be as in Theorem 6. Suppose that $f(x)$ is irreducible over $\mathbb{Z}[x]$. Then $W(C; c)$ ($2 \leq c < \infty$) is not finitely automatic with respect to any basis $U \in GL(s; \mathbb{Z})$

of the lattice $\mathbb{Z}^s$, and to any base $\boldsymbol{b} \in (\mathbb{N} \setminus \{0,1\})^s$ of an automton $(@, K, G(\boldsymbol{b}), \zeta)$.

Can we show that $W(A;c)$ $(1 < c < \infty)$ is not automatic with respect to any basis $U \in GL(s; \mathbb{Z})$, and to any base $\boldsymbol{b}$ for A mentioned in Remark 8 ? If so, what will happen when we extend the conception of automaticity ? For example, we may consider a representation ρ of numbers i given by a sum of terms of a linear recurrence sequence instead of the base-b expansion $\rho_b(i)$. We may consider $\rho(\boldsymbol{x})$ $(\boldsymbol{x} \in \mathbb{N}^s)$ coming from s linear recurrence sequences. In such a case, the set $\{\rho(\boldsymbol{x}); \ \boldsymbol{x} \in \mathbb{N}^s\}$ is retricted to a subset of $G(\boldsymbol{b})^*$; for instance, if we take the Fibonacci representation ρ with dimension $s = 2$, then $\rho(\boldsymbol{x}) \in R$, where

$$R := \{\boldsymbol{x} \in (\{0,1\}^*)^{(2)}; \ \boldsymbol{x} = \boldsymbol{x}_n \cdots \boldsymbol{x}_0, \ \boldsymbol{x}_j = {}^T(x_{1j}, x_{2j}) \in \{0,1\}^{(2)},$$
$$0 \leqq j \leqq n, \ \boldsymbol{x}_n \neq \boldsymbol{o}, \ x_{in} \cdots x_{i0} \notin \{0,1\}^* 11 \{0,1\}^*, \ i = 1, 2\}.$$

We may define an automatic class of words as a class of words W generated by an automaton $\mathcal{M} = (@, K, G(\boldsymbol{b}), \zeta)$ together with a representation ρ and a projection $\pi : K \to F$:

$$W = (w_{\boldsymbol{x}})_{\boldsymbol{x} \in \mathbb{Z}^s}, \ w_{T \otimes \boldsymbol{x}} := \pi({}^T \zeta(@, \rho(\boldsymbol{x}))) \quad (\boldsymbol{x} \in \mathbb{N}^s).$$

Can we find a matrix $A \notin \mathcal{D}$ such that $W(A;c)$ becomes automatic in this sense ?

We say $A, B \in M_r(s; \mathbb{Z})$ have the same behavior if $H((\det A)^n A^{-n}) = H((\det B)^n B^{-n})$ holds for all $n \geq 0$. If the behaviors of $A, B \in (\text{Bdd})$ are the same, then $W(A;c) = W(B;c)$ follows. We remark that, in general, any two matrices among A, UA, AU, $U^{-1}AU$, and ${}^T A$ do not have the same behavior for $U \in GL(s; \mathbb{Z})$. On the other hand, for instance, the behaviors for $[2, 1//2, 2]$, $[2, 3//2, 2]$, $[2, 2//3, 2]$, and $[4, -7//-6, 10]$ coincide. Recall the example (vii) in Section 3; the matrices A_n (resp., B_n) have the same behavior for all $n \in \mathbb{Z}$. Can we find an effective criterion by which we can see whether two given matrices have the same behavior ?

Concerning the non-Σ-periodicity, it is remarkable that $p(n) = n + 1$ is the minimum complexity for the non-Σ-periodic words on $\mathbb{Z}$, and on $\mathbb{N}$, cf. Remark 7. Here, in general, for a word of W on a subset of $\mathbb{Z}^s$, $p(n_1, \ldots, n_s) = p(n_1, \ldots, n_s; W)$ counts the number of the subwords of size $n_1 \times \cdots \times n_s$ of W. Can we find the minimum complexity for the non-Σ-periodic words on $\mathbb{Z}^s$, or on $\mathbb{N}^s$? Note that the existence of minimum complexity is not clear.

Conjecture 2: The word $W(A;c)$ corresponding to the partition (2) is non-Σ-periodic for any $A \in (\text{Bdd})$, $1 < c < \infty$.

Note that the conjecture implies that all the tessellations $\Omega(A, c; i)$ $(0 \leq i < c)$ of $\mathbb{R}^s$ are non-Σ-periodic for any $A \in$ (Bdd), and finite $c > 1$, cf. Theorem 4.

It will be interesting to consider the partitions of the form

$$\bigcup_{0 \leq j < c} A^j \Gamma = \mathbb{Z}^s \setminus \{x \in \mathbb{Z}^s; \text{ind }_A(x) = \infty\}, \quad s \geq 3$$

instead of (2) for non-singular matrices $A \notin$ (Bdd).

Acknowledgement: I would like to thank Hiroshi Kumagai for making the TEXversion of my preprint, and to Maki Furukado for her help in making calculation of the sequences $\{t_n\}_{n=1,2,...}$, $\{u_n\}_{n=0,1,...}$ in Remark 8 for a considerably wide range by a computer. I am grateful to Boris Solomyak for helpful discussions related to Remark 6. I also thank an anonymous referee for some valuable suggestions, in particular, the assertion (i) in Theorem 2 came from her/him. I would like to express my deep gratitude to Kohji Matsumoto, who kindly made so many correspondences, beyond the usual range of duties of an editor, among Kluwer Academic Publishings, World Scientific Co. Pte. Ltd., and the author.

References

[1] B. Mossé, *Puissances de mots et reconnaissabilité des points fixes d'une substitution*, Theoret. Comput. Sci. **99:2** (1992), 327–334.

[2] J.-I. Tamura, *Certain partitions of the set of positive integers, and non-periodic fixed points in various senses*, Chapter 5 in Introduction to Finite Automata and Substitution Dynamical Systems, Editors: V. Berthé, S. Ferenczi, C. Mauduit, and A. Siegel, Springer–Verlag, to appear.

[3] J.-I. Tamura, *Certain partitions of a lattice*, from Crystal to Chaos —— Proceedings of the Conference in Honor of Gérard Rauzy on His 60th Brithday, Editors: J.-M. Gambaudo, P. Hubert, P. Tisseur and S. Vaienti, World Scientific Co. Pte. Ltd., 2000, pp. 199–219.

DETERMINATION OF ALL Q-RATIONAL CM-POINTS IN MODULI SPACES OF POLARIZED ABELIAN SURFACES

Atsuki UMEGAKI

Department of Mathematical Science, School of Science and Engineering,
Waseda University, 3-4-1, Okubo, Shinjuku-ku, Tokyo, 169-8555, Japan
umegaki@gm.math.waseda.ac.jp

Keywords: polarized abelian surface, moduli space, complex multiplication

Abstract   It is known that in the moduli space $\mathcal{A}_1$ of elliptic curves, there exist precisely 9 Q-rational points corresponding to the isomorphism class of elliptic curves with complex multiplication by the ring of integers of an imaginary quadratic field. We consider the same determination problem in the case of dimension two. In the moduli space of principally polarized abelian surfaces, this problem was solved in [3]. In this paper, we shall determine all Q-rational points corresponding to the isomorphism class of abelian surfaces whose endomorphism ring is isomorphic to the ring of integers of a quartic CM-field in the non-principal case.

2000 Mathematics Subject Classification: 11G15, 11G10, 14K22, 14K10.

1.   INTRODUCTION

Let E be an elliptic curve defined over the complex number field $\mathbf{C}$ whose endomorphism ring is isomorphic to the ring of integers of an imaginary quadratic field K. It is well known that

the j-invariant of E is contained in the rational number field $\mathbf{Q}$ (1.1)
if and only if the class number of K is one,

and that

This work was partially supported by the Ministry of Education, Science, Sports and Culture, Grant-in-Aid for Encouragement of JSPS Research Fellows.

C. Jia and K. Matsumoto (eds.), Analytic Number Theory, 349–357.
© 2002 *Kluwer Academic Publishers. Printed in the Netherlands.*

there are 9 imaginary quadratic fields $\mathbf{Q}(\sqrt{d})$, $d < 0$, having   (1.2)
class number equal to one:
$$-d = 1, 2, 3, 7, 11, 19, 43, 67, 163.$$

Let $\mathcal{A}_1$ be the moduli space of elliptic curves. This is represented by

$$\mathcal{A}_1 = \mathrm{SL}_2(\mathbf{Z}) \backslash \mathfrak{H}_1,$$

where $\mathfrak{H}_1 = \{\tau \in \mathbf{C} \mid \mathrm{Im}(\tau) > 0\}$. From the above facts, we can see that in $\mathcal{A}_1$ there exist precisely 9 $\mathbf{Q}$-rational points corresponding to the isomorphism class of elliptic curves with complex multiplication by the ring of integers of an imaginary quadratic field.

Now we consider the same questions in the case of dimension two. Let $(A, \mathcal{C})$ be a polarized abelian surface over $\mathbf{C}$. We can regard it analytically as a pair $(\mathbf{C}^2/\Lambda, E)$ of a complex torus and a non-degenerate Riemann form. By choosing a suitable basis $(u_1, \ldots, u_4)$ of the lattice Λ, the matrix $(E(u_i, u_j))_{i,j=1,\ldots,4}$ can be reduced to the form

$$\begin{pmatrix} 0 & D \\ -D & 0 \end{pmatrix},$$

where $D = \begin{pmatrix} d_1 & 0 \\ 0 & d_2 \end{pmatrix}$, $d_1 \mid d_2$. Moreover we can assume that $d_1 = 1$,

i.e. $D = \begin{pmatrix} 1 & 0 \\ 0 & d \end{pmatrix}$, $d = \dfrac{d_2}{d_1}$, since two polarized abelian surfaces are isomorphic. In the following, we call such a surface $(A, \mathcal{C})$ a d-polarized abelian surface for the sake of convenience. For $d \in \mathbf{Z}_{\geq 1}$, put

$$D(d) := \begin{pmatrix} 1 & 0 \\ 0 & d \end{pmatrix}, \quad W(d) := \begin{pmatrix} -D(d) & 0 \\ 0 & 1 \end{pmatrix}, \quad T(d) := \begin{pmatrix} 0 & D(d) \\ -D(d) & 0 \end{pmatrix}.$$

Denote the Siegel upper half space of degree 2 by

$$\mathfrak{H}_2 := \{\tau \in M_4(\mathbf{C}) \mid {}^t\tau = \tau, \ \mathrm{Im}(\tau) > 0\}$$

and define a subgroup $\Gamma(d)$ of $\mathrm{GL}_4(\mathbf{Z})$ by

$$\Gamma(d) = \{G \in \mathrm{GL}_4(\mathbf{Z}) \mid G \cdot T(d) \cdot {}^tG = T(d)\}$$

for any d. For $G \in \Gamma(d)$, we put

$$\psi(G) := W(d)^{-1} G W(d).$$

Then we can regard the moduli space of d-polarized abelian surfaces as

$$\mathcal{A}_2(d) = \psi(\Gamma(d))\backslash\mathfrak{H}_2.$$

So we consider how many **Q**-rational points $\mathcal{A}_2(d)$ includes for each d. In the case of $d = 1$, $\Gamma(1) = \mathrm{Sp}_2(\mathbf{Z})$, so $\mathcal{A}_2(1)$ means the moduli space of principally polarized abelian surfaces. In this case, we have already obtained an answer (cf. [3]), namely it was proved that in $\mathcal{A}_2(1)$ there exist precisely 19 **Q**-rational points whose endomorphism ring is isomorphic to the ring of integers of a quartic CM-field.

In this paper we determine all **Q**-rational points corresponding to the isomorphism class of abelian surfaces whose endomorphism ring is isomorphic to the ring of integers of a quartic CM-field in $\mathcal{A}_2(d)$ for any d.

Acknowledgments

The author expresses sincere thanks to Professor Naoki Murabayashi for valuable talks with him.

2.   PRELIMINARIES ON THE THEORY OF COMPLEX MULTIPLICATION

Let K be a quartic CM-field and F the maximal real subfield of K. We consider a structure $P = (A, \mathcal{C}, \theta)$ formed by an abelian variety A of dimension two defined over **C**, a polarization $\mathcal{C}$ of A, and an injection θ of K into $\mathrm{End}^\circ(A) := \mathrm{End}(A) \otimes_{\mathbf{Z}} \mathbf{Q}$ such that $\theta^{-1}(\mathrm{End}(A)) = \mathcal{O}_K$, where $\mathcal{O}_K$ is the ring of integers of K.

We always assume the following condition:

$\theta(K)$ is stable under the Rosati involution of $\mathrm{End}^\circ(A)$ (2.1)
determined by $\mathcal{C}$.

Let $\varphi : K \longrightarrow \mathrm{End}_{\mathbf{C}}(\mathrm{Lie}(A))$ be the representation on the Lie algebra of A with respect to θ. Then there exist two injections σ_1, σ_2 of K into **C** such that φ is equivalent to the direct sum $\sigma_1 \oplus \sigma_2$. We put $\Phi = \{\sigma_1, \sigma_2\}$. Then Φ must satisfy the following condition:

$\{\sigma_1, \sigma_2\} \cup \{\rho \circ \sigma_1, \rho \circ \sigma_2\}$ coincides with the set of all injections   (2.2)
of K into **C**,

where ρ denotes the complex conjugation in **C**. This pair (K, Φ) is called the *CM-type* of P. We define an isomorphism $\tilde{\Phi}$ of **R**-linear spaces by

$$\tilde{\Phi}: K \otimes_{\mathbf{Q}} \mathbf{R} \longrightarrow \mathbf{C}^2, \quad \alpha \otimes a \mapsto (a\alpha^{\sigma_1}, a\alpha^{\sigma_2}).$$

For each $\alpha \in \mathcal{O}_K$, we set $S_{\Phi}(\alpha) = \begin{pmatrix} \alpha^{\sigma_1} & 0 \\ 0 & \alpha^{\sigma_2} \end{pmatrix}$. Then there exist a fractional ideal $\mathfrak{a}$ of K and an analytic isomorphism

$$\lambda \colon \mathbf{C}^2/\tilde{\Phi}(\mathfrak{a}) \xrightarrow{\sim} A$$

such that the following diagram commutes for any $\alpha \in \mathcal{O}_K$:

$$
\begin{array}{ccc}
\mathbf{C}^2/\tilde{\Phi}(\mathfrak{a}) & \xrightarrow{\lambda} & A \\
S_{\Phi}(\alpha) \downarrow & \widetilde{} & \downarrow \theta(\alpha) \\
\mathbf{C}^2/\tilde{\Phi}(\mathfrak{a}) & \xrightarrow[\lambda]{\sim} & A
\end{array}
\tag{2.3}
$$

Take a basic polar divisor in $\mathcal{C}$ and consider its Riemann form $E(x, y)$ on $\mathbf{C}^2$ with respect to λ. Then there is an element η of K such that

$$E(\tilde{\Phi}(x), \tilde{\Phi}(y)) = \mathrm{Tr}_{K/\mathbf{Q}}(\eta x y^{\rho})$$

for any $(x, y) \in K \times K$. The element η must satisfy the following conditions:

$$\eta^{\rho} = -\eta, \quad \mathrm{Im}(\eta^{\sigma_i}) > 0 \ (i = 1, 2); \tag{2.4}$$

$$\mathrm{Tr}_{K/\mathbf{Q}}(\eta \mathfrak{a} \mathfrak{a}^{\rho}) = \mathbf{Z}. \tag{2.5}$$

Then the structure P is said to be *of type* $(K, \Phi; \eta, \mathfrak{a})$. Conversely for any $\Phi, \mathfrak{a}, \eta$ satisfying the conditions (2.2), (2.4) and (2.5), we can construct $(A, \mathcal{C}, \theta)$ of type $(K, \Phi; \eta, \mathfrak{a})$.

Let X be a basic polar divisor in the class of $\mathcal{C}$. We can consider the isogeny

$$\phi_X \colon A \longrightarrow \mathrm{Pic}^0(A), \quad v \mapsto [X_v - X]$$

from A onto its Picard variety, where X_v is the translation of X by v. Then there exists an ideal $\mathfrak{f}$ of F which is uniquely determined by the property

$$\ker(\phi_X) = \{y \in A \mid \theta(\mathfrak{f}) \, y = 0\}. \tag{2.6}$$

The ideal $\mathfrak{f}$ satisfies

$$\mathfrak{f}\mathcal{O}_K = \eta \mathfrak{d} \mathfrak{a} \mathfrak{a}^{\rho}, \tag{2.7}$$

where $\mathfrak{d}$ is the different of $K/\mathbf{Q}$.

For any subfield L of K, one can prove that there exists a subfield M_L of $\mathbf{C}$ which has the property:

for any automorphism σ of $\mathbf{C}$, σ is the identity map on M_L (2.8)
if and only if there exists an isomorphism ι of A to A^σ such
that $\iota(\mathcal{C}) = \mathcal{C}^\sigma$ and $\iota \circ \theta(\alpha) = \theta^\sigma(\alpha) \circ \iota$ for all $\alpha \in L$.

M_L is called the *field of moduli* of $(A, \mathcal{C}, \theta|_L)$. The field of moduli of $(A, \mathcal{C})$, which is equal to $M_{\mathbf{Q}}$, can not be understood only by the theory of complex multiplication. But we have the following result proved by Murabayashi (cf. Theorem 4.12 in [2]):

Theorem 2.1. *Let K be a quartic CM-field and $(A, \mathcal{C})$ a polarized abelian surfaces such that $\mathcal{O}_K \subseteq \mathrm{End}(A)$, where $\mathcal{O}_K$ denotes the ring of integers of K. Then $\mathcal{O}_K \cong \mathrm{End}(A)$ and the field of moduli of $(A, \mathcal{C})$ coincides with $\mathbf{Q}$ if and only if the following three conditions hold:*

(a) K *has an expression of the form*

$$K = \mathbf{Q}\left(\sqrt{-q_1 \cdots q_t \varepsilon_0 \sqrt{p}}\right) \quad (t \geq 0),$$

where ε_0 is a fundamental unit of $F = \mathbf{Q}(\sqrt{p})$ such that $\varepsilon_0 > 0$ and $p, q_1, \ldots, q_t$ are distinct prime numbers which satisfy one of the following conditions:

(a_1) $p \equiv 5 \pmod 8$. *Moreover if $t \geq 1$, then*

$$q_i \equiv 1 \pmod 4 \quad \text{and} \quad \left(\frac{p}{q_i}\right) = -1 \quad (i = 1, \ldots, t);$$

(a_2) $p \equiv 5 \pmod 8$, $t \geq 1$, $q_1 = 2$. *Moreover if $t \geq 2$, then*

$$q_i \equiv 1 \pmod 4 \quad \text{and} \quad \left(\frac{p}{q_i}\right) = -1 \quad (i = 2, \ldots, t);$$

(a_3) $p = 2$. *Moreover if $t \geq 1$, then*

$$q_i \equiv 5 \pmod 8 \quad (i = 1, \ldots, t).$$

(b) $h_K^- = 2^t$, *where h_K^- denotes the relative class number of K.*

(c) $\mathfrak{f} = \mathfrak{f}^\tau$ *for any $\tau \in \mathrm{Gal}(F/\mathbf{Q})$.*

Remark 2.2. In the statement of Theorem 4.12 in [2], we assume the condition (2.1). But we can ignore this assumption (see Remark 1.2 in [3]).

3.   FIELDS WHICH SATISFY NECESSARY CONDITIONS

We have the following theorem which has been proved by Louboutin in [1]:

Theorem 3.1. *Let K be an imaginary cyclic number field of 2-power degree $2n = 2^m \geq 4$ with conductor f_K and discriminant d_K. Then we have*

$$h_K^- \geq \frac{2\varepsilon_K}{e(2n-1)} \left(\frac{\sqrt{f_K}}{\pi(\log f_K + 0.05)} \right)^n,$$

where

$$\varepsilon_K = 1 - \frac{2\pi n e^{\frac{1}{n}}}{d_K^{\frac{1}{2n}}} \quad or \quad \frac{2}{5}\exp\left(-\frac{2n\pi}{d_K^{\frac{1}{2n}}} \right).$$

By using this result, we can determine quartic CM-fields satisfying the conditions (a) and (b) in Theorem 2.1 (cf. [3]):

Theorem 3.2. *There are exactly 13 quartic CM-fields satisfying the conditions (a) and (b) in Theorem 2.1. Namely, the ones with class number h given as follows:*

$h = 1$ $\mathbf{Q}\left(\sqrt{-(2+\sqrt{2})} \right)$ $h = 2$ $\mathbf{Q}\left(\sqrt{-5(2+\sqrt{2})} \right)$

$\mathbf{Q}\left(\sqrt{-(5+2\sqrt{5})} \right)$ $\mathbf{Q}\left(\sqrt{-(5+\sqrt{5})} \right)$

$\mathbf{Q}\left(\sqrt{-(13+2\sqrt{13})} \right)$ $\mathbf{Q}\left(\sqrt{-13(5+2\sqrt{5})} \right)$

$\mathbf{Q}\left(\sqrt{-(29+2\sqrt{29})} \right)$ $\mathbf{Q}\left(\sqrt{-17(5+2\sqrt{5})} \right)$

$\mathbf{Q}\left(\sqrt{-(37+6\sqrt{37})} \right)$ $\mathbf{Q}\left(\sqrt{-(13+3\sqrt{13})} \right)$

$\mathbf{Q}\left(\sqrt{-(53+2\sqrt{53})} \right)$ $\mathbf{Q}\left(\sqrt{-5(13+2\sqrt{13})} \right)$

$\mathbf{Q}\left(\sqrt{-(61+6\sqrt{61})} \right)$

Remark 3.3. Note that the above fields are cyclic extensions over $\mathbf{Q}$ and have an expression of the form

$$K = \mathbf{Q}\left(\sqrt{-n\varepsilon_0\sqrt{p}} \right) \quad (n = q_1 \cdots q_t)$$

which is given by the condition (a) in Theorem 2.1:

K	p	ε_0	n
$\mathbf{Q}\left(\sqrt{-(2+\sqrt{2})}\,\right)$	2	$1+\sqrt{2}$	1
$\mathbf{Q}\left(\sqrt{-(5+2\sqrt{5})}\,\right)$	5	$\frac{1+\sqrt{5}}{2}$	1
$\mathbf{Q}\left(\sqrt{-(13+2\sqrt{13})}\,\right)$	13	$\frac{3+\sqrt{13}}{2}$	1
$\mathbf{Q}\left(\sqrt{-(29+2\sqrt{29})}\,\right)$	29	$\frac{5+\sqrt{29}}{2}$	1
$\mathbf{Q}\left(\sqrt{-(37+6\sqrt{37})}\,\right)$	37	$6+\sqrt{37}$	1
$\mathbf{Q}\left(\sqrt{-(53+2\sqrt{53})}\,\right)$	53	$\frac{7+\sqrt{53}}{2}$	1
$\mathbf{Q}\left(\sqrt{-(61+6\sqrt{61})}\,\right)$	61	$\frac{39+5\sqrt{61}}{2}$	1
$\mathbf{Q}\left(\sqrt{-5(2+\sqrt{2})}\,\right)$	2	$1+\sqrt{2}$	5
$\mathbf{Q}\left(\sqrt{-(5+\sqrt{5})}\,\right)$	5	$\frac{1+\sqrt{5}}{2}$	2
$\mathbf{Q}\left(\sqrt{-13(5+2\sqrt{5})}\,\right)$	5	$\frac{1+\sqrt{5}}{2}$	13
$\mathbf{Q}\left(\sqrt{-17(5+2\sqrt{5})}\,\right)$	5	$\frac{1+\sqrt{5}}{2}$	17
$\mathbf{Q}\left(\sqrt{-(13+3\sqrt{13})}\,\right)$	13	$\frac{3+\sqrt{13}}{2}$	2
$\mathbf{Q}\left(\sqrt{-5(13+2\sqrt{13})}\,\right)$	13	$\frac{3+\sqrt{13}}{2}$	5

4.   RESULTS

Proposition 4.1. *Let K be a quartic CM-field. Assume that (A,C) is a polarized abelian surface such that $\mathrm{End}(A) \cong o_K$ and its field of moduli coincides with $\mathbf{Q}$. Then (A,C) has either principally polarization or p-polarization, where p is the prime number which appears in the expression of K as in Theorem 2.1 (a).*

Proof. Let (A,C) be an abelian surface satisfying the assumption and $\mathfrak{f}$ the ideal of F determined by (2.6). From the condition (c) of Theorem 2.1, we can write

$$\mathfrak{f} = (r)(\sqrt{p})^{\delta}, \quad r \in \mathbf{Q}, \ \delta = 0, 1, \tag{4.1}$$

since F has class number 1. Then one can ignore the gap of r, since it causes an isomorphism (cf. [4], § 14.2). So the polarization C is determined only by the value of δ. By the calculation of the Riemann form, it is a principally polarization if $\delta = 0$, but otherwise a p-polarization. This completes the proof.   ∎

Theorem 4.2. *Let $\mathcal{A}_2(d)$ be the moduli space of d-polarized abelian surfaces and*

$$n(d) = \begin{cases} 19 & \text{if } d = 1, \\ 7 & \text{if } d = 5, \\ 5 & \text{if } d = 13, \\ 3 & \text{if } d = 2, \\ 1 & \text{if } d = 29, 37, 53, 61, \\ 0 & \text{otherwise.} \end{cases}$$

Then in $\mathcal{A}_2(d)$ there exist exactly $n(d)$ $\mathbf{Q}$-rational points corresponding to the abelian surface whose endomorphism ring is isomorphic to the ring of integers of a quartic CM-field.

Proof. Let K be a quartic cyclic CM-field and A an abelian surface with $\text{End}(A) \cong \mathcal{O}_K$. We fix an injection $\iota\colon K \hookrightarrow \mathbf{C}$, a generator σ of $\text{Gal}(K/\mathbf{Q})$ and an isomorphism $\theta\colon K \xrightarrow{\sim} \text{End}^\circ(A)$. Let $\varphi\colon K \longrightarrow \text{End}_{\mathbf{C}}(\text{Lie}(A))$ be the representation corresponding to θ. Then φ is equivalent to $\iota \circ \sigma^i \oplus \iota \circ \sigma^j$ for some i, j $(0 \le i < j \le 3)$. Since the condition (2.2) holds, we have $(i, j) = (0, 1), (0, 3), (1, 2)$ or $(2, 3)$. Changing θ by

$$\theta' := \begin{cases} \theta & \text{if } (i, j) = (0, 1), \\ \theta \circ \sigma & \text{if } (i, j) = (0, 3), \\ \theta \circ \sigma^3 & \text{if } (i, j) = (1, 2), \\ \theta \circ \sigma^2 & \text{if } (i, j) = (2, 3), \end{cases}$$

we may assume that φ is equivalent to $\iota \oplus \iota \circ \sigma$. Put $\Phi = \{\iota, \iota \circ \sigma\}$. Then A is isomorphic to $\mathbf{C}^2/\tilde{\Phi}(\mathfrak{a})$ for some $\mathfrak{a} \in I_K$. It holds that for any $\mathfrak{a}, \mathfrak{b} \in I_K$, $\mathbf{C}^2/\tilde{\Phi}(\mathfrak{a})$ and $\mathbf{C}^2/\tilde{\Phi}(\mathfrak{b})$ are isomorphic if and only if $\mathfrak{a}$ and $\mathfrak{b}$ are in the same ideal class. Therefore we obtain 19 abelian surfaces from the fields in Theorem 3.2. For each of these abelian surfaces, a principal polarization has been constructed in [3]. By the similar calculation, we can construct a p-polarization by giving η in §2 concretely. In our case, it holds that $N_{F/\mathbf{Q}}(\varepsilon_0) = -1$. So any totally positive unit of F is of the form $\varepsilon_0^{2n} = \varepsilon_0^n \cdot (\varepsilon_0^n)^\rho$. Hence any two d-polarizations are equivalent for each $d = 1, p$ (see § 14 in [4]). From Proposition 4.1, this completes the proof.   ∎

Remark 4.3. We can determine the points in $\mathfrak{H}_2$ corresponding to these $\mathbf{Q}$-rational points.

References

[1] S. Louboutin, CM-fields with cyclic ideal class groups of 2-power orders, *J. Number Theory* **67** (1997), 11–28.

[2] N. Murabayashi, The field of moduli of abelian surfaces with complex multiplication, *J. reine angew. Math.* **470** (1996), 1–26.

[3] N. Murabayashi and A. Umegaki, Determination of all **Q**-rational CM-points in the moduli space of principally polarized abelian surfaces, *J. Algebra* **235** (2001), 267–274.

[4] G. Shimura, *Abelian varieties with complex multiplication and modular functions*, Princeton university press, 1998.

ON FAMILIES OF CUBIC THUE EQUATIONS

Isao WAKABAYASHI

Faculty of Engineering, Seikei University, Musashino-shi, Tokyo 180-8633, Japan
wakaba@ge.seikei.ac.jp

Keywords: cubic Thue equation, Padé approximation

Abstract   We survey results on families of cubic Thue equations. We also state new results on the family of cubic Thue inequalities $|x^3 + axy^2 + by^3| \leq k$, where a, k are positive integers and b is an integer: We give upper bounds for the solutions of these inequalities when a is larger than a certain value depending on b, and as an example, for the case $b = 1, 2$, $a \geq 1$, and $k = a + b + 1$, we solve these inequalities completely. Our method is based on Padé approximations.

1991 Mathematics Subject Classification: 11D.

1.   INTRODUCTION

The study of cubic Thue equations has a long history. We survey some of the results on families of cubic Thue equations. We also explain the Padé approximation method. There are also many results on various kinds of single cubic Thue equation, and also on the number of solutions, however we will not mention these subjects except a few results. We also state new results.

In 1909, A. Thue [Thu2] proved that if $F(x, y)$ is a homogeneous irreducible polynomial with integer coefficients, of degree at least 3, and k is an integer, then the Diophantine equation

$$F(x, y) = k \tag{1}$$

has only a finite number of integer solutions. After this important work, equations of this type are called *Thue equations*. His method also yields an estimate for the number of solutions. However, his result is ineffective and does not provide an algorithm for finding all solutions.

For a Thue equation, we call the solutions which can be found easily the *trivial solutions*. This is a vague terminology, but will be well un-

359

C. Jia and K. Matsumoto (eds.), Analytic Number Theory, 359–377.
© 2002 *Kluwer Academic Publishers. Printed in the Netherlands.*

derstood in each situation. The problem is usually to show that a given Thue equation has no other solution than the trivial solutions. For a family of Thue equations, it is usually expected that, if some suitable parameters are large, then there would be no non-trivial solution.

To actually solve a Thue equation or to give an effective upper bound for the solutions, there are mainly four methods: 1. Algebraic method; 2. Baker's method; 3. Padé approximation method; and 4. Bombieri's method. In the following we discuss results obtained by these methods. (Some of the results are stated in the form of Theorem, and some are not. But no meaning is laid on this distinction.)

Let θ be a real number. If there exist positive constants c and λ such that for any integers p, q $(q > 0)$, we have

$$\left| \theta - \frac{p}{q} \right| > \frac{1}{cq^\lambda}, \tag{2}$$

then we call the number λ an *irrationality measure* for θ. If c is effective, then we say that θ has an *effective* irrationality measure. For equation (1), if we can obtain for every real root of $F(x, 1)$ an effective irrationality measure smaller than the degree of F, then we can obtain, by an elementary estimation, an upper bound for the solutions of (1).

Let us consider the family of cubic Thue inequalities

$$|x^3 + axy^2 + by^3| \le k \tag{3}$$

where a, b, and k are integers. This is a general form of cubic Thue equation or inequality in the sense that the solutions of any given cubic Thue equation are contained in the set of solutions of an inequality as (3) with suitable a, b and k. In Section 7 we state new results on (3): We give an upper bound for the solutions of (3) when a is positive and larger than a certain value depending on b, and as an example, for the case $b = 1, 2$, $a \ge 1$, and $k = a + b + 1$, we solve these inequalities completely. Our method is based on Padé approximations. We give a sketch of the proof in Section 8.

The author expresses his gratitude to Professors Yann Bugeaud and Attila Pethő for valuable comments and information.

2.   BAKER'S METHOD

In 1968, A. Baker [Ba2] succeeded to give an effective upper bound for the solutions of any Thue equation. The method is based on his result on estimate of lower bound for linear forms in the logarithms of algebraic numbers. See [Ba3, Chapter 4] for this method. The upper bounds obtained by Baker's method are usually very large. However, since the result of Baker appeared, estimate of lower bounds for linear forms in the logarithms of algebraic numbers has been largely improved. See for example M. Waldschmidt [Wal] for the case of n logarithms, and M. Laurent, M. Mignotte, and Y. Nesterenko [LMN] for the case of two logarithms. The result in [LMN] provides us, in some cases, a fairly good estimate close to an estimate obtained by the Padé approximation method. Using these improvements and other computation techniques, we can nowadays actually solve a single Thue equation by aid of computer if the coefficients are concretely given, and if they are small. For example Y. Bilu and G. Hanrot [BH] provide a method using computation techniques and number theoretical datas for finding actually all solutions of such Thue equations or Thue inequalities. In fact this method is implemented by them in the software called *KANT* [Dal] and also in the system called *PARI*. These softwares return us all solutions of a Thue equation if we just input to the computer the values of the coefficients of the equation.

In 1990, E. Thomas [Tho1] investigated for the first time a family of Thue equations by using Baker's method. Since then, various families of Thue equations or inequalities of degree 3 or 4 or even of higher degree have been solved. See also C. Heuberger, A. Pethő, and R. F. Tichy [HPT] for a survey on families of Thue equations. Heuberger [H] is also a nice survey on families of Thue equations, and describes recent development of Baker's method for solving them. As for cubic Thue equations or inequalities, the following have been solved partly or completely.

(i) $|x^3 - (a - 1)x^2y - (a + 2)xy^2 - y^3| \leq k.$

Thomas [Tho1] proved that this family with $k = 1$ has only the trivial solutions $(0,0), (\pm1,0), (0,\pm1)$ and $\pm(1,-1)$ if $a \geq 1.365 \times 10^7$. Mignotte [M1] in 1993 solved the remaining cases with $k = 1$.

For general k, Mignotte, Pethő, and F. Lemmermeyer [MPL] in 1996 gave an upper bound for the solutions in the case $a \geq 1650$, and solved completely this family for $k = 2a + 1$. In 1999, G. Lettl, Pethő, and P. Voutier [LPV], using the Padé approximation method, gave an upper bound for the solutions in the case $a \geq 31$, and $k \geq 1$. This upper bound is much better than the former one.

(ii) $x^3 - ax^2y - (a + 1)xy^2 - y^3 = 1.$

Mignotte and N. Tzanakis [MT] proved that this family has only 5 trivial solutions if $a \geq 3.67 \cdot 10^{32}$. Recently, Mignotte [M2] solved this family completely.

(iii) $x(x - a^b y)(x - a^c y) \pm y^3 = 1$.

Thomas [Tho2] proved that if $0 < b < c$ and $a \geq (2 \cdot 10^6(b + 2c))^{4.85/(c-b)}$, then the equation has only the trivial solutions. For the case $b = 1$, $c \geq 143$, $a \geq 2$, he completely solved the equation. He conjectures that, if $d \geq 3$ and $p_1(t), \cdots, p_{d-1}(t)$ are polynomials with integer coefficients satisfying $\deg p_1 < \cdots < \deg p_{d-1}$, then the family $x(x - p_1(a)y) \cdots (x - p_{d-1}(a)y) \pm y^d = 1$ has only the trivial solutions if a is sufficiently large. He calls this a *split family*.

These were all treated by Baker's method, except the work of [LPV] on (i). To apply this method, it is necessary to determine a fundamental unit system of the associated algebraic number field or of the associated order. Therefore, to be able to treat a family of Thue equations by this method, it is strongly required that the associated algebraic number field or the associated order has a fundamental unit system whose form does not vary much depending on the values of parameters, or at least that we are able to obtain a good estimate for a fundamental unit system. If the right-hand side k of a Thue equation is not ± 1 as in family (i), then it is also necessary to determine a system of representatives of algebraic integers whose norm is equal to k. Therefore, when Baker's method is used, usually the case with $k = \pm 1$ is treated.

3.   PADÉ APPROXIMATION METHOD

The idea of Padé approximation method is originally due to Thue [Thu1], [Thu4]. This method can be applied only to certain types of equations. Moreover, when we apply this method to a family of Thue equations, it fails usually for the case where the values of the parameters of the family are small, and we must treat the remaining equations by Baker's method. However, if this method is applicable, then it usually provides a better estimate than Baker's method. Therefore, it often happens that the number of remaining equations is finite and small, and in such a case we can solve completely the family. An example is mentioned above concerning equation (i). Moreover, to apply this method, it is not necessary to know the structure of the associated number field, and we can easily treat Thue inequalities at the same time.

(iv) $|ax^d - by^d| \leq k$, $d \geq 3$.

In 1918 Thue [Thu4] succeeded to give an upper bound for the solutions of this inequality using one good solution under the assumption that it exists. When one can find easily a good solution, then his result gives actually an upper bound for the other solutions. As an example he gave an upper bound for the solutions of

$$|(a+1)x^3 - ay^3| = 1 \tag{4}$$

when $a \geq 37$.

To our knowledge, this is the first time that an upper bound for the solutions of a family of Diophantine equations was obtained. His idea is really original. He constructed, already in his earlier work [Thu1], polynomials $P_n(x)$ and $Q_n(x)$ of degree at most n for any positive integer n such that $xP_n(x^d) - Q_n(x^d) = (x-1)^{(d-2)n}R_n(x)$. Actually, these polynomials P_n and Q_n are some hypergeometric polynomials, namely $P_n(x) = F(-1/d - n, -n, -1/d + 1; 1/x)x^n$ and $Q_n(x) = F(-1/d-n, -n, -1/d+1; x)$, even though he does not mention this fact and this is observed by C. L. Siegel [S1]. Taking a good solution (x_0, y_0), Thue puts $x = \sqrt[d]{a/b}x_0/y_0$, and obtains the result. He works only with polynomials, but this is essentially equivalent to Padé approximations. See Section 6 for further explanation.

Using the Padé approximation method, Siegel [S1] in 1929 showed that the number of solutions of any cubic Thue equation as (1) with positive discriminant is at most 18 if the discriminant is larger than a certain value depending on k. To construct Padé approximations he as well as Thue uses hypergeometric polynomials, so this Padé approximation method is also called the *hypergeometric method*. In a student paper from 1949, A. E. Gel'man showed that 18 can be replaced by 10. For a proof of this, see B. N. Delone and D. K. Faddeev [DF, Chapter 5]. J.-H. Evertse [E2] proved that the number of solutions of any cubic Thue equation with positive discriminant (not assuming it is large) and with $k = 1$ is at most 12, and M. A. Bennett [Be1] proved that this number is at most 10. Recently (in 2000) R. Okazaki announced at a conference held at Debrecen, Hungary that the number of solutions of any cubic Thue equation with sufficiently large positive discriminant and with $k = 1$ is at most 7. On the other hand, based on extensive computer search, Pethő [P] formulated a conjecture on the number of solutions of cubic Thue equations with positive discriminant and with $k = 1$, by classifying cases. For the general case, he conjectures that the number of solutions is at most 5. His conjecture is refined by F. Lippok [Li]. For the equation $|ax^d - by^d| = k$, Evertse [E1] gave an upper bound for the number of solutions, by the Padé approximation method and congruence consideration. The paper [Be1] includes also many references on Thue equations.

Siegel [S2] refined the above work of Thue on (iv), and proved that if $|ab|$ is larger than a certain value depending on d and k, then inequality (iv) has at most one positive coprime solution. This is a nice result since it asserts that, if we find one solution, then there is no more solution. Siegel gave interesting examples: 1. For $d = 7, 11, 13$, the equation $33x^d - 32y^d = 1$ has only the solution $(1, 1)$. 2. The equation $(a+1)x^3 - ay^3 = 1$ has only the solution $(1, 1)$ if $a \geq 562$. Some special cases of Siegel's results are contained in earlier results of Delaunay, T. Nagell, and V. Tartakovskij obtained by the algebraic method. See (viii) and (ix) below.

Refining the Thue–Siegel method, Baker [Ba1] proved that $\sqrt[3]{2}$ has irrationality measure $\lambda = 2.955$ with $c = 10^6$, and obtained an upper bound for the solutions of (iv) in the case $a = 1, b = 2$, and $d = 3$, and completely solved (iv) in the case $a = 1, b = 2, d = 3$, and $k = 1$.

Baker and C. L. Stewart [BaS] treated (iv) in the case $b = 1, a \geq 2$ and $d = 3$ by Baker's method, and proved that the solutions are bounded by $(c_1 k)^{c_2}$, where $c_1 = \epsilon^{(50 \log \log \epsilon)^2}$, $c_2 = 10^{12} \log \epsilon$, and ϵ is the fundamental unit > 1 of the field $\mathbf{Q}(\sqrt[3]{a})$.

An irrationality measure for the number $\sqrt[d]{a/b}$ is obtained by Bombieri's method. See Section 4.

After an extensive work done by Bennett and B. M. M. de Weger [BedW], recently Bennett [Be2] proved, by using both the Padé approximation method and Baker's method, that for $d \geq 3$ the equation $ax^d - by^d = 1$ has at most one positive solution. This can be considered to be a final result about family (iv) with $k = 1$.

There are some other families of Thue equations of degree > 3 which were treated by the Padé approximation method.

(v) $|x^4 - ax^3 y - 6x^2 y^2 + axy^3 + y^4| \leq k$.

Chen Jianhua and P. M. Voutier [CheV] solved this family with $k = 1$ for $a \geq 128$. They use an important lemma of Thue [Thu3] which provides a possibility to apply the Padé approximation method to some special type of Thue equations of degree ≥ 3. For general k, Lettl, Pethő, and Voutier [LPV] gave an upper bound for the solutions in the case $a \geq 58$, and proved that, in the case $k = 6a + 7$ and $a \geq 58$, the only solutions with $|x| \leq y$ are $(0, 1), (\pm 1, 1), (\pm 1, 2)$. Historically, before [CheV], by using Baker's method Lettl and Pethő [LP] had already solved completely the corresponding family of Thue equations (not inequalities) for $k = 1$ and 4.

(vi) $|x^6 - 2ax^5 y - (5a+15)x^4 y^2 - 20x^3 y^3 + 5ax^2 y^4 + (2a+6)xy^5 + y^6| \leq k$.

Lettl, Pethő, and Voutier [LPV] gave an upper bound for the solutions in the case $a \geq 89$, and solved the inequality completely in the case $k = 120a + 323$ and $a \geq 89$. Families (i), (v) and (vi) are called *simple families* since the solutions of the corresponding algebraic equation $F(x, 1) =$

0 are permuted transitively by a fractional linear transformation with rational coefficients (see [LPV] for the definition).

(vii) $|x^4 - a^2x^2y^2 - by^4| \le k$.

For $b = -1$, the author [Wak1] gave an upper bound for the solutions in the case $a \ge 8$, and solved the inequality when $k = a^2 - 2$ and $a \ge 8$. When $b \ge 1$ and a is greater than a certain value depending on b, an upper bound for the solutions was obtained by [Wak2], and the case $b = 1, 2$, $a \ge 1$, and $k = a^2 + b - 1$ was completely solved. The method is based on Padé approximations. In the Padé approximation method, one usually constructs, for a real solution θ of the corresponding algebraic equation, linear forms in $1, \theta$. In these two works however, linear forms in $1, \theta, \theta^2$ were used, and it seems that linear forms of this kind were used for the first time to solve Thue equations.

4.   BOMBIERI'S METHOD

The proof of the finiteness theorem of Thue on equation (1) is based on the idea that if there is a rational number which is sufficiently close to a given real number θ, then other rational numbers can not be too close to θ. In fact, Thue first assumes existence of a rational number p_0/q_0 having sufficiently large denominator and exceptionally close to θ. Then, taking another p/q close to θ, and using the box principle, he constructs an auxiliary polynomial $P(x, y)$ which vanishes at (θ, θ) to a high order. And he proves that P vanishes at $(p_0/q_0, p/q)$ only to a low order. To prove this, he needs the assumption that q_0 is sufficiently large. However, this assumption for p_0/q_0 is too strong, and no pair $(\theta, p_0/q_0)$ having the required property is known at present, and by this reason no effective result has been found using this method.

E. Bombieri [Bo1] however succeeded, by using Dyson's lemma, to remove the requirement that q_0 should be large. This lemma asserts in fact that P vanishes at $(p_0/q_0, p/q)$ only to a low order, by using information on the degree of P and the vanishing degree at (θ, θ) only. Hence it is free from the size of q_0, and it allows to use an approximation p_0/q_0 to θ even if its denominator is small. Thus, Bombieri could find examples of good pairs $(\theta, p_0/q_0)$, and obtained effective irrationality measures for certain algebraic numbers.

Example [Bo1]. Let $d \ge 40$, and let θ be the positive root of $x^d - ax^{d-1} + 1$, where $a \ge A(d)$ and $A(d)$ is effectively computable. Then, θ has effective irrationality measure $\lambda = 39.2574$.

Refining the above method, Bombieri and J. Mueller [BoM] obtained the following.

Theorem ([BoM]). *Let* $d \geq 3$, *and let* a *and* b *be coprime positive integers, and put* $\mu = \dfrac{\log|a - b|}{\log b}$. *Let* $\theta \in \mathbf{Q}(\sqrt[d]{a/b})$ *be of degree* d. *If* $\mu < 1 - 2/d$, *then for any* $\varepsilon > 0$, θ *has effective irrationality measure*

$$\lambda = \frac{2}{1 - \mu} + 6 \left(\frac{d^5 \log d}{\log b} \right)^{1/3} + \varepsilon.$$

Pursuing this direction, Bombieri [Bo2] in 1993 succeeded to create a new method for obtaining an effective irrationality measure for any algebraic number. This method applies to all algebraic numbers, hence it is a very general theorem as well as Baker's theorem. The method is completely different from that of Baker.

Bombieri, A. J. van der Poorten, and J. D. Vaaler [BvPV] applied Bombieri's method to algebraic numbers of degree three over an algebraic number field, and under a certain assumption, obtained an irrationality measure with respect to the ground field. For simplicity, we state their result for cubic numbers over the rational number field.

Theorem ([BvPV]). *Let* a $(\neq 0)$ *and* b *be integers* (*positive or negative*), *and* $f(x) = x^3 + ax + b$ *be irreducible. Let* θ *be the real root of* f *whose absolute value is the smallest among the three roots of* f. *Assume that* $|a| > e^{1000}$ *and* $|a| \geq b^2$. *Then,* θ *has effective irrationality measure*

$$\lambda = \frac{2 \log |a|^3}{\log(|a|^3/b^2)} + \frac{14}{(\log(|a|^3/b^2))^{1/3}}.$$

Example. For $|a| = [e^{1000}] + 1$ and $|b| = 1$ we have $\lambda = 2.971$.

Remark. If $a > 0$, then f has only one real root, and its absolute value is smaller than that of the complex roots of f.

5.   ALGEBRAIC METHOD

After Thue proved his theorem on the finiteness of the number of solutions, the maximal number of solutions for some special families of Thue equations of low degrees was determined by the algebraic method. This method uses intensively properties of the units of the associated number field or of the associated order. Therefore, only the cases where the units group has rank 1 have been treated. However, this method gives very precise results on the maximal number of solutions, and it also gives an alternative proof of Thue's finiteness theorem for these special cases. Nagell [N4], and Delone and Faddeev [DF] are good books on this subject. Families (viii) and (ix) are special cases of (iv).

(viii) $ax^3 + by^3 = k$.

Case $b = 1$ and $k = 1$. Delaunay and Nagell independently proved that this equation has at most one solution with $xy \neq 0$. For the proof, see [De1], [N1] and [N2].

Case $k = 1$ or 3. Nagell [N1] [N2] proved that this equation has at most one solution with $xy \neq 0$ except the equation $2x^3 + y^3 = 3$ which has two such solutions $(1, 1)$ and $(4, -5)$.

(ix) $x^4 - ay^4 = 1$.

Tartakovskij [Ta] proved that if $a \neq 15$ then this equation has at most one positive solution. Actually, one can see by using KANT that the same holds for $a = 15$ also.

(x) $F(x, y) = 1$ *where F is a homogeneous cubic polynomial with negative discriminant.*

The assumption that the discriminant is negative implies that $F(x, 1)$ has only one real zero and the associated unit group has rank 1. Delaunay [De2] and Nagell [N3] independently proved that, if the discriminant is not equal to $-23, -31, -44$, then this equation has at most 3 solutions.

(xi) $x^3 + axy^2 + y^3 = 1$, $a \geq 2$.

The only solutions of this equation are $(1, 0), (0, 1)$ and $(1, -a)$. This is a corollary of the above general result of Delaunay and Nagell on equation (x) since the discriminant $-4a^3 - 27$ is negative by the assumption and the equation has already three solutions. The case $a = 1$ corresponds to one of the exceptional cases.

6. PRINCIPLE OF THE PADÉ APPROXIMATION METHOD

We explain here how we apply Padé approximations to solve Thue equations. For this purpose, let us consider equation (4), namely the example treated by Thue and Siegel. The principle is as follows. In order to solve this equation, we need to obtain properties concerning rational approximations to the algebraic number $\alpha = \sqrt[d]{(a + 1)/a}$ with $d = 3$. Let us note that this number is written as $\alpha = \sqrt[d]{1 + 1/a}$. In order to obtain properties of this number, we first consider the binomial function $\sqrt[d]{1 + x}$, and obtain some property concerning Padé approximations to this function. Then, putting $x = 1/a$ and using the property of this function, we obtain properties of the number α.

In order to explain what Padé approximations are, we state the following proposition.

Proposition. *Let $f(x)$ be a Taylor series at the origin with rational coefficients. Then, for every positive integer n, there exist polynomials $P_n(x)$ and $Q_n(x)$ $(Q_n \neq 0)$ with rational coefficients and of degree at most n such that*

$$P_n(x) - Q_n(x)f(x) = c_n x^{2n+1} + \cdots$$

holds, namely, such that the Taylor expansion of the left-hand side begins with the term of degree at least $2n + 1$.

We call the rational functions $P_n(x)/Q_n(x)$ or the pairs $(P_n(x), Q_n(x))$ *Padé approximations* to $f(x)$.

Proof. The necessary condition for P_n and Q_n is written as a system of linear equations in their coefficients. Comparing the number of equations and the number of unknowns, we find a non-trivial solution.

By this proposition we see that Padé approximations to a given function exist always. For application to Diophantine equations, this existence theorem is not sufficient, and it is very important to know properties of the polynomials $P_n(x)$ and $Q_n(x)$. Namely, it is necessary to know the size of the denominators of their coefficients, and upper bounds for values of these polynomials and the right-hand side. Therefore, it is important to be able to construct Padé approximations concretely.

Example (Thue-Siegel). Padé approximations to the binomial function $\sqrt[d]{1+x}$ are given by

$$F(-1/d - n, -n, -2n; -x) - F(1/d - n, -n, -2n; -x)\sqrt[d]{1+x}$$
$$= c_n x^{2n+1} + \cdots,$$

where F denotes the hypergeometric function of Gauss (in this case they are hypergeometric polynomials).

We put $x = 1/a$ into this formula, and we multiply the relation by the common denominators of the coefficients of these hypergeometric polynomials and also by a^n. Then we obtain

$$p_n - q_n \sqrt[d]{1 + \frac{1}{a}} = c_n' \left(\frac{1}{a}\right)^{n+1} + \cdots,$$

$$p_n, q_n \in \mathbf{Z}.$$

Suppose a is sufficiently large. Then we can verify that if n tends to the infinity, then the right-hand side tends to zero. From this we obtain a sequence of rational numbers p_n/q_n which approach to the number $\alpha = \sqrt[d]{1 + 1/a}$. Since we know well about hypergeometric functions, we can obtain necessary information about p_n, q_n, and the right-hand side of the above relation. Then, comparing a rational number p/q close to the number α with the sequence p_n/q_n, we obtain a result on rational approximations to α. (Actually we should construct two sequences approaching to α.) This final process is based on the following lemma. Its idea is due to Thue.

Lemma. *Let θ be a non-zero real number. Suppose there are positive numbers ρ, P, l, L with $L > 1$, and further there are, for each integer $n \geq 1$, two linear forms*

$$p_{in} + q_{in}\theta = l_{in} \quad (i = 1, 2)$$

with integer coefficients p_{in} and q_{in} satisfying the following conditions:
   (i) the two linear forms are linearly independent;
   (ii) $|q_{in}| \leq \rho P^n$; and
   (iii) $|l_{in}| \leq l/L^n$.
Then θ has irrationality measure

$$\lambda = 1 + \frac{\log P}{\log L} \quad with \quad c = 2\rho P(\max\{2l, 1\})^{(\log P)/(\log L)}.$$

Proof. See for example [R, Lemma 2.1].

If we can obtain some information about the behavior of the principal convergents to the algebraic number $\alpha = \sqrt[d]{1 + 1/a}$, it would be extremely nice. But, since α is an algebraic number of degree greater than 2, we do not know any information at all about its principal convergents. Even though the sequence p_n/q_n obtained by Padé approximations converges to α not so strongly as the principal convergents, we any way know some information about its behavior, and this provides us with information about rational approximations to α by the above lemma. This is the principle of the Padé approximation method.

Remark. As explained in Section 4, in the proof of the finiteness theorem of Thue, the denominator of an exceptional approximation p_0/q_0 is required to be large. Contrary to this, in the Padé approximation method, it is not required for a good solution (or equivalently for an exceptional approximation) to have large size.

In the above lemma, the smaller the constant P is, the better result is obtained. G. V. Chudnovsky [Chu] estimated more precisely the common denominators of the coefficients of the hypergeometric polynomials than Siegel [S2] and Baker [Ba1] did. Thus he improved for example Baker's result on irrationality measure for $\sqrt[3]{2}$, and obtained the value $\lambda = 2.42971$. He also gave irrationality measures for general cubic numbers under a certain assumption [Chu, Main Theorem]. As an example, he obtained the following (see [Chu, p.378]).

Theorem ([Chu]). *Let a be an integer (positive or negative) with $a \equiv -3 \pmod{9}$, and let θ be the real zero of $f(x) = x^3 + ax + 1$ whose absolute value is the smallest among the three zeros of f. Then, for any $\varepsilon > 0$, θ has effective irrationality measure*

$$\lambda = \frac{\log\left(\left(G_1 + \sqrt{G_1^2 + D_1}\right)^2 / |D_1|\right)}{\log\left(\left(G_1 + \sqrt{G_1^2 + D_1}\right) / |D_1|\gamma\right)} + \varepsilon,$$

where

$$D_1 = -\frac{4}{27}a^3 - 1, \quad G_1 = \frac{2}{27}a^6 + \frac{2}{3}a^3 + 1, \quad \gamma = e^{\sqrt{3}\pi/6}/\sqrt{3}.$$

Taking ε small, we have $\lambda < 3$. When $|a|$ is large, we have asymptotically

$$\lambda \sim 3 - \frac{4.5\log 3 - 2\log 2 - \sqrt{3}\pi/2}{3\log|a| + 0.5\log 3 - \sqrt{3}\pi/6}.$$

7.   THUE INEQUALITIES GIVEN BY (3)

Hereafter, we suppose $a > 0$. We give new results on the family of cubic Thue inequalities given by (3). Our method is based on Padé approximations. We give an outline of the proof in Section 8. The full proof will appear elsewhere.

As mentioned above, there are preceding results concerning this family: Equation (xi) was solved by algebraic method; irrationality measures for the associated algebraic numbers were given by Bombieri's method; Chudnovsky gave irrationality measures for the case $b = 1$ and $a \equiv -3 \pmod 9$. (In the last two cases, the results hold for $a < 0$ also.)

Put $f(x) = x^3 + ax + b$. From the assumption that $a > 0$, the algebraic equation $f(x) = 0$ has only one real solution and the other two solutions are complex. Let us denote by θ this real solution. Further we put

$R = 4a^3 + 27b^2$. Note that R is equal to the discriminant of f multiplied by -1. Our results are as follows.

Theorem 1. *Suppose* $a > 2^{2/3}3^4|b|^{8/3}\left(1 + \dfrac{1}{390|b|^3}\right)^{2/3}$. *Then for any integers* p, q $(q > 0)$, *we have*

$$\left|\theta - \frac{p}{q}\right| > \frac{1}{1.73 \cdot 10^6 a^4 b^4 q^{\lambda(a,b)}},$$

where

$$\lambda(a, b) = 1 + \frac{\log(4\sqrt{R} + 12\sqrt{6}|b|)}{\log((1 - 27b^2/R)\sqrt{R}/(27b^2))} < 3.$$

Further, $\lambda(a, b)$ *is a decreasing function of* a *and tends to* 2 *when* a *tends to* ∞.

Remark. The above assumption on the size of a is imposed in order to obtain the inequality $\lambda(a, b) < 3$ which is the essential requirement for application. For example, we obtain $\lambda(a, b) < 3$ for $b = 1$ and $a \geq 129$, and for $b = 2$ and $a \geq 817$. This shows that for small b we have $\lambda(a, b) < 3$ for relatively small a. Compare with the assumption in the theorem of Bombieri, van der Poorten, and Vaaler mentioned in Section 4. Moreover, in their result, λ behaves asymptotically for large a

$$\lambda(a, b) \sim 2 + \frac{14}{\sqrt[3]{3}(\log a)^{1/3}},$$

while ours behaves

$$\lambda(a, b) \sim 2 + \frac{4\log|b| + 2\log 108}{3\log a},$$

and behaves better. However we should note that their result holds for cubic algebraic numbers over any algebraic number field. Compare also with the behavior of λ in Chudnovsky's theorem.

As an easy consequence of Theorem 1, we obtain the following.

Theorem 2. *Under the same assumption as in Theorem 1, we have, for any solution* (x, y) *of the Thue inequality* (3),

$$|y| < (1.73 \cdot 10^6 a^3 b^4 k)^{\frac{1}{3 - \lambda(a,b)}}$$

with the same $\lambda(a, b)$ *as in Theorem 1.*

Since we have obtained an upper bound for the solutions of (3), we may consider that (3) is solved in a sense under the assumption. In order to find all solutions of (3) completely, we need to specify the value k. Let $b > 0$. Taking into account that for $(x, y) = (1, 1)$ the left-hand side of (3) is equal to $a + b + 1$, let us put $k = a + b + 1$, and let us consider the Thue inequalities

$$|x^3 + axy^2 + by^3| \leq a + b + 1. \tag{5}$$

Theorem 3. *Let $b > 0$ and $a \geq 3000b^4$. Then the only solutions of (5) with $y \geq 0$ and $\gcd(x, y) = 1$ are*

$$(0, 0), \ (\pm 1, 0), \ (0, 1), \ (-1, 1), \ (1, 1), \ (-b/d, a/d), \quad \text{where} \ \ d = \gcd(a, b).$$

Let us call these solutions the *trivial solutions* of (5) with $y \geq 0$.
For $b = 1, 2$ we can solve (5) completely.

Theorem 4. *Let $b = 1$ or $b = 2$, and let $a \geq 1$. Then the only solutions of (5) with $y \geq 0$ and $\gcd(x, y) = 1$ are the trivial solutions except the cases $b = 1$, $1 \leq a \leq 3$ and $b = 2$, $1 \leq a \leq 7$. Further, we can list up all solutions for the exceptional cases.*

8.   SKETCH OF THE PROOF

Proof of Theorem 1. We use the Padé approximation method explained in Section 6. As explained above, in order to solve equation (4) or the more general equation (iv), it was necessary to obtain a result on rational approximations to the associated algebraic number. Since those equations have diagonal form, namely they contain only the terms x^d and y^d, the associated algebraic number is a root of a rational number. Therefore the binomial function $\sqrt[d]{1 + x}$ was used. However, in our case, equation (3) has not diagonal form. But this inconvenience can be overcome by transforming the equation into an equation of diagonal form as follows. The transformation itself is an easy consequence of the syzygy theorem for cubic forms in invariant theory, namely the relation $4\mathcal{H}^3 = \mathcal{D}\mathcal{F}^2 - \mathcal{J}^2$ among a cubic form $\mathcal{F}$, its discriminant $\mathcal{D}$, and its covariants $\mathcal{H}$ and $\mathcal{J}$ of degree 2 and 3 respectively.

Lemma 1. *The polynomial f can be written as*

$$x^3 + ax + b = \frac{1}{2}(\alpha(x + \beta)^3 + \overline{\alpha}(x + \overline{\beta})^3),$$

where

$$\alpha, \overline{\alpha} = 1 \pm \frac{3\sqrt{3b}}{\sqrt{R}}, \qquad \beta, \overline{\beta} = \frac{3b}{2a} \mp \frac{\sqrt{3R}}{6a}.$$

Proof. We can verify the formula by a simple calculation starting from the right-hand side.

In order to obtain the form of the formula, we just set $\alpha = s + t\sqrt{R}$ and $\beta = u + v\sqrt{R}$ with unknowns s, t, u, v, and put them into the right-hand side, and we compare the coefficients. Then we obtain the formula.

This fact was used also in Siegel [S1] to determine the number of solutions of cubic Thue equations, and also it was used in Chudnovsky [Chu]. This made possible to obtain results on general cubic numbers of wide class by considering only the binomial function $\sqrt[3]{1+x}$ and its Padé approximations.

We use this formula in a different way from theirs. Since θ is a real zero of $f(x)$, we have

$$\sqrt[3]{\frac{\overline{\alpha}}{\alpha}} = -\frac{\theta + \beta}{\theta + \overline{\beta}}. \tag{6}$$

The key point of our method is to consider the number on the left-hand side, and to apply to this number Padé approximations to the function $\sqrt[3]{(1-x)/(1+x)}$. To construct Padé approximations, we use Rickert's integrals [R].

Lemma 2. *For $n \geq 1$, $i = 1, 2$ and small x, let*

$$I_{in}(x) = \frac{1}{2\pi\sqrt{-1}} \int_{\gamma_i} \frac{(1+xz)^{n+1/3}}{(z^2-1)^{n+1}} dz,$$

where γ_i $(i = 1, 2)$ is a small simple closed counter-clockwise curve enclosing the point 1 (resp. -1). Then these integrals are written as

$$I_{1n}(x) = P_n(x)\sqrt[3]{1+x}, \qquad I_{2n}(x) = -P_n(-x)\sqrt[3]{1-x},$$

where $P_n(x)$ is a polynomial of degree at most n with rational coefficients. Further we have

$$P_n(x) - P_n(-x)\sqrt[3]{\frac{1-x}{1+x}} = c_n x^{2n+1} + \cdots .$$

Now we put $x = 3\sqrt{3}b/\sqrt{R}$ into these Padé approximations. Then using (6) we rewrite the left-hand side in terms of θ, β and $\overline{\beta}$, and we multiply by $\theta + \overline{\beta}$. Then we observe that the left-hand side does not contain any square root. Thus we obtain linear forms

$$p_n + q_n \theta = l_n$$

with rational coefficients.

Since the Padé approximations are given by integrals, we can obtain necessary information about the linear forms by residue calculus and by estimation from above of the values of the integrals. In a similar way, we obtain another set of linear forms $p'_n + q'_n\theta = l'_n$ by replacing the integrand by $z(1 + xz)^{n+1/3}/(z^2 - 1)^{n+1}$. Using these linear forms and the lemma in Section 6, we obtain Theorem 1.

Proof of Theorem 3. We use the well-known classical theorem of Legendre which says that, for any real number ξ and integers p and q with $q > 0$, if the inequality $|\xi - p/q| < 1/(2q^2)$ holds, then p/q is one of the principal convergents to ξ. We see easily that the quotient x/y of any non-trivial solution (x, y) of (5) satisfies this condition for θ, and hence it is one of the principal convergents to θ. Regarding this, we calculate first several convergents to θ, and verify that these do not give solutions of (5) except the trivial solutions. This implies that, for any non-trivial solution (x, y) of (5) with $y > 0$, y is larger than the denominators of these convergents. Thus we obtain a lower bound for y. Moreover, we can verify that, if a is sufficiently large compared with b, then this lower bound is larger than the upper bound given by Theorem 2, which implies that there is no non-trivial solution. Actually, in order to obtain a lower bound for y as large as possible, we generalize Legendre's theorem to a form applicable to generalized continued fractions containing rational numbers in their denominators.

Proof of Theorem 4. For the case $b = 1$, the proof is as follows.

Case $a \geq 3000$. It is nothing but Theorem 3.

Case $129 \leq a < 3000$. By Theorem 2 we have an upper bound for solutions of (5). So, for each a we calculate by computer convergents to θ until the denominator becomes greater than this upper bound, and we verify that these convergents do not satisfy (5) except the trivial solutions. From Legendre's theorem, we see that there is no non-trivial solution. Actually, we did this only for $140 \leq a < 3000$ because of the capacity of the computer. For $129 \leq a < 140$ we used KANT as in the next case.

Case $1 \leq a < 129$. For this case our method does not give any information about θ. So, we use Baker's method. Actually, we use the software KANT. This works since the number of the remaining equations is not large.

References

[Ba1]   A. Baker, Rational approximation to $\sqrt[3]{2}$ and other algebraic numbers, *Quart. J. Math. Oxford*, **15** (1964), 375–383.

[Ba2] ———, Contributions to the theory of Diophantine equations: I. On the representation of integers by binary forms, *Phil. Trans. Royal Soc.*, A **263** (1968), 173–191.

[Ba3] ———, *Transcendental Number Theory*, Cambridge Univ. Press, 1975.

[BaS] A. Baker and C. L. Stewart, On effective approximations to cubic irrationals, *in* "New Advances in Transcendence Theory", A. Baker, Ed., pp.1–24, Cambridge Univ. Press, 1988.

[Be1] M. A. Bennett, On the representation of unity by binary cubic forms, to appear in *Trans. Amer. Math. Soc.*

[Be2] ———, Rational approximation to algebraic numbers of small height: The Diophantine equation $|ax^n - by^n| = 1$, to appear in *J. Reine Angew. Math.*

[BedW] M. A. Bennett and B. M. M. de Weger, On the Diophantine equation $|ax^n - by^n| = 1$, *Math. Comp.*, **67** (1998), 413–438.

[BH] Y. Bilu and G. Hanrot, Solving Thue equations of high degree, *J. Number Theory*, **60** (1996), 373–392.

[Bo1] E. Bombieri, On the Thue-Siegel-Dyson theorem, *Acta Math.*, **148** (1982), 255–296.

[Bo2] ———, Effective Diophantine approximation on G_m, *Ann. Scuola Norm. Sup. Pisa Cl. Sci.*, (4) **20** (1993), 61–89.

[BoM] E. Bombieri and J. Mueller, On effective measures of irrationality for $\sqrt[r]{a/b}$ and related numbers, *J. Reine Angew. Math.*, **342** (1983), 173–196.

[BvPV] E. Bombieri, A. J. van der Poorten, and J. D. Vaaler, Effective measures of irrationality for cubic extensions of number fields, *Ann. Scuola Norm. Sup. Pisa Cl. Sci.*, **23** (1996), 211–248.

[CheV] Chen Jianhua and P. M. Voutier, Complete solution of the Diophantine equation $X^2 + 1 = dY^4$ and a related family of quartic Thue equations, *J. Number Theory*, **62** (1997), 71–99.

[Chu] G. V. Chudnovsky, On the method of Thue-Siegel, *Ann. of Math.*, **117** (1983), 325–382.

[Dal] M. Daberkow, C. Fieker, J. Klüners, M. Pohst, K. Roegner, and K. Wildanger, KANT V4, *J. Symbolic Comp.*, **24** (1997), 267–283.

[Del] B. Delaunay, Vollständige Lösung der unbestimmten Gleichung $X^3 q + Y^3 = 1$ in ganzen Zahlen, *Math. Z.*, **28** (1928), 1–9.

[De2] ———, Über die Darstellung der Zahlen durch die binären kubische Formen von negativer Diskriminante, *Math. Z.*, **31** (1930), 1–26.

[DF]   B. N. Delone and D. K. Faddeev, *The Theory of Irrationalities of the Third Degree*, Translations of Math. Monographs, Vol. 10, Amer. Math. Soc., 1964.

[E1]   J.-H. Evertse, On the equation $ax^n - by^n = c$, *Compositio Math.*, **47** (1982), 289–315.

[E2]   ———, On the representation of integers by binary cubic forms of positive discriminant, *Invent. Math.*, **73** (1983), 117–138.

[H]   C. Heuberger, On general families of parametrized Thue equations, in "Algebraic Number Theory and Diophantine Analysis", F. Halter-Koch and R. F. Tichy, Eds., Proc. Conf. Graz, 1988, W. de Gruyter, Berlin (2000), 215–238.

[HPT]   C. Heuberger, A. Pethő, and R. F. Tichy, Complete solution of parametrized Thue equations, *Acta Math. Inform. Univ. Ostraviensis*, **6** (1998), 93–113.

[LMN]   M. Laurent, M. Mignotte, and Y. Nesterenko, Formes linéaires en deux logarithmes et déterminants d'interpolation, *J. Number Theory*, **55** (1995), 285–321.

[LP]   G. Lettl and A. Pethő, Complete solution of a family of quartic Thue equations, *Abh. Math. Sem. Univ. Hamburg*, **65** (1995), 365–383.

[LPV]   G. Lettl, A. Pethő, and P. Voutier, Simple families of Thue inequalities, *Trans. Amer. Math. Soc.*, **351** (1999), 1871–1894.

[Li]   F. Lippok, On the representation of 1 by binary cubic forms of positive discriminant, *J. Symbolic Computation*, **15** (1993), 297–313.

[M1]   M. Mignotte, Verification of a conjecture of E. Thomas, *J. Number Theory*, **44** (1993), 172–177.

[M2]   ———, Pethő's cubics, *Publ. Math. Debrecen*, **56** (2000), 481–505.

[MPL]   M. Mignotte, A. Pethő, and F. Lemmermeyer, On the family of Thue equations $x^3 - (n-1)x^2y - (n+2)xy^2 - y^3 = k$, *Acta Arith.*, **76** (1996), 245–269.

[MT]   M. Mignotte and N. Tzanakis, On a family of cubics, *J. Number Theory*, **39** (1991), 41–49.

[N1]   T. Nagell, Solution complète de quelques équations cubiques à deux indéterminées, *J. Math. Pures. Appl.*, **6** (1925), 209–270.

[N2]   ———, Über einige kubische Gleichungen mit zwei Unbestimmten, *Math. Z.*, **24** (1926), 422–447.

[N3] ——, Darstellung ganzer Zahlen durch binäre kubische Formen mit negativer Diskriminante, *Math. Z.*, **28** (1928), 10–29.

[N4] ——, *L'analyse indéterminée de dergré supérieur*, Mémorial Sci. Math. Vol. 39, Gauthier-Villars, Paris, 1929.

[P] A. Pethő, On the representation of 1 by binary cubic forms with positive discriminant, *in* "Number Theory, Ulm 1987 Proceedings," H. P. Schlickewei and E. Wirsing, Eds., Lecture Notes Math. Vol. 1380, Springer (1989), 185–196.

[R] J. H. Rickert, Simultaneous rational approximations and related Diophantine equations, *Math. Proc. Cambridge Philos. Soc.*, **113** (1993), 461–472.

[S1] C. L. Siegel, Über einige Anwendungen diophantischer Approximationen, *Abh. Preuss. Akad. Wiss. Phys. Math. Kl. No.1*, (1929) ; Gesam. Abh. I, 209–266.

[S2] ——, Die Gleichung $ax^n - by^n = c$, *Math. Ann.*, **114** (1937), 57–68 ; Gesam. Abh. II, 8–19.

[Ta] V. Tartakovskij, Auflösung der Gleichung $x^4 - \rho y^4 = 1$, *Bull. Acad. Sci. de l'URSS*, 1926, 301–324.

[Tho1] E. Thomas, Complete solutions to a family of cubic Diophantine equations, *J. Number Theory*, **34** (1990), 235–250.

[Tho2] ——, Solutions to certain families of Thue equations, *J. Number Theory*, **43** (1993), 319–369.

[Thu1] A. Thue, Bemerkungen über gewisse Näherungsbrüche algebraischer Zahlen, *Kristiania Vidensk. Selsk. Skrifter.*, I, Mat. Nat. Kl., 1908, No.3.

[Thu2] ——, Über Annäherungswerte algebraischer Zahlen, *J. Reine Angew. Math.*, **135** (1909), 284–305.

[Thu3] ——, Ein Fundamentaltheorem zur Bestimmung von Annähe-rungswerten aller Wurzeln gewisser ganzer Funktionen, *J. Reine Angew. Math.*, **138** (1910), 96–108.

[Thu4] ——, Berechnung aller Lösungen Gewisser Gleichungen von der Form $ax^r - by^r = f$, *Kristiania Vidensk. Selsk. Skrifter.*, I, Mat. Nat. Kl., 1918, No.4.

[Wak1] I. Wakabayashi, On a family of quartic Thue inequalities I, *J. Number Theory*, **66** (1997), 70–84.

[Wak2] ——, On a family of quartic Thue inequalities II, *J. Number Theory*, **80** (2000), 60–88.

[Wal] M. Waldschmidt, Minorations de combinaisons linéaires de logarithmes de nombres algébriques, *Canad. J. Math.*, **45** (1993), 176–224.

TWO EXAMPLES OF ZETA-REGULARIZATION

Masami YOSHIMOTO

Research Institute for Mathematical Sciences, Kyoto University
Kyoto, 606-8502, Japan
yosimoto@kurims.kyoto-u.ac.jp

Keywords: zeta-regularization, Kronecker limit formula

Abstract   In this paper we shall exhibit the close mutual interaction between zeta-regularization theory and number theory by establishing two examples; the first gives the unified Kronecker limit formula, the main feature being that stated in terms of zeta-regularization, the second limit formula is informative enough to entail the first limit formula, and the second example gives a generalization of a series involving the Hurwitz zeta-function, which may have applications in zeta-regularization theory.

1991 Mathematical Subject Classification: 11M36, 11M06, 11M41, 11F20

1.   INTRODUCTION

In this paper we shall give two theorems, Theorems 1 and 2, which have their genesis in zeta-regularization. One (Theorem 1) is the second limit formula [16] of Kronecker in which effective use is made of the zeta-regularization technique, and the other (Theorem 2) is a generalization of a formula of Erdélyi [6] (in the setting of Hj. Mellin [13]) which gives a generalized zeta regularized sequence (Elizalde et al [4]).

First, we shall give the definition of the zeta-regularization.

Definition.   Let $\{\lambda_k\}_{k=1,2,...}$ be a sequence of complex numbers such that

$$\sum_{k=1}^{\infty} |\lambda_k|^{-\alpha} < \infty$$

for some positive α.

Define $Z(s)$ by $Z(s) = \sum_{k=1}^{\infty} \lambda_k^{-s}$ for s such that Re $s = \sigma > \alpha$ and call it the *zeta-function* associated to the sequence $\{\lambda_k\}$.

C. Jia and K. Matsumoto (eds.), Analytic Number Theory, 379–393.

We suppose that $Z(s)$ can be continued analytically to a region containing the origin.

Then we define the *regularized product*, denoted by $\prod_z \lambda_k$, of $\{\lambda_k\}$ as

$$\prod_z \lambda_k = e^{-\text{FP } Z'(0)},$$

where FP $f(s_0)$ indicates the constant term in the Laurent expansion of $f(s)$ at $s = s_0$. If this definition is meaningful, the sequence $\{\lambda_k\}$ is called *zeta-regularizable*.

Similarly, if $Z(s)$ can be continued analytically to a region containing $s = -1$, the zeta-regularized sum is defined by

$$\sum_z \lambda_k = \log\left\{ \prod_z \exp(\lambda_k) \right\} = \text{FP } Z(-1).$$

Giving a meaning in this way to the otherwise divergent series or products by interpreting them as special values of suitable zeta-functions (or derivatives thereof) is called Zeta-regularization.

Remark 1. i) As $\{\lambda_k\}$ we consider mostly the discrete spectrum of a differential operator.

ii) Formal differentiation gives

$$-Z'(s) = \sum_k \lambda_k^{-s} \log \lambda_k = \sum_k \log \lambda_k^{\lambda_k^{-s}},$$

or

$$\prod_k \lambda_k^{\lambda_k^{-s}} = e^{-Z'(s)},$$

so that this gives a motivation for interpreting the formal infinite product $\prod_{k=1}^{\infty} \lambda_k$ as $e^{-\text{FP } Z'(0)}$.

iii) The merit of the zeta-regularization method lies in the fact that by only formal calculation one can get the expressions wanted, save for the main term, which is given as a residual function (the sum of the residues), for more details, see Bochner [3]. To calculate the residual function one has to appeal to classical methods. Cf. Remark 2 and Remark 3, below.

We shall now state the background of our results more precisely. Regarding Kronecker's limit formulas, J. R. Quine et al [15] were the first who used the zeta regularized products to derive Kronecker's first limit formula in the form

$$\prod_{\substack{m,n\in\mathbb{Z}}}^{'} |m+nz|^2 = (2\pi)^2 |\eta(z)|^4, \tag{1}$$

where the prime "'" indicates that the pair $(m,n) = (0,0)$ should be excluded from the product, $\eta(z)$ denotes the Dedekind eta-function defined by

$$\eta(z) = e^{\frac{\pi i z}{12}} \prod_{m=1}^{\infty} \left(1 - e^{2\pi i m z}\right), \quad \operatorname{Im} z > 0. \tag{2}$$

On the other hand, Quine et al [15], Formula (53), states Kronecker's second limit formula in the form

$$\prod_{z}(\gamma + w) = i\eta^{-1}(z) \exp\left(-\frac{\pi i z}{6} - \pi i w\right) \vartheta_1(w, z), \tag{3}$$

where γ runs through all elements $m + n\tau$ of the lattice with basis 1, τ ($-\pi \le \arg\gamma < \pi$) and $\vartheta_1(w, z)$ is one of Jacobi's theta-function given by

$$\vartheta_1(w, z) = \sum_{n=-\infty}^{\infty} \exp\left(\pi i(n + \frac{1}{2})^2 z + 2\pi i(n + \frac{1}{2})(w - \frac{1}{2})\right)$$

$$= 2e^{\frac{\pi i z}{6}}\eta(z) \sin \pi w \prod_{n=1}^{\infty} \left(1 - e^{2\pi i(nz+w)}\right)\left(1 - e^{2\pi i(nz-w)}\right). \tag{4}$$

Quine et al [15] proved (3) using a generalization of Voros' theorem [19] on the ratio of the zeta-regularized product and the Weierstrass product to the case of zeta regularizable sequences (Theorem 2 [15]).

It is rather surprising that they reached (3) without knowing the first form of Kronecker's second limit formula (from the references they gave they do not seem to know Siegel's most famous book [16] on Kronecker limit formulas) as (3) does not immediately lead to it (the conjugate decomposition property does not necessarily hold for complex sequences γ).

Siegel's book [16] had so much impact on the development of Kronecker's limit formulas and their applications in algebraic number theory that the general understanding before Berndt's paper [1] was that the first and the second limit formulas of Kronecker should be treated separately, and may not be unified into one (see e.g. Lang [11] and others). Berndt [1] was the first who unified these two limit formulas into one, but this paper does not seem to be well-known for its importance probably because it was published under a rather general title "Identities involving the coefficients of a class of Dirichlet series".

It is also possible to unify the investigation of Chowla and Selberg with Siegel to deduce a unified version of Kronecker's limit formula (Kumagai [10]), but this is superseded by Berndt's theorem in the sense that in Berndt's one is to take only $w = 0$ while in Kumagai's paper, the residual function has not been extracted out.

To restore the importance of Berndt's paper and clarify the situation surrounding Kronecker's limit formulas we shall prove the unified Kronecker limit formula. The main feature is that, stated in terms of zeta-regularized product, the second limit formula is informative enough to entail the first limit formula:

$$\prod_{\substack{m,n\in\mathbb{Z}}}^{z} |m + nz + w|^2 = e^{-2\pi u^2 y} \left|\frac{\vartheta_1(w, z)}{\eta(z)}\right|^2, \quad w \neq 0. \tag{5}$$

Indeed, noting that

$$\vartheta_1(w, z) = 2\pi\eta^3(z)w + O(|w|^3) \quad |w| \to 0,$$

we can take the limit in (5) as $w \to 0$, and

$$\prod_{\substack{m,n\in\mathbb{Z}}}^{z}{}' |n + mz|^2 = \lim_{w\to 0}\frac{1}{|w|^2} \prod_{\substack{m,n\in\mathbb{Z}}}^{z} |n + mz + w|^2$$

$$= 4\pi^2|\eta(z)|^4.$$

To state Theorem 1 we fix the following notation.

Let $Q(\xi, \eta) = a\xi^2 + 2b\xi\eta + c\eta^2$ be a positive definite quadratic form with $a > 0$, and discriminant $d = ac - b^2 > 0$ which we decompose as

$$Q(\xi, \eta) = \sqrt{d}y^{-1}|\xi + \eta z|^2, \quad z = x + iy = \frac{b + \sqrt{d}i}{a}.$$

For $0 \leq u, v < 1$ we define the Epstein (-Lerch) zeta-function associated with $Q(u, v)$ by

$$\zeta_Q(s; u, v) := \sum_{m,n=-\infty}^{\infty}{}' \frac{e^{2\pi i(mu+nv)}}{Q(m, n)^s}, \quad \sigma > 1. \tag{6}$$

Also define the classical Epstein zeta-function by

$$\zeta_Q(s) = \zeta_Q(s; u, v) = \sum_{m,n}{}' \frac{1}{Q(m, n)^s}, \quad \sigma > 1. \tag{7}$$

We can now state the unified Kronecker limit formula.

Theorem 1. i) $\zeta_Q(s; u, v)$ *has the expansion at $s = 1$,*

$$\zeta_Q(s; u, v) = \frac{c_{-1}}{s - 1} + c_0 + O(s - 1), \tag{8}$$

where

$$c_{-1} = \frac{\pi}{\sqrt{d}} \varepsilon(u, v), \tag{9}$$

and

$$
c_0 = \frac{2\pi}{\sqrt{d}} \left\{ \varepsilon(u, v) \left(\gamma + \log \pi - \log \sqrt{\sqrt{d} y} \right) + 2\pi u^2 y \right.
$$
$$
\left. - \log \left| \frac{\vartheta_1(v - uz, z)}{(v - uz)\eta(z)} \right| - (1 - \varepsilon(u, v)) \log |v - uz| \right\}, \tag{10}
$$

where $\varepsilon(u, v)$ is defined by

$$
\varepsilon(u, v) = \begin{cases} 1, & (u, v) = (0, 0) \\ 0, & \text{otherwise.} \end{cases}
$$

ii) *Under the convention*

$$(1 - \varepsilon(u, v)) \log |v - uz| = 0 \text{ for } (u, v) = (0, 0),$$

we can rewrite (8) as Kronecker's first limit formula stated in Siegel [16].

$$
\lim_{s \to 1} \left(\zeta_Q(s) - \frac{\pi/\sqrt{d}}{s - 1} \right) = \frac{2\pi}{\sqrt{d}} \left\{ \gamma - \log 2 - \log \sqrt{\sqrt{d} y} |\eta(z)|^2 \right\}, \tag{8'}
$$

where γ is Euler constant and $\zeta(s) = \zeta(s, 1) = \sum_{n=1}^{\infty} n^{-s}$ denotes the Riemann zeta-function.

We now turn to the second example, which is discussed in Elizalde et al [4] under the title "Zeta-regularization generalized". Our main object of study is the function

$$F(s, \tau, a) = F_\alpha(s, \tau, a) := \sum_{n=0}^{\infty} \frac{e^{-(n+a)^\alpha \tau}}{(n + a)^{\alpha s}}, \tag{11}$$

where the series, extended over $n \neq -a$, is absolutely convergent for $\alpha\sigma > 1$ and $\tau \geq 0$.

Theorem 2. *For $0 < \alpha \leq 1$, $a > 0$ and $|\tau| < 2\pi$, we have the expansion*

$$F(s, \tau) = F(s, \tau, a) = \sum_{\substack{n=0 \\ \alpha(s-n) \neq 1}}^{\infty} \frac{(-1)^n}{n!} \zeta(\alpha(s - n), a)\tau^n + P(s, \tau), \tag{12}$$

where $\zeta(s,v) = \sum_{n=0}^{\infty}(n+v)^{-s}$ denotes the Hurwitz zeta-function and $P(s,\tau)$ the generalized residual function (see e.g. Bochner [3])

$$P(s,\tau) = P(s,\tau,a) = \frac{1}{2\pi i}\int_{|z+s-\frac{1}{\alpha}|=r}\Gamma(z)\zeta(\alpha(s+z),a)\tau^{-z}dz, \quad (13)$$

and where r is a small positive number so as to enclose the pole of the integrand as follows:

$$r < \begin{cases} |t|, & t \neq 0 \\ \min\left(\left\{-\sigma+\frac{1}{\alpha}\right\}, 1-\left\{-\sigma+\frac{1}{\alpha}\right\}\right), & -\sigma+\frac{1}{\alpha} \notin \mathbb{Z},\ t=0 \\ 1, & s-\frac{1}{\alpha} \in \mathbb{Z}. \end{cases}$$

In particular, if $s - \frac{1}{\alpha} \notin \mathbb{Z}$,

$$P(s,\tau) = \frac{1}{\alpha}\Gamma\left(\frac{1}{\alpha}-s\right)\tau^{s-\frac{1}{\alpha}}.$$

The series on the right-hand side of (12) is absolutely convergent for any τ or $|\tau| < 2\pi$ according as $0 < \alpha < 1$ or $\alpha = 1$.

In the case $\alpha = 1$, (12) corresponds to Formula (8) in Erdélyi [6, p.29]: $|\log z| < 2\pi$, $s \neq 1, 2, 3, \ldots$, $v \neq 0, -1, -2, \ldots$

$$\sum_{n=0}^{\infty}\frac{z^n}{(n+v)^s} = \frac{\Gamma(1-s)}{z^v}(\log 1/z)^{s-1} + z^{-v}\sum_{r=0}^{\infty}\zeta(s-r,v)\frac{(\log z)^r}{r!}.$$

Theorem 2 is a companion formula to Theorem 1 of Katsurada [8] which gives an asymptotic expansion for the function

$$G(s,\tau,a) = \sum_{n>\operatorname{Re}s+1}\zeta(\alpha(n-s),a)\frac{(-\tau)^n}{n!}$$

(Katsurada considered the special case of $\alpha = 1$, $a = 1$: $G_\nu(\tau) = G(\nu,\tau) = \sum_{n>\operatorname{Re}\nu+1}\zeta(n-\nu)\frac{(-\tau)^n}{n!}$).
We are as yet unable to obtain a counterpart result for G.

Acknowledgment
The author would like to express his hearty thanks to the referee for his thorough scrutinizing the paper and for many useful comments which improved the earlier version completely, resulting in this improver version.

2.   PROPERTIES OF ZETA-REGULARIZATION

We quote some basic properties of the zeta-regularized products which we use in the proof of Theorem 1 (see e.g. Quine, et al [15], Song [17]).

We call the least integer h such that the series for $Z(h+1)$ converges absolutely the convergence index.

I. Partition Property. Let $\{\lambda_k^{(1)}\}$ and $\{\lambda_k^{(2)}\}$ be zeta-regularizable sequences and let $\{\lambda_k\} = \{\lambda_k^{(1)}\} \cup \{\lambda_k^{(2)}\}$ (disjoint union). Then

$$\prod_{k}{}_z \lambda_k = \prod_{k}{}_z \lambda_k^{(1)} \prod_{k}{}_z \lambda_k^{(2)}.$$

II. Splitting Property. If $\{\lambda_k\}$ is zeta-regularizable and $\prod_{k} a_k$ is convergent absolutely, then

$$\prod_{k}{}_z a_k \lambda_k = \prod_{k} a_k \prod_{k}{}_z \lambda_k.$$

III. Conjugate Decomposition Property. If $\{\lambda_k\}$ is a real zeta-regularizable sequence with convergence index $h < 2$, then

$$\prod_{k}{}_z |\lambda_k - \lambda|^2 = \prod_{k}{}_z (\lambda_k - \lambda) \prod_{k}{}_z (\lambda_k - \bar{\lambda}).$$

IV. Iteration Property. Suppose $\Lambda = \{\lambda_{mn}\}$ is a double-indexed zeta-regularizable sequence, i.e. $\prod_{m,n}{}_z \lambda_{mn}$ is meaningful. Then for $\{\Lambda_m\} = \{ \prod_{n}{}_z \lambda_{mn}\}$, $\prod_{m}{}_z \Lambda_m$ is meaningful, and we have

$$\prod_{m,n}{}_z \lambda_{mn} = \prod_{m}{}_z \Lambda_m = \prod_{m}{}_z \prod_{n}{}_z \lambda_{mn}$$

(for details, cf. Song [17], p.31).

We shall now give two examples. The first example interpreting the otherwise divergent series as special values of the Riemann zeta-function, is originally due to Euler [7] and is an example of various regular methods of summation or a consequence of the functional equation, while the second example enabling us to interpret the otherwise divergent products as the zeta-regularized products is due to Lerch [12, p.13].

Example 1. $\zeta(-n)$, $n = 0, 1, 2, \ldots$ (see Berndt [2, pp.133–136] for more details about Ramanujan's ideas)

$$1^0 + 2^0 + 3^0 + 4^0 + \cdots = -\frac{1}{2},$$

$$1^1 + 2^1 + 3^1 + 4^1 + \cdots = -\frac{1}{12},$$

$$1^2 + 2^2 + 3^2 + 4^2 + \cdots = 0,$$

$$1^3 + 2^3 + 3^3 + 4^3 + \cdots = -\frac{1}{120}.$$

Example 2. $\exp(-\zeta'(-n))$, $\exp(-\zeta'(-n, \alpha))$

$$\prod_{n=1}^{\infty} {}_z\, n = \sqrt{2\pi},$$

$$\prod_{n=0}^{\infty} {}_z\, (n + \alpha) = \frac{\sqrt{2\pi}}{\Gamma(\alpha)}, \quad \alpha \in \mathbb{C} - [-\infty, 0], \tag{14}$$

$$\prod_{n=1}^{\infty} {}_z\, n^n = A e^{-\frac{1}{12}},$$

where $A = 1.282427130\ldots$ is the Glaisher-Kinkelin constant defined by

$$A = \lim_{n \to \infty} \frac{1^1 2^2 3^3 \cdots n^n}{n^{n^2 + \frac{n}{2} + \frac{1}{12}}} e^{\frac{n^2}{4}} = e^{-\zeta'(-1) + \frac{1}{12}}.$$

3.   PROOF OF THEOREMS

Proof of Theorem 1. Let Q^* be the reciprocal of Q given by

$$Q^*(\xi, \eta) = \frac{1}{d} Q(\eta, -\xi).$$

If we define the Epstein-Hurwitz zeta-function ϕ_Q by

$$\phi_Q(s; u, v) = \sideset{}{'}\sum_{m,n \in \mathbb{Z}} \frac{1}{Q(m + u, n + v)^s}, \quad \sigma > 1, \tag{15}$$

then ζ_Q satisfies the functional equation

$$d^{\frac{1}{4}} \pi^{-s} \Gamma(s) \zeta_Q(s; u, v) = (d^{-1})^{\frac{1}{4}} \pi^{-(1-s)} \Gamma(1 - s) \phi_{Q^*}(1 - s; u, v) \tag{16}$$

(see e.g. Epstein [5] or Berndt [1]).

In particular, $\phi_{Q^*}(0; u, v) = -\varepsilon(u, v)$ follows from this.

We use the functional equation (16) to expand ζ_Q around $s = 1$ as follows:

$$\zeta_Q(s; u, v) = \frac{\pi d^{-1/2}}{s - 1} \varepsilon(u, v)$$
$$+ \pi d^{-\frac{1}{2}} \phi'_{Q^*}(0; u, v) + 2\pi d^{-\frac{1}{2}} \varepsilon(u, v)(\gamma + \log \pi)$$
$$+ O(s - 1). \tag{17}$$

Thus we are led, after Stark (see e.g. [18]), to compute $\phi'_{Q^*}(s; u, v)$ at $s = 0$ (rather than at $s = 1$) which we do by the zeta-regularized product

$$e^{-\phi'_{Q^*}(0; u, v)} = (\sqrt{dy})^{\varepsilon(u, v)} \prod_{m, n \in \mathbb{Z}}{}^{'}{}_z |n + mz + w|^2, \tag{18}$$

where $w = v - uz$.

We compute $\prod = \prod_{m, n \in \mathbb{Z}}{}_z |n + mz + w|^2$. By Property IV,

$$\prod = \prod_{m \in \mathbb{Z}}{}_z \prod_{n \in \mathbb{Z}}{}_z |n + mz + w|^2.$$

For $w \neq 0$, by Properties I and III, the inner product can be expressed as

$$\prod_{n \in \mathbb{Z}}{}_z |n + mz + w|^2 = \prod_{n=0}^{\infty}{}_z (n + mz + w) \prod_{n=0}^{\infty}{}_z (n + m\bar{z} + \bar{w})$$
$$\times \prod_{n=0}^{\infty}{}_z (n + 1 - mz - w) \prod_{n=0}^{\infty}{}_z (n + 1 - m\bar{z} - \bar{w}).$$

By Lerch's formula $\prod_{n=0}^{\infty}{}_z (n + \alpha) = \frac{\sqrt{2\pi}}{\Gamma(\alpha)}$, we may express each product in terms of the gamma function. Since these in pairs are of the form

$$\frac{\sqrt{2\pi}}{\Gamma(mz + w)} \frac{\sqrt{2\pi}}{\Gamma(1 - mz - \bar{w})} \quad \text{and} \quad \frac{\sqrt{2\pi}}{\Gamma(m\bar{z} + \bar{w})} \frac{\sqrt{2\pi}}{\Gamma(1 - m\bar{z} - \bar{w})},$$

we can rewrite them, using the reflection property of the gamma function, in sines to get

$$\prod_{n \in \mathbb{Z}}{}_z |n + mz + w|^2 = |2 \sin \pi(mz + w)|^2.$$

On the right-hand of this formula we express the sine $|2 \sin \pi(mz+w)|^2$ by exponentials

$$\left|e^{-\pi i(lz+w)}\right|^2 \left|1 - e^{2\pi i(w+lz)}\right|^2 \quad \text{or} \quad \left|e^{-\pi i(lz-w)}\right|^2 \left|1 - e^{-2\pi i(w-lz)}\right|^2$$

according as $m = l$ or $m = -l$, $l \in \mathbb{N}$.

We multiply these over l, $-l$, and 0, using Properties I and II, we have

$$\prod = |2 \sin \pi w|^2 \prod_{l=1}^{\infty} \left|e^{-\pi i(lz+w)}\right|^2 \left|e^{-\pi i(lz-w)}\right|^2$$

$$\times \prod_{l=1}^{\infty} \left|1 - e^{2\pi i(w+lz)}\right|^2 \left|1 - e^{-2\pi i(w-lz)}\right|^2.$$

The infinite product part can be rewritten as

$$\frac{e^{\frac{\pi y}{3}}}{|2 \sin \pi w|^2} \left|\frac{\vartheta_1(w, z)}{\eta(z)}\right|^2$$

and the zeta-product part as

$$e^{-2\pi y(u^2 - \frac{1}{6})},$$

whence we conclude (5):

$$\prod = \prod_{m,n \in \mathbb{Z}} |m + nz + w|^2 = e^{-2\pi u^2 y} \left|\frac{\vartheta_1(w, z)}{\eta(z)}\right|^2.$$

Hence by (5), the product in (18) becomes

$$\prod_{m,n \in \mathbb{Z}}' |m + nz + w|^2 = e^{-2\pi u^2 y} \left|\frac{\vartheta_1(w, z)}{w\eta(z)}\right|^2,$$

or

$$-\phi_{Q^*}'(0; u, v) = \log\left\{ (\sqrt{dy})^{\varepsilon(u,v)} e^{-2\pi u^2 y} \left|\frac{\vartheta_1(w, z)}{w\eta(z)}\right|^2 \right\}. \tag{18'}$$

Substituting (18)' in (17) proves the assertion. ∎

Remark 2. Deduction of (18)' in this formal way is the advantage of the zeta-regularization method (cf. the ingenious proof of Siegel [16]).

Proof of Theorem 2. Suppose first that $0 < \tau < 2\pi$. Then by the Mellin inversion for fixed $s = \sigma + it$,

$$F(s,\tau) = \frac{1}{2\pi i} \int_{(c)} \Gamma(z)\zeta(\alpha(s+z),a)\tau^{-z}dz,$$

where the integral is the Bromwich integral extended over a vertical line (c): $z = c + iy$, $-\infty < y < \infty$, with $c > 0$.

We shift the line of integration to (c') with $c' = -N - \kappa$, where $0 < \kappa < 1$ and N is a large positive integer satisfying

$$N > \sigma - \frac{1}{\alpha}, \quad N \gg |t|, \tag{19}$$

which we may, because for the Hurwitz zeta-function we have the well-known estimate

$$|\zeta(u+iv,\alpha)| \ll \left|\frac{v}{2\pi}\right|^{\frac{1}{2}-u} \quad \text{for } u < 0, \tag{20}$$

and for the gamma function we have

$$|\Gamma(u+iv)| \sim \sqrt{2\pi}|v|^{u-\frac{1}{2}}e^{-\frac{\pi}{2}|v|} \tag{21}$$

valid in any strip $u_1 \le u \le u_2$, a consequence of Stirling's formula

$$\log\Gamma(z) = \left(z - \frac{1}{2}\right)\log z - z + \log\sqrt{2\pi} + O(z^{-1}) \tag{22}$$

for $|\arg z| < \pi$, $|z| \to \infty$.

Hence by the residue theorem, we find that

$$F(s,\tau) = \sum_{\substack{0 \le k < -c' \\ \alpha(s-k) \ne 1}} \operatorname*{Res}_{z=-k} \Gamma(z)\zeta(\alpha(s+z),a)\tau^{-z}$$

$$+ P(s,\tau) + R_N(s,\tau)$$

say, where $P(s,\tau)$ is defined by (13) and

$$R_N(s,\tau) = R_N(s,\tau,a) = \frac{1}{2\pi i} \int_{(c')} \Gamma(z)\zeta(\alpha(s+z),a)\tau^{-z}dz. \tag{23}$$

Hence

$$F(s,\tau) = \sum_{0 \le k \le N}' \frac{(-1)^k}{k!}\zeta(\alpha(s-k),a)\tau^k + P(s,\tau) + R_N(s,\tau). \tag{24}$$

We wish to determine the radius of absolute convergence of the series on the right-hand side of (12) whose partial sum appears on the right-hand side of (24). For this purpose we estimate the coefficients.

If $z = x + iy$, $|\arg z| < \pi$, and $|z| \to \infty$, then by (22) we have

$$|\Gamma(z)| \sim \sqrt{2\pi}|z|^{x-\frac{1}{2}}e^{-y\arg z-x}. \tag{25}$$

Using (25) and the functional equation

$$\zeta(s,a) = \frac{2\Gamma(1-s)}{(2\pi)^{1-s}} \sum_{m=1}^{\infty} \frac{\sin(2\pi ma + \frac{\pi s}{2})}{m^{1-s}} \qquad \text{for } \sigma < 0,$$

we infer that

$$\begin{aligned}
|\zeta(\alpha(s-z),a)| &\leq 2(2\pi)^{\alpha(\sigma-x)-1}e^{\frac{\pi\alpha}{2}|t-y|} \\
&\quad \times |\Gamma(\alpha(x-\sigma)+1-i\alpha(t-y)))|\zeta(\alpha(k-\sigma)+1) \\
&\ll (2\pi)^{-\alpha(\sigma-x)-1}e^{\frac{\pi\alpha}{2}|t-y|} \\
&\quad \times (\alpha(x-\sigma)+1)^{\alpha(x-\sigma)+\frac{1}{2}}e^{-\alpha(x-\sigma)-1} \\
&\ll e^{-\alpha x(\log 2\pi)}e^{-\alpha x}(\alpha x)^{\alpha x+\frac{1}{2}-\alpha\sigma}e^{\frac{\pi\alpha}{2}|t-y|} \tag{26}
\end{aligned}$$

for $x \to \infty$.

Thus we conclude that by the Cauchy-Hadamard Theorem, (25) and (26), the series on the right-hand side of (12) is absolutely convergent for $|\tau| < \infty$ or $|\tau| < 2\pi$ according as $0 < \alpha < 1$ or $\alpha = 1$, whence we conclude that (12) gives the Taylor expansion for $F(s,\tau) - P(s,\tau)$ in the respective range of τ if $\lim_{N\to\infty} R_N(s,\tau) = 0$.

It remains to prove that

$$\lim_{N\to\infty} R_N(s,\tau) = 0 \tag{27}$$

for any τ if $0 < \alpha < 1$ and for $|\tau| < 2\pi$ if $\alpha = 1$.

With

$$c(\alpha) = \frac{\pi\alpha}{\pi\alpha + 2},$$

let

$$T = \frac{1}{c(\alpha)}(N + \kappa). \tag{28}$$

Then $N = [c(\alpha)T]$ and

$$T \gg |t| \tag{29}$$

in view of (19). Divide the integration path of $R_N(s,\tau)$ into three parts: $v = \text{Im } z \in (-\infty, -T)$, $[-T, T)$, $[T, \infty)$ and denote the corresponding integral by $I^{(-)}$, $I^{(0)}$ and $I^{(+)}$, respectively, where

$$
\begin{cases}
I^{(-)} = \dfrac{1}{2\pi} \displaystyle\int_T^\infty \Gamma(-N - \kappa - iv)\zeta(\alpha(s - N - \kappa - iv), a)\tau^{N+\kappa+iv}dv, \\[2mm]
I^{(0)} = \dfrac{1}{2\pi} \displaystyle\int_0^T \Gamma(-N - \kappa - iv)\zeta(\alpha(s - N - \kappa - iv), a)\tau^{N+\kappa+iv}dv \\[2mm]
\qquad + \dfrac{1}{2\pi} \displaystyle\int_0^T \Gamma(-N - \kappa + iv)\zeta(\alpha(s - N - \kappa + iv), a)\tau^{N+\kappa-iv}dv, \\[2mm]
I^{(+)} = \dfrac{1}{2\pi} \displaystyle\int_T^\infty \Gamma(-N - \kappa + iv)\zeta(\alpha(s - N - \kappa + iv), a)\tau^{N+\kappa-iv}dv.
\end{cases}
\tag{30}
$$

To estimate $I^{(\pm)}$ we apply (20) and (21) and then the Cauchy-Schwarz inequality to get

$$
\begin{aligned}
I^{(-)} &\ll \left(\frac{\alpha}{2\pi}\right)^{\frac{1}{2}+\alpha(N+\kappa-\sigma)} \tau^{N+\kappa} \\
&\quad \times \int_T^\infty e^{-\frac{\pi}{2}v}v^{-N-\kappa-\frac{1}{2}}(|t| + v)^{\frac{1}{2}+\alpha(N+\kappa-\sigma)}dv \\
&\ll \left(\frac{\alpha}{2\pi}\right)^{\frac{1}{2}+\alpha(N+\kappa-\sigma)} \tau^{N+\kappa} \\
&\quad \times \left(\int_T^\infty e^{-\frac{\pi}{2}v}v^{-2N-2\kappa-1}dv\right)^{\frac{1}{2}} \\
&\quad \times \left(\int_T^\infty e^{-\frac{\pi}{2}v}(|t| + v)^{1+2\alpha(N+\kappa-\sigma)}dv\right)^{\frac{1}{2}} \\
&\ll \left(\frac{\alpha}{2\pi}\right)^{\frac{1}{2}+\alpha(N+\kappa-\sigma)} \tau^{N+\kappa}e^{-\frac{\pi}{2}T}T^{-N-\kappa-\frac{1}{2}}(|t| + T)^{\frac{1}{2}+\alpha(N+\kappa-\sigma)}.
\end{aligned}
$$

Hence by (29)

$$
I^{(-)} \ll \left(\frac{\alpha}{2\pi}\right)^{\alpha c(\alpha)T} \tau^{c(\alpha)T}e^{-\frac{\pi}{2}T}T^{-(1-\alpha)c(\alpha)T}T^{-\alpha\sigma},
\tag{31}
$$

and similarly,

$$
I^{(+)} \ll \left(\frac{\alpha}{2\pi}\right)^{\alpha c(\alpha)T} \tau^{c(\alpha)T}e^{-\frac{\pi}{2}T}T^{-(1-\alpha)c(\alpha)T}T^{-\alpha\sigma}.
\tag{32}
$$

For the estimation of $I^{(0)}$ we apply (25) and (26),

$$
I^{(0)} \ll e^{-c_1(\alpha)T}T^{-(1-\alpha)c(\alpha)T-\alpha\sigma}e^{\frac{\pi\alpha}{2}|t|}\tau^{c(\alpha)T},
\tag{33}
$$

where

$$
c_1(\alpha) = c(\alpha)\{\alpha\log 2\pi - \alpha\log\alpha + (1 - \alpha)(\log c(\alpha) - 1)\}.
$$

Thus, from (31), (32) and (33), it follows that

$$R_N(s, \tau) \ll e^{-c_2(\alpha)T} T^{-(1-\alpha)c(\alpha)T} T^{-\alpha\sigma} \tau^{c(\alpha)T}$$
$$\ll e^{-c_2'(\alpha)N} N^{-(1-\alpha)(N+\kappa)} N^{-\alpha\sigma} \tau^{N+\kappa}, \tag{34}$$

with $c_2(\alpha) = c_1(\alpha)$, $c_2'(\alpha) = \alpha(\log 2\pi - \log \alpha + 1) - 1$.
Hence, as long as $0 < \alpha < 1$, we have

$$R_N(s, \tau) \ll N^{-(1-\alpha)N} \to 0 \tag{35}$$

as $N \to \infty$ and for $\alpha = 1$, $c_2'(1) = \log 2\pi > 0$, we have

$$R_N(s, \tau) \ll e^{-(\log 2\pi - \log \tau)N} \to 0 \tag{35}'$$

as $N \to \infty$.
By (35) and (35)' we conclude (27). This completes the proof. ∎

Remark 3. We transform formally $F(s, \tau)$ by the method of zeta-regularization to obtain

$$F(s, \tau) = \sum_{n=0}^{\infty} \frac{e^{-(n+a)^\alpha \tau}}{(n+a)^{\alpha s}} = \sum_{n=0}^{\infty} \sum_{k=0}^{\infty} \frac{(-\tau)^k}{k!} \frac{1}{(n+a)^{\alpha(s-k)}}$$
$$= \sum_{k=0}^{\infty} \frac{(-\tau)^k}{k!} \zeta(\alpha(s-k), a) + (\text{a correction term})$$

with $s - \frac{1}{\alpha} \neq 0, 1, 2, \ldots$
In fact $P(s, \tau)$ is the correction term to be calculated by another method.

References

[1] B. C. Berndt, *Identities involving the coefficients of a class of Dirichlet series. VI*, Trans. Amer. Math. Soc. **160** (1971), 157–167.

[2] B. C. Berndt, *Ramanujan's notebooks. Part I*, Springer-Verlag, New York-Berlin, 1985.

[3] S. Bochner, *Some properties of modular relations*, Ann. of Math. (2) **53** (1951), 332–363.

[4] E. Elizalde, et al, *Zeta regularization techniques with applications*, World Scientific Publishing Co. Pte. Ltd., 1994.

[5] P. Epstein, *Zur Theorie allgemeiner Zetafunctionen*, Math. Ann. **56** (1903), 615–644.

[6] A. Erdélyi, et al, *Higher transcendental functions*, McGraw-Hill, 1953.

[7] L. Euler, *Introduction to the analysis of the infinite*, Springer Verlag, 1988.

[8] M. Katsurada, *On Mellin-Barnes type of integrals and sums associated with the Riemann zeta-function*, Publ. Inst. Math. (Beograd) (N.S.) **62(76)** (1997), 13–25.

[9] H. Kumagai, *The determinant of the Laplacian on the n-sphere*, Acta Arith. **91** (1999), 199–208.

[10] H. Kumagai, *On unified Kronecker limit formula*, Kyushu J. Math., to appear.

[11] S. Lang, *Elliptic functions*, Addison-Wesley, 1973.

[12] M. Lerch, *Dalši studie v oboru Malmsténovských řad*, Rozpravy České Akad. **3** (1894) no.28, 1–61.

[13] Hj. Mellin, *Die Dirichletschen Reihen, die zahlentheoretischen Funktionen und die unendlichen Produkte von endlichem Geschlecht*, Acta Soc. Sci. Fennicæ **31** (1902), 3–48.

[14] T. Orloff, *Another proof of the Kronecker limit formulae*, Number theory (Montreal, Que., 1985), 273–278, CMS Conf. Proc., **7**, Amer. Math. Soc., Providence, R. I., 1987.

[15] J. R. Quine, S. H. Heydari and R. Y. Song, *Zeta regularized products*, Trans. Amer. Math. Soc. **338** (1993), 213–231.

[16] C. L. Siegel, *Lectures on advanced analytic number theory*, Tata Inst. 1961, 2nd ed., 1980.

[17] R. Song, *Properties of zeta regularized products*, Thesis, Florida State University, 1993.

[18] H. M. Stark, *L-Functions at $s = 1$. II*, Advances in Math. **17** (1975), 60–92.

[19] A. Voros, *Spectral functions, special functions and the Selberg zeta function*, Comm. Math. Phys. **110** (1987), 439–465.

A HYBRID MEAN VALUE FORMULA OF DEDEKIND SUMS AND HURWITZ ZETA-FUNCTIONS

ZHANG Wenpeng
Department of Mathematics, Northwest University, Xi'an, Shaanxi, P. R. China

Keywords: Dedekind sums, Hurwitz zeta-function, The hybrid mean value

Abstract   The main purpose of this paper is using the mean value theorem of the Dirichlet L-functions and the estimates for character sums to study the mean value distribution of Dedekind sums with a weight of Hurwitz zeta-function, and give an interesting asymptotic formula.

2000 Mathematics Subject Classification: 11N37.

1.   INTRODUCTION

For a positive integer k and an arbitrary integer h, the Dedekind sum $S(h,k)$ is defined by

$$S(h,k) = \sum_{a=1}^{k} \left(\left(\frac{a}{k}\right)\right) \left(\left(\frac{ah}{k}\right)\right),$$

where

$$((x)) = \begin{cases} x - [x] - \frac{1}{2} & \text{if } x \text{ is not an integer;} \\ 0 & \text{if } x \text{ is an integer.} \end{cases}$$

Various properties of $S(h,k)$ were investigated by many authors. For example, Carlitz [3], Mordell [5] and Rademacher [6] obtained an important reciprocity formula for $S(h,k)$. Conrey et al. [4] studied the mean value distribution of $S(h,k)$, and gave an interesting asymptotic formula. The main purpose of this paper is to study the asymptotic

This work is supported by the P.N.S.F. and N.S.F. of P. R. China.

C. Jia and K. Matsumoto (eds.), Analytic Number Theory, 395–408.

properties of the hybrid mean value

$$\sum_{a=1}^{q}{}' \zeta^n(\frac{1}{2}, \frac{a}{q}) S^m(a, q), \tag{1}$$

where $\zeta(s, \alpha)$ is Hurwitz zeta-function, $\sum_{a}{}'$ denotes the summation over

all integers a coprime to q.

Regarding (1), it seems that it has not yet been studied, at least I have not seen expressions like (1) before. The problem is interesting because it can help us to find some new relationship between Dedekind sums and Hurwitz zeta-functions. In this paper, we shall give a sharper asymptotic formula for (1). The constants implied by the O-symbols and the symbols $\ll$ used in this paper do not depend on any parameter, unless otherwise indicated. By using the estimates for character sums and the mean value Theorems of Dirichlet L-functions, we shall prove the following main conclusion:

Theorem. *Let q be an integer with $q \geq 3$, χ be any Dirichlet character modulo q. Then for any fixed positive integer m and n, we have the asymptotic formula*

$$\sum_{a=1}^{q} \chi(a)\zeta^n(\frac{1}{2}, \frac{a}{q}) S^m(a, q)$$

$$= \frac{q^{m+\frac{n}{2}}}{(12)^m} L\left(m + \frac{n}{2}, \chi\right) + O\left(q^{m+\frac{n-1}{2}} \exp\left(\frac{(m+n+2)\log q}{\log\log q}\right)\right),$$

where $L(s, \chi)$ is Dirichlet L-function and $\exp(y) = e^y$.

From this Theorem we may immediately deduce the following two Corollaries:

Corollary 1. *Let q be an integer with $q \geq 3$. Then for any fixed positive integer m and n, we have the asymptotic formula*

$$\sum_{a=1}^{q}{}' \zeta^n(\frac{1}{2}, \frac{a}{q}) S^m(a, q) = \frac{q^{m+\frac{n}{2}}}{(12)^m} \zeta\left(m + \frac{n}{2}\right) \prod_{p|q}\left(1 - \frac{1}{p^{m+\frac{n}{2}}}\right) +$$

$$+ O\left(q^{m+\frac{n-1}{2}} \exp\left(\frac{(m+n+2)\log q}{\log\log q}\right)\right),$$

where $\sum_{a}{}'$ denotes the summation over all a such that $(q, a) = 1$, $\prod_{p|q}$ denotes the product over all prime divisors of q, and $\zeta(s)$ is the Riemann zeta-function.

Corollary 2. *Let p be an odd prime, $\mathcal{A} = \mathcal{A}(p)$ denotes the set of all quadratic residues modulo p in the interval $[1, p-1]$. Then the asymptotic formula*

$$
\sum_{a \in \mathcal{A}} \zeta^n(\tfrac{1}{2}, \tfrac{a}{p}) S^m(a, p) = \frac{1}{2} \frac{p^{m+\frac{n}{2}}}{(12)^m} \zeta\left(m + \frac{n}{2}\right) \left[1 + \prod_q^* \frac{q^{m+\frac{n}{2}} - 1}{q^{m+\frac{n}{2}} + 1}\right] +
$$

$$
+ O\left(p^{m+\frac{n-1}{2}} \exp\left(\frac{(m+n+2)\log p}{\log\log p}\right)\right)
$$

holds uniformly for any fixed positive integers m and n, where $\prod_q^*$ de-notes the product over all prime q such that $\left(\frac{q}{p}\right) = -1$.

2.   SOME LEMMAS

To complete the proof of the theorem, we need the following Lemmas 1, 2 and 3.

Lemma 1. *Let q be an integer with $q \geq 3$, $(a, q) = 1$. Then we have*

$$
S(a, q) = \frac{1}{\pi^2 q} \sum_{d \mid q} \frac{d^2}{\phi(d)} \sum_{\substack{\chi \bmod d \\ \chi(-1) = -1}} \chi(a) \, |L(1, \chi)|^2,
$$

where $\phi(d)$ is the Euler function, χ denotes a Dirichlet character mod-ulo d with $\chi(-1) = -1$, and $L(s, \chi)$ denotes the Dirichlet L-function corresponding to χ.

Proof. (See reference [9]).

Lemma 2. *Let q be any integer with $q \geq 3$, χ denotes an odd Dirichlet character modulo d with $d \mid q$, χ_1 be any Dirichlet character modulo q. Then for any fixed positive integer m, we have the asymptotic formula*

$$
\sum_{\substack{\chi \bmod d \\ \chi(-1) = -1}} L(m - \tfrac{1}{2}, \chi_1\chi) |L(1, \chi)|^2
$$

$$
= \frac{\pi^2}{12} \phi(d) L(m + \tfrac{1}{2}, \chi_1) \prod_{p \mid d} \left(1 - \frac{1}{p^2}\right) + O\left(\frac{q}{\sqrt{d}} \exp\left(\frac{2\log d}{\log\log d}\right)\right).
$$

Proof. For the sake of simplicity we only prove that Lemma 2 holds for $m = 1$. Similarly we can deduce other cases. For any character

χ_q modulo q, let $A(y, \chi_q) = \sum\limits_{d < a \le y} \chi_q(a)$, and χ_q^0 denote the principal character modulo q. Then for any character $\chi_q(\ne \chi_q^0)$ modulo q and Re $s > 0$, applying Abel's identity we have

$$L(s, \chi_q) = \sum_{1 \le n \le d} \frac{\chi_q(n)}{n^s} + s \int_d^\infty \frac{A(y, \chi_q)}{y^{s+1}} dy$$

$$= S(s, \chi_q) + R(s, \chi_q).$$

So that

$$\sum_{\substack{\chi \bmod d \\ \chi(-1) = -1 \\ \chi\chi_1 \ne \chi_q^0}} L(\tfrac{1}{2}, \chi\chi_1) |L(1, \chi)|^2$$

$$= \sum_{\substack{\chi \bmod d \\ \chi(-1) = -1 \\ \chi\chi_1 \ne \chi_q^0}} \left\{ S(\tfrac{1}{2}, \chi\chi_1) + R(\tfrac{1}{2}, \chi\chi_1) \right\} \times$$

$$\times \{ S(1, \chi) + R(1, \chi) \} \cdot \{ S(1, \overline{\chi}) + R(1, \overline{\chi}) \} . \tag{2}$$

Note that for $(lmn, d) = 1$, from the orthogonality relation for character sums modulo d we have the identity

$$\sum_{\substack{\chi \bmod d \\ \chi(-1) = -1}} \chi(l)\chi(n)\overline{\chi}(m) = \begin{cases} \tfrac{1}{2}\phi(d), & \text{if } ln \equiv m \bmod d; \\ -\tfrac{1}{2}\phi(d), & \text{if } ln \equiv -m \bmod d; \\ 0, & \text{otherwise.} \end{cases} \tag{3}$$

Now we use (3) to estimate each term on the right side of (2). First we have

$$\sum_{\substack{\chi \bmod d \\ \chi(-1) = -1}} S(\tfrac{1}{2}, \chi\chi_1) S(1, \chi) S(1, \overline{\chi})$$

$$= \frac{\phi(d)}{2} \sum_{1 \le l < d} {\sum_{1 \le m < d}}' {\sum_{\substack{1 \le n < d \\ ln \equiv m (\bmod d)}}}' \frac{\chi_1(l)}{\sqrt{l}\, mn} - \frac{\phi(d)}{2} \sum_{1 \le l < d} {\sum_{1 \le m < d}}' {\sum_{\substack{1 \le n < d \\ ln \equiv -m (\bmod d)}}}' \frac{\chi_1(l)}{\sqrt{l}\, mn}$$

$$= \frac{\phi(d)}{2} \sum_{1 \le l < d} {\sum_{1 \le m < d}}' {\sum_{\substack{1 \le n < d \\ ln = m}}}' \frac{\sqrt{l}\,\chi_1(l)}{lmn} +$$

$$+ \frac{\phi(d)}{2} \sum_{\substack{1 \le l < d \\ ln \equiv m(\bmod d) \\ ln > d}} \sideset{}{'}\sum_{1 \le m < d} \sideset{}{'}\sum_{1 \le n < d} \frac{\sqrt{l}\,\chi_1(l)}{lmn} -$$

$$- \frac{\phi(d)}{2} \sum_{\substack{1 \le l < d \\ ln = d-m}} \sideset{}{'}\sum_{1 \le m < d} \sideset{}{'}\sum_{1 \le n < d} \frac{\sqrt{l}\,\chi_1(l)}{lmn} -$$

$$- \frac{\phi(d)}{2} \sum_{\substack{1 \le l < d \\ ln \equiv -m(\bmod d) \\ ln > d}} \sideset{}{'}\sum_{1 \le m < d} \sideset{}{'}\sum_{1 \le n < d} \frac{\sqrt{l}\,\chi_1(l)}{lmn}$$

$$= X_1 + X_2 - X_3 - X_4. \tag{4}$$

Here and in what follows each $\sum'$ indicates that the summation index runs through all integers coprime to d in the respective range. Note that the asymptotic estimates

$$X_1 = \frac{\phi(d)}{2} \sideset{}{'}\sum_{t=1}^{d} \frac{\sum_{n|t} \sqrt{n}\,\chi_1(n)}{t^2}$$

$$= \frac{\phi(d)}{2} \sideset{}{'}\sum_{t=1}^{\infty} \frac{\sum_{n|t} \sqrt{n}\,\chi_1(n)}{t^2} + O\left(\phi(d) \sum_{t=d}^{\infty} \frac{\sum_{n|t} 1}{t^{3/2}}\right)$$

$$= \frac{\phi(d)}{2} \left(\sum_{t=1}^{\infty} \frac{\chi_1(t)}{t^{3/2}}\right) \left(\sideset{}{'}\sum_{n=1}^{\infty} \frac{1}{n^2}\right) + O\left(\frac{\phi(d)}{\sqrt{d}}\right)$$

holds, and hence

$$X_1 = \frac{\pi^2}{12}\phi(d)L(\tfrac{3}{2},\chi_1) \prod_{p|d}\left(1 - \frac{1}{p^2}\right) + O\left(\frac{\phi(d)}{\sqrt{d}}\right). \tag{5}$$

Next we have

$$X_2 = \frac{\phi(d)}{2} \sum_{\substack{1 \le u < d \\ ln = ud+m}} \sum_{1 \le l < d} \sideset{}{'}\sum_{1 \le m < d} \sideset{}{'}\sum_{1 \le n < d} \frac{\sqrt{l}\,\chi_1(l)}{lmn}$$

$$= O\left(\phi(d) \sum_{1 \le u < d} \sum_{1 \le m < d} \frac{\sqrt{d}}{(ud+m)m}\right)$$

$$= O\left(\frac{\phi(d)}{\sqrt{d}} \sum_{1\le u<d} \sum_{1\le m<d} \frac{1}{mu}\right).$$

Hence

$$X_2 = O\left(\frac{\phi(d)}{\sqrt{d}} \log^2 d\right). \tag{6}$$

Also,

$$X_3 = \frac{\phi(d)}{2} \sum_{\substack{1\le l<d\ 1\le n<d \\ 1\le ln<\frac{d}{2}}} {\sum}' \frac{\chi_1(l)}{(d-ln)n\sqrt{l}} + \frac{\phi(d)}{2} \sum_{\substack{1\le l<d\ 1\le n<d \\ \frac{d}{2}\le ln<d}} {\sum}' \frac{\sqrt{l}\chi_1(l)}{ln(d-ln)}$$

$$= O\left(\frac{\phi(d)}{d} \sum_{1\le l<d}\sum_{1\le n<d} \frac{1}{n\sqrt{l}}\right) + O\left(\frac{\phi(d)}{d} \sum_{1\le u<d/2} \frac{\sqrt{d}\,\tau(d-u)}{u}\right)$$

$$= O\left(\frac{\phi(d)}{\sqrt{d}} \log^2 d\right) + O\left(\frac{\phi(d)}{\sqrt{d}} \exp\left(\frac{2\log d}{\log\log d}\right)\right),$$

where $\tau(d)$ is the divisor function and the estimate $\tau(d) \ll \exp\left(\frac{\log d}{\log\log d}\right)$ has been applied. Hence

$$X_3 = O\left(\frac{\phi(d)}{\sqrt{d}} \exp\left(\frac{2\log d}{\log\log d}\right)\right). \tag{7}$$

Similarly,

$$X_4 = \frac{\phi(d)}{2} \sum_{2\le u<d}\sum_{1\le m<d} {\sum}' \frac{\sqrt{l}\,\chi_1(l)}{(ud-m)m}$$

$$= O\left(\phi(d) \sum_{1\le u<d}\sum_{1\le m<d} \frac{\sqrt{d}}{umd}\right).$$

Hence

$$X_4 = O\left(\frac{\phi(d)}{\sqrt{d}} \log^2 d\right). \tag{8}$$

Combining (4) – (8) we obtain

$$\sum_{\substack{\chi \bmod d \\ \chi(-1)=-1 \\ \chi\chi_1\ne\chi_q^0}} S(\tfrac{1}{2},\chi\chi_1)S(1,\chi)S(1,\overline{\chi})$$

$$= \sum_{\substack{\chi \bmod d \\ \chi(-1)=-1}} S(\tfrac{1}{2}, \chi\chi_1) S(1, \chi) S(1, \overline{\chi}) + O\left(\sqrt{d} \log^2 d\right)$$

$$= \frac{\pi^2}{12} \phi(d) L(\tfrac{3}{2}, \chi_1) \prod_{p|d} \left(1 - \frac{1}{p^2}\right) + O\left(\frac{\phi(d)}{\sqrt{d}} \exp\left(\frac{2\log d}{\log\log d}\right)\right). \quad (9)$$

We next estimate the terms involving $R(s, \chi_q)$'s on the right-hand side of (2). For this we first note that $\chi\chi_1$ is a character modulo q, and hence we may assume $d < y < q$ in the following. Then from (3) and the properties of characters we have

$$\sum_{\substack{\chi \bmod d \\ \chi(-1)=-1 \\ \chi\chi_1 \neq \chi_q^0}} S(1, \chi) S(1, \overline{\chi}) A(y, \chi\chi_1)$$

$$= {\sum_{1 \leq n < d}}' \; {\sum_{1 \leq m < d}}' \; {\sum_{d < l \leq y}}' \frac{\chi_1(l)}{mn} \sum_{\substack{\chi \bmod d \\ \chi(-1)=-1 \\ \chi\chi_1 \neq \chi_q^0}} \chi(ln) \overline{\chi}(m)$$

$$\ll \phi(d) \sum_{\substack{1 \leq l < q \\ ln \equiv m(\bmod d)}} \sum_{1 \leq n < d} \sum_{1 \leq m < d} \frac{1}{mn} + \sum_{1 \leq l < q} \sum_{1 \leq n < d} \sum_{1 \leq m < d} \frac{\chi_q^0(l)}{mn}$$

$$\ll \frac{q}{d} \phi(d) \sum_{1 \leq n < d} \sum_{1 \leq m < d} \frac{1}{mn} + \phi(q) \log^2 d$$

$$\ll \frac{q\phi(d)}{d} \log^2 d + \phi(q) \log^2 d \ll q \log^2 d. \quad (10)$$

Similarly, for $d < y_j < q$ $(j = 1, 2, 3)$ we can also get the estimates

$$\sum_{\substack{\chi \bmod d \\ \chi(-1)=-1 \\ \chi\chi_1 \neq \chi_q^0}} S(1, \chi) A(y_1, \chi\chi_1) A(y_2, \overline{\chi})$$

$$= {\sum_{1 \leq n < d}}' \; {\sum_{d < a \leq y_1}}' \; {\sum_{d < b \leq y_2}}' \frac{\chi_1(a)}{n} \sum_{\substack{\chi \bmod d \\ \chi(-1)=-1 \\ \chi\chi_1 \neq \chi_q^0}} \chi(an) \overline{\chi}(b)$$

$$\ll \phi(d) \sum_{\substack{1 \leq n < d \\ an \equiv b(\bmod d)}} \sum_{1 \leq a < q} \sum_{1 \leq b < d} \frac{\chi_q^0(a)}{n} + {\sum_{1 \leq n < d}}' \; {\sum_{1 \leq a < q}}' \; {\sum_{1 \leq b < d}}' \frac{1}{n}$$

$$\ll \phi(d) \sum_{\substack{1\leq a<q \\ (a,q)=1}} \sum_{1\leq n<d} \frac{1}{n} + \phi(q)\phi(d)\log d$$

$$\ll \phi(q)\phi(d)\log d \ll qd\log d, \tag{11}$$

and the estimates

$$\sum_{\substack{\chi \bmod d \\ \chi(-1)=-1 \\ \chi\chi_1\neq\chi_q^0}} A(y_1,\chi\chi_1)A(y_2,\chi)A(y_3,\overline{\chi})$$

$$= \sideset{}{'}\sum_{d<a\leq y_1} \sideset{}{'}\sum_{d<b\leq y_2} \sideset{}{'}\sum_{d<c\leq y_3} \chi_1(a) \sum_{\substack{\chi \bmod d \\ \chi(-1)=-1 \\ \chi\chi_1\neq\chi_q^0}} \chi(ab)\overline{\chi}(c)$$

$$\ll \phi(d) \sideset{}{'}\sum_{\substack{1\leq a<q \\ ab\equiv c(\bmod d)}} \sideset{}{'}\sum_{1\leq b<d} \sideset{}{'}\sum_{1\leq c<d} \chi_q^0(a) + \sum_{\substack{1\leq a<q \\ (a,q)=1}} \sum_{\substack{1\leq b<d \\ (b,d)=1}} \sum_{\substack{1\leq c<d \\ (c,d)=1}} 1$$

$$\ll \phi(d) \sum_{\substack{1\leq a<q \\ (a,q)=1}} \sum_{\substack{1\leq b<d \\ (b,d)=1}} 1 + \phi(q)\phi^2(d) \ll \phi(q)\phi^2(d) \ll qd^2. \tag{12}$$

Hence from (10) we get

$$\sum_{\substack{\chi \bmod d \\ \chi(-1)=-1 \\ \chi\chi_1\neq\chi_q^0}} S(1,\chi)S(1,\overline{\chi})R(\tfrac{1}{2},\chi\chi_1)$$

$$= \int_d^\infty \frac{1}{y^{3/2}} \left(\sum_{\substack{\chi \bmod d \\ \chi(-1)=-1 \\ \chi\chi_1\neq\chi_q^0}} S(1,\chi)S(1,\overline{\chi})A(y,\chi\chi_1) \right) dy$$

$$= O\left(\int_d^\infty \frac{q\log^2 d}{y^{3/2}} dy \right) = O\left(\frac{q}{\sqrt{d}}\log^2 d \right). \tag{13}$$

Similarly from (11) we deduce that

$$\sum_{\substack{\chi \bmod d \\ \chi(-1)=-1 \\ \chi\chi_1\neq\chi_q^0}} S(1,\chi)R(\tfrac{1}{2},\chi\chi_1)R(1,\overline{\chi}) = O\left(\frac{q}{\sqrt{d}}\log d \right), \tag{14}$$

and from (12) we obtain

$$\sum_{\substack{\chi \bmod d \\ \chi(-1)=-1 \\ \chi\chi_1 \neq \chi_q^0}} R(\frac{1}{2},\chi\chi_1)R(1,\chi)R(1,\overline{\chi}) = O\left(\frac{q}{\sqrt{d}}\right). \qquad (15)$$

Using the same method of proving (13) we also have

$$\sum_{\substack{\chi \bmod d \\ \chi(-1)=-1 \\ \chi\chi_1 \neq \chi_q^0}} S(\frac{1}{2},\chi\chi_1)S(1,\chi)R(1,\overline{\chi})$$

$$= \int_d^\infty \frac{1}{y^2}\left(\sum_{1\leq l<d}\sideset{}{'}\sum_{1\leq n<d}\sideset{}{'}\sum_{d<m\leq y}\frac{\chi_1(l)}{nl^{\frac{1}{2}}}\sum_{\substack{\chi \bmod d \\ \chi(-1)=-1 \\ \chi\chi_1 \neq \chi_q^0}}\chi(ln)\overline{\chi}(m)\right)dy$$

$$= O\left(\frac{\phi(d)}{d}\sum_{1\leq l<d}\sum_{1\leq n<d}\sum_{\substack{1\leq m<d \\ ln\equiv m(\bmod d)}}\frac{1}{nl^{\frac{1}{2}}}\right)$$

$$+ O\left(\frac{1}{d}\sum_{1\leq l<d}\sum_{1\leq n<d}\sum_{1<m\leq d}\frac{\chi_q^0(lmn)}{nl^{\frac{1}{2}}}\right)$$

$$= O\left(\frac{\phi(d)}{d}\sum_{1\leq l<d}\sum_{1\leq n<d}\frac{1}{nl^{\frac{1}{2}}}\right) + O\left(\frac{\phi(d)}{\sqrt{d}}\log d\right)$$

$$= O\left(\frac{\phi(d)}{\sqrt{d}}\log d\right). \qquad (16)$$

$$\sum_{\substack{\chi \bmod d \\ \chi(-1)=-1 \\ \chi\chi_1 \neq \chi_q^0}} S(\frac{1}{2},\chi\chi_1)R(1,\chi)R(1,\overline{\chi}) = O\left(\sqrt{d}\right). \qquad (17)$$

Now combining (2), (9), (13), (14), (15), (16) and (17) we obtain

$$\sum_{\substack{\chi \bmod d \\ \chi(-1)=-1}} L(\frac{1}{2},\chi_1\chi)|L(1,\chi)|^2$$

$$= \sum_{\substack{\chi \bmod d \\ \chi(-1)=-1 \\ \chi\chi_1\neq\chi_q^0}} L(\tfrac{1}{2},\chi_1\chi)|L(1,\chi)|^2 + O\left(q^{\frac{1}{4}}\log^2 d\right)$$

$$= \frac{\pi^2}{12}\phi(d)L(\tfrac{3}{2},\chi_1)\prod_{p|d}\left(1-\frac{1}{p^2}\right) + O\left(\frac{q}{\sqrt{d}}\exp\left(\frac{2\log d}{\log\log d}\right)\right).$$

This proves Lemma 2.

Lemma 3. *Let q and m be positive integers with $q \geq 3$, χ be any Dirichlet character modulo q. Then we have*

$$\sum_{a=1}^{q}\chi(a)\zeta(\tfrac{1}{2},\tfrac{a}{q})S^m(a,q)$$

$$= \frac{q^{m+\frac{1}{2}}}{(12)^m}L\left(m+\frac{1}{2},\chi\right) + O\left(q^m\exp\left(\frac{(m+2)\log q}{\log\log q}\right)\right).$$

Proof. For any complex number $s = \sigma + it$ with $s \neq 1$, from [1] we know that

$$L(s,\chi) = \frac{1}{q^s}\sum_{a=1}^{q}\chi(a)\zeta(s,\tfrac{a}{q}).$$

Applying this identity and Lemma 1 we may immediately get

$$\sum_{a=1}^{q}\chi(a)\zeta(s,\tfrac{a}{q})S^m(a,q)$$

$$= \frac{1}{\pi^{2m}q^m}\sum_{d_1|q}\cdots\sum_{d_m|q}\frac{d_1^2\cdots d_m^2}{\phi(d_1)\cdots\phi(d_m)}\times$$

$$\times \sum_{\substack{\chi_1 \bmod d_1 \\ \chi_1(-1)=-1}}\cdots\sum_{\substack{\chi_m \bmod d_m \\ \chi_m(-1)=-1}}\left(\sum_{a=1}^{q}\chi(a)\chi_1(a)\cdots\chi_m(a)\zeta(s,\tfrac{a}{q})\right)\times$$

$$\times |L(1,\chi_1)|^2\cdots|L(1,\chi_m)|^2$$

$$= \frac{q^{s-m}}{\pi^{2m}}\sum_{d_1|q}\cdots\sum_{d_m|q}\frac{d_1^2\cdots d_m^2}{\phi(d_1)\cdots\phi(d_m)}\times$$

$$\times \sum_{\substack{\chi_1 \bmod d_1 \\ \chi_1(-1)=-1}}\cdots\sum_{\substack{\chi_m \bmod d_m \\ \chi_m(-1)=-1}}L(s,\chi\chi_1\cdots\chi_m)\times$$

$$\times |L(1,\chi_1)|^2\cdots|L(1,\chi_m)|^2. \tag{18}$$

Note that the estimates

$$\sum_{\substack{\chi \bmod d \\ \chi(-1)=-1}} |L(1,\chi)|^2 \le \frac{\pi^2}{12}\phi(d), \qquad q\sum_{d|q} \frac{d^{3/2}}{\phi(d)} \ll q^{\frac{3}{2}} \exp\left(\frac{\log q}{\log\log q}\right)$$

and the identity

$$\sum_{d|q} d^2 \prod_{p|d} \left(1 - \frac{1}{p^2}\right) = q^2.$$

Applying (18) and Lemma 2 repeatedly we have

$$\sideset{}{'}\sum_{a=1}^{q} \chi(a)\zeta(\frac{1}{2}, \frac{a}{q})S^m(a,q)$$

$$= \frac{q^{\frac{1}{2}-m}}{\pi^{2m}} \sum_{d_1|q} \cdots \sum_{d_{m-1}|q} \frac{d_1^2 \cdots d_{m-1}^2}{\phi(d_1)\cdots\phi(d_{m-1})} \times$$

$$\times \sum_{\substack{\chi_1 \bmod d_1 \\ \chi_1(-1)=-1}} \cdots \sum_{\substack{\chi_{m-1} \bmod d_{m-1} \\ \chi_{m-1}(-1)=-1}} \sum_{d_m|q} \frac{d_m^2}{\phi(d_m)} \times$$

$$\times \left(\sum_{\substack{\chi_m \bmod d_m \\ \chi_m(-1)=-1}} L(\frac{1}{2}, \chi\chi_1\cdots\chi_m)|L(1,\chi_m)|^2\right) \times$$

$$\times |L(1,\chi_1)|^2 \cdots |L(1,\chi_{m-1})|^2$$

$$= \frac{q^{\frac{1}{2}-m}}{\pi^{2m}} \sum_{d_1|q} \cdots \sum_{d_{m-1}|q} \frac{d_1^2 \cdots d_{m-1}^2}{\phi(d_1)\cdots\phi(d_{m-1})} \times$$

$$\times \sum_{\substack{\chi_1 \bmod d_1 \\ \chi_1(-1)=-1}} \cdots \sum_{\substack{\chi_{m-1} \bmod d_{m-1} \\ \chi_{m-1}(-1)=-1}} |L(1,\chi_1)|^2 \cdots |L(1,\chi_{m-1})|^2 \times$$

$$\times \left[\frac{\pi^2}{12}q^2 L(\frac{3}{2}, \chi\chi_1\cdots\chi_{m-1}) + O\left(q^{\frac{3}{2}}\exp\left(\frac{3\log q}{\log\log q}\right)\right)\right]$$

$$= \cdots\cdots\cdots$$

$$= \frac{q^{m+\frac{1}{2}}}{(12)^m} L(m+\frac{1}{2}, \chi) + O\left(q^m \exp\left(\frac{(m+2)\log q}{\log\log q}\right)\right).$$

This proves Lemma 3.

3.   PROOF OF THE THEOREM

In this section, we shall prove the Theorem by mathematical induction. First from Lemma 3 we know that the Theorem is true for $n = 1$. Then we assume that the Theorem is true for $n = k \geq 1$. That is,

$$\sum_{a=1}^{q} \chi(a)\zeta^{k}(\frac{1}{2}, \frac{a}{q})S^{m}(a, q) = \frac{q^{m+\frac{k}{2}}}{(12)^{m}} L\left(m + \frac{k}{2}, \chi\right) +$$

$$+ O\left(q^{m+\frac{k-1}{2}} \exp\left(\frac{(m+k+2)\log q}{\log\log q}\right)\right). \tag{19}$$

Now we prove that the Theorem is true for $n = k + 1$. From the orthogonality relation for character sums modulo q and the inductive assumption (19) we have

$$\sum_{a=1}^{q} \chi(a)\zeta^{k+1}(\frac{1}{2}, \frac{a}{q})S^{m}(a, q)$$

$$= \frac{1}{\phi(q)} \sum_{\chi_1 \bmod q} \left(\sum_{a=1}^{q} \chi(a)\chi_1(a)\zeta^{k}(\frac{1}{2}, \frac{a}{q})S^{m}(a, q)\right) \left(\sum_{b=1}^{q} \overline{\chi_1}(b)\zeta(\frac{1}{2}, \frac{b}{q})\right)$$

$$= \frac{\sqrt{q}}{\phi(q)} \sum_{\chi_1 \bmod q} \left(\sum_{a=1}^{q} \chi(a)\chi_1(a)\zeta^{k}(\frac{1}{2}, \frac{a}{q})S^{m}(a, q)\right) L\left(\frac{1}{2}, \overline{\chi_1}\right)$$

$$= \frac{\sqrt{q}}{\phi(q)} \sum_{\chi_1 \bmod q} \left[\frac{q^{m+\frac{k}{2}}}{(12)^{m}} L\left(m + \frac{k}{2}, \chi\chi_1\right) + \right.$$

$$\left. + O\left(q^{m+\frac{k-1}{2}} \exp\left(\frac{(m+k+2)\log q}{\log\log q}\right)\right)\right] L\left(\frac{1}{2}, \overline{\chi_1}\right)$$

$$= \frac{q^{m+\frac{k+1}{2}}}{(12)^{m}\phi(q)} \sum_{\chi_1 \bmod q} L\left(m + \frac{k}{2}, \chi\chi_1\right) L\left(\frac{1}{2}, \overline{\chi_1}\right) +$$

$$+ O\left(\frac{q^{m+\frac{k}{2}}}{\phi(q)} \sum_{\chi_1 \bmod q} \left|L\left(\frac{1}{2}, \overline{\chi_1}\right)\right| \exp\left(\frac{(m+k+2)\log q}{\log\log q}\right)\right). \tag{20}$$

Using the method of proving Lemma 2 we can easily get the asymptotic formula

$$\sum_{\chi_1 \bmod q} L\left(m + \frac{k}{2}, \chi\chi_1\right) L\left(\frac{1}{2}, \overline{\chi_1}\right) = \phi(q)L\left(m + \frac{k+1}{2}, \chi\right) + O\left(q^{\frac{1}{2}}\right)$$

$$\tag{21}$$

and the estimate

$$\sum_{\chi_1 \bmod q} \left| L\left(\frac{1}{2}, \overline{\chi_1}\right) \right| \ll \phi(q)\sqrt{\log q}. \tag{22}$$

Combining (20), (21) and (22) we obtain

$$\sum_{a=1}^{q} \chi(a)\zeta^{k+1}(\frac{1}{2}, \frac{a}{q})S^m(a,q) = \frac{q^{m+\frac{k+1}{2}}}{(12)^m} L\left(m + \frac{k+1}{2}, \chi\right) +$$
$$+ O\left(q^{m+\frac{k}{2}} \exp\left(\frac{(m+k+3)\log q}{\log\log q}\right)\right).$$

This completes the proof of the Theorem.

Proof of the Corollaries. Taking $\chi = \chi_q^0$ (the principal character mod q) in the Theorem and noting the identity

$$L(s, \chi_q^0) = \zeta(s) \prod_{p|q}\left(1 - \frac{1}{p^s}\right),$$

we may immediately get the assertion of Corollary 1.

Next we prove Corollary 2. Let p be an odd prime, $\left(\dfrac{a}{p}\right)$ denotes the Legendre symbol. Then note that the identity

$$\sum_{a \in \mathcal{A}} \zeta^n(\frac{1}{2}, \frac{a}{p})S^m(a,p) = \frac{1}{2}\sum_{a=1}^{p-1}\left(1 + \left(\frac{a}{p}\right)\right) \zeta^n(\frac{1}{2}, \frac{a}{p})S^m(a,p)$$

holds. From Theorem and Corollary 1 we may deduce Corollary 2.

Note. It is clear that using the method of proving the Theorem we can also get the following more general conclusion:
   For any $\frac{1}{2} \le \sigma < 1$, we have

$$\sum_{a=1}^{q}{}' \zeta^n(\sigma, \frac{a}{q})S^m(a,q) = \frac{q^{m+n\sigma}}{(12)^m}\zeta(m+n\sigma) \prod_{p|q}\left(1 - \frac{1}{p^{m+n\sigma}}\right) +$$
$$+ O\left(q^{m+(n-1)\sigma} \exp\left(\frac{(m+n+2)\log q}{\log\log q}\right)\right).$$

Acknowledgments

The author expresses his gratitude to the referee for very helpful and detailed comments.

References

[1] Tom M. Apostol, Introduction to Analytic Number Theory, Springer-Verlag, New York, 1976.

[2] Tom M. Apostol, Modular Functions and Dirichlet Series in Number Theory, Springer-Verlag, New York, 1976.

[3] L. Carlitz, The reciprocity theorem of Dedekind Sums, Pacific J. of Math., **3** (1953), 513–522.

[4] J. B. Conrey, Eric Fransen, Robert Klein and Clayton Scott, Mean values of Dedekind sums, J. Number Theory, **56** (1996), 214–226.

[5] L. J. Mordell, The reciprocity formula for Dedekind sums, Amer. J. Math., **73** (1951), 593–598.

[6] H. Rademacher, On the transformation of $\log \eta(\tau)$, J. Indian Math. Soc., **19** (1955), 25–30.

[7] H. Rademacher, "Dedekind Sum", Carus Mathematical Monographs, Math. Assoc. Amer., Washington D.C., 1972.

[8] H. Walum, An exact formula for an average of L-series, Illinois J. Math., **26** (1982), 1–3.

[9] Wenpeng Zhang, On the mean values of Dedekind Sums, J. Théorie des Nombres de Bordeaux, **8** (1996), 429–442.